MYCOLOGY IN THE TROPICS

DEVELOPMENTS IN MICROBIOLOGY

MYCOLOGY IN THE TROPICS

UPDATES ON PHILIPPINE FUNGI

Edited by

JONATHAN JAIME G. GUERRERO
College of Medicine, University of the Philippines Manila, Manila, Philippines

TERESITA U. DALISAY
Institute of Weed Science, Entomology and Plant Pathology, College of Agriculture and Food Science,
University of the Philippines Los Baños, Laguna, Philippines

MARIAN P. DE LEON
Microbial Culture Collection, Museum of Natural History,
University of the Philippines Los Baños, Laguna, Philippines

MARK ANGELO O. BALENDRES
Institute of Plant Breeding, College of Agriculture and Food Science,
University of the Philippines Los Baños, Laguna, Philippines

KIN ISRAEL R. NOTARTE
Faculty of Medicine & Surgery, University of Santo Tomas, Manila, Philippines
Department of Pathology, Johns Hopkins University School of Medicine, Baltimore, Maryland, United States

THOMAS EDISON E. DELA CRUZ
Department of Biological Sciences, College of Science, University of Santo Tomas, Manila, Philippines

Academic Press is an imprint of Elsevier
125 London Wall, London EC2Y 5AS, United Kingdom
525 B Street, Suite 1650, San Diego, CA 92101, United States
50 Hampshire Street, 5th Floor, Cambridge, MA 02139, United States
The Boulevard, Langford Lane, Kidlington, Oxford OX5 1GB, United Kingdom

Copyright © 2023 Elsevier Inc. All rights reserved.

No part of this publication may be reproduced or transmitted in any form or by any means, electronic or mechanical, including photocopying, recording, or any information storage and retrieval system, without permission in writing from the publisher. Details on how to seek permission, further information about the Publisher's permissions policies and our arrangements with organizations such as the Copyright Clearance Center and the Copyright Licensing Agency, can be found at our website: www.elsevier.com/permissions.

This book and the individual contributions contained in it are protected under copyright by the Publisher (other than as may be noted herein).

Notices
Knowledge and best practice in this field are constantly changing. As new research and experience broaden our understanding, changes in research methods, professional practices, or medical treatment may become necessary.

Practitioners and researchers must always rely on their own experience and knowledge in evaluating and using any information, methods, compounds, or experiments described herein. In using such information or methods they should be mindful of their own safety and the safety of others, including parties for whom they have a professional responsibility.

To the fullest extent of the law, neither the Publisher nor the authors, contributors, or editors, assume any liability for any injury and/or damage to persons or property as a matter of products liability, negligence or otherwise, or from any use or operation of any methods, products, instructions, or ideas contained in the material herein.

ISBN: 978-0-323-99489-7

For information on all Academic Press publications visit our website at
https://www.elsevier.com/books-and-journals

Publisher: Stacy Masucci
Acquisitions Editor: Linda Versteeg-Buschman
Editorial Project Manager: Mica Ella Ortega
Production Project Manager: Kiruthika Govindaraju
Cover Designer: Greg Harris

Typeset by TNQ Technologies

Contents

Contributors   ix
Editors' Biographies   xiii
Preface   xv
About the Cover   xvii

1. A brief introduction to Philippine mycology

Thomas Edison E. dela Cruz, Jonathan Jaime G. Guerrero and Angeles M. De Leon

1. Introduction   1
2. Fungal research in Philippine higher education and research institutions   2
3. Professional societies and local journals for information dissemination on fungi   4
4. Looking forward … *updates on Philippine fungi*   11
References   11

I

Fungi in aquatic and terrestrial habitats

2. Diversity and applications of fungi associated with mangrove leaves in the Philippines

Thomas Edison E. dela Cruz, Reuel M. Bennett and Carlo Chris S. Apurillo

1. Introduction   17
2. Early studies on Philippine mangrove fungi   18
3. Fungi associated with healthy mangrove leaves   20
4. Fungi associated with decaying mangrove leaves   26

5. Strategies to isolate and identify mangrove leaf-associated fungi   29
6. Concluding remarks and future direction   30
Acknowledgments   30
References   30

3. Fungi and fungus-like microorganisms in Philippine marine ecosystems

Irish Emmanuel P. Agpoon, Mark Kevin P. Devanadera, Kimberly D. Neri and Gina R. Dedeles

1. Introduction   33
2. Marine yeasts, mangrove-derived fungi, and marine fungal endophytes   34
3. Marine oomycetes   65
4. Biosynthesis, isolation, analysis, and applications of fatty acids from Philippine marine oomycetes   75
5. Concluding remarks   80
Acknowledgments   81
References   81

4. Species and functional diversity of forest fungi for conservation and sustainable landscape in the Philippines

Nelson M. Pampolina, Edwin R. Tadiosa, Jessa P. Ata, Janine Kaysee R. Soriano, Jason A. Parlucha and Jennifer M. Niem

1. Introduction   89
2. Forest formations in the Philippines   90
3. Ecological functions and substrates of forest fungi   92
4. Fungal growth requirements   94
5. Fungal diversity and mycological collection across forest formations and landscape   94
6. Challenges in mycology for conservation and sustainable landscape   128
Acknowledgments   129
References   129
Further reading   134

5. Bioluminescent mushrooms of the Philippines

Carlo Oliver M. Olayta and Thomas Edison E. dela Cruz

1. Introduction   137
2. Bioluminescence in fungi   138
3. Bioluminescent fungi in Southeast Asia   139
4. Bioluminescent fungi in the Philippines   143
5. Uses and applications of bioluminescent fungi   143
6. Concluding remarks   144
Acknowledgments   144
References   144

6. Lichens in the Philippines: diversity and applications in natural product research

Thomas Edison E. dela Cruz, Jaycee Augusto G. Paguirigan and Krystle Angelique A. Santiago

1. Introduction   147
2. Early studies on Philippine lichens   150
3. Taxonomic diversity of Philippine lichens   150
4. Philippine lichens in natural product research   151
5. Pioneering studies on endolichenic fungi from the Philippines   156
6. Concluding remarks and future direction   157
Acknowledgments   157
References   158

II

Fungi in agriculture, health, and environment

7. Plant diseases caused by fungi in the Philippines

Mark Angelo O. Balendres

1. Introduction   163
2. Genera of plant pathogenic fungi in the Philippines   164
3. Plant diseases caused by fungi in Phylum Ascomycota   166

4. Plant diseases caused by fungi in Phylum Basidiomycota   177
5. Plant diseases caused by fungi in Phylum Blastocladiomycota, Chytridiomycota, and Mucoromycota   179
6. Concluding remarks and future direction   181
Acknowledgments   182
References   182

8. Epidemiology of fungal plant diseases in the Philippines

Ireneo B. Pangga, John Bethany M. Macasero and Joselito E. Villa

1. Introduction   189
2. Epidemiological studies   191
3. Crop loss studies   194
4. Survey studies   196
5. Plant disease modeling   197
6. Disease-yield loss modeling   204
7. Climate change and plant diseases   205
8. Concluding remarks and future directions   206
References   207

9. Mycosis in the Philippines: Epidemiology, clinical presentation, diagnostics and interventions

Kin Israel R. Notarte, Adriel M. Pastrana, Abbygail Therese M. Ver, Jacqueline Veronica L. Velasco, Ma. Margarita Leticia D. Gellaco and Melissa H. Pecundo

1. Overview of mycosis   213
2. Common fungal infections   215
3. Concluding remarks   228
Acknowledgments   229
References   229
Further reading   233

10. Environmental mycology in the Philippines

Jonathan Jaime G. Guerrero, Charmaine A. Malonzo, Ric Ryan H. Regalado and Arnelyn D. Doloiras-Laraño

1. Introduction: the growing environmental concern in Southeast Asia   235
2. Mycoremediation   236

Contents **vii**

3. Mycoremediation of heavy metals 238
4. Fungi as bioremediation agents for pesticides 242
5. Fungi as biocontrol agents in agriculture 245
6. Plastic degrading fungi in the Philippines 253
7. Mycoremediation of hydrocarbons 256
8. Moving forward 258
Acknowledgment 258
References 259
Further reading 267

III

Fungal conservation and management

11. Edible mushrooms of the Philippines: traditional knowledge, bioactivities, mychochemicals, and in vitro cultivation

Thomas Edison E. dela Cruz and Angeles M. De Leon

1. Introduction 271
2. Ethnomycology in the Philippines 272
3. Species listing of wild edible mushrooms in the Philippines 273
4. In vitro culture of Philippine wild edible mushrooms 275
5. Mycochemicals and bioactivities of cultivated wild edible mushrooms 285
6. Concluding remarks and future direction 288
Acknowledgments 288
References 288

12. Culture collections and herbaria: Diverse roles in mycological research in the Philippines

Marian P. De Leon and Maria Auxilia T. Siringan

1. Introduction 293
2. Microbial culture collections 294
3. Fungal resources of Philippine MCCs 305

4. Opportunities and challenges in microbial culture collections 310
5. Mycological herbarium 313
6. Opportunities and challenges in mycological herbaria 317
7. Concluding remarks and future direction 319
Acknowledgments 320
References 320
Further reading 324

13. Fungal plant pathogens of quarantine importance in the Philippines

Lilia A. Portales, Jonar I. Yago and Amor C. Dimayacyac

1. History, plant quarantine protocols and current policy 325
2. Fungal pathogens with high impact in Philippine agriculture 328
3. Practices in controlling entry, exit and epidemics 331
4. Related administrative orders 335
References 342
Further reading 342

14. Innovative learning activities for teaching mycology in the Philippines

Thomas Edison E. dela Cruz, Reuel M. Bennett, Marilen P. Balolong, Bryna Thezza D. Leaño, Angeles M. De Leon, James Kennard S. Jacob, Joel C. Magday, Jr. Almira Deanna Lynn C. Valencia, Maria Feliciana Benita M. Eloreta, Jocelyn E. Serrano, Jayzon G. Bitacura, Carlo Chris S. Apurillo, Judee N. Nogodula, Melissa H. Pecundo, Krystle Angelique A. Santiago and Jeane V. Aril-dela Cruz

1. Introduction 344
2. Teaching fungi to kids in preschool and elementary levels 345
3. Learning activities on fungi for high school students 350
4. Innovative teaching strategies on fungi for undergraduate courses 354
5. Safety when working with fungi 363
6. Concluding remarks 364
References 364

Index **367**

Contributors

Irish Emmanuel P. Agpoon The Graduate School, University of Santo Tomas, Manila, Metro Manila, Philippines; Research Center for the Natural and Applied Sciences, University of Santo Tomas, Manila, Metro Manila, Philippines

Carlo Chris S. Apurillo Center for Research in Science and Technology (CReST), Philippine Science High School - Eastern Visayas Campus, Palo, Leyte, Philippines

Jeane V. Aril-dela Cruz Institute of Biology, Freie Universität Berlin, Berlin, Germany

Jessa P. Ata Department of Forest Biological Science, College of Forestry and Natural Resources, University of the Philippines Los Baños, College, Laguna, Philippines

Mark Angelo O. Balendres Plant Pathology Laboratory, Institute of Plant Breeding, College of Agriculture and Food Science, University of the Philippines Los Baños, Laguna, Philippines

Marilen P. Balolong Department of Biology, College of Arts and Science, University of the Philippines Manila, Manila, Philippines

Reuel M. Bennett Department of Biological Sciences, College of Science, University of Santo Tomas, Manila, Philippines

Jayzon G. Bitacura Department of Biological Sciences, Visayas State University, Baybay City, Leyte, Philippines

Gina R. Dedeles The Graduate School, University of Santo Tomas, Manila, Metro Manila, Philippines; Research Center for the Natural and Applied Sciences, University of Santo Tomas, Manila, Metro Manila, Philippines; Department of Biological Sciences, College of Science, University of Santo Tomas, Manila, Metro Manila, Philippines

Thomas Edison E. dela Cruz Department of Biological Sciences, College of Science, University of Santo Tomas, Manila, Philippines; Fungal Biodiversity, Ecogenomics and Systematics (FBeS) Group, Research Center for the Natural and Applied Sciences, University of Santo Tomas, Manila, Philippines; The Graduate School, University of Santo Tomas, Manila, Philippines

Marian P. De Leon Microbial Culture Collection, Museum of Natural History, University of the Philippines Los Baños, College, Laguna, Philippines

Angeles M. De Leon Department of Biological Sciences, College of Science, Central Luzon State University, Science City of Muñoz, Philippines

Mark Kevin P. Devanadera Research Center for the Natural and Applied Sciences, University of Santo Tomas, Manila, Metro Manila, Philippines; Department of Biochemistry, Faculty of Pharmacy, University of Santo Tomas, Manila, Metro Manila, Philippines

Amor C. Dimayacyac Bureau of Plant Industry, Central Post-Entry Quarantine Station, Los Banos, Laguna, Philippines

Arnelyn D. Doloiras-Laraño Graduate School of Science and Engineering, Ehime University, Matsuyama, Ehime, Japan

Maria Feliciana Benita M. Eloreta Philippine Science High School - Bicol Region Campus, Goa, Camarines Sur, Philippines

Jonathan Jaime G. Guerrero College of Medicine, University of the Philippines Manila, Manila, Philippines

James Kennard S. Jacob Department of Biological Sciences, College of Arts and Sciences, Isabela State University, Echague, Isabela, Philippines

Bryna Thezza D. Leaño Department of Biology, College of Science, De la Salle University, Manila, Philippines

Ma. Margarita Leticia D. Gellaco Faculty of Medicine & Surgery, University of Santo Tomas, Manila, Philippines; University of Santo Tomas Hospital, Manila, Philippines

Almira Deanna Lynn C. Valencia Department of Biology, College of Arts and Sciences, Partido State University, Goa, Camarines Sur, Philippines

John Bethany M. Macasero Institute of Weed Science, Entomology and Plant Pathology, College of Agriculture and Food Science, University of the Philippines Los Baños, College, Laguna, Philippines

Joel C. Magday, Jr. Philippine Science High School - Cagayan Valley Campus, Bayombong, Nueva Vizcaya, Philippines

Charmaine A. Malonzo Department of Biology, College of Science, Bicol University, Legazpi City, Philippines

Kimberly D. Neri The Graduate School, University of Santo Tomas, Manila, Metro Manila, Philippines; Research Center for the Natural and Applied Sciences, University of Santo Tomas, Manila, Metro Manila, Philippines

Jennifer M. Niem UPLB Museum of Natural History, University of the Philippines Los Baños, College, Laguna, Philippines

Judee N. Nogodula Department of Natural Sciences, College of Arts and Sciences, University of Southeastern Philippines, Davao City, Philippines

Kin Israel R. Notarte Faculty of Medicine & Surgery, University of Santo Tomas, Manila, Philippines; Department of Pathology, Johns Hopkins University School of Medicine, Baltimore, Maryland, United States

Carlo Oliver M. Olayta The Graduate School, University of Santo Tomas, Manila, Philippines; Laboratory Equipment and Supplies Office, University of Santo Tomas, Manila, Philippines

Jaycee Augusto G. Paguirigan Department of Biological Sciences, College of Science, University of Santo Tomas, Manila, Philippines; Korean Lichen Research Institute, Sunchon National University, Suncheon, Korea

Nelson M. Pampolina Department of Forest Biological Science, College of Forestry and Natural Resources, University of the Philippines Los Baños, College, Laguna, Philippines; UPLB Museum of Natural History, University of the Philippines Los Baños, College, Laguna, Philippines

Ireneo B. Pangga Institute of Weed Science, Entomology and Plant Pathology, College of Agriculture and Food Science, University of the Philippines Los Baños, College, Laguna, Philippines

Jason A. Parlucha Department of Wood Science and Technology, College of Forestry and Environmental Science, Central Mindanao University, Maramag, Bukidnon, Philippines

Adriel M. Pastrana Faculty of Medicine & Surgery, University of Santo Tomas, Manila, Philippines

Melissa H. Pecundo Research Center for Natural and Applied Sciences, University of Santo Tomas, Manila, Philippines

Lilia A. Portales Bureau of Plant Industry, National Plant Quarantine Services Division, Manila, Philippines

Ric Ryan H. Regalado National Institute of Molecular Biology and Biotechnology, College of Science, University of the Philippines, Diliman, Philippines

Krystle Angelique A. Santiago School of Science, Monash University Malaysia, Bandar Sunway, Selangor Darul Ehsan, Malaysia

Jocelyn E. Serrano Department of Biology, College of Science, Bicol University, Legazpi, Albay, Philippines

Maria Auxilia T. Siringan University of the Philippines Culture Collection, Microbiological Research and Service Laboratory, Natural Sciences Research Institute, University of the Philippines Diliman, Quezon City, Philippines

Janine Kaysee R. Soriano Department of Forest Biological Science, College of Forestry and Natural Resources, University of the Philippines Los Baños, College, Laguna, Philippines

Edwin R. Tadiosa Science Department, College of Sciences, Bulacan State University, Malolos, Bulacan, Philippines

Jacqueline Veronica L. Velasco Faculty of Medicine & Surgery, University of Santo Tomas, Manila, Philippines

Abbygail Therese M. Ver Faculty of Medicine & Surgery, University of Santo Tomas, Manila, Philippines

Joselito E. Villa Institute of Weed Science, Entomology and Plant Pathology, College of Agriculture and Food Science, University of the Philippines Los Baños, College, Laguna, Philippines

Jonar I. Yago Bureau of Plant Industry, Office of the Assistant Director for Operations, Technical and Production Services, Manila, Philippines

Editors' Biographies

Jonathan Jaime G. Guerrero

Jonathan Jaime G. Guerrero obtained his Master of Science in Plant Pathology, majoring in Mycology with cognates in Soil Science (Soil Microbiology) at the University of the Philippines Los Baños, graduating as the Salutatorian of the batch. He pursued nanotechnology and mycology research at the Kasetsart University Faculty of Agriculture in Bangkok, Thailand as a Southeast Asian Regional Center for Graduate Study and Research in Agriculture (SEARCA) scholar. He served as the Director of the Publication and Knowledge Management Division under the Office of the Vice President for Research, Development, and Extension of Bicol University (BU) in the Philippines, and concurrently became the Editor-in-chief of the BU R&D Journal. Currently, he is pursuing a Doctor of Medicine at the University of the Philippines Manila and is passionately training during his spare time to become a licensed pilot.

Teresita U. Dalisay

Teresita U. Dalisay, Professor 8 at the Institute of Weed Science, Entomology and Plant Pathology, University of the Philippines Los Baños (UPLB), is an accomplished plant pathologist with mycology as a specialization that is rooted back in her Bachelor's degree to her Ph.D. with impressive mentorship by professors and advisers. Her being an educator, researcher, and extensionist broadened her knowledge and experience in certain aspects of plant pathology (biological control, endophytic fungi, taxonomic works, and disease management), thus was conferred as a national awardee as Outstanding Plant Pathologist and holder of One U.P. Professorial Chair Awards. She has several published papers, book chapters, pamphlets, and brochures to her name and is frequently being tapped to review papers/articles for publications, project proposals, and ongoing projects. Having reached the peak of her career is influenced greatly by her passion to work on fungi with the exemplary guidance of her advisers namely: Dr. Delfin B. Lapis and Dr. Tricita H. Quimio, retired faculty members of UPLB, and Dr. Kevin D. Hyde, former faculty of the University of Hong Kong.

Marian P. De Leon

Marian Pulido De Leon earned her Doctor of Engineering in Material and Life Sciences from Osaka University, Osaka, Japan and has been appointed in 2018 as Visiting Associate Professor in the International Center for Biotechnology of the same university. She has served in different capacities in professional and scientific organizations including the Philippine

Society for Microbiology, Philippine Network of Microbial Culture Collections, Philippine Society for Lactic Acid Bacteria, and the Mycological Society of the Philippines. Currently, she is the Chair of the Biological Sciences Division of the National Research Council of the Philippines and Curator of the Microbial Culture Collection, of the University of the Philippines Los Baños Museum of Natural History, which she also serves as the Director. She served as a reviewer in local and international scientific journals, including the PLOS One Journal, Philippine Journal of Science, Microbial Drug Resistance, International Journal of Life Sciences and Biotechnology, and Applied Microbiology and Biotechnology. She is happily married to a civil engineer and a university professor, Mario with whom she is blessed with a son, Matthew Dominic, and a daughter, Madeline Dorothy.

Mark Angelo O. Balendres

Mark Angelo Balendres obtained his Doctor of Philosophy in Agricultural Science (Plant Pathology) from the University of Tasmania, Australia. He has more than fifty publications in plant pathology, plant disease diagnostics, plant-pathogen interaction, and plant disease resistance. He is a reviewer for more than fifteen international journals and a fellow of the Global Young Academy. His extension works include promoting scholarly writing and awareness of the importance of plant diseases, their causes, and their management. He likes bush walking and studies ancient philosophy and psychology in his free time.

Kin Israel R. Notarte

Kin Israel Notarte completed a Doctor of Medicine and Master of Science in Microbiology from the University of Santo Tomas. He is currently pursuing Ph.D. in Pathobiology (Microbiology and Infectious Disease) at the Johns Hopkins University School of Medicine, Baltimore, Maryland. He has more than thirty publications in research areas encompassing microbiology, medicinal chemistry, pharmacology, public health, and evidence-based medicine. He is a reviewer in several journals, including the New England Journal of Medicine, the Journal of Infection and Public Health, and the Current Research in Microbial Sciences. He is also a fellow of the Royal Society for Public Health. In his spare time, he loves to travel, watch anime, and read fictional books.

Thomas Edison E. Dela Cruz

Thomas Edison E. dela Cruz is a professor of microbiology at the Department of Biological Sciences, College of Science at the University of Santo Tomas (UST) in Manila, Philippines. He is a faculty researcher and head of the Fungal Biodiversity, Ecogenomics and Systematics (FBeS) group at the UST Research Center for the Natural and Applied Sciences. He has published more than 125 papers on fungal diversity, ecology, and natural products including research on lichens and myxomycetes. He completed his BSc Microbiology and MSc Biological Sciences at UST, his Diploma in Science Teaching at UP Open University, and his Doctor of Natural Sciences (Dr. Rer. Nat.) at Technical University Braunschweig in Germany. He likes to travel with his family, explore nature and city sights, and take photos.

Preface

The science of mycology in the Philippines is relatively young, yet dynamically evolving. In the last 10 years, we have seen a rapid increase in interest and research on fungi, their biodiversity and ecology, and their varied applications in medicine, agriculture, and the environment.

This first book fully puts together the literature on Philippine fungi. It draws readers on how the science of mycology started in the country and evolved to what we know of it today. It highlights the changing educational landscape with the goal of finally making mycology as an independent and integral course in the biology and microbiology programs. It also puts into perspective our understanding of their medical, environmental, and agricultural uses with emphasis on local standpoint. Its different chapters are useful in supplementing and concretizing mycological concepts found in other international references.

Each chapter in this book focuses on a particular field of mycology, that is, from basic to applied aspects, from perspectives in medicine and environment to agriculture, and from specialized topics on culture collection, ethnomycology, and education. These have pooled together recent literature authored by Filipinos or by studies conducted mainly in the Philippines. Readers need to become familiar with terms commonly used in mycology and see examples of local studies to better appreciate this field.

About the Cover

The Philippines is home to a variety of fungal species, and many of these have significantly contributed to the sociocultural and economic sectors of the country. There is also an increasing interest on fungi along science and technology to find long-lasting, sustainable, and eco-friendly solutions to challenges in health, environment, and agriculture. The cover celebrates the Kingdom Fungi's simplicity and complexity while acknowledging its vital role in nature and in the country's growth and development.

By **James Julian G. Guerrero**

A brief introduction to Philippine mycology

Thomas Edison E. dela Cruz[1], Jonathan Jaime G. Guerrero[2] and Angeles M. De Leon[3]

[1]Department of Biological Sciences, College of Science, University of Santo Tomas, Manila, Philippines; [2]College of Medicine, University of the Philippines Manila, Manila, Philippines; [3]Department of Biological Sciences, College of Science, Central Luzon State University, Science City of Muñoz, Philippines

1. Introduction

The Philippines is blessed with an enormous biodiversity. Owing to its geographic isolation for millions of years and to its tropical climate, the country is among those areas with the highest concentration of flora and fauna. In fact, the Convention on Biological Diversity ranked the Philippines fifth in the number of plant species with a very high endemism covering more than 25 plant genera (https://www.cbd.int). It is also not surprising that the country has a very high fungal diversity. To date, about 4698 fungal species belonging to 1031 genera are described in the Philippines (Quimio, 2002). But this number is certainly lower than the actual number as the species listing of Quimio was 20 years old and covers only species reported in the country from 1806 to 2001. If we look at the case of slime molds, Reynold's annotated list identified 107 species of myxomycetes for the country (Reynolds, 1981), a number that was also included in Quimio's list, but now numbers to 162 (Dagamac & dela Cruz, 2015, 2019; Macabago et al., 2020). Yulo and dela Cruz (2011) reported 13 species of dictyostelids in the country, two more records from the previous list of 11 species. Other recent updates on the species list of Philippine fungi are the papers of Paguirigan et al. (2020) which annotated the Philippine lichens and noted 1262 published taxa with 1234 validated species names and of Bennett and Thines (2020) which accounted 68 species of Philippine oomycetes. Certainly, there is a need to update the species list of fungi so far recorded in the Philippines.

Studies on fungi in the Philippines began as early as the 18th century through reports and expeditions made by Spanish botanists, particularly the Augustinian friars. Tadiosa (2012) had written a detailed account of the early history of Philippine mycology. He noted some of the early contributors to Philippine mycology, namely, Adelbert von Chamisso and Thaddeus Haenke, whose discoveries are reported in Nees von Esenback's *Horae Physicae Berolinensis* published in 1820. Father Francisco Manuel Blanco (1778–1845) in his monumental book, *Flora de Filipinas*, mentioned some species of fungi in the country. During the 19th to the 20th century, development in the field was due to the contributions made by American and European botanists and mycologists. Tadiosa (2012) enumerated some of these well-recognized mycologists, Paul W. Graff, Paul Christoph Hennings, George Edward Massee, William A. Murrill, Pier Andrea Saccardo, Hans Sydow, Paul Sydow, Roland Thaxter, to name a few. Among Filipino mycologists, the works of Teodoro (1937) is considered as one of the earliest. He published the *Enumeration of Philippine Fungi* which listed all fungi reported in the country up to the year 1935. This book is considered as one of the most valuable pieces of work to date on this topic. The 21st century also saw the establishment of local herbarium dedicated to the conservation of fungi, the Gerardo O. Ocfemia Memorial Herbarium, which is housed in the University of the Philippines Los Baños. Surely Philippine mycology has gone a long way since then. In this chapter, we briefly present some of the recent research studies on fungi by leading higher education and research institutions in the country. While we may not be able to describe in full details all the research conducted by these institutions, we aim to give a snapshot of the types of fungal research being done in the Philippines. We also introduce some professional scientific organizations and local journals in the country where fungal research is often disseminated. Finally, we introduce this book as an update and a compilation of studies done on Philippine fungi.

2. Fungal research in Philippine higher education and research institutions

Universities remain one of the main generators of knowledge through research. Here we list down some of the leading universities in the country and describe some of their research on fungi. This may not highlight all research activities in all institutions, but it can give ideas on the direction of mycology in the Philippines. The universities listed here are also not exhaustive but are based on recent publications. We did not include unpublished undergraduate and graduate theses related to fungi which are impossible to document in the absence of a common database.

University of the Philippines. The University of the Philippines (UP), the premier state university in the country, certainly has a strong tradition on fungal research. The Gerardo O. Ocfemia Memorial Herbarium, the first mycological herbarium in the country, and the Philippine National Collection of Microorganisms (PNCM) and the Museum of Natural History (MNH), all based in the UP Los Baños, are testament to the enormous diversity of fungi, both as preserved and living cultures, in the Philippines. Researchers at MNH have also pioneered the study of fungi from Philippine caves (Niem & Baldovino, 2015; Abris et al., 2020). The University of the Philippines Los Baños offers some of the early courses in mycology and has strong research on plant pathogenic fungi, specifically from the Institute of Weed Science, Entomology and Plant Pathology and the Institute of Plant Breeding of the College of Agriculture and Food

Science. A few examples of recent published research on fungal diseases in the Philippines by these institutes are the papers of dela Cueva et al. (2017) and Aguilar-Hawod et al. (2020) on the geographic distribution, genetic distribution, and race of *Fusarium oxysporum* Schltdl., on the geographic distribution of *Magnaporthe oryzae* (Lopez et al. 2019), and on the population structure of *Pseudocercospora fijiensis* (M. Morelet) Deighton (Mendoza & Ardales, 2019). Other fungal pathogens recently studied in the Philippines included several species of *Colletotrichum*, that is, *C. acutatum* J.H. Simmonds (dela Cueva et al., 2018), *C. fructicola* Prihast., L. Cai & K.D. Hyde (Evallo et al., 2022), *C. musae* (Berk. & M.A. Curtis) Arx (Balendres et al., 2019), *C. tropicale* E.I. Rojas, S.A. Rehner & Samuels, and *C. theobromicola* Delacr (dela Cueva et al., 2021), *Corynespora cassiicola* (Berk. & M.A. Curtis) C.T. Wei (Dimayacyac & Balendres, 2021), and *Lasiodiplodia theobromae* (Pat.) Griffon & Maubl (Evallo et al., 2022). A more comprehensive discussion on fungal diseases of plants in the Philippines is treated in Chapters 7 and 8 of this book. The National Institute of Molecular Biology and Biotechnology (BIOTECH) at UP Los Baños has also research on fungi, from fermenting yeasts (Madigal et al., 2019) to agricultural applications of arbuscular mycorrhizal fungi (Aggangan & Jomao-as, 2019; Cortes et al., 2021). Other UP units have also reported studies on fungi. For example, fungal endophytes and soil filamentous fungi were studied by researchers from UP Baguio (Hipol et al., 2019, 2020; Pablo et al., 2020), UP Diliman (Baluyot et al., 2022), and UP Visayas (Calabon et al., 2019). At UP Manila, known for its health research, Ablola and Bungay (2020) isolated and identified indoor fungi from a public tertiary hospital.

University of Santo Tomas. At the University of Santo Tomas (UST), fungal research is led by the Fungal Biodiversity, Ecogenomics and Systematics (FBeS) group of the Research Center for the Natural and Applied Sciences (https://ustfungalbiodiversitylab.wordpress.com/). In the last 15 years, the group has published more than 90 technical papers on Philippine fungi, distributed as follows: on filamentous fungi including fungal endophytes, marine fungi, and mangrove-associated fungi (31 papers), mushrooms (7 papers), lichens (6 papers), and on the fungus-like protists, the slime molds or myxomycetes (50 papers). The group has also spearheaded some of the first and extensive fungal studies in the Philippines, e.g., the first study on fungal endophytes from Philippine *Pandanus* (Bungihan et al., 2010), the first ethnomycological study in the country (De Leon et al., 2012), the first interactive database on myxomycetes (Dagamac et al., 2011), and the most comprehensive comparative ecological studies on myxomycetes (Dagamac et al., 2012) as well as updating the species list of fungi and fungus-like organisms in the country, see Dagamac and dela Cruz (2015, 2019) on myxomycetes and Paguirigan et al. (2020) on lichens. The group has established a one-stop, online information hub on Philippine myxomycetes (https://philmyxos.wordpress.com/). Recent research of the group focused on applications of Philippine fungi in bioremediation (de Padua & dela Cruz, 2021), agriculture (De Mesa et al., 2020; dela Cruz et al., 2021), and drug discovery (dela Cruz et al., 2020; Santiago, Edrada-Ebel et al., 2021), and on the role of endolichenic fungi in the lichen symbiosis (Galinato et al., 2021; Santiago, Edrada-Ebel et al., 2021; Santiago, dela Cruz, & Ting, 2021) and on the impact of disturbances to microorganisms (Cabutaje et al., 2021). Other research groups at UST have also focused on other fungal groups, for example, on estuarine oomycetes and thraustochytrids (Bennett & Thines, 2020; Perez et al., 2020), and on their application as sources of fatty acids (Caguimbal et al., 2019; Devanadera et al., 2019).

Central Luzon State University. At Central Luzon State University (CLSU), research on fungi generally centers on the historical uses of mushrooms by indigenous communities and how mushrooms influence their culture and tradition, in a field known as ethnomycology. Among the pioneering works on this topic at CLSU is the work of De Leon et al. (2012) that serves as the most extensive ethnomycological study focusing on the Aetas of Central Luzon. Other most recent ethnomycological studies are the Kalanguya communities in Carranglan, Nueva Ecija (De Leon et al., 2016), the Ifugao communities in Banaue, Hungduan, and Mayoyao, Ifugao Province (De Leon et al., 2018, 2019), and the indigenous communities from Paracelis, Mountain Province (De Leon et al., 2021). In addition to the documentation of their knowledge about mushrooms, collected specimens during mycological expeditions in these areas were identified to assess species diversity and distribution, for example, for the macrofungi in three provinces in Central Luzon (De Leon et al., 2013), and in Ifugao (De Leon et al., 2019) and Mountain Province (De Leon et al., 2021). Mushrooms are considered as functional foods, and at CLSU, local mushroom species were evaluated for their proximate nutritive compositions and other medicinal benefits. To date, at least 18 locally grown mushrooms were evaluated for these beneficial properties. Among these local mushrooms that were recently evaluated are *Lentinus strigosus* Fr (Dulay et al., 2018), *Pleurotus florida* Singer (Zacarias et al., 2021), and *Xylaria papulis* Llyod (De Leon et al., 2020). CLSU has also invested efforts for the cultivation of these edible mushrooms through the development of new technologies that increase yield. Studies on the cultivation of at least 16 local mushrooms were conducted in the university through its Center for Tropical Mushroom Research and Development (https://cos.clsu-ovpaa.edu.ph/research-centers/ctmrd/).

Bicol University*.* Early mycological research at Bicol University (BU) centers on biodiversity and plant pathology as undergraduate research of the Department of Biology, College of Science (Guerrero, 2020) and from various departments at the College of Agriculture and Forestry. Fungal biodiversity studies were also recently conducted on microfungi and macrofungi (Guerrero et al., 2020) and on fungal endophytes associated with medicinal plants and mangroves (Guerrero et al., 2019). Other studies on fungal endophytes focused on the economically important and endemic plant in the region, *Canarium ovatum* Engl. (General & Guerrero, 2017; Guerrero & Dalisay, 2018).

Other Higher Education and Research Institutions. Other institutions across the country have likewise contributed to the growing interest on Philippine mycology. These included the study on species diversity of macrofungi in mountain forest ecosystems in southern Luzon at Bulacan State University (Parlucha et al., 2021), on the interaction between agroeconomically important fungi and bats in Mindanao by researchers from the University of Southern Mindanao (Tanalgo, 2017), on the heavy metal tolerance of filamentous fungi at Visayas State University (Eliseo & Bitacura, 2021), and on the biopharmaceutical and biocontrol properties of fungal endophytes at Isabela State University (Campos et al., 2019; Campos & Jacob, 2021). Studies on medically important yeasts, specifically on *Candida albicans* (C.-P. Robin) Berkhout, were conducted by researchers in De la Salle University in Manila (Moron & Cabrera, 2018, 2019).

3. Professional societies and local journals for information dissemination on fungi

Research outputs are often disseminated first through scientific meetings and conferences organized by learned societies. These professional organizations have primarily educational and informational functions through publication of journals and conduct of meetings and

3. Professional societies and local journals for information dissemination on fungi

congresses, developing professional excellence, raising public awareness, and granting awards that promote and set standards for the field (Institute of Medicine, 2005). Here we describe Philippine professional societies that can serve as venues to share research on fungi.

Mycological Society of the Philippines. The Mycological Society of the Philippines (MSP) Inc. is the leading professional organization focusing on fungal research in the country and the representative society to the Asian Mycological Association of the International Mycological Association. Organized on October 19, 1996, MSP started from a conference on tropical microbial biodiversity held at the University of the Philippines Visayas in Iloilo Province. During this event, a small group of mycologists attending this conference has decided to form an initial organizing committee that will start the society. Dr. Tricita H. Quimio, then the Philippine representative to the International Mycological Association Committee for Asia (IMACA, now the Asian Mycological Association) and a professor and curator of the Mycology Herbarium at the UP Los Baños MNH was appointed as chairman together with Dr. Resurreccion B. Sabada and Dr. Maria Vicenta V. Gacutan, both from UP Visayas, serving as vice chairman and secretary, respectively. Subsequently, 11 Filipino mycologists formed the society at UPLB and served as its first board of directors: Tricita H. Quimio, Teresita U. Dalisay, Lourdes M. Tapay, Marilyn B. Brown, Nenita L. Opina, Elsa M. Luis, Silvino D. Merca, Ernesto P. Militante, Nelly S. Aggangan, and Juanito M. Dangan. The first scientific meeting was held on April 8, 1999, with 167 charter members. The society now held its 22nd annual scientific meetings and first virtual symposium in 2021. MSP aims to promote and nurture interest on fungi. It regularly offers pre-symposium workshops that hone technical skills on fungi. We listed in Table 1.1 the theme, dates, and venues of the different annual scientific meetings and symposium of the Mycological Society of the Philippines from its documented souvenir programs. Through this we can get a glimpse on the progress of mycology in the country.

Aside from MSP, there are other professional scientific organizations that can also serve as venue to disseminate information, particularly research outputs, related to fungi. While these professional societies have a general field as its focus, mycological researches are occasionally presented either as oral or poster presentation. We listed below some of these professional societies in the Philippines that cater and sustain interests on fungi.

Philippine Phytopathological Society. The Philippine Phytopathological Society (PPS) Inc., established on May 11, 1963, aims to support and advance the learning of infection management and to build up a successful working connection between people engaged in plant pathology. It is one of the five associations listed under the umbrella of the Pest Management Council of the Philippines (PMCP) Inc. PPS also confers the G. O. Ocfemia Outstanding Plant Pathologist Awards annually to Filipino plant pathologists who contributed significantly to research, instruction, and extension.

Philippine Society for Microbiology. The Philippine Society for Microbiology (PSM) Inc. was established on October 14, 1971, with the aim of promoting scientific knowledge in the field of microbiology and other related fields. The organization, in addition to its annual convention, conducts two cluster symposia and supports regional meetings of its four regional chapters. PSM also supports the Philippine Academy of Microbiology (PAM), which certifies microbiologists in the country either as Registered or Specialist Microbiologist.

Philippine Network on Microbial Culture Collection. The Philippine Network on Microbial Culture Collection (PNMCC) Inc., a network of culture collections in the Philippines, was established in 1996 and organizes annual symposium and training courses on culture collection and related fields. It also serves as an interagency arm of the Philippine National Collection of

6
1. A brief introduction to Philippine mycology

TABLE 1.1 General information on the annual scientific meetings and symposia of the Mycological Society of the Philippines, Inc.

Annual scientific meeting and symposium	Date	Venue	Theme
1st	April 8, 1999	APEC Building, UPLB, Laguna	Philippine mycology faces the challenges of the next millennium
2nd	April 28, 2000	Institute of Plant Breeding, UPLB, Laguna	Mycology at the threshold of the 21st century
3rd	April 20, 2001	Plant Pathology Auditorium, UPLB, Laguna	Role of mycology in boosting food security
4th	April 12, 2002	Plant Pathology Auditorium, UPLB, Laguna	Fungal diversity: conservation and utilization
5th	April 11, 2003	Plant Pathology Auditorium, UPLB, Laguna	Fungal biotechnology for food security
6th	April 16, 2004	Ecosystems Research and Development Bureau, UPLB, Laguna	Fungi and the environment
7th	April 8, 2005	Ecosystems Research and Development Bureau, UPLB, Laguna	Recent trends in fungal research
8th	—	Ecosystems Research and Development Bureau, UPLB, Laguna	Fungi as nutraceuticals
9th	April 19, 2007	Central Luzon State University, Nueva Ecija	Fungi and health
10th	April 15–16, 2008	Benguet State University, Benguet	Fungi in organic agriculture
11th	April 29, 2009	Cavite State University, Cavite	Fungi in agriculture and alternative medicine
12th	April 16, 2010	Nueva Viscaya State University, Nueva Vizcaya	Fungi in microbiology education and industry
13th	April 15, 2011	SEARCA, UPLB, Laguna	Fungi and climate change
14th	April 24, 2012	PhilRice, Nueva Ecija	Doing business with fungi
15th	April 11, 2013	Rizal Technological University, Mandaluyong City	Fungi in human and animal health
16th	April 23, 2014	University of Santo Tomas, Manila City	Fungi in industrial biotechnology
17th	April 25, 2015	Southern Luzon State University, Quezon	Fungi for organic agriculture

TABLE 1.1 General information on the annual scientific meetings and symposia of the Mycological Society of the Philippines, Inc.—cont'd

Annual scientific meeting and symposium	Date	Venue	Theme
18th	April 15, 2016	Batangas State University, Batangas	The bad, the good, the amazing fungi
19th	August 26, 2017	Benguet State University, Benguet	The role of fungi in restoring degraded ecosystems
20th	August 24, 2018	Institute of Biological Sciences, UPLB, Laguna	Mycology and global change: capitalizing on fungal research and education in building resilient communities
21st [together with the 6th annual conference and iCollect workshop of the ASEAN network on microbial utilization (AnMicro)]	April 27, 2019	National Institute of Molecular Biology and Biotechnology, UPLB, and SEARCA, Laguna	Strengthening ASEAN linkages in fungal diversity conservation and utilization
22nd	June 16–17, 2021	Virtual Meeting	Mycological innovations for a resilient and progressive new normal

Microorganisms (PNCM), the national repository of microorganisms in the country. PNMCC publishes a directory of microbial strains in the country which includes fungi.

Association of Systematic Biologists of the Philippines. Established in 1982, the Association of Systematic Biologists of the Philippines (ASBP) Inc. aims to promote the science and practice of taxonomy, systematics, and natural history in the country. ASBP publishes the Philippine Journal of Systematic Biology, an annual, peer-reviewed journal in the English language that reports high quality original research and reviews on taxonomy, systematics, ecology and conservation biology of plants, animals, and microorganisms including fungi in the Philippines.

Biodiversity Conservation Society of the Philippines. Formerly as the Wildlife Conservation Society of the Philippines (WCSP), the Biodiversity Conservation Society of the Philippines (BCSP) Inc. aims to promote the advancement of wildlife research and conservation in the Philippines. The Society was officially registered as WCSP in 1993 and later as BCSP in 2014. Occasionally, fungal conservation studies are presented in their annual Philippine biodiversity symposium.

Biology Teachers Association of the Philippines. The Biology Teachers Association of the Philippines (BIOTA) Inc. promotes the science of biology in all fields in the country. Established in February 1966, the society is one of the oldest professional organizations in the country and has now more than 20 chapter-members across the country. Presentations in their annual convention cover both research in biology and biology education.

In addition to oral and poster presentations delivered during annual scientific meetings and conferences that showcases fungal research, local based journals are also important venues to disseminate information about Philippine fungi. We have listed under Table 1.2 some of these local journals where submission of mycological research is very much welcome.

TABLE 1.2 Some Philippine-based journals that publish research on fungi.

Journal	Publisher	Frequency of publication	Scope of the journal	Types of accepted papers	Number of published papers in mycology for the last 5 years (2017−2021)	Link to journal website
Acta Manilana	University of Santo Tomas	Annual	Research and development in all areas of natural and applied sciences	Original research paper, short communication, reviews	4	https://actamanilana.ust.edu.ph/
Acta Medica Philippina	University of the Philippines Manila	Quarterly	Original scientific papers in the field of basic and clinical medical or health-related research	Original papers, case reports, current reviews, commentaries	4	https://actamedicaphilippina.upm.edu.ph/
Asian Journal of Biodiversity	Liceo de Cagayan University	Annual	New discoveries in genetic, species, and ecological diversity, biodiversity education	Original research articles	1	https://asianscientificjournals.com/new/publication/index.php/ajob
Bicol University R&D Journal	Bicol University	Biannual	Concerned with varied disciplines such as sciences and technology, languages, humanities, social sciences, management, and pedagogy	Original research articles	2	https://journal.bicol-u.edu.ph/
CLSU International Journal of Science and Technology	Central Luzon State University	Biannual	Research publication that seeks solutions to the problems of science and technology of the developing world	Original research articles	6	https://clsu-ijst.org/
Laksambuhay The UPLB Journal of Natural History	University of the Philippines Los Baños	Annual	Research on systematics, taxonomy, ecology, evolution, biogeography, and on natural history and tropical biology	Original research articles, notes and brief communications	1	https://sites.google.com/up.edu.ph/laksambuhay/home

3. Professional societies and local journals for information dissemination on fungi

Journal	Publisher	Frequency	Scope	Article types	No.	Website
Journal of Environmental Science and Management	University of the Philippines Los Baños	Biannual	Research that employs integrated methods resulting to analyses that provide new insights in environmental science	Original research articles	1	https://jesam.sesam.uplb.edu.ph/
Manila Journal of Science	De La Salle University	Biannual	Researches in biology, chemistry, mathematics, statistics, physics, computer science, and science education	Original research articles	3	https://www.dlsu.edu.ph/research/publishing-house/journals/manila-journal-of-science/
Philippine Agricultural Scientist	University of the Philippines Los Baños	Quarterly	Tropical agricultural science and related areas including environmental science, food science, engineering, biotechnology, economics, extension, rural sociology, development communication, agroforestry and silviculture, and marine and fishery sciences	Original research paper, short communication, reviews	8	https://pas.cafs.uplb.edu.ph/
Philippine Journal of Crop Science	Crop Science Society of the Philippines and Federation of Plant Science Societies of the Philippines	Annual	All aspects of crop science and technology (research, instruction, training, or extension)	Research report, review, lecture or policy statement	4	https://philippjcropsci.com/
Philippine Journal of Science	Department of Science and Technology	Quarterly	Natural sciences, engineering, mathematics, and social sciences	Original research articles, research note, reviews	14	https://philjournalsci.dost.gov.ph/
Philippine Journal of Health Research and Development	University of the Philippines Manila	Quarterly	All aspects of environmental and human health	Systematic reviews, original research, case reports	3	https://pjhrd.upm.edu.ph/

(Continued)

TABLE 1.2 Some Philippine-based journals that publish research on fungi.—cont'd

Journal	Publisher	Frequency of publication	Scope of the journal	Types of accepted papers	Number of published papers in mycology for the last 5 years (2017–2021)	Link to journal website
Philippine Journal of Systematic Biology	Association of Systematic Biologists in the Philippines	Annual	Taxonomy and systematics, ecology, conservation biology	Original research paper, reviews	13	https://asbp.org.ph/journals/
Philippine Society of Nature Studies Journal	Philippine Society for the Study of Nature	Biannual	Natural and applied sciences (agriculture, biology and its allied fields including studies on social sciences as related to nature and the environment)	Original research paper, short communication, reviews	1	https://www.journalofnaturestudies.org/
Science Diliman	University of the Philippines Diliman	Biannual	Studies on pure and applied sciences, interdisciplinary research	Original research articles, short communication	2	https://journals.upd.edu.ph/
Science & Engineering Journal (formerly Philippine Science Letters)	Philippine-American Academy of Science & Engineering	Biannual	Studies on natural, physical, mathematical, computational and social Sciences	Regular research articles, commentaries, letters, position papers, popular science articles	7 (incl. special issue on micro biology—2018–2021)	https://scienggj.org/
Silliman Journal	Silliman University	Biannual	Investigation in the humanities, social sciences, and sciences	Original research articles, notes, reviews	1	https://su.edu.ph/sillimanjournal/

4. Looking forward … *updates on Philippine fungi*

The growing interest and sustained enthusiasm on fungal research in the country has led to the publication of this very first book solely devoted to Philippine fungi. This book, *Mycology in the Tropics: Updates on Philippine Fungi*, is the first single leading reference that encapsulates the many facets of mycology in the Philippines and comprehensively discuss the current state and historical developments of Philippine mycology. Presented in three parts, the book highlights local studies in its 14 chapters. Chapter 1, this book chapter, briefly introduces the history of Philippine mycology and highlights some of the latest fungal research in some of the leading universities in the country. In Part I—Fungi in Aquatic and Terrestrial Habitats, five chapters outline the diversity and distribution of fungi in the country, from marine and estuarine habitats to terrestrial forest ecosystems, and from filamentous fungi to mushrooms and lichens. While this part primarily addresses the species list of fungi so far recorded in the country, it also highlights their applications, particularly in drug discovery and other agricultural uses. Part II—Fungi in Agriculture, Health, and the Environment has four chapters and details the unique roles of fungi in various aspects of our society. It covers fungal diseases in plants and in humans and discusses fungi in treating environmental pollution. The last part, Part III—Fungal Conservation and Management, with its four chapters, introduces the rising field of ethnomycology and innovative learning activities related to fungi. It also presents the importance of culture collection and quarantine guidelines in fungal research. With this pioneering book, it is hoped that more Filipinos will be interested to study Philippine fungi.

References

Ablola, F. B., & Bungay, A. A. C. (2020). Isolation of fungi in indoor air environment of selected air-conditioned and non-airconditioned wards in a public tertiary hospital in Metro Manila, Philippines. *The Philippine Journal of Health Research and Development, 24*(1), 27−38.

Abris, M. I., Palanca, M. G., De Leon, M. P., & Banaay, C. G. B. (2020). Microbial air and water quality assessment of a freshwater limestone cave in the Philippines and its implications for ecotourism management. *Journal of Nature Studies, 19*(1), 33−48.

Aggangan, N. S., & Jomao-as, J. G. (2019). Biochar from sugarcane bagasse and arbuscular mycorrhizal fungi on growth and nutrient status of Cacao (*Theobroma cacao* L.) seedlings under nursery conditions. *The Philippine Journal of Science, 148*(4), 647−657.

Aguilar-Hawod, K. G. I., de la Cueva, F. M., & Cumagun, C. J. R. (2020). Genetic diversity of *Fusarium oxysporum* f.sp. *cubense* causing Panama Wilt of Banana in the Philippines. *Pathogens, 9*, 32. https://doi.org/10.3390/pathogens9010032

Balendres, M. A., Mendoza, J., & dela Cueva, F. (2019). Characteristics of *Colletotrichum musae* PHBN0002 and the susceptibility of popular banana cultivars to postharvest anthracnose. *Indian Phytopathology, 73*(1), 57−64.

Baluyot, J. C., Santos, H. K., Batoctoy, D. C. R., Torreno, V. P. M., Ghimire, L. B., Joson, S. E. A., Obusan, M. C. M., Yu, E. T., Bela-ong, D. B., Gerona, R. R., & Velarde, M. C. (2022). *Diaporthe/Phomopsis longicolla* degrades an array of bisphenol analogues with secreted laccase. *Microbiological Research, 257*, 126973.

Bennett, R. M., & Thines, M. (2020). An overview on Philippine estuarine oomycetes. *The Philippine Journal of Systematic Biology, 14*(1). https://doi.org/10.26757/pjsb2020a14007

Bungihan, M. E., Tan, M. A., Kogure, N., Kitajima, M., dela Cruz, T. E. E., Takayama, H., & Nonato, M. G. (2010). A new isocoumarin compound from an endophytic fungus *Guignardia* sp. isolated from *Pandanus amaryllifolius* Roxb. *The ACGC Chemical Research Communications, 24*, 13−16.

Cabutaje, E. M., Pecundo, M. H., & dela Cruz, T. E. E. (2021). Diversity of myxomycetes in typhoon-prone areas: A case study in beach and inland forests of Aurora and Quezon province, Philippines. *Sydowia, 73*, 113–132.

Caguimbal, N. A., Devanadera, M. K., Bennett, R. M., Arafiles, K. H. V., Watanabe, K., Aki, T., & Dedeles, G. R. (2019). Fatty acid profiles of *Halophytophthora vesicula* and *Salispina spinosa* from Philippine mangrove leaves. *Letters in Applied Microbiology, 69*(3), 221–228.

Calabon, M. S., Sadaba, R. B., & Campos, W. L. (2019). Fungal diversity of mangrove-associated sponges from New Washington, Aklan, Philippines. *Mycology, 10*(1), 6–21.

Campos, R. P. G., & Jacob, J. K. S. (2021). Biocontrol potential of endophytic *Aspergillus* spp. against *Fusarium verticillioides*. *Biotropia, 28*(2), 141–148.

Campos, R. P. G., Jacob, J. K. S., Ramos, H. C., & Temanel, F. B. (2019). Mycopharmacological properties of endophytic fungi isolated from Cuban oregano (*Plectranthus amboinicus* Lour.) leaves. *Asian Journal of Biological and Life sciences, 8*(3), 103–110.

Cortes, A. D., Aggangan, N. S., & Opulencia, R. B. (2021). Taxonomic microbiome profiling and abundance patterns in the Cacao (*Theobroma cacao* L.) rhizosphere treated with arbuscular mycorrhizal fungi and bamboo biochar. *Philippine Agricultural Scientist, 104*(1), 19–33.

dela Cruz, T. E. E., Din, H. J. F., & Aril-dela Cruz, J. V. (2021) Microbes for sustainable agriculture: Isolation and identification of beneficial soil- and plant-associated microorganisms. SEARCA Professorial Chair Lecture Monograph No. 6. SEARCA, Los Baños, Laguna, Philippines.

dela Cruz, T. E. E., Notarte, K. I. R., Apurillo, C. C. S., Tarman, K., & Bungihan, M. E. (2020). Biomining fungal endophytes from tropical plants and seaweeds for drug discovery. In M. Ozturk, D. Egamberdieva, & M. Pesic (Eds.), *Biodiversity and biomedicine our future*. Elsevier Academic Press, ISBN 978-0-12-819541-3.

dela Cueva, F. M., Dalisay, T. U., Silva, F. F. M. A., De Castro, A. M., Pozon, A. P. M., & Molina, A. B. (2017). Geographic distribution and race identification of Fusarium wilt disease in the Philippines. *Philippine Journal of Crop Science, 42*, 74–75.

dela Cueva, F. M., Laurel, N. R., Dalisay, T. U., & Sison, M. L. J. (2021). Identification and characterisation of *Colletotrichum fructicola, C. tropicale* and *C. theobromicola* causing mango anthracnose in the Philippines. *Arch Phytopathol Plant Prot, 54*, 1989–2006.

dela Cueva, F. M., Mendoza, J. S., & Balendres, M. A. (2018). A new *Colletotrichum* species causing anthracnose of chilli in the Philippines and its pathogenicity to chilli cultivar Django. *Crop Protection, 112*, 264–268.

Dagamac, N. H. A., & dela Cruz, T. E. E. (2015). Myxomycetes research in the Philippines: Updates and opportunities. *Mycosphere, 6*(6), 784–795.

Dagamac, N. H. A., & dela Cruz, T. E. E. (2019). The Philippine slime molds after Dogma's 1975 list—how far have we been? *The Philippine Journal of Systematic Biology, 13*(2), 58–65.

Dagamac, N. H. A., Pangilinan, M. V. B., Stephenson, S. L., & dela Cruz, T. E. E. (2011). List of species collected and interactive database of myxomycetes (plasmodial slime molds) for Mt. Arayat National Park, Pampanga, Philippines. *Mycosphere, 2*(4), 449–455.

Dagamac, N. H. A., Stephenson, S. L., & dela Cruz, T. E. E. (2012). Occurrence, distribution, and diversity of myxomycetes (plasmodial slime molds) along two transects in Mt. Arayat, National Park, Pampanga, Philippines. *Mycology, 3*(2), 119–126.

De Leon, A. M., Cruz, A. S., Evangelista, A. B. B., Miguel, C. M., Pagoso, E. J. A., dela Cruz, T. E. E., Nelsen, D. J., & Stephenson, S. L. (2019). Species listing of macrofungi found in Ifugao indigenous community in Ifugao Province, Philippines. *Philippine Agricultural Scientist, 102*(2), 118–131.

De Leon, A. M., Diego, E. O., Domingo, L. K. F., & Kalaw, S. P. (2020). Mycochemical screening, antioxidant evaluation and assessment of bioactivities of *Xylaria papulis*: A newly reported macrofungi from Paracelis, Mountain Province, Philippines. *Current Research in Environmental & Applied Mycology (Journal of Fungal Biology), 10*(1), 300–318.

De Leon, A. M., Fermin, S. M. C., Rigor, R. P. T., Kalaw, S. P., dela Cruz, T. E. E., & Stephenson, S. L. (2018). Ethnomycological report on the macrofungi utilized by the indigenous community in Ifugao, Province, Philippines. *Philippine Agricultural Scientist, 101*(1), 84–92.

De Leon, A. M., Kalaw, S. P., Dulay, R. M., Undan, J. R., Alfonso, D. O., Undan, J. Q., & Reyes, R. G. (2016). Ethnomycological survey of the Kalanguya indigenous community in Carangalan, Nueva Ecija, Philippines. *Current Research in Environmental & Applied Mycology (Journal of Fungal Biology), 6*(2), 61–66.

De Leon, A. M., Luangsa-ard, J. J. D., Karunarathna, S. C., Hyde, K. D., Reyes, R. G., & dela Cruz, T. E. E. (2013). Species listing, distribution, and molecular identification of macrofungi in six Aeta tribal communities in Central Luzon, Philippines. *Mycosphere, 4*(3), 478–494.

De Leon, A. M., Pagaduan, M. A. Y., Panto, B. E., & Kalaw, S. P. (2021). Species listing of macrofungi found Paracelis, Mountain Province, Philippines. *CLSU International Journal of Science and Technology, 5*(2), 22–40.

De Leon, A. M., Reyes, R. G., & dela Cruz, T. E. E. (2012). An ethnomycological survey of the macrofungi utilized by the Aeta communities in Central Luzon, Philippines. *Mycosphere, 3*(2), 251–259.

De Mesa, R.B.C., Espinosa, I.R., Agcaoili, M.C.R.R., Calderon, M.A.T., Pangilinan, M.V.B., De Padua, J.C., dela Cruz, T.E.E. (2020) Antagonistic activities of needle-leaf fungal endophytes against *Fusarium* spp. MycoAsia 2020/06.

De Padua, J. C., & dela Cruz, T. E. E. (2021). Isolation and characterization of nickel-tolerant *Trichoderma* strains from marine and terrestrial environments. *Journal of Fungi, 7*(8), 591. https://doi.org/10.3390/jof7080591

Devanadera, M. K., Bennett, R. M., Santiago, M., Ramos, C., Watanabe, K., Aki, T., & Dedeles, G. R. (2019). Marine oomycetes (*Halophytophthora* and *Salispina*): A potential source of fatty acids with cytotoxic activity against breast adeno-carcinoma cells (MCF7). *Journal of Oleo Science, 68*(12), 1163–1174.

Dimayacyac, D. A., & Balendres, M. A. (2021). *Commelina benghalensis* harbors *Corynespora cassiicola*, the tomato target spot pathogen. *International Journal of Pest Management*, 1–7. https://doi.org/10.1080/09670874.2021.1980246

Dulay, R. M. R., Pamiloza, D. G., & Ramirez, R. L. (2018). Toxic and teratogenic effects of mycelia and fruiting body extracts of *Lentinus strigosus* (BIL 1324) in zebra fish (*Danio rerio*) embryo. *International Journal of Biosciences, 13*(5), 205–211.

Eliseo, R. M. M., & Bitacura, J. G. (2021). Heavy metal tolerance of filamentous fungi from the sediments of Visayas State University wastewater pond. *Annals of Tropical Research, 43*(1), 88–101.

Evallo, E., Taguiam, J., Bengoa, J., & Balendres, M. A. (2022). First report of *Lasiodiplodia theobromae* causing stem canker of *Selenicereus monacanthus* in the Philippines. *The Journal of Plant Pathology*. https://doi.org/10.1007/s42161-022-01042-0

Galinato, M. G. M., Bungihan, M. E., Santiago, K. A. A., Sangvichien, E., & dela Cruz, T. E. E. (2021). Antioxidant activities of fungi inhabiting *Ramalina peruviana*: Insights on the role of endolichenic fungi in the lichen symbiosis. *Current Research in Environmental & Applied Mycology (Journal of Fungal Biology), 11*(1), 119–136.

General, M. A., & Guerrero, J. J. G. (2017). Records of fungal endophytes from *Canarium ovatum* Engl. (Family Burseraceae) leaves. *The Philippine Journal of Science, 146*(1), 1–5.

Guerrero, J. J. (2020). Insights and prospects toward the undergraduate mycological researches of Bicol University. *The Philippine Journal of Science, 149*(2), 405–413.

Guerrero, J. J., Banares, E. N., General, M. A., & Imperial, J. T. (2020). Rapid survey of macro-fungi within an urban forest fragment in Bicol, Eastern Philippines. *Österreichische Zeitschrift fur Pilzkunde, 28*, 37–43.

Guerrero, J. J. G., & Dalisay, T. U. (2018). Fungal endophytes across tissue layers of *Canarium ovatum* (Burseraceae) fruit. *Österreichische Zeitschrift für Pilzkunde, 27*, 11–21.

Guerrero, J. J. G., Imperial, J. T., General, M. A., Arena, E. A. A., & Bernal, M. B. R. (2019). Antibacterial activities of secondary metabolites of endophytic *Aspergillus fumigatus, Aspergillus* sp. and *Diaporthe* sp. isolated from medicinal plants. *Österreichische Zeitschrift für Pilzkunde, 28*, 53–61.

Hipol, R. M., Baldelomar, J. A., Bolinget, K. C., & Solis, A. F. F. (2019). The soil fungi producing siderophores of Mt. Yangbew, Tawang, La Trinidad, Benguet. *Studies in Fungi, 4*(1), 1–13.

Hipol, R. M., Hipol, R. B., Fabian, M. C. P., Sasotona, J. S., & Hernandez, C. C. (2020). HMG-CoA reductase inhibitory activity of leaf-associated fungi. *Acta Medica Philippina, 54*(5), 498–502.

Institute of Medicine. (2005). *Facilitating interdisciplinary research*. The National Academies Press. https://doi.org/10.17226/11153

Lopez, A. L. C., Yli-Matilla, T., & Cumagun, C. J. R. (2019). Geographic distribution of avirulence genes of the rice blast fungus *Magnaporthe oryzae* in the Philippines. *Microorganisms, 7*(1), 23.

Macabago, S. A. B., Dagamac, N. H. A., dela Cruz, T. E. E., & Stephenson, S. L. (2020). Myxomycetes of the Caramoan Islands and update on the myxomycetes of the Bicol Peninsula in the Philippines. *The Philippine Journal of Systematic Biology, 14*, 1. https://doi.org/10.26757/pjsb2019a13010

Madigal, J. P. T., Simbahan, J. F., Lantican, N. B., Agrupis, S., & Elegado, F. B. (2019). Yeast and bacterial community profiling of fermenting Nipa (*Nypa fruticans*) sap in two Philippine locations and fermentation characteristics of selected yeasts. *Philippine Agricultural Scientist (Philippines), 102*(3), 220–229.

Mendoza, M. J. C., & Ardales, E. Y. (2019). Population structure of the banana Black Sigatoka pathogen [*Pseudocercospora fijiensis* (M. Morelet) Deighton] in Luzon, Philippines. *Philippine Agricultural Scientist (Philippines), 102*(3), 211—219.

Moron, L. S., & Cabrera, E. C. (2018). Detection of azole resistance and ERG11 point mutations in *Candida albicans* isolates from tertiary hospitals in the Philippines. *Current Research in Environmental & Applied Mycology (Journal of Fungal Biology), 8*(3), 298—305.

Moron, L. S., & Cabrera, E. C. (2019). ABC genotyping and putative virulence factors of *Candida albicans* clinical isolates. *Malaysian Journal of Microbiology, 15*(5), 400—407.

Niem, J. M., & Baldovino, M. M. (2015). Initial checklist of macrofungi in the karst area of Cavinti, Laguna. *Museum Publications in Natural History, 4*.

Pablo, C. H. D., Pagaduan, J. R. R., Langres, H. K. C., & Hipol, R. M. (2020). Plant growth-promoting characteristics of root fungal endophytes isolated from a traditional Cordillera rice landrace. *Studies in Fungi, 5*(1), 536—549.

Paguirigan, J. A. G., dela Cruz, T. E. E., Santiago, K. A. A., Gerlach, A., & Aptroot, A. (2020). A checklist of lichens known from the Philippines. *Current Research in Environmental & Applied Mycology (Journal of Fungal Biology), 10*(1), 319—376.

Parlucha, J. A., Soriano, J. K. R., Yabes, M. D., Pampolina, N. M., & Tadiosa, E. R. (2021). Species and functional diversity of macrofungi from protected areas in mountain forest ecosystems of Southern Luzon, Philippines. *Tropical Ecology, 62*, 359—367.

Perez, J. G. A., Mappala, A. K. A., Icaro, C. K. L., Estrada, A. M. T., Arafiles, K. H. V., & Dedeles, G. R. (2020). Occurrence of thraustochytrids on fallen mangrove leaves from Pagbilao mangrove Park, Quezon province. *The Philippine Journal of Systematic Biology, 14*(1). https://doi.org/10.26757/pjsb2020a14006

Quimio, T. H. (2002). *Checklist and database of Philippine Fungi (1806—2001)*. ASEAN Regional Center for Biodiversity Conservation.

Reynolds, D. R. (1981). Southeast Asian myxomycetes: II. Philippines. *Kalikasan. Philippine Journal of Biology, 10*(2—3), 127—150.

Santiago, K. A. A., dela Cruz, T. E. E., & Ting, A. (2021). Diversity and bioactivity of endolichenic fungi in *Usnea* lichens of the Philippines. *Czech Mycol, 73*(1), 1—19.

Santiago, K. A. A., Edrada-Ebel, R., dela Cruz, T. E. E., Cheow, Y. L., & Ting, A. S. Y. (2021). Biodiscovery of potential antibacterial diagnostic metabolites from the endolichenic fungus *Xylaria venustula* using LC—MS-based metabolomics. *Biology, 10*, 191.

Tadiosa, E. R. (2012). The growth and development of mycology in the Philippines. *Fungal Conservation, 2*, 18—22.

Tanalgo, K. C. T. (2017). Presence of important agro-economic fungi in common frugivorous bats from southcentral Mindanao, Philippines. *Current Research in Environmental & Applied Mycology (Journal of Fungal Biology), 7*(2), 73—81.

Teodoro, N. G. (1937). An Enumeration of Philippine Fungi. Commonwealth of the Philippines, Department of Agriculture and Commerce, Technical Bulletin 4. Bureau of Printing, Manila.

Yulo, P. R. J., & dela Cruz, T. E. E. (2011). Cellular slime molds isolated from Lubang Island, Occidental Mindoro, Philippines. *Mycosphere, 2*(5), 565—573.

Zacarias, R. C., Kalaw, S. P., & De Leon, A. M. (2021). Evaluation of angiosuppressive activity of *Pleurotus florida* on developing chick embryo using chorioallantoic membrane assay. *Studies in Fungi, 6*(1), 334—341.

PART I

Fungi in aquatic and terrestrial habitats

CHAPTER 2

Diversity and applications of fungi associated with mangrove leaves in the Philippines

Thomas Edison E. dela Cruz[1,2], Reuel M. Bennett[1] and Carlo Chris S. Apurillo[3]

[1]Department of Biological Sciences, College of Science, University of Santo Tomas, Manila, Philippines; [2]Fungal Biodiversity, Ecogenomics and Systematics (FBeS) Group, Research Center for the Natural and Applied Sciences, University of Santo Tomas, Manila, Philippines; [3]Center for Research in Science and Technology (CReST), Philippine Science High School - Eastern Visayas Campus, Palo, Leyte, Philippines

1. Introduction

Mangrove forests along the coastlines are among the most biologically diverse ecosystems on the planet, with its root systems submerged in waters harboring a variety of fishes, crustaceans, and other water-dwelling animals, and its above-ground parts being home to diverse species of birds, snakes, and even mammals. Mangroves are also the most productive and most carbon-rich forests in the tropics, with an average of 1023 Mg carbon per hectare (Donato et al., 2011). They are found in 123 countries worldwide but cover only an estimated area of 152,000 sq km, that is, less than 1% of all tropical forests (UNEP, 2014). This ecosystem is unique as the plants living in this area need to adapt and cope with changing salinity, oxygen-deprived soil, and regular tidal inundation. About 80 plant species spread across 18 families are known to thrive in mangrove forests (Duke et al., 2007). However, the uniqueness and remarkable traits of mangrove forests resulted in a variety of goods and services that human communities can benefit from. Among the important ecosystem services provided by the mangroves include provisions for food, fuelwood, fiber, fodder, timber, and medicines (UNEP, 2014). As further outlined by UNEP (2014), mangroves also regulate climate, protect coastlines, particularly during storm surges, prevent soil erosion, maintain water quality, and

Mycology in the Tropics
https://doi.org/10.1016/B978-0-323-99489-7.00006-8

recycle nutrients. Socio-culturally, mangroves are also important for ecotourism, recreation, aesthetics, and other human-related activities. However, we have also seen a decline in mangrove forest cover. An estimated 20%, equivalent to roughly 3.6 million hectares of mangroves, were lost between 1980 and 2005, with many Asian and Pacific regions registering this percentage within this period (FAO, 2007). Among the identified drivers of this decline is the conversion of mangrove forests to aquaculture or agriculture areas, coastal development, overexploitation, pollution, and climate change or extreme weather events (UNEP, 2014). In Southeast Asia, conversion to aquaculture or agriculture, coastal development, and overexploitation are the leading causes of mangrove loss. Recently, through satellite mapping, it was reported that 62% of the global mangrove losses between 2000 and 2016 was due to land-use change, primarily through conversion to aquaculture and agriculture, with up to 80% of these human-driven losses occurring in Southeast Asian nations (Goldberg et al., 2020). In the Philippines, Long and Giri (2011) estimated the total area of our mangrove forests to be 256,185 ha. Thirty-nine species of mangroves belonging to 14 families and around 20—30 species of shrubs and vines classified as mangrove associates are also found in the Philippines (Primavera, 2000). Recently, Garcia et al. (2013) reported 60 species of mangroves belonging to 23 genera and 14 families for the country.

Mangroves are also colonized by fungi. Mangrove fungi, also known as manglicolous fungi, include both lower fungi, exemplified by oomycetes and thraustochytrids, and higher fungi, that is, ascomycetes and basidiomycetes, which include mostly marine fungi and a small group of terrestrial fungi (Thatoi et al., 2013). Recently, fungus-like protists, the slime molds or myxomycetes, have been reported from substrates collected in mangrove ecosystems (Lim et al., 2021). Mangrove fungi are important ecologically either as saprophytes, parasites, or symbionts. Their economic uses include as sources of novel bioactive metabolites (Thatoi et al., 2013) or as a producer of enzymes with biotechnological applications (Torres & dela Cruz, 2013). In this paper, we explored early studies on mangrove fungi in the Philippines. We then documented diversity studies on higher and lower fungi colonizing mangrove leaves. We also discussed the different applications or uses of these mangrove-associated fungi and outlined strategies for their study.

2. Early studies on Philippine mangrove fungi

One of the early studies on marine fungi found in mangrove forests in the Philippines was that of Gacutan and Uyenco (1983). They noted *Clavariopsis bulbosa* Anastasiou, *Lulworthia* sp., *Halosphaeria quadricornuta* Cribb & J.W. Cribb, *Torpedospora radiata* Meyers, *Didymosphaeria enalia* Kohlm., *Nia vibrissa* R.T. Moore & Meyers, and *Halocyphina villosa* Kohlm. & E. Kohlm. from sugarcane bagasse and rice straw which were baited in Batan Bay, Aklan. This is followed by the study of Jones et al. (1988) on the driftwoods found in mangrove intertidal zones in Pagbilao, Quezon which reported 31 marine fungi. Among the most common species recorded in this study are *Massarina velatospora* K.D. Hyde & Borse, *Savoryella lignicola* E.B.G. Jones & R.A. Eaton, *Zalerion varia* Anastasiou, and *Rosellinia* sp. Other species recorded are *Halosarpheia marina* (Cribb & J.W. Cribb) Kohlm., *Trichocladium achrasporum* (Meyers & R.T. Moore) M. Dixon ex Shearer & J.L. Crane, *Dactylospora haliotrepha* (Kohlm. & E. Kohlm.) Hafellner, *Cirrenalia tropicalis* Kohlm., *Didymosphaeria rhizophorae* Kohlm. & E. Kohlm, *Lulworthia grandispora* Meyers, *Aniptodera chesapeakensis* Shearer & M.A. Mill., *Aniptodera mangrovii*

K.D. Hyde, *Dendryphiella arenaria* Nicot, *Didymosphaeria enalia* Kohlm., *Halocyphina villosa* Kohlm. & E. Kohlm, *Helicascus kanaloanus* Kohlm., *Halosarpheia ratnagiriensis* S.D. Patil & Borse, *Halosarpheia retorquens* Shearer & J.L. Crane, *Halosarpheia viscosa* (I. Schmidt) Shearer & J.L. Crane, *Leptosphaeria australiensis* (Cribb & J.W. Cribb) G.C. Hughes, and *Lignincola laevis* Höhnk. They also reported unidentified species of *Acrocordiella, Lulworthia, Dictyosporium, Didymosphaeria, Leptosphaeria, Microthelia,* and *Monodictys*. Jones et al. (1996) then redescribed a novel genus, *Trisporella beccariana* comb. nov., which was isolated from a decomposing leaf petiole or rachis of *Nypa fruticans* Wurmb that was collected in Malaysia and the Philippines. Another early report of mangrove-associated fungi was that of Alias et al. (1999). Their study collected mangrove decayed woods and driftwoods from Boracay Island in Aklan, Taklong Island in Guimaras, and Pagbilao in Quezon and documented 50 intertidal fungi, 21 of which were new records for the country, and with one species, *Acrocordiopsis sphaerica* Alias & E.B.G. Jones as new to science. Among the new records for the country as stated by this study are *Trematosphaeria mangrovis* Kohlm., *Lophiostoma mangrovei* Kohlm. & Vittal, *Biatriospora marina* K.D. Hyde & Borse, *Acrocordiopsis sphaerica* Alias & E.B.G. Jones, *Acrocordiopsis patilii* Borse & K.D. Hyde, *Salsuginea ramicola* K.D. Hyde, *Quintaria lignatilis* (Kohlm.) Kohlm. & Volkm.-Kohlm., *Trichocladium alopallonellum* (Meyers & R.T. Moore) Kohlm. & Volkm.-Kohlm., *Trichocladium achrasporum* (Meyers & R.T. Moore) M. Dixon, *Lignincola longirostris* (Cribb & J.W. Cribb) Kohlm., *Xylomyces* sp., *Antennospora salina* (Meyers) Yussoff, E.B.G. Jones & S.T. Moss, *Helicascus kanaloanus* Kohlm., *Adomia* sp., *Aigialus parvus* S. Schatz & Kohlm., *Arenariomyces trifurcatus* Höhnk, *Cucullosporella mangrovei* (K.D. Hyde & E.B.G. Jones) K.D. Hyde & E.B.G. Jones, *Lignincola tropica* Kohlm., *Dendryphiella salina* (G.K. Sutherl.) Pugh & Nicot, *Marinosphaera mangrovei* K.D. Hyde, and *Tetraploa aristata* Berk. & Broome. It is interesting to note that four species, *Trematosphaeria mangrovis, Lophiostoma mangrovei, Cucullosporella mangrovei,* and *Marinosphaera mangrovei,* had species epithet derived from mangroves. *Helicascus kanaloanus* and *Trichocladium achrasporum* were also previously reported by Jones et al. (1988) and thus, should not be considered anymore as new records. Besitulo et al. (2010) later documented the fungi associated with the mangroves in Del Carmen, Siargao Island in the province of Surigao del Norte. Their study reported a high fungal diversity, with 66 identified taxa, including 57 ascomycetes, 2 basidiomycetes, and 7 anamorphic fungi. These were recorded from 400 wood samples collected from *Rhizophora apiculata* Blume, *Xylocarpus granatum* K.D. Koenig, *Nypa fruticans,* and various driftwoods. Besitulo et al. (2010) also noted 46 species as new records for the Philippines, with the most frequent species as follows: *Linocarpon appendiculatum* K.D. Hyde, *Microthyrium* sp., *Coronopapilla mangrovei* (K.D. Hyde) Kohlm. & Volkm.-Kohlm., *Astrosphaeriella striatispora* (K.D. Hyde) K.D. Hyde, *Swampomyces* sp., *Phialophorophoma litoralis* Linder, *Dactylospora haliotrepha* (Kohlm. & E. Kohlm.) Hafellner, and *Oxydothis nypicola* K.D. Hyde. At this point, it is essential to note that these studies focus mainly on driftwoods and decayed mangrove woods as substrates for the isolation of Philippine mangrove fungi. A more recent study has also explored sponges found in mangrove forests. In that study, a total of 110 species of sponge-associated fungi belonging to 22 genera of ascomycetes, 2 genera of basidiomycetes, 21 morphospecies of *Mycelia sterilia,* 1 unidentified yeast species, and 11 unidentified hyphomycetes were isolated from four species of mangrove sponges (Calabon et al., 2019). However, other plant parts of mangroves could also be used for the isolation of mangrove-associated fungi (Moron et al., 2018). The following section described these studies reporting fungal endophytes from mangrove leaves, their diversity, and various reported applications.

3. Fungi associated with healthy mangrove leaves

Diversity. Mangrove ecosystems are "hot spots" for fungal diversity (Shearer et al., 2007). In the Philippines, fungi isolated from healthy mangrove leaves have been reported from the following provinces: Zambales, Batangas, Cavite, Quezon in Central and Southern Luzon (Ramirez et al., 2020), Albay, Catanduanes, Sorsogon, Masbate, Camarines Norte, and Camarines Sur in the Bicol Region (Guerrero et al., 2018), and Leyte and Samar in Eastern Visayas (Apurillo et al., 2019). These studies focused on endophytic fungi or fungal endophytes, which are described as fungi that reside in the host plants without causing any harm (Petrini, 1991). Healthy mangroves are said to be host to more than 200 species of fungal endophytes and are often described as belonging to the genera *Alternaria, Aspergillus, Cladosporum, Colletotrichum, Fusarium, Paecilomyces, Penicillium, Pestalotiopsis, Phoma, Phomopsis, Phyllosticta,* and *Trichoderma* (Liu et al., 2007). In the studies conducted in mangroves from different areas in Luzon and the Visayas, the commonly isolated fungi from healthy mangrove leaves belong to the genera *Aspergillus, Colletotrichum, Fusarium, Penicillium,* and *Pestalotiopsis*. However, it is worth noting that there are very few similar fungal species reported between the provinces despite the presence of similar host mangrove species in these sites. This bolsters the hypothesis that the diversity of fungal endophytes varies between host mangroves and their geographical origin.

There are also many unique genera reported only in a specific province. For instance, *Phialophora* was reported from *Sonneratia alba* Sm. and *Rhizophora mucronata* Lam. collected in Zambales (Ramirez et al., 2020). Due to the large number of mangrove species sampled in the Bicol region, a number of unique fungal genera were reported in this area, and these included *Alternaria, Botryosphaeria, Ceriporia, Chaetomium, Curvularia, Daldinia, Fomes, Glomerella, Hyalotiella, Lasiodoplodia, Lentinus, Lichtheimia, Microascus, Nigrospora, Paecilomyces, Rhizopus, Rigidoporus, Schizophyllum,* and *Trametes* (Guerrero et al., 2018). *Guignardia, Marasmiellus, Phaeosphaeriopsis, Cytospora,* and *Verticillium* were reported from mangroves collected in Samar (Apurillo et al., 2019). Table 2.1 lists the 65 fungal species isolated from different species of mangroves from various areas in the Philippines. It is interesting to note that of the 40−60 species of mangroves identified in the country, roughly half or only 28 mangrove species have been studied so far for foliar fungal endophytes. *Rhizophora mucronata* has been consistently observed to have the most number of distinct fungal species (Apurillo et al., 2019; Guerrero et al., 2018; Ramirez et al., 2020).

Many of the isolated fungi from healthy mangrove leaves are known as plant pathogens, e.g., *Colletotrichum gloeosporioides* (Penz.) Penz. & Sacc., *Fusarium solani* (Mart.) Sacc., *Marasmiellus palmivorus* (Sharples) Desjardin, and *Verticillium nigrescens* Pethybr. These species, despite their pathogenicity to other plants, may have existed as asymptomatic endophytes within their mangrove hosts (Apurillo et al., 2019). The relationship between an endophyte and its host is usually of a balanced antagonism (Schulz et al., 1999). Since these fungi exist as endophytes in mangroves without causing harm, this only means that the mangrove hosts may have evolved mechanisms to prevent these fungi from becoming pathogenic to their hosts. Some mangrove fungal endophytes are shown in Fig. 2.1.

While most studies on mangrove fungal endophytes focused on leaves, one study looked at the fungi associated with the stems and roots of 13 mangrove hosts (*Avicennia officinalis,*

TABLE 2.1 List of fungal endophytes from healthy mangrove leaves in the Philippines.

Mangrove species	Fungi from leaves	Site	Reference
Acanthus ebracteatus Vahl	*Aspergillus ochraceus* K. Wilh. *Colletotrichum* sp. *Nigrospora* sp.	Batangas Quezon	Ramirez et al. (2020)
Aegiceras corniculatum (L.) Bianco	*Cladosporium* sp. *Nigrospora* sp.	Quezon	Ramirez et al. (2020)
Aegiceras floridum Roem. & Schult.	*Cladosporium* sp. *Phialophora* sp.	Zambales Quezon	Ramirez et al. (2020)
	Verticillium nigrescens Pethybr.	Samar	Apurillo et al. (2019)
Avicennia alba Blume	*Aspergillus* sp. *Fomes* sp. *Hyalotiella rubi* Senan., Camporesi & K.D. Hyde	Bicol region	Guerrero et al. (2018)
Avicennia marina (Forssk.) Vierh.	*Aspergillus aculeatus* Iizuka *Glomerella* sp. *Lentinus squarrosulus* Mont. *Pestalotiopsis cocculi* (Guba) G.C. Zhao & N. Li *Trametes cubensis* (Mont.) Sacc.	Bicol region	Guerrero et al. (2018)
	Aspergillus niger Tiegh. *Aspergillus ochraceus* K. Wilh. *Cladosporium* sp. *Colletotrichum* sp. *Nigrospora* sp. *Penicillium* sp. *Pestalotiopsis* sp. *Phialophora* sp.	Zambales Batangas Cavite Quezon	Ramirez et al. (2020)
	Colletotrichum fructicola (Prihastuti, L. Cai & K.D. Hyde) *Colletotrichum queenslandicum* B.S. Weir & P.R. Johnston *Colletotrichum tropicale* Rojas, Rehner & Samuels *Diaporthe* sp.	Leyte	Apurillo et al. (2019)
Avicennia officinalis L.	*Aspergillus niger* Tiegh. *Phialophora* sp.	Quezon	Ramirez et al. (2020)
Avicennia rumphiana Hallier f.	*Aspergillus minisclerotigenes* Vaamonde, Frisvad & Samson *Aspergillus niger* Tiegh. *Aspergillus* sp. *Ceriporia lacerata* N. Maek, Suhara & R. Kondo *Fomes* sp. *Hyalotiella rubi* Senan, Camporesi & K.D. Hyde *Peniophora* sp. *Xylaria feejeensis* (Berk.) Fr.	Bicol region	Guerrero et al. (2018)
Bruguiera cylindrica (L.) Blume	*Aspergillus aculeatus* Iizuka *Aspergillus flavus* Link *Aspergillus niger* Tiegh. *Botryosphaeria rhodina* (Berk. & M.A. Curtis) Arx *Ceriporia lacerata* N. Maek., Suhara & R. Kondo *Trametes cubensis* (Mont.) Sacc.	Bicol region	Guerrero et al. (2018)

(Continued)

TABLE 2.1 List of fungal endophytes from healthy mangrove leaves in the Philippines.—cont'd

Mangrove species	Fungi from leaves	Site	Reference
Bruguiera gymnorhiza (L.) Lam	*Daldinia eschscholtzii* (Ehrenb.) Rehm *Lichtheimia ramose* (Zopf) Vuill. *Pestalotiopsis cocculi* (Guba) G.C. Zhao & N. Li	Bicol region	Guerrero et al. (2018)
Bruguiera parviflora (Roxb.) Wight & Am. Ex Griff.	*Aspergillus flavus* Link *Aspergillus minisclerotigenes* Vaamonde, Frisvad & Samson *Ceriporia lacerata* N. Maek, Suhara & R. Kondo *Colletotrichum* sp.	Bicol region	Guerrero et al. (2018)
Bruguiera sexangula (Lour.) Poir.	*Aspergillus flavus* Link *Aspergillus fumigatus* Fresen. *Penicillium citrinum* Thom *Pestalotiopsis uvicola* (Speg.) Bissett *Trametes cubensis* (Mont.) Sacc.	Bicol region	Guerrero et al. (2018)
Ceriops decandra (Griff.) Ding Hou	*Aspergillus* sp. *Fomes* sp. *Rigidoporus vinctus* (Berk.) Ryvarden *Trametes ljubarskyi* Pilát	Bicol region	Guerrero et al. (2018)
	Penicillium sp. *Phialophora* sp.	Quezon	Ramirez et al. (2020)
Ceriops tagal (Pers.) C.B. Rob.	*Nigrospora oryzae* (Berk. & Broome) Petch	Bicol region	Guerrero et al. (2018)
	Penicillium sp. *Phialophora* sp.	Batangas Quezon	Ramirez et al. (2020)
Ceriops zippeliana Blume	*Rhizopus microsporus* Tiegh.	Bicol region	Guerrero et al. (2018)
Excoecaria agallocha (L.)	*Ceriporia lacerata* N. Maek., Suhara & R. Kondo *Chaetomium globosum* Kunze *Curvularia verruculosa* Tandon & Bilgrami ex M.B. Ellis	Bicol region	Guerrero et al. (2018)
	Penicillium sp. *Phialophora* sp.	Zambales & Quezon	Ramirez et al. (2020)
Lumnitzera littorea (Jack) Voigt	*Lasiodiplodia theobromae* (Pat.) Griffon & Maubl.	Bicol region	Guerrero et al. (2018)
Lumnitzera racemosa Wild., Nuee Schriften Ges. Naturf. Freunde Berlin iv.	*Aspergillus aculeatus* Iizuka *Aspergillus japonicus* Salto *Aspergillus minisclerotigenes* Vaamonde, Frisvad & Samson *Aspergillus* sp. *Ceriporia lacerata* N. Maek., Suhara & R. Kondo *Colletotrichum kahawae* J.M. Waller & Bridge *Fomes* sp. *Nigrospora oryzae* (Berk. & Broome) Petch *Nigrospora sphaerica* (Sacc.) E.W. Mason *Pestalotiopsis cocculi* (Guba) G.C. Zhao & N. Li *Rigidoporus vinctus* (Berk.) Ryvarden *Trametes maxima* (Mont.) A. David & Rajchenb.	Bicol region	Guerrero et al. (2018)
Nypa fruticans Wurmb	*Nigrospora* sp. *Phialophora* sp.	Zambales	Ramirez et al. (2020)
Osbornia octodonta F. Muell.	*Penicillium* sp. *Phialophora* sp.	Zambales Quezon	Ramirez et al. (2020)

TABLE 2.1 List of fungal endophytes from healthy mangrove leaves in the Philippines.—cont'd

Mangrove species	Fungi from leaves	Site	Reference
Pemphis acidula J.R. Forst & G. Forst	*Nigrospora sphaerica* (Sacc.) E.W. Mason	Bicol region	Guerrero et al. (2018)
Rhizophora apiculata Blume	*Aspergillus aculeatus* Iizuka *Aspergillus fumigatus* Fresen. *Aspergillus minisclerotigenes* *Colletotrichum gloeosporoides* (Penz.) Penz. & Sacc. *Pestalotiopsis uvicola* (Speg.) Bissett *Pestalotiopsis* sp. *Trametes cubensis* (Mont.) Sacc.	Bicol region	Guerrero et al. (2018)
Rhizophora mucronata Lam.	*Alternaria* sp. *Aspergillus flavus* Link *Ceriporia lacerata* N. Maek, Suhara & R. Kondo *Hyalotiella rubi* Senan. Camporesi & K.D. Hyde *Lichtheimia ramose* (Zopf) Vuill. *Nigrospora oryzae* (Berk. & Broome) Petch *Nigrospora* sp. *Paecilomyces variotii* Bainier *Penicillium griseofulvum* Dierckx *Pestalotiopsis mangiferae* (Henn.) Steyaert *Pestalotiopsis cocculi* Guba) G.C. Zhao & N. Li *Pestalotiopsis microspora* (Speg.) Bat. & Peres *Phlebiopsis crassa* (Lév) Floudas & Hibbett *Schizophyllum commune* Fr. *Trametes cubensis* (Mont.) Sacc.	Bicol region	Guerrero et al. (2018)
	Aspergillus oryzae (Ahlb.) Cohn *Phomopsis pittospori* (Cooke & Harkn.) Grove	Leyte	Apurillo et al. (2019)
	Phaeosphaeriopsis musae M. Arzanlou & Crous *Pestalotiopsis adusta* (Ellis & Everh.) Steyaert *Xylaria cubensis* (Mont.) Fr. *Diaporthe siamensis* Udayanga, Xing Z. Liu & K.D. Hyde *Cytospora rhizophorae* Kohlm. & E. Kohlm. *Phomopsis pittospori* (Cooke & Harkn.) Grove	Samar	Apurillo et al. (2019)
	Cladosporum sp. *Colletotrichum* sp. *Fusarium* sp. *Nigrospora* sp. *Penicillium* sp. *Pestalotiopsis* sp. *Phialophora* sp. *Trichoderma* sp.	Zambales Batangas Cavite Quezon	Ramirez et al. (2020)
Rhizophora stylosa Griff.	*Aspergillus flavus* Link *Aspergillus minisclerotigenes* Vaamonde, Frisvad & Samson *Aspergillus versicolor* (Vuillemin) Tiraboschi *Ceriporia lacerata* N. Maek, Suhara & R. Kondo *Pestalotiopsis cocculi* (Guba) G.C. Zhao & N. Li *Pestalotiopsis uvicola* (Speg.) Bissett *Trametes cubensis* (Mont.) Sacc. *Valsa eugeniae* Nutman & F.M. Roberts	Bicol region	Guerrero et al. (2018)
	Cladosporium sp.	Quezon	Ramirez et al. (2020)

(Continued)

24 2. Diversity and applications of fungi associated with mangrove leaves in the Philippines

TABLE 2.1 List of fungal endophytes from healthy mangrove leaves in the Philippines.—cont'd

Mangrove species	Fungi from leaves	Site	Reference
Scyphiphora hydrophylacea C.F. Gaerth	*Diaporthe longicolla* (Hobbs) J.M. Santos, Vrandecic & A.J.L. Phillips *Fomes* sp. *Hyalotiella rubi* Senan, Camporesi & K.D. Hyde *Rhizopus microsporus* Tiegh.	Bicol region	Guerrero et al. (2018)
Sonneratia alba Sm.	*Aspergillus aculeatus* Iizuka *Aspergillus flavus* Link *Aspergillus minisclerotigenes* Vaamonde, Frisvad & Samson *Colletotrichum coffeanum* F. Noack *Fomes* sp. *Lasiodiplodia theobromae* (Pat.) Griffon & Maubl. *Microascus cinereus* Cursi *Paecilomyces formosus* Sakag., May. Inoue & Tada ex Houbraken & Samson *Pestalotiopsis cocculi* (Guba) G.C. Zhao & N. Li *Trametes cubensis* (Mont.) Sacc. *Trametes elegans* (Spreng.) Fr.	Bicol region	Guerrero et al. (2018)
	Valsa brevispora G.C. Adams & Jol. Roux	Leyte	Apurillo et al. (2019)
	Guignardia mangiferae A.J. Roy *Marasmiellus palmivorus* (Sharples) Desjardin *Aspergillus nidulans* (Eidam) G. Winter	Samar	Apurillo et al. (2019)
	Phialophora sp. *Cladosporum* sp. *Penicillium* sp.	Zambales	Ramirez et al. (2020)
	Penicillium sp.	Batangas	Ramirez et al. (2020)
	Penicillium sp.	Quezon	Ramirez et al. (2020)
Sonneratia caseolaris (L.) Engl.	*Fusarium solani* (Mart.) Sacc. *Glomerella cingulata* (Stoneman) Spauld. & H. Schrenk *Penicillium citrinum* Thom *Pestalotiopsis cocculi* Guba) G.C. Zhao & N. Li *Rhizopus microsporus* Tiegh. *Trametes cubensis* (Mont.) Sacc. *Trametes ljubarskyi* Pilát	Bicol region	Guerrero et al. (2018)
Sonneratia ovalis (L.) Engl.	*Aspergillus minisclerotigenes* Vaamonde, Frisvad & Samson *Fomes* sp. *Pestalotiopsis cocculi* (Guba) G.C. Zhao & N. Li *Phomopsis longicolla* Hobbs *Rigidoporus vinctus* (Berk.) Ryvarden *Trametes cubensis* (Mont.) Sacc.	Bicol region	Guerrero et al. (2018)
Xylocarpus granatum K.D. Koenig	*Aspergillus ochraceus* K. Wilh. *Phialophora* sp.	Zambales Quezon	Ramirez et al. (2020)

I. Fungi in aquatic and terrestrial habitats

3. Fungi associated with healthy mangrove leaves

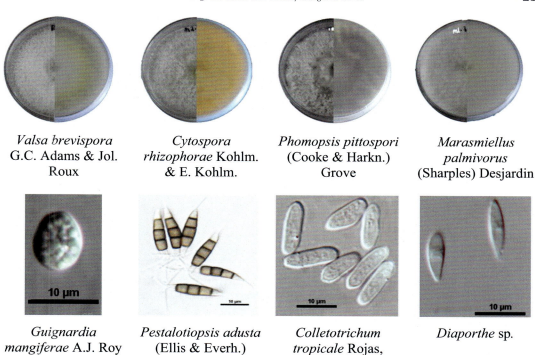

FIGURE 2.1 Colonies (front and back view) and conidia of mangrove fungal endophytes grown on Potato Dextrose Agar plates at room temperature. *Images were originally published in Apurillo et al. (2019).*

Avicennia marina, Avicennia sp., *A. corniculatum, A. rumphiana, Aegiceras corniculatum, Bruguiera gymnorrhiza, Camptostemon philippinense, Excoecaria agallocha, Lumnitzera litorea, Rhizophora apiculata, R. stylosa,* and *Sonneratia alba*) that were collected from Camarines Sur, Pangasinan, and Zambales (Moron et al., 2018). In this study, 40 morphospecies of mangrove fungal endophytes were identified as belonging to the genera *Fusarium, Xylaria, Colletotrichum, Aspergillus, Lasiodiplodia, Trichoderma, Penicillium, Paecilomyces, Pestalotiopsis, Alternaria,* and *Phomopsis*. It was also noted that the isolated fungal endophytes were similar with those reported from other plants, which indicates that these endophytic fungi are not host-specific to mangroves. They were also reported to possess antimicrobial activities (Moron et al., 2018).

Applications. Mangrove fungal endophytes have also been studied for their economic importance, particularly for their bioactivities. A fungal endophyte isolated from *Rhizophora annamalayana* Kathiresan was reported to produce taxol, a potent anticancer metabolite (Elavarasi et al., 2012). Mangrove-derived fungi belonging to the genera *Xylaria* and *Cladosporium* had potent antibacterial activities (Bhimba et al., 2012; Chaeprasert et al., 2010). Bioactivities of mangrove foliar fungal endophytes in the Philippines have also been reported. For example, *Phyllosticta* sp. isolated from a mangrove collected in Luzon Island produced Tyrosol C and Cytosporone B, which were reported to have antimicrobial and antioxidant activities, respectively (Tan et al., 2015). The mangrove fungal endophytes *Phialophora* sp.,

Cladosporium sp., and *Penicillium* sp. were very active against *Staphylococcus aureus* and *Micrococcus luteus* (Ramirez et al., 2020), while *P. adusta* from *R. mucronata* showed potent antibacterial activity against *P. aeruginosa* (Apurillo et al., 2019). Other mangrove foliar fungal endophytes such as *C. queenslandicum* and *C. tropicale* also showed antibacterial activity against *S. aureus*. In addition, *Xylaria cubensis* from *R. mucronata* exhibited cytotoxic activity against K562 myelogenous leukemia cells (Apurillo et al., 2019). *Pestalotiopsis microspora* isolated from mangrove stems showed strong antimicrobial activities against gram-positive bacteria, while species of *Lasiodiplodia* were very active against *Saccharomyces cerevisiae* (Moron et al., 2018). Aside from pharmaceutical applications, endophytic fungi from mangroves, *Fomitopsis* sp. and *Aspergillus tubingensis* Mosseray, produced xylanase, endoglucanase, and β-glucosidase with high enzymatic indices (Bacal & Yu, 2017). These studies showed the potential of Philippine mangrove fungal endophytes as rich sources of bioactive components that can be further explored for drug discovery.

4. Fungi associated with decaying mangrove leaves

Diversity. Fungi could also be isolated from dead and decaying mangrove leaves but remain underexplored in the Philippines. Most research in the country focuses on the isolation of fungal endophytes from healthy leaves of various mangrove species (e.g., Apurillo et al., 2019; Guerrero et al., 2018; Ramirez et al., 2020) and from stems and roots (Moron et al., 2018). To date, only the study of Torres and dela Cruz (2013) reported the occurrence of several mitosporic fungi from moribund mangrove leaves. This includes species of *Aspergillus*, *Aureobasidium*, *Colletotrichum*, *Fusarium*, *Guignardia*, *Paecilomyces*, *Penicillium*, and *Phomopsis* (Table 2.2). These species are ubiquitous, either as allochthonous or autochthonous, and can colonize a wide range of habitats and substrates as saprotrophs. Nakagiri et al. (1989) proposed that *Fusarium* sp., and *Penicillium* sp. are initial colonizers of fallen senescent mangrove leaves and thrive throughout the entire leaf-decay stages.

However, despite the limited studies on mitosporic fungi from decaying mangrove leaves, there have been several reports related to the occurrence of saprotrophic lower fungi, the

TABLE 2.2 Genera of fungi isolated from decaying leaves of Philippine mangroves.

Phylum	Order	Decaying leaves
Ascomycota	*incertae sedis* (Diaporthales ?)	*Phomopsis*
	Eurotiales	*Aspergillus* *Paecilomyces* *Penicillium*
	Dothideales	*Aureobasidium*
	Glomerellales	*Colletotrichum*
	Botryosphaeriales	*Guignardia*
Labyrinthulomycota	Thraustrochytriales	*Schizochytrium* *Thraustochytrium*

thraustochytrids, in dead mangrove leaves (Table 2.2). Leaño (2001) reported one of the early studies that identified the occurrence of *Schizochytrium mangrovei* Raghuk. and several *Thraustochytrium* species from dead mangrove leaves. It was followed by Arafiles et al. (2011), a decade after Leaño (2001), where they reported *Thraustochytrium* sp. and *Schizochytrium* sp. from Subic Bay, Philippines. The genera *Schizochytrium* and *Thraustochytrium* were considered as common isolates at that time. However, with the advent of molecular phylogenetics, several genera within the "Thraustochytrid" clade were proposed separately from Thraustochytriaceae, that is, *Aurantiochytrium, Botryochytrium, Hondaea, Monorhizochytrium, Oblongichytrium, Parietichytrium, Stellarchytrium* (Bennett et al., 2017). This led to the idea that the previously reported *Thraustochytrium* spp. from several mangrove areas in the country may be congeners of the recently proposed genera. Since morphology appears to be a limitation in species delineation, Perez et al. (2020) refrained from assigning their thraustochytrid isolates to a definite genus, although presented similar thalli morphology to *Aurantiochytrium, Hondaea,* or *Monorhizochytrium*, and proposed that combined molecular and morphological analyses should be presented. Fig. 2.2 shows some of the thraustochytrids isolated from decayed mangrove leaves collected in the Philippines.

Applications. The Philippine mangrove fungi reported from dead and decaying leaves were also explored for biotechnological applications (Table 2.3). The mangrove fungal isolates studied by Torres and dela Cruz (2013) produced xylanase, an enzyme currently used in wastewater management (Collins et al., 2005), and the paper and pulp bleaching process (Jiang et al., 2010). Specifically, *Fusarium* sp. (strain KAWIT-A) produced xylanase on the second day of incubation, whereas *Phomopsis* sp. (strain MACA-J) exhibited maximum xylanase production after 6 days of incubation. These fungi were also screened for enzymatic pretreatment of recycled paper pulp. The data presented by Torres and dela Cruz (2013) showed the promising application of mangrove fungi from decayed leaves in the paper industry and other enzyme-based technologies.

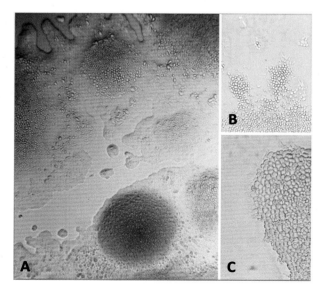

FIGURE 2.2 Thraustochytrids colonizing dead leaves of Philippine mangroves. (A) Typical colonies of thraustochytrids. (B−C) Typical ameboid form of cells for some congeners of thraustochytrids.

28   2. Diversity and applications of fungi associated with mangrove leaves in the Philippines

Thraustochytrids are also known economically as oleaginous organisms. They are currently tapped for their ability to produce high amounts of fatty acids. In the Philippines, several strains of thraustochytrids were reported to produce high amounts of docosahexaenoic acid (DHA), eicosapentaenoic acid (EPA), arachidonic acid, myristic acid, palmitic acid, stearic acid, oleic acid, linoleic acid, and linolenic acid (Arafiles et al., 2011; Leaño et al., 2003). An advantage of thraustochytrid cells is their ability to produce high biomass, which can be used as an alternative to the increasing demand for limited and nonrenewable oil supply. Apart from the ability of thraustochytrids to produce fatty acids, some strains were explored for their carotenogenic capabilities. Atienza et al. (2012) analyzed the potential of *Thraustochytrium* sp. strain SB04 and *Schizochytrium* sp. strain SB11 for the production of carotenoids. Their findings could lead to its potential as fish feed to commercially farmed fishes such as tilapia (*Oreochromis niloticus*), a common fish available in the Philippine market. Interestingly, fish fed with the thraustochytrid cells yield better average body length and weight than those fed with commercial feeds. Furthermore, the reported isolated thraustochytrids produce astaxanthin, echinenone, and lutein. Mangrove oomycetes *Halophytophthora vesicula* AK1YB2, *H. vesicula* PQ1YB3 and *Salispina spinosa* were also found to produce different fatty acids such as arachidonic acid, linoleic acid, and vaccenic acid (Caguimbal et al., 2019) (Table 2.3).

TABLE 2.3   Uses of fungi isolated from mangrove leaves in the Philippines.

Applications	Fungi	Hosts
Antimicrobial activity	*Phyllosticta* sp.	Not specified
	Phialophora sp.	*R. mucronata*
	Cladosporium sp.	*R. stylosa*
	Penicillium sp.	*R. mucronata*
	P. adusta	*R. mucronata*
	C. queenslandicum	*A. marina*
	C. tropicale	*A. marina*
	P. microspora	*R. apiculata*
	Lasiodiploida sp.	*R. stylosa*
Cytotoxic activity	*X. cubensis*	*R. mucronata*
Enzyme production	*Fusarium* sp.	Decayed mangrove leaves
	Fomitopsis sp.	*A. marina*
	Aspergillus tubingensis	*S. caseolaris*
Carotenoid production	*Thraustochytrium* sp. strain SB04	Decayed mangrove leaves
	Schizochytrium sp. strain SB11	Decayed mangrove leaves
Fatty acid production	*Halophytophthora vesicula* AK1YB2	Decayed mangrove leaves
	H. vesicula PQ1YB3	Decayed mangrove leaves
	Salispina spinosa	Decayed mangrove leaves

I. Fungi in aquatic and terrestrial habitats

5. Strategies to isolate and identify mangrove leaf-associated fungi

The limited studies on mangrove-associated fungi and the high diversity of Philippine mangrove species certainly merit continuous exploration of our mangrove forests. Herewith, we describe some strategies that can be employed to isolate these mangrove-associated fungi from mangrove leaves.

Fungal endophytes. Fungal endophytes can be easily isolated from healthy mangrove leaves. Leaf discs of about 6 mm in diameter are cut with a one-hole puncher, placed in sterile Petri plates, and surface-sterilized by sequential washing with 75% ethyl alcohol (1 min), 0.5% sodium hypochlorite (3 min), and 75% ethyl alcohol (30 s), with final washing of sterile distilled water (3 times, 3 min) (Apurillo et al., 2019). The leaf explants are placed on Petri plates prefilled with a suitable medium such as Potato Dextrose Agar or Malt Extract Agar. To prevent growth of mangrove-associated bacteria, antibiotics are added to the culture media, for example, 500 mg/L streptomycin sulfate. The culture plates are incubated at room temperature and observed daily for a week which can be extended to a month. Fungi that grow out of the leaf explants are subsequently subcultured for isolation. Identification can be made through a polyphasic approach of morphological characterization and molecular techniques. For molecular identification, fragments of the nuclear ribosomal internal transcribed spacer (ITS) are primarily used for species identification. However, it may be necessary to use other gene markers for isolates whose identification cannot be resolved using ITS. These include *translation elongation factor, partial actin gene* (ACT), *partial beta-tubulin gene, partial histone gene, partial calmodulin gene* (CAL), *partial glyceraldehyde-3-phosphate dehydrogenase genes* (GAPDH), and the *Apn2/MAT* locus (ApMAT) (Apurillo et al., 2019). Phylogenetic analysis using maximum likelihood, maximum parsimony, or Bayesian posterior probability is conducted to confirm the identities of the fungal isolates.

Thraustochythrids. For the isolation of thraustochytrids, leaf disks are prepared from senescent mangrove leaves as described by Leaño (2001) and Arafiles et al. (2011). No surface sterilization with alcohol or NaOCl is done with the leaf samples; however, it is necessary to wash the leaf disks three times for an hour, each with 50% sterile natural seawater (NSW) (Fan & Kamlandee, 2003). The washed leaf disks are placed on glucose-yeast extract peptone seawater (GYPS) agar which is composed of 3.0 g glucose, 1.25 g yeast extract, 1.25 g peptone, 1.0 L 50% natural seawater (NSW) at pH 6.0, 15.0 g agar, and supplemented with 300 mg/L streptomycin and penicillin G (Leaño et al., 2003; Perveen et al., 2006). A small amount of sterile 50% NSW is dropped onto the leaf disks to promote primary sporulation of the organisms. The culture plates are incubated at 25−28°C for 2−3 days, and thraustochytrid cells growing around the leaf disks are subcultured on GYPS agar. Identification is done through morphological characterization in combination with the sequencing of 18S rRNA genes (Bennett et al., 2017). However, as the systematics and phylogeny of Thraustochytrids, and Labyrinthulomycota in general, are still in flux, the concept of a standard barcode is yet to be established.

I. Fungi in aquatic and terrestrial habitats

6. Concluding remarks and future direction

Many mangrove-associated fungi have already been reported in the Philippines. However, these fungi were isolated only from a small portion of the total area of mangrove ecosystems in the country. In fact, most studied mangrove forests are found in Luzon Island, with few reports in the Visayas and none so far, in Mindanao. It would be interesting to compare the diversity of mangrove fungi from other mangrove ecosystems in the country and study the possible influence of geographical location on their diversity and distribution. An early attempt on this comparative diversity was done with mangrove-associated fungi from the Philippines and Indonesia (dela Cruz et al., 2020). Another interesting line of research would be to compare fungal diversity between different mangrove species. It has been shown that mangroves from different areas harbored high fungal diversity. This would mean a higher probability of isolating novel fungal species if other mangrove areas or mangrove species in the Philippines will be explored. The use of polyphasic approach, that is, a combined application of morphological and molecular techniques, is an essential tool in identifying these fungal species and to determine the novel taxa. While a few studies were reported on mangrove fungi associated with healthy mangrove leaves, stems and roots, the diversity of fungi associated with dead or senescent mangrove leaves, particularly on lower fungi such as thrautochytrids, is a gap in Philippine mycology. From an ecological perspective, studying these fungi, either those considered as higher or lower fungi, will certainly shed light on fungal succession and distribution. From an evolutionary standpoint, one can also understand how habitat transition and physiological adaptation impacts colonization and speciation. As many of the species reported in Philippine mangrove forests are identified only up to the genus level, it is also equally important to update the current species listing and contribute to the refinement of fungal nomenclature. Lastly, mangrove fungi are ideal and promising microorganisms for biotechnological and biopharmaceutical applications. It has been previously shown that marine microbes produce diverse and biologically active secondary metabolites (Schulz et al., 2008). This will certainly warrant full attention due to the complex substances they produce in response to their unique environment, which can be helpful in future drug discovery program.

Acknowledgments

The authors acknowledge the numerous authors whose papers are cited in this paper.

References

Alias, S. A., Jones, E. B. G., & Torres, J. (1999). Intertidal fungi from the Philippines, with a description of *Acrocordiopsis sphaerica* sp. nov. *Fungal Divers, 2*, 35–41.

Apurillo, C. C. S., Cai, L., & dela Cruz, T. E. E. (2019). Diversity and bioactivities of mangrove fungal endophytes from Leyte and Samar, Philippines. *Philippine Science Letters, 12*, 33–48.

Arafiles, K. H. V., Alcantara, J. C. O., Cordero, P. R. F., Batoon, J. A. L., Galura, F. S., Leaño, E. M., & Dedeles, G. R. (2011). Cultural optimization of thraustochytrids for biomass and fatty acid production. *Mycosphere, 2*(5), 521–531.

Atienza, G. A. M. V., Arafiles, K. H. V., Carmona, M. C. M., Garcia, J. P. C., Macabago, A. M. B., Peñacerrada, B. J. D. C., Cordero, P. R. F., Bennett, R. M., & Dedeles, G. R. (2012). Carotenoid analysis of locally

References

isolated thraustochytrids and their potential as an alternative fish feed for *Oreochromis niloticus* (Nile Tilapia). *Mycosphere, 3*(4), 420−428.

Bacal, C. J. O., & Yu, E. T. (2017). Cellulolytic activities of a novel *Fomitopsis* sp. and *Aspergillus tubingensis* isolated from Philippine mangroves. *Philippine Journal of Science, 146*(4), 403−410.

Bennett, R. M., Honda, D., Beakes, G. W., & Thines, M. (2017). Labyrinthulomycota. In J. M. Archibald, A. G. B. Simpson, C. H. Slamovits, L. M. Margulis, M. Melkonian, D. J. Chapman, & J. O. Corliss (Eds.), *Handbook of the protists* (pp. 507−542). Springer International Publishing.

Besitulo, A., Moslem, M. A., & Hyde, K. D. (2010). Occurrence and distribution of fungi in a mangrove forest on Siargao Island, Philippines. *Botanica Marina, 53*(6), 535−543.

Bhimba, B. V., Franco, D. A., Matthew, J., Jose, G. M., Joel, E. L., & Thangaraj, M. (2012). Anticancer and antimicrobial activity of mangrove-derived fungi *Hypocrea lixii VB1. Chinese Journal of Natural Medicines, 10*(1), 0077−0080.

Caguimbal, N. A. L. E., Devanadera, M. K. P., Bennett, R. M., Arafiles, K. H. V., Watanabe, K., Aki, T., & Dedeles, G. R. (2019). Growth and fatty acid profiles of *Halophytophthora vesicula* and *Salispina spinosa* from Philippine mangrove leaves. *Letters in Applied Microbiology, 69*(3), 221−228.

Calabon, M., Sadaba, R. B., & Campos, W. L. (2019). Fungal diversity of mangrove-associated sponges from New Washington, Aklan, Philippines. *Mycology, 10*(1), 6−21.

Chaeprasert, S., Piapukiew, J., Whalley, A. J., & Sihanonth, P. (2010). Endophytic fungi from mangrove plant species of Thailand: Their antimicrobial and anticancer potentials. *Botanica Marina, 53*, 555−564.

Collins, T., Gerday, C., & Feller, G. (2005). Xylanases, xylanase families and thermophilic xylanases. *FEMS Microbiology Reviews, 29*, 3−23.

dela Cruz, T. E. E., Notarte, K. I. R., Apurillo, C. C. S., Tarman, K., & Bungihan, M. E. (2020). Biomining fungal endophytes from tropical plants and seaweeds for drug discovery. In M. Ozturk, D. Egamberdieva, & M. Pesic (Eds.), *Biodiversity and biomedicine our future* (pp. 51−62). Elsevier Academic Press. ISBN: 978-0-12-819541-3.

Donato, D. C., Kauffman, J. B., Murdiyarso, D., Kurnianto, S., Stidham, M., & Kanninen, M. (2011). Mangroves among the most carbon-rich forests in the Tropics. *Nature Geoscience, 4*(5), 293−297.

Duke, N. C., Meynecke, J. O., Dittmann, S., Ellison, A. M., Anger, K., Berger, U., Cannicci, S., Diele, K., Ewel, K. C., Field, C. D., Koedam, N., Lee, S. Y., Marchand, C., Nordhaus, I., & Dahdouh-Guebas, F. (2007). A world without mangroves? *Science, 317*, 41−42.

Elavarasi, A., Rahna, G. S., & Kalaiselvam, M. (2012). Taxol producing mangrove endophytic fungi *Fusarium oxysporum* from *Rhizophora annamalayana. Asian Pacific Journal of Tropical Biomedicine, 2*(2), 1081−1085.

Fan, K. W., & Kamlandee, N. (2003). Polyunsaturated fatty acids production by *Schizochytrium* sp. isolated from mangrove. *Journal of Science and Technology, 25*, 643−650.

FAO. (2007). *The world's mangroves 1980−2005.* FAO Forestry Paper 153.

Gacutan, V. C., & Uyenco, F. R. (1983). Marine fungi from Batan, Aklan. *Natural and Applied Science Bulletin, 35*, 1−16.

Garcia, K. B., Malabrigo, P. L., & Gevaña, D. T. (2013). Philippines' mangrove ecosystem: Status, threats and conservation. *Mangrove Ecosystems of Asia*, 81−94. https://doi.org/10.1007/978-1-4614-8582-7_5

Goldberg, L., Lagomasino, D., Thomas, N., & Fatoyinbo, T. (2020). Global declines in human-driven mangrove loss. *Global Change Biology, 26*(10), 5844−5855.

Guerrero, J. J., General, M. A., & Serrano, J. E. (2018). Culturable foliar fungal endophytes of mangrove species in Bicol Region, Philippines. *Philippine Journal of Science, 147*(4), 563−574.

Jiang, Z., Cong, Q., Yan, Q., Kumar, N., & Du, X. (2010). Characterization of a thermostable xylanase from *Chaetomium* sp. and its application in Chinese steamed bread. *Food Chemistry, 120*, 457−462.

Jones, E. B. G., Hyde, K. D., Alias, S. A., & Moss, S. T. (1996). *Tirisporella* gen. novo an ascomycete from the mangrove palm *Nypa fruiticans. Canadian Journal of Botany, 74*, 1487−1495.

Jones, E. B. G., Uyenco, F. R., & Follosco, M. P. (1988). Fungi on driftwood collected in the intertidal zone from the Philippines. *Asian Marine Biology, 5*, 103−106.

Leaño, E. M. (2001). Straminipilous organisms from fallen mangrove leaves from Panay Island, Philippines. *Fungal Divers, 6*, 75−81.

Leaño, E. M., Gapasin, R. S. J., Polohan, B., & Vrijmoed, L. L. P. (2003). Growth and fatty acid production of thraustochytrids from Panay mangroves, Philippines. *Fungal Divers, 12*, 111−122.

Lim, A. R. U., Silva, R. M. N., Lesaca, G. R. E., Mapalo, V. J. C., Pecundo, M. H., & dela Cruz, T. E. E. (2021). First survey of myxomycetes in successional and mangrove forests of Negros Oriental, Philippines. *Slime Molds, 1*, V1A7.

I. Fungi in aquatic and terrestrial habitats

Liu, A., Wu, X., & Xu, T. (2007). Research advances in endophytic fungi of mangrove. *Chinese Journal of Applied Ecology, 18*(4), 912–918.

Long, J. B., & Giri, C. (2011). Mapping the Philippines' mangrove forests using landsat imagery. *Sensors, 11*(3), 2972–2981. https://doi.org/10.3390/s110302972

Moron, L. S., Lim, Y., & dela Cruz, T. E. E. (2018). Antimicrobial activities of crude culture extracts from mangrove fungal endophytes collected in Luzon island, Philippines. *Philippine Science Letters, 11*, 28–36.

Nakagiri, A., Tokumasu, S., Araki, H., Koreeda, S., & Tubaki, K. (1989). Succession of fungi in decomposing mangrove leaves in Japan. In T. Hattori, Y. Ishida, Y. Maruyama, R. Y. Morita, & A. Ochida (Eds.), *Recent advances in microbial ecology* (pp. 297–301). Japan Science Society Press.

Perez, J. G. A., Mappala, A. K. A., Icaro, C. K. L., Estrada, A. M. T., Arafiles, K. H. V., & Dedeles, G. R. (2020). Occurrence of thraustochytrids on fallen mangrove leaves from Pagbilao mangrove Park, Quezon province. *Philippine Journal of Systematic Biology, 14*, 1–8.

Perveen, Z., Ando, H., Ueno, A., Ito, Y., Yamamoto, Y., Yamada, Y., Takagi, T., Kaneko, T., Kogame, K., & Okuyama, H. (2006). Isolation and characterization of a novel thraustochytrid-like microorganisms that efficiently produces docosahexaenoic acid. *Biotechnology Letters, 28*, 197–202.

Petrini, O. (1991). Fungal endophytes of tree leaves. In J. H. Andrews, & S. S. Hirano (Eds.), *Microbial ecology of leaves* (pp. 179–197). Springer.

Primavera, J. H. (2000). Development and conservation of Philippine mangroves: Institutional issues. *Ecological Economics, 35*, 91–106.

Ramirez, C. S. P., Notarte, K. I. R., & dela Cruz, T. E. E. (2020). Antibacterial activities of mangrove leaf endophytic fungi from Luzon Island, Philippines. *Studies in Fungi, 3*, 320–331.

Schulz, B., Draeger, S., dela Cruz, T. E., Rheinheimer, J., Siems, K., Loesgen, S., Bitzer, J., Schloerke, O., Zeeck, A., Koch, I., Hussain, H., Dai, J., & Krohn, K. (2008). Screening strategies for obtaining novel, biologically active, fungal secondary metabolites from marine habitats. *Botanica Marina, 51*, 219–234.

Schulz, B., Römmert, A.-K., Dammann, U., Aust, H.-J., & Strack, D. (1999). The endophyte-host interaction: A balanced antagonism. *Mycological Research, 103*, 1275–1283.

Shearer, C. A., Descals, E., Kohlmeyer, J., Marvanova, L., Pedgett, D., Porter, D., Raja, H. A., Schmit, J. P., Thorton, H. A., & Voglymayr, H. (2007). Fungal diversity in aquatic habitats. *Biodiversity and Conservation, 16*, 49–67.

Tan, M. A., dela Cruz, T. E. E., Apurillo, C. C. S., & Proksch, P. (2015). Chemical constituents from a Philippie mangrove endophytic fungi *Phyllosticta* sp. *Der Pharma Chemica, 7*(2), 43–45.

Thatoi, H., Behera, B. C., & Mishra, R. R. (2013). Ecological role and biotechnological potential of mangrove fungi: A review. *Mycology, 4*(1), 54–71.

Torres, J. M. O., & dela Cruz, T. E. E. (2013). Production of xylanases by mangrove fungi from the Philippines and their application in enzymatic pretreatment of recycled paper pulps. *World Journal of Microbiology and Biotechnology, 29*, 645–655.

UNEP. (2014). In J. van Bochove, E. Sullivan, & T. Nakamura (Eds.), *The importance of mangroves to People: A call to action* (p. 128). United Nations Environment Programme World Conservation Monitoring Centre.

CHAPTER 3

Fungi and fungus-like microorganisms in Philippine marine ecosystems

Irish Emmanuel P. Agpoon[1,2], *Mark Kevin P. Devanadera*[2,3], *Kimberly D. Neri*[1,2] *and Gina R. Dedeles*[1,2,4]

[1]The Graduate School, University of Santo Tomas, Manila, Metro Manila, Philippines; [2]Research Center for the Natural and Applied Sciences, University of Santo Tomas, Manila, Metro Manila, Philippines; [3]Department of Biochemistry, Faculty of Pharmacy, University of Santo Tomas, Manila, Metro Manila, Philippines; [4]Department of Biological Sciences, College of Science, University of Santo Tomas, Manila, Metro Manila, Philippines

1. Introduction

At the onset of these marine fungal studies, this section describes the yeast strains and other organisms isolated from salt marshes to marine mangroves. The Philippines, as a tropical country, abound with several mangrove forests all over its archipelago. There are about 47 species of true mangroves and associated species belonging to 26 families (Primavera, 2009). Studies on marine fungi and fungal-like organisms isolated from different substrates like wood, leaves, and roots of mangrove trees, macroalgae (seaweeds), seagrasses, and even sponges have been reported and described (Apurillo et al., 2019; Besitulo et al., 2010; Calabon et al., 2019; Guerrero et al., 2018; Moron et al., 2018; Notarte et al., 2018). These microorganisms thrive in the marine environment where they exist as saprophytes, parasites, and symbionts on algae, vascular plants, invertebrates, and fishes. They are also found on matrices such as sediments, logs, and water (Johnson & Sparrow, 1961). Marine fungi can be classified either as obligate that complete their lifecycle in marine environments, or facultative that can grow in freshwater, terrestrial, and marine habitats (Kohlmeyer & Kohlmeyer, 1979). Despite the abundance of mangrove ecosystems, the identification of unicellular yeasts

34   3. Fungi and fungus-like microorganisms in Philippine marine ecosystems

is still lacking, likewise with the documentation of oomycetes, which superficially resemble fungi in mycelial growth and mode of nutrition. This section hopes to provide valuable information about these fungal species, particularly their involvement in a variety of ecological functions. With the country's type of climate and geography, the occurrence and distribution of marine microorganisms are assumed to be abundant and diverse.

2. Marine yeasts, mangrove-derived fungi, and marine fungal endophytes

2.1 Yeasts from the marine environment

Yeasts are unicellular, heterotrophic, eukaryotic microorganisms classified in the kingdom of Fungi (Shurson, 2018). They are considered a polyphyletic group of basidiomycetous and ascomycetous fungi (Kutty & Philip, 2008). Generally, the cells are about 3–4 μm in size, nonphotosynthetic, but with prominent nuclear membrane and cell walls. They reproduce asexually through budding and fission, where they are distinguished from filamentous fungi by their unique unicellular growth (Kurtzman & Fell, 1998; Paulino et al., 2017). Yeasts are facultative anaerobes in which they can propagate in the presence or absence of oxygen. Cells secrete proteolytic, glycolytic, or lipolytic enzymes that facilitate the digestion of organic matter and absorb amino acids and monosaccharides through their cell walls (Shurson, 2018).

Yeasts thrive in various aquatic habitats such as lakes, rivers, estuaries, seas, and oceans (Kutty & Philip, 2008). A wide range of substrates such as barks, leaves, soil, and mangrove water offer a microhabitat for yeasts (Liu & Chi, 2012; Nasr et al., 2017). The existence of yeasts has been implicated in ecologically significant processes, including decomposition of plant substrates, biogeochemical cycling, and detrital food web, and can act as a food source for some marine invertebrates and zooplanktons (Kutty & Philip, 2008; Liu & Chi, 2012; Paulino et al., 2017). It was suggested that yeasts found in the aquatic environments are generally asporogenous and weakly fermentative or oxidative (Pitt & Miller, 1970). In ocean waters, they are involved in the degradation of macrophytes, including seaweeds and seagrasses (Raghukumar, 2008).

Globally, reports on yeasts' distribution in the terrestrial and aquatic environment are well documented. In other countries, the diversity of marine yeasts on seaweeds (Francis et al., 2016), occurrence in mangroves (Liu & Chi, 2012; Nasr et al., 2017), yeasts associated with coral reefs (Paulino et al., 2017), and seawater and marine sediments (Chi et al., 2008; Gao et al., 2007) have been investigated. In the Philippines, only a few papers discussed the isolation of yeasts from the marine environment. For instance, Ramirez et al. (2010) isolated and characterized marine yeasts from substrates, which include seawater, marine sediments, decaying seagrasses (*Enhalus acoroides* (Linnaeus f.) Royle, 1839), and seaweeds (*Kappaphycus* sp. Doty, 1988) collected in Calatagan, Batangas. Their analyses identified Ascomycetous yeasts such as *Candida tropicalis* (Castell.) Berkhout (1923) and *Pichia caribbica* Vaughan-Martini (2005).Their presence in the marine environment is suggestive of their role in the degradation of these substrates.

From seawater samples collected from Manila Bay, Sia Su et al. (2014) have only identified yeast through microscopic and germ tube techniques. Calabon et al. (2019) reported the diversity of fungi of mangrove-associated sponges. They identified the following yeasts

I. Fungi in aquatic and terrestrial habitats

species: *Candida famata* (F.C. Harrison) S.A. Mey. & Yarrow, *Candida guilliermondii* (Castellani) Langeron & Guerra, *Cryptococcus* sp. Vuill., *Geotrichum* sp., *Hortaea werneckii* (Horta) Nishim. & Miyaji, *Kloeckera* sp. Janke, and *Pichia angusta* (Teun., H.H. Hall & Wick.) Kurtzman. However, identification through molecular sequencing was not performed. In contrast, the first report of *C. tropicalis* from edible freshwater fish *Glossogobius giuris* Hamilton (tank goby) in Lake Danao was reported by De Jesus Milanez et al. (2020). This study and other findings show that yeasts are involved in the decomposition of marine substrates (Ramirez et al., 2010) or form a sponge-symbiotic relationship (Calabon et al., 2019). As yeasts are widely distributed in the environment, Calabon et al. (2019) asserted that further research is needed to provide strong evidence and identify the origin of the yeast and its association with the marine host. De Jesus Milanez et al. (2020) also support that an intensive study is also needed to identify whether yeast species isolated from the aquatic environment are pathogenic or not. These present studies suggest that further research on the distribution of yeasts in the Philippine marine environment is necessary.

2.2 Fungi from mangrove forests

The mangrove ecosystem is generally productive, rich in nutrients, and organic matter that offers a habitat to numerous microorganisms such as fungi (Besitulo et al., 2010; Holguin et al., 2001). Some factors influence the colonization of fungi in mangroves, which includes salinity, the temperature of seawater, position in the intertidal region, nature of the mangrove floor (mud or sand), pH, oceanic region, tidal exposure, microbial interactions, and type of substratum (Alias & Jones, 2000, 2009; Hyde & Jones, 1988). Studies have reported that fungi in mangroves may act as endophytes to colonize the internal plant tissues without causing any damage. But depending on the substrates, some have reported them as pathogens or saprobes. The nutrients released from decomposed dead plants and animal remains are either recycled in the environment or used to support plant growth or as a nutrient source for some marine invertebrates (Hyde & Lee, 1995; Liu & Chi, 2012).

The occurrence of fungi in select mangrove locations in the Philippines is listed in Table 3.1. Besitulo et al. (2010) studied the occurrence of manglicolous fungi in the mangrove woody tissue. A total of 66 fungal species were recorded from wood and driftwood samples from the mangroves of Del Carmen, Siargao Island, Philippines. Most species collected are ascomycetous obtained from the driftwood and suggest that they degrade lignocellulose (Bucher et al., 2004). This is followed by the basidiomycetes, which are involved in lignin mineralization and the production of lignin-modifying enzymes (Mouzouras et al., 1986).

Fungal endophytes in healthy leaves were studied by Guerrero et al. (2018) and identified 172 isolates that belonged to Ascomycota (75.47%), Basidiomycota (20.75%), and Zygomycota (3.77%). However, their study did not discuss the exact role of fungi as an endophyte in plants (pathogenic or saprophytic). The fungal endophytes from healthy mangrove leaves were reported by Ramirez et al. (2020), while stems and roots were used as substrates by Moron et al. (2018). Results showed a high colonization rate in roots, indicating a preference for tissue types (Carrol & Petrini, 1983). Another factor that can influence fungal colonization is the leaf age. More endophytes can be isolated from old/decaying leaves (Osono, 2008) and host-preference of the different fungal taxa (Calabon et al., 2019). In contrast, Rafael and Calumpong (2019) identified foliar diseases in mangrove leaves, namely, Brown Leaf Spot

36 3. Fungi and fungus-like microorganisms in Philippine marine ecosystems

TABLE 3.1 Records of Philippine fungi from aquatic habitats.

Species	Substrate/host	Habitat	Locality	References
Acremonium kiliense Grütz	Sponge-associated fungi, in mangroves habitat	Brackish	New Washington, Aklan, Panay	Calabon et al. (2019)
Acremonium Link	Saprophytic, with dye decolorization bioactivities; Sponge-associated fungi, in mangroves habitat	Marine; Brackish	Coastal Road, Pasay City; Navotas Fish Port, Navotas; Calampit Ternate, Cavite; New Washington, Aklan, Panay	Torres et al. (2011), Calabon et al. (2019)
Acremonium strictum W. Gams	Sponge-associated fungi, in mangroves habitat	Brackish	New Washington, Aklan, Panay	Calabon et al. (2019)
Acrocordiopsis patilii Borse and K.D. Hyde	Saprophytic	Marine	Boracay, Aklan	Alias, Jones, and Torres (1999)
Acrocordiopsis patilii Borse & K.D. Hyde	Saprophytic	Brackish, Mangroves swamp	Del Carmen, Siargao Island	Besitulo et al. (2010)
Acrocordiopsis sphaerica Alias and E.B.G. Jones	Saprophytic	Marine	Boracay, Aklan	Alias, Jones, and Torres (1999)
Acrodictys liputii Cai, Zhang, McKenzie, Ho & Hyde	Saprophytic	Freshwater	Liput River, Alegria, Negros Occidental	Cai et al. (2003)
Acrodictys sacchari M.B. Ellis	Saprophytic	Freshwater	Liput River, Alegria, Negros Occidental	Cai et al. (2003)
Acrodontium crateriforme (J.F.H. Beyma) de Hoof	Sponge-associated fungi, in mangroves habitat	Brackish	New Washington, Aklan, Panay	Calabon et al. (2019)
Acrogenospora sphaerocephala (Berk. & Broome) M.B. Ellis	Saprophytic	Freshwater	Liput River, Alegria, Negros Occidental	Cai et al. (2003)
Adomia S. Schatz	Saprophytic	Marine, Brackish	Boracay, Aklan; Pagbilao Pangasinan	Alias, Jones, and Torres (1999)
Aigialus grandis Kohlm. & S. Schatz	Saprophytic	Brackish, Mangroves swamp	Del Carmen, Siargao Island	Besitulo et al. (2010)
Aigialus mangrovei Borse	Saprophytic	Brackish, Mangroves swamp	Del Carmen, Siargao Island	Besitulo et al. (2010)

I. Fungi in aquatic and terrestrial habitats

2. Marine yeasts, mangrove-derived fungi, and marine fungal endophytes   37

TABLE 3.1 Records of Philippine fungi from aquatic habitats.—cont'd

Species	Substrate/host	Habitat	Locality	References
Aigialus parvus Schatz and Kohlm.	Saprophytic	Marine, Brackish	Boracay, Aklan; Pagbilao Pangasinan	Alias, Jones, and Torres (1999)
Alternaria Nees	Saprophytic, with dye decolorization bioactivities; Mangrove (*Rhizophora stylosa*) associated endophytes	Marine, Brackish	Calatagan, Batangas; Camarines Sur	Torres et al. (2011), Moron et al. (2018)
Aniptodera chesapeakensis Shearer & M.A. Mill	Saprophytic	Marine	Boracay, Aklan; Pagbilao Pangasinan	Alias, Jones, and Torres (1999)
Aniptodera inflatiascigera K.M. Tsui, K.D. Hyde & Hodgkiss	Saprophytic	Freshwater	Liput River, Alegria, Negros Occidental	Cai et al. (2003)
Aniptodera intermedia K.D. Hyde & Alias	Saprophytic	Brackish, Mangroves swamp	Del Carmen, Siargao Island	Besitulo et al. (2010)
Aniptodera lignicola K.D. Hyde, W.H. Ho & K.M. Tsui	Saprophytic	Freshwater	Liput River, Alegria, Negros Occidental	Cai et al. (2003)
Aniptodera longispora K.D. Hyde	Saprophytic	Brackish, Mangroves swamp	Del Carmen, Siargao Island	Besitulo et al. (2010)
Aniptodera mangrovei K.D. Hyde	Saprophytic	Marine	Boracay, Aklan; Pagbilao Pangasinan	Alias, Jones, and Torres (1999)
Aniptodera (Halosarpheia) marina (Cribb and Cribb) Kohlm.	Saprophytic	Marine, Brackish	Taklong, Guimaras; Boracay, Aklan; Pagbilao, Pangasinan	Alias, Jones, and Torres (1999)
Annulatascus liputii L. Cai & K.D. Hyde	Saprophytic	Freshwater	Liput River, Alegria, Negros Occidental	Cai et al. (2003)
Annulatascus K.D. Hyde	Saprophytic	Freshwater	Liput River, Alegria, Negros Occidental	Cai et al. (2003)
Annulatascus velatisporus K.D. Hyde	Saprophytic	Freshwater	Liput River, Alegria, Negros Occidental	Cai et al. (2003)

(Continued)

I. Fungi in aquatic and terrestrial habitats

38   3. Fungi and fungus-like microorganisms in Philippine marine ecosystems

TABLE 3.1   Records of Philippine fungi from aquatic habitats.—cont'd

Species	Substrate/host	Habitat	Locality	References
Antennospora quadricornuta (Cribb and J.W. Cribb) T.W. Johnson	Saprophytic	Marine	Taklong, Guimaras; Boracay, Aklan; Pagbilao, Pangasinan	Alias, Jones, and Torres (1999)
Antennospora quadricornuta (Cribb and J.W. Cribb) T.W. Johnson	Saprophytic	Brackish, Mangroves swamp	Del Carmen, Siargao Island	Besitulo et al. (2010)
Antennospora salina (Meyers) Yusoff, E.B.G. Jones and S.T. Moss	Saprophytic	Marine, Brackish	Taklong, Guimaras; Boracay, Aklan; Pagbilao, Pangasinan	Alias, Jones, and Torres (1999)
Anthostomella nypae K.D. Hyde, B.S. Lu & Alias	Saprophytic	Brackish, Mangroves swamp	Del Carmen, Siargao Island	Besitulo et al. (2010)
Anthostomella nypensis K.D. Hyde, Alias & B.S. Lu	Saprophytic	Brackish, Mangroves swamp	Del Carmen, Siargao Island	Besitulo et al. (2010)
Aniptodera chesapeakensis Shaearer & M.A. Mill	Saprophytic	Brackish, Mangroves swamp	Del Carmen, Siargao Island	Besitulo et al. (2010)
Aquaphila albicans Goh, K.D. Hyde & W.H. Ho	Saprophytic	Freshwater	Liput River, Alegria, Negros Occidental	Cai et al. (2003)
Arenariomyces trifurcatus Hönk	Saprophytic	Marine	Boracay, Aklan; Pagbilao Pangasinan	Alias, Jones, and Torres (1999)
Arthrinium phaeospermum (Corda) M.B. Ellis	Mangrove (*Rhizophora stylosa)* associated endophytes	Brackish	Camarines Sur	Moron et al. (2018)
Ascocratera manglicola Kohlm.	Saprophytic	Brackish, Mangroves swamp	Del Carmen, Siargao Island	Besitulo et al. (2010)
Aspergilloides	Sponge-associated fungi, in mangroves habitat	Brackish	New Washington, Aklan, Panay	Calabon et al. (2019)
Aspergillus candidus Link	Sponge-associated fungi, in mangroves habitat	Brackish	New Washington, Aklan, Panay	Calabon et al. (2019)

I. Fungi in aquatic and terrestrial habitats

2. Marine yeasts, mangrove-derived fungi, and marine fungal endophytes 39

TABLE 3.1 Records of Philippine fungi from aquatic habitats.—cont'd

Species	Substrate/host	Habitat	Locality	References
Aspergillus fumisynnematus Y. Horie, Miyaji, Nishim., Taguchi & Udagawa	Sponge-associated fungi, in mangroves habitat	Brackish	New Washington, Aklan, Panay	Calabon et al. (2019)
Aspergillus novofumigatus S.B. Hong, Frisvad & Samson	Sponge-associated fungi, in mangroves habitat	Brackish	New Washington, Aklan, Panay	Calabon et al. (2019)
Aspergillus penicilloides Speg.	Sponge-associated fungi, in mangroves habitat	Brackish	New Washington, Aklan, Panay	Calabon et al. (2019)
Aspergillus flavus Link	Saprophytic	Marine	Calatagan, Batangas, Manila Bay	Solis et al. (2010), Sia Su et al. (2014)
Aspergillus fumigatus Fresen	Saprophytic	Marine	Manila Bay	Sia Su et al. (2014)
Aspergillus nidulans (Eidam) Winter	Fungal endophytes, on mangroves (*Sonneratia alba*)	Brackish, Marine	Marabut, Samar	Apurillo et al. (2019)
Aspergillus niger Tiegh.	Sponge-associated fungi, in mangroves habitat	Brackish	New Washington, Aklan, Panay	Calabon et al. (2019)
Aspergillus niveus Blochwitz	Sponge-associated fungi, in mangroves habitat	Brackish	New Washington, Aklan, Panay	Calabon et al. (2019)
Aspergillus ochraceus K. Wilh.	Saprophytic; likely hemibiotrophic casing *ice-ice* disease on its host macroalgae (*Kappaphycus*); Sponge-associated fungi, in mangroves habitat; Mangrove (*Rhizophora apiculata*) associated endophytes	Marine, Brackish	Calatagan, Batangas; New Washington, Aklan, Panay; Zambales	Solis et al. (2010), Moron et al. (2018), Calabon et al. (2019)
Aspergillus oryzae (Ahlb.) Cohn	Fungal endophytes, on mangroves (*Rhizophora mucronata*)	Brackish, Marine	Babatngon, Leyte	Apurillo et al. (2019)
Aspergillus restrictus G.Sm.	Sponge-associated fungi, in mangroves habitat	Brackish	New Washington, Aklan, Panay	Calabon et al. (2019)
Aspergillus sclerotiorum G.A. Huber	Sponge-associated fungi, in mangroves habitat	Brackish	New Washington, Aklan, Panay	Calabon et al. (2019)

(*Continued*)

I. Fungi in aquatic and terrestrial habitats

40   3. Fungi and fungus-like microorganisms in Philippine marine ecosystems

TABLE 3.1   Records of Philippine fungi from aquatic habitats.—cont'd

Species	Substrate/host	Habitat	Locality	References
Aspergillus P. Micheli ex Haller	Fungal endophytes; Saprophytic; Mangrove associated; Fungal endophytes on mangroves	Marine, Brackish	New Washington, Aklan, Panay; Manila Bay; Bais, Alcantara; Albay; Catanduanes; Sorsogon; Masbate; Camarines Sur; Camarines Norte; Pangangan Island	Calabon et al. (2019), Sia Su et al. (2014), Guerrero et al. (2018), Rafael and Calumpong (2019)
Aspergillus sydowii (Bainier & Sartory) Thom & Church	Saprophytic; Sponge-associated fungi, in mangroves habitat	Marine, Brackish	Calatagan, Batangas; New Washington, Aklan, Panay	Solis et al. (2010), Calabon et al. (2019)
Aspergillus tamarii Kita	Sponge-associated fungi, in mangroves habitat	Brackish	New Washington, Aklan, Panay	Calabon et al. (2019)
Aspergillus terreus Thom	Saprophytic; likely hemibiotrophic casing *ice-ice* disease on its host macroalgae (*Kappaphycus*); Sponge-associated fungi, in mangroves habitat	Marine, Brackish	Calatagan, Batangas; New Washington, Aklan, Panay	Solis et al. (2010), Calabon et al. (2019)
Astrosphaeriella mangrovei (Kohlm. & Vittal) Aptroot & K.D. Hyde	Saprophytic	Brackish, Mangroves swamp	Del Carmen, Siargao Island	Besitulo et al. (2010)
Astrosphaeriella papillata Hyde & J. Fröhl.	Saprophytic	Freshwater	Liput River, Alegria, Negros Occidental	Cai et al. (2003)
Astrosphaeriella stellata (Pat.) Sacc.	Saprophytic	Freshwater	Liput River, Alegria, Negros Occidental	Cai et al. (2003)
Astrosphaeriella striatispora (K.D. Hyde) K.D. Hyde	Saprophytic	Brackish, Mangroves swamp	Del Carmen, Siargao Island	Besitulo et al. (2010)
Astrosphaeriella tornata (Berk. & Curtis) Hawksw. & Boise	Saprophytic	Freshwater	Liput River, Alegria, Negros Occidental	Cai et al. (2003)
Bactrodesmium longisporum M.B. Ellis	Saprophytic	Freshwater	Liput River, Alegria, Negros Occidental	Cai et al. (2003)
Bactrodesmium stilboideum R.F. Castañeda & G.R.W. Arnold	Saprophytic	Freshwater	Liput River, Alegria, Negros Occidental	Cai et al. (2003)

I. Fungi in aquatic and terrestrial habitats

2. Marine yeasts, mangrove-derived fungi, and marine fungal endophytes

41

TABLE 3.1 Records of Philippine fungi from aquatic habitats.—cont'd

Species	Substrate/host	Habitat	Locality	References
Bathyascus grandisporus K.D. Hyde	Saprophytic	Brackish, Mangroves swamp	Del Carmen, Siargao Island	Besitulo et al. (2010)
Beauveria Vuill.	Sponge-associated fungi, in mangroves habitat	Brackish	New Washington, Aklan, Panay	Calabon et al. (2019)
Biatriospora marina K.D. Hyde & Borse	Saprophytic	Marine	Boracay, Aklan	Alias, Jones, and Torres (1999)
Blastocladia angusta Lund	Saprophytic on fruits materials in stagnant water	Freshwater	Los Baños, Laguna	Dogma (1986)
Blastocladia globosa Kanouse	Saprophytic on fruits materials in stagnant water	Freshwater	Los Baños, Laguna	Dogma (1986)
Blastocladia gracilis Kanouse	Saprophytic on fruits materials in stagnant water	Freshwater	Los Baños, Laguna	Dogma (1986)
Blastocladia incrassata Indoh	Saprophytic on fruits materials in stagnant water	Freshwater	Los Baños, Laguna	Dogma (1986)
Blastocladia pringsheimii Reinsch	Saprophytic, occurring in tufts on various rosaceous fruits and twigs	Freshwater	Los Baños, Laguna	Dogma (1986)
Blastocladia ramosa Thaxt.	Saprophytes of fruits and twigs	Freshwater	Los Baños, Laguna	Dogma (1986)
Blastocladia sparrowii Indoh	Saprophytic on fruits materials in stagnant water	Freshwater	Los Baños, Laguna	Dogma (1986)
Boerlagiomyces grandisporus S.J. Stanley & K.D. Hyde	Saprophytic	Freshwater	Liput River, Alegria, Negros Occidental	Cai et al. (2003)
Camposporium fusisporum Whitton, McKenzie & K.D. Hyde	Saprophytic	Freshwater	Liput River, Alegria, Negros Occidental	Cai et al. (2003)
Camposporium quercicola Mercado, Heredia & J. Mena	Saprophytic	Freshwater	Liput River, Alegria, Negros Occidental	Cai et al. (2003)

(*Continued*)

I. Fungi in aquatic and terrestrial habitats

42　　　　3. Fungi and fungus-like microorganisms in Philippine marine ecosystems

TABLE 3.1　Records of Philippine fungi from aquatic habitats.—cont'd

Species	Substrate/host	Habitat	Locality	References
Candelabrum brocchiatum Tubaki	Saprophytic	Freshwater	Liput River, Alegria, Negros Occidental	Cai et al. (2003)
Candida famata (F.C. Harrison) S.A. Mey. & Yarrow	Sponge-associated fungi, in mangroves habitat	Brackish	New Washington, Aklan, Panay	Calabon et al. (2019)
Candida guilliermondii (Castell.) Langeron & Guerra	Sponge-associated fungi, in mangroves habitat	Brackish	New Washington, Aklan, Panay	Calabon et al. (2019)
Catenaria anguillulae Sorokin	Saprophyte (on keratinic and chitinic materials) and endoparasitic (eggs and adults nematodes, mites, rotifers, *Daphnia*, liver fluke (*Fasciola hepatica*))	Freshwater	La Trinidad, Benguet; Daraga, Albay; Montalban, Rizal; Baguio City; Calamba, Laguna; Lutayan, Sultan Kudarat; Davao City	Dogma (1986)
Carinispora nypae K.D. Hyde	Saprophytic	Brackish, Mangroves swamp	Del Carmen, Siargao Island	Besitulo et al. (2010)
Caryospora callicarpa (Curry) Nitschke ex Fuckel	Saprophytic	Freshwater	Liput River, Alegria, Negros Occidental	Cai et al. (2003)
Caryospora minima Jeffers	Saprophytic	Freshwater	Liput River, Alegria, Negros Occidental	Cai et al. (2003)
Caryosporella rhizophorae Kohlm.	Saprophytic	Brackish, Mangroves swamp	Del Carmen, Siargao Island	Besitulo et al. (2010)
Chaetospermum Sacc.	Saprophytic	Brackish, Mangroves swamp	Del Carmen, Siargao Island	Besitulo et al. (2010)
Chloridium cylindrosporum Gams & Hol.-Jech.	Saprophytic	Freshwater	Liput River, Alegria, Negros Occidental	Cai et al. (2003)
Chrysosporium Corda	Saprophytic	Marine	Manila Bay	Sia Su et al. (2014)
Chytridium lagenaria Schenk	Hemibiotrophic on moribund filaments of *Oedogonium* sp.	Freshwater	Los Baños, Laguna	Dogma (1986)

I. Fungi in aquatic and terrestrial habitats

2. Marine yeasts, mangrove-derived fungi, and marine fungal endophytes **43**

TABLE 3.1 Records of Philippine fungi from aquatic habitats.—cont'd

Species	Substrate/host	Habitat	Locality	References
Chytridium oedogoniarum Dogma	Ectoparasitic on young filaments of green alga *Oedogonium*, and especially its oogonia and oospores	Freshwater	La Trinidad Valley, Benguet	Dogma (1986)
Chytridium palmelloidea Dogma	Parasitic on encysted (palmella stage) *Chlamydomonas* sp.	Freshwater	Daet, Camarines Norte	Dogma (1986)
Chytridium schenkii (P.A. Dang.) Scherff.	Epi-endobiotic, weakly parasites of *Oedogonium*, *Closterium*, *Bulbochaete*, *Spirogyra*, *Zygnema*, and *Cladophora*	Freshwater	Los Baños, Laguna	Dogma (1986)
Chytriomyces annulatus Dogma	Weakly parasitic on desmids; also live as saprophytes on pine pollens and snake skin baits	Freshwater	Pilar, Sorsogon	Dogma (1986)
Chytriomyces appendiculatus Karling	Epibiotic saprophytes of chitinic substrata (e.g., shrimp skin, termite wings)	Freshwater	Atimonan, Quezon; Mt. Makiling, Quezon	Dogma (1986)
Chytriomyces tabellariae (J. Schrot) Canter	Parasitic on the diatom *Tabellaria* sp.	Freshwater	Molawin creek, Los Baños, Laguna	Dogma (1986)
Cirrenalia Meyers & R.T. Moore	Saprophytic	Marine	Taklong, Guimaras; Boracay, Aklan; Pagbilao, Pangasinan	Alias, Jones, and Torres (1999)
Cirrenalia tropicalis Kohlm.	Saprophytic	Marine, Brackish	Boracay, Aklan; Pagbilao Pangasinan	Alias, Jones, and Torres (1999)
Cladochytrium replicatum Karling	Saprophyte of cellulosic materials	Freshwater	Molawin Creek, Los Baños, Laguna; Montalban, Rizal;	Dogma (1986)
Cladosporium cladosporioides (Fresen.) G.A. de Vries	Sponge-associated fungi, in mangroves habitat	Brackish	New Washington, Aklan, Panay	Calabon et al. (2019)
Cladosporium Link	Sponge-associated fungi, in mangroves habitat: Saprophytic with dye decolorization bioactivities	Marine; Brackish	Calatagan, Batangas; New Washington, Aklan, Panay; Manila Bay	Solis et al. (2010), Torres et al. (2011), Sia Su et al. (2014), Calabon et al. (2019)

(Continued)

I. Fungi in aquatic and terrestrial habitats

44
3. Fungi and fungus-like microorganisms in Philippine marine ecosystems

TABLE 3.1 Records of Philippine fungi from aquatic habitats.—cont'd

Species	Substrate/host	Habitat	Locality	References
Cladosporium sphaerospermum Penz.	Sponge-associated fungi, in mangroves habitat	Brackish	New Washington, Aklan, Panay	Calabon et al. (2019)
Clavatospora bulbosa (Anast.) Nakagiri & Tubaki	Saprophytic	Marine, Brackish	Taklong, Guimaras; Boracay, Aklan; Pagbilao, Pangasinan	Alias, Jones, and Torres (1999)
Coelomomyces indiana A.V.V. Iyengar	Endoparasitic on *Anopheles subpictus indefinitus*	Freshwater	Sta. Veronica, San Pablo City, Laguna	Dogma (1986)
Coelomomyces quadrangulatus Couch var. *lamborni* Couch	Endoparasitic on *Aedes aegypti* larvae	Freshwater	Molawin Creek, Los Baños, Laguna	Dogma (1986)
Coelomomyces stegomyiae Keilin	Endoparasitic on *Aedes aegypti* larvae	Freshwater	Los Baños, Laguna	Dogma (1986)
Coelomycetes Grove	Saprophytic	Freshwater	Liput River, Alegria, Negros Occidental	Cai et al. (2003)
Colletotrichum fructicola Prihastuti, L. Cai & K.D. Hyde	Fungal endophytes, on mangroves (*Avicennia marina*)	Brackish, Marine	Babatngon, Leyte	Apurillo et al. (2019)
Colletotrichum queenslandicum B.S Weir & P.R. Johnston	Fungal endophytes, on mangroves (*Avicennia marina*)	Brackish, Marine	Babatngon, Leyte	Apurillo et al. (2019)
Colletotrichum siamense Prihastuti, L. Cai & K.D. Hyde	Mangrove (*Rhizophora stylosa*) associated endophytes	Brackish	Camarines Sur	Moron et al. (2018)
Colletotrichum Corda	Mangrove (*Avicennia marina*) associated endophytes; Fungal endophytes on mangroves	Brackish	Camarines Sur; Albay; Catanduanes; Sorsogon; Masbate; Camarines Sur; Camarines Norte	Moron et al. (2018), Guerrero et al. (2018)
Colletotrichum tropicale Rojas, Rehner & Samuels	Fungal endophytes, on mangroves (*Avicennia marina*)	Brackish, Marine	Babatngon, Leyte	Apurillo et al. (2019)
Colletotrichum tropicale Rojas, Rehner & Samuels	Mangrove (*Rhizophora stylosa, Avicennia corniculatum*) associated endophytes	Brackish	Camarines Sur	Moron et al. (2018)

I. Fungi in aquatic and terrestrial habitats

2. Marine yeasts, mangrove-derived fungi, and marine fungal endophytes

TABLE 3.1 Records of Philippine fungi from aquatic habitats.—cont'd

Species	Substrate/host	Habitat	Locality	References
Coniothyrium Corda	Saprophytic, with dye decolorization bioactivities	Marine	Carpuran, Calatagan, Batangas	Torres et al. (2011)
Cordana abramovii Semen & Davydkina	Saprophytic	Freshwater	Liput River, Alegria, Negros Occidental	Cai et al. (2003)
Corollospora Wederm.	Saprophytic	Marine	Pagbilao, Pangasinan	Alias, Jones, and Torres (1999)
Coronopapilla mangrovei (K.D. Hyde) Kohlm. et Volkm.-Kohlm.	Saprophytic	Brackish, Mangroves swamp	Del Carmen, Siargao Island	Besitulo et al. (2010)
Cryptococcus Vuill.	Sponge-associated fungi, in mangroves habitat	Brackish	New Washington, Aklan, Panay	Calabon et al. (2019)
Cucullosporella mangrovei K.D. Hyde & E.B.G. Jones	Saprophytic	Marine, Brackish, Mangroves swamp	Boracay, Aklan; Pagbilao Pangasinan; Del Carmen, Siargao Island	Alias, Jones, and Torres (1999), Besitulo et al. (2010)
Curvularia intermedia Boedijn	Saprophytic	Marine	Calatagan, Batangas	Solis et al. (2010)
Cylindrochytridium Johnstonii Karling	Saprophytic on vegetables materials	Freshwater	Quezon National Park, Atimonan, Quezon	Dogma (1986)
Cytospora rhizophorae Kolm. & E. Kolm.	Fungal endophytes, on mangroves (*Rhizophora mucronata*)	Brackish, Marine	Marabut, Samar	Apurillo et al. (2019)
Dactylaria africana (S. Hughes) G.C. Bhatt & W.B. Kendr.	Saprophytic	Freshwater	Liput River, Alegria, Negros Occidental	Cai et al. (2003)
Dactylaria longidentata Cazau, Aramb. & Cabello	Saprophytic	Freshwater	Liput River, Alegria, Negros Occidental	Cai et al. (2003)
Dactylospora halioptrepha (Kohlm. & E. Kohlm) Hafellner	Saprophytic	Marine	Taklong, Guimaras; Pagbilao, Pangasinan	Alias, Jones, and Torres (1999)
Dactylospora haliotrepha ((Kohlm. & E. Kohlm) Hafellner	Saprophytic	Marine, Brackish, Mangroves swamp	Boracay, Aklan; Del Carmen, Siargao Island	Alias, Jones, and Torres (1999), Besitulo et al. (2010)

(Continued)

I. Fungi in aquatic and terrestrial habitats

46 3. Fungi and fungus-like microorganisms in Philippine marine ecosystems

TABLE 3.1 Records of Philippine fungi from aquatic habitats.—cont'd

Species	Substrate/host	Habitat	Locality	References
Dendryphiella salina (G.K. Suth.) Pugh and Nicot	Saprophytic	Marine	Taklong, Guimaras; Pagbilao, Pangasinan	Alias, Jones, and Torres (1999)
Diaporthe siamensis Udayanga, Xing Z., Liu & K.D. Hyde	Fungal endophytes, on mangroves (*Rhizophora mucronata*)	Brackish, Marine	Marabut, Samar	Apurillo et al. (2019)
Diaporthe Nitschke	Saprophytic, Fungal endophytes, on mangroves (*Avicennia marina*)	Marine, Brackish, Freshwater	Liput River, Alegria, Negros Occidental; Babatngon, Leyte	Cai et al. (2003), Apurillo et al. (2019)
Dictyochaeta curvispora L. Cai, McKenzie & K.D. Hyde	Saprophytic	Freshwater	Liput River, Alegria, Negros Occidental	Cai et al. (2003)
Dictyochaeta plovercovensis Goh & K.D. Hyde	Saprophytic	Freshwater	Liput River, Alegria, Negros Occidental	Cai et al. (2003)
Dictyochaeta Speg.	Saprophytic	Freshwater	Liput River, Alegria, Negros Occidental	Cai et al. (2003)
Dictyomorpha dioica (Couch) Mullins	Endoparasitic on the zoosporangia of *Achlya Americana*	Freshwater	Los Baños, Laguna	Reynolds (1970), Dogma (1986)
Dictyosporium heptasporum (Garov.) Damon	Saprophytic	Freshwater	Liput River, Alegria, Negros Occidental	Cai et al. (2003)
Didymella aptrootii K.D. Hyde & S.W. Wong	Saprophytic	Freshwater	Liput River, Alegria, Negros Occidental	Cai et al. (2003)
Didymella Sacc.	Saprophytic	Brackish, Mangroves swamp	Del Carmen, Siargao Island	Besitulo et al. (2010)
Digitodesmium bambusicola L. Cai, K.Q. Zhang, McKenzie, W.H. Ho & K.D. Hyde	Saprophytic	Freshwater	Liput River, Alegria, Negros Occidental	Cai et al. (2003)
Diplophlyctis asteroidea Dogma	Saprophyte of chitinous substrata (e.g., insect exuviae)	Freshwater	Quezon	Dogma (1976), Dogma (1986)

I. Fungi in aquatic and terrestrial habitats

2. Marine yeasts, mangrove-derived fungi, and marine fungal endophytes

47

TABLE 3.1 Records of Philippine fungi from aquatic habitats.—cont'd

Species	Substrate/host	Habitat	Locality	References
Diplophlyctis complicata (Willoughby) Dogma	Saprophyte of chitinous substrata	Freshwater	Mt. Makiling (Calamba side) Laguna	Dogma (1974)
Ellisembia adscendens (Berk.) Subram.	Saprophytic	Freshwater	Liput River, Alegria, Negros Occidental	Cai et al. (2003)
Ellisembia Subram.	Saprophytic	Freshwater	Liput River, Alegria, Negros Occidental	Cai et al. (2003)
Ellisembia vaginata McKenzie	Saprophytic	Freshwater	Liput River, Alegria, Negros Occidental	Cai et al. (2003)
Engyodontium album (Limber) de Hoog	Saprophytic; likely hemibiotrophic casing *ice-ice* disease on its host macroalgae (*Kappaphycus*)	Marine	Calatagan, Batangas	Solis et al. (2010)
Entophlyctis confervae-glomeratae (Cienk.) Sparrow	Hemibiotrophic on filaments of *Cladophora* sp.	Freshwater	Los Baños, Laguna	Dogma (1974)
Eupenicillium javanicum (J.F.H. Beyma) Stolk & D.B. Scott	Sponge-associated fungi, in mangroves habitat	Brackish	New Washington, Aklan, Panay	Calabon et al. (2019)
Eurotium Link	Saprophytic; likely hemibiotrophic casing *ice-ice* disease on its host macroalgae (*Kappaphycus*)	Marine	Calatagan, Batangas	Solis et al. (2010)
Fluminicola bipolaris S.W. Wong, K.D. Hyde & E.B.G. Jones	Saprophytic	Freshwater	Liput River, Alegria, Negros Occidental	Cai et al. (2003)
Fusarium chlamydosporum Wollenw. & Reinking	Mangrove (*Rhizophora apiculata*) associated endophytes	Brackish	Pangasinan	Moron et al. (2018)
Fusarium oxysporum Schltdl.	Mangrove (*Rhizophora apiculate, Avicennia corniculatum*) associated endophytes	Brackish	Camarines Sur	Moron et al. (2018)

(Continued)

I. Fungi in aquatic and terrestrial habitats

48   3. Fungi and fungus-like microorganisms in Philippine marine ecosystems

TABLE 3.1   Records of Philippine fungi from aquatic habitats.—cont'd

Species	Substrate/host	Habitat	Locality	References
Fusarium proliferatum (Matsushima) Nirenberg ex Gerlach & Nirenberg	Mangrove (*Rhizophora stylosa, Lumnitzera litorea*) associated endophytes	Brackish	Camarines Sur, Zambales	Moron et al. (2018)
Fusarium solani (Mart.) Sacc.	Saprophytic; Mangrove (*Rhizophora apiculata*) associated endophytes	Marine; Brackish	Calatagan, Batangas; Pangasinan	Solis et al. (2010), Moron et al. (2018)
Fusarium Link	Saprophytic; Mangrove (*Rhizophora apiculata*) associated endophytes	Marine; Brackish	Camarines Sur; Zambales; Calatagan, Batangas; Manila Bay	Solis et al. (2010), Sia Su et al. (2014), Moron et al. (2018)
Gaeumannomyces Arx & D.L. Olivier	Saprophytic	Freshwater	Liput River, Alegria, Negros Occidental	Cai et al. (2003)
Geotrichum Link	Sponge-associated fungi, in mangroves habitat	Brackish	New Washington, Aklan, Panay	Calabon et al. (2019)
Gliomastix Gueg.	Sponge-associated fungi, in mangroves habitat	Brackish	New Washington, Aklan, Panay	Calabon et al. (2019)
Gonapodya polymorpha Thaxt.	Saprophytic on fruits materials (e.g., apple) in stagnant water	Freshwater	Los Baños, Laguna	Dogma (1986)
Gonapodya prolifera (Cornu) A. Fischer	Saprophytes of fruits and twigs	Freshwater	Los Baños, Laguna	Dogma (1986)
Guignardia mangiferae A.J. Roy	Fungal endophytes, on mangroves (*Sonneratia alba*)	Brackish, Marine	Marabut, Samar	Apurillo et al. (2019)
Haematonectria haematococca (Berk. & Broome) Samuels & Nirenberg	Saprophytic	Freshwater	Liput River, Alegria, Negros Occidental	Cai et al. (2003)
Halocyphina Kohlm & E. Kohlm.	Saprophytic	Brackish, Mangroves swamp	Del Carmen, Siargao Island	Besitulo et al. (2010)
Halocyphina villosa Kohlm. and Kohlm.	Saprophytic	Marine, Brackish, Mangroves swamp	Taklong, Guimaras; Boracay, Aklan; Pagbilao, Pangasinan; Del Carmen, Siargao Island	Alias, Jones, and Torres (1999), Besitulo et al. (2010)

I. Fungi in aquatic and terrestrial habitats

2. Marine yeasts, mangrove-derived fungi, and marine fungal endophytes

49

TABLE 3.1 Records of Philippine fungi from aquatic habitats.—cont'd

Species	Substrate/host	Habitat	Locality	References
Halomassarina thalassiae (Kohlm. & Volkm.-Kohlm.) Suetrong et al.	Saprophytic	Brackish, Mangroves swamp	Del Carmen, Siargao Island	Besitulo et al. (2010)
Halorosellinia oceanica (S. Schatz) Whalley, E.B.G. Jones, K.D. Hyde & Laessoe	Saprophytic	Brackish, Mangroves swamp	Del Carmen, Siargao Island	Besitulo et al. (2010)
Halosarpheia heteroguttulata S.W. Wong, K.D. Hyde & E.B.G Jones	Saprophytic	Freshwater	Liput River, Alegria, Negros Occidental	Cai et al. (2003)
Halosarpheia lotica Shearer	Saprophytic	Freshwater	Liput River, Alegria, Negros Occidental	Cai et al. (2003)
Halosarphiea marina (Cribb. & J. Cribb.) Kohlm.	Saprophytic	Brackish, Mangroves swamp	Del Carmen, Siargao Island	Besitulo et al. (2010)
Helicascus kanaloanus Kohlm.	Saprophytic	Marine	Taklong, Guimaras; Boracay, Aklan; Pagbilao, Pangasinan	Alias, Jones, and Torres (1999)
Helicorhoidion nypicola K.D. Hyde & Goh	Saprophytic	Brackish, Mangroves swamp	Del Carmen, Siargao Island	Besitulo et al. (2010)
Helicosporium gigasporum C.K.M. Tsui, Goh, K.D. Hyde & Hodgkiss	Saprophytic	Freshwater	Liput River, Alegria, Negros Occidental	Cai et al. (2003)
Hortaea werneckii (Horta) Nishim. & Miyajo	Sponge-associated fungi, in mangroves habitat	Brackish	New Washington, Aklan, Panay	Calabon et al. (2019)
Hydea pygmea (Kohlm.) K.L. Pang & E.B.G. Jones	Saprophytic	Brackish, Mangroves swamp	Del Carmen, Siargao Island	Besitulo et al. (2010)
Hypoxylon oceanicum Schatz	Saprophytic	Marine, Brackish	Pagbilao, Pangasinan	Alias, Jones, and Torres (1999)
Ityorhoptrum verruculosum (M.B. Ellis) P.M. Kirk	Saprophytic	Freshwater	Liput River, Alegria, Negros Occidental	Cai et al. (2003)

(Continued)

I. Fungi in aquatic and terrestrial habitats

50   3. Fungi and fungus-like microorganisms in Philippine marine ecosystems

TABLE 3.1   Records of Philippine fungi from aquatic habitats.—cont'd

Species	Substrate/host	Habitat	Locality	References
Jahnula seychellensis K.D. Hyde & S.W. Wong	Saprophytic	Freshwater	Liput River, Alegria, Negros Occidental	Cai et al. (2003)
Kallichroma tethys (Kohlm. & E. Kohlm.) Kohlm. & Volkm.-Kohlm.	Saprophytic	Brackish, Mangroves swamp	Del Carmen, Siargao Island	Besitulo et al. (2010)
Kirschsteiniothelia elaterascus Shearer	Saprophytic	Freshwater	Liput River, Alegria, Negros Occidental	Cai et al. (2003)
Kloeckera Janke	Sponge-associated fungi, in mangroves habitat	Brackish	New Washington, Aklan, Panay	Calabon et al. (2019)
Lasiodiplodia Ellis & Everh.	Mangrove (*Rhizophora apiculate, R. stylosa*) associated endophytes	Brackish	Camarines Sur	Moron et al. (2018)
Lasiodiplodia theobromae (Pat.) Griffon & Maubl.	Mangrove (*Avicennia lanata*) associated endophytes	Brackish	Zambales	Moron et al. (2018)
Leptosphaeria Ces. & De Not.	Saprophytic	Marine	Boracay, Aklan; Pagbilao Pangasinan	Alias, Jones, and Torres (1999)
Lichtheimia ramosa (Zopf) Vuill.	Fungal endophytes on mangroves	Brackish	Alba; Catanduanes; Sorsogon; Masbate; Camarines Sur; Camarines Norte	Guerrero et al. (2018)
Lignincola longirostris (Cribb & J.W. Cribb) Kohlm.	Saprophytic	Marine, Brackish	Taklong, Guimaras; Boracay, Aklan; Pagbilao, Pangasinan	Alias, Jones, and Torres (1999)
Lignincola laevis Hönk	Saprophytic	Marine, Brackish	Taklong, Guimaras; Boracay, Aklan; Pagbilao, Pangasinan	Alias, Jones, and Torres (1999)
Lignincola nypae K.D. Hyde & Alias	Saprophytic	Brackish, Mangroves swamp	Del Carmen, Siargao Island	Besitulo et al. (2010)
Lignincola tropica Kohlm.	Saprophytic	Marine; Brackish, Mangroves swamp	Taklong, Guimaras; Pagbilao, Pangasinan; Del Carmen, Siargao Island; Boracay Aklan	Besitulo et al. (2010), Alias, Jones, and Torres (1999)

I. Fungi in aquatic and terrestrial habitats

2. Marine yeasts, mangrove-derived fungi, and marine fungal endophytes **51**

TABLE 3.1 Records of Philippine fungi from aquatic habitats.—cont'd

Species	Substrate/host	Habitat	Locality	References
Lineolata rhizophorae (Kohlm. & E. Kohlm.) Kohlm. et Volkm.-Kohlm.	Saprophytic	Brackish, Mangroves swamp	Del Carmen, Siargao Island	Besitulo et al. (2010)
Linocarpon angustatum K.D. Hyde & Alias	Saprophytic	Brackish, Mangroves swamp	Del Carmen, Siargao Island	Besitulo et al. (2010)
Linocarpon appendiculatum K.D. Hyde	Saprophytic	Brackish, Mangroves swamp	Del Carmen, Siargao Island	Besitulo et al. (2010)
Linocarpon bambusicola L. Cai & K.D. Hyde	Saprophytic	Freshwater	Liput River, Alegria, Negros Occidental	Cai et al. (2003)
Lophiostoma bipolare (K.D. Hyde) E.C.Y. Liew, Aptroot & K.D. Hyde	Saprophytic	Freshwater	Liput River, Alegria, Negros Occidental	Cai et al. (2003)
Lophiostoma mangrovei Kohlm. & Vittal	Saprophytic	Marine	Boracay, Aklan	Alias, Jones, and Torres (1999)
Lulworthia grandispora Meyers	Saprophytic	Marine, Brackish, Mangroves swamp	Taklong, Guimaras; Boracay, Aklan; Pagbilao, Pangasinan; Del Carmen, Siargao Island	Alias, Jones, and Torres (1999), Besitulo et al. (2010)
Lulworthia G.K. Sutherl.	Saprophytic	Brackish, Mangroves swamp	Del Carmen, Siargao Island	Besitulo et al. (2010)
Mammaria Ces.	Sponge-associated fungi, in mangroves habitat	Brackish	New Washington, Aklan, Panay	Calabon et al. (2019)
Mangrovispora pemphii K.D. Hyde. & Nakagiri	Saprophytic	Brackish, Mangroves swamp	Del Carmen, Siargao Island	Besitulo et al. (2010)
Marasmiellus palmivorus (Sharples) Desjardin	Fungal endophytes, on mangroves (*Sonneratia alba*)	Brackish, Marine	Marabut, Samar	Apurillo et al. (2019)

(Continued)

I. Fungi in aquatic and terrestrial habitats

52   3. Fungi and fungus-like microorganisms in Philippine marine ecosystems

TABLE 3.1   Records of Philippine fungi from aquatic habitats.—cont'd

Species	Substrate/host	Habitat	Locality	References
Marinosphaera mangrovei K.D. Hyde	Saprophytic	Marine, Brackish, Mangroves swamp	Boracay, Aklan; Pagbilao Pangasinan; Del Carmen, Siargao Island	Alias, Jones, and Torres (1999), Besitulo et al. (2010)
Marinosphaera K.D. Hyde	Saprophytic	Brackish, Mangroves swamp	Del Carmen, Siargao Island	Besitulo et al. (2010)
Massarina Sacc.	Saprophytic	Brackish, Mangroves swamp	Del Carmen, Siargao Island	Besitulo et al. (2010)
Massarina thalassioidea Hyde & Aptroot	Saprophytic	Freshwater	Liput River, Alegria, Negros Occidental	Cai et al. (2003)
Microthyrium Desm.	Saprophytic	Brackish, Mangroves swamp	Del Carmen, Siargao Island	Besitulo et al. (2010)
Monodictys levis (Wiltshire) S. Hughes	Saprophytic	Freshwater	Liput River, Alegria, Negros Occidental	Cai et al. (2003)
Monodictys monilicellularis Matsush.	Saprophytic	Freshwater	Liput River, Alegria, Negros Occidental	Cai et al. (2003)
Monodictys S. Hughes	Saprophytic	Marine, Brackish	Taklong, Guimaras; Boracay, Aklan; Pagbilao, Pangasinan	Alias, Jones, and Torres (1999)
Monotosporella microaquatica (Tubaki) S.V. Nilsson	Saprophytic	Freshwater	Liput River, Alegria, Negros Occidental	Cai et al. (2003)
Morosphaeria ramunculicola (K.D. Hyde) Suetrong et al.	Saprophytic	Brackish, Mangroves swamp	Del Carmen, Siargao Island	Besitulo et al. (2010)
Morosphaeria velatispora (K.D. Hyde & Borse) Suetrong et al.	Saprophytic	Brackish, Mangroves swamp	Del Carmen, Siargao Island	Besitulo et al. (2010)
Mucor sp.	Saprophytic	Marine	Manila Bay	Sia Su et al. (2014)
Myrothecium Tode	Saprophytic, with dye decolorization bioactivities	Marine	Navotas Fish Port, Navotas	Torres et al. (2011)

I. Fungi in aquatic and terrestrial habitats

2. Marine yeasts, mangrove-derived fungi, and marine fungal endophytes

TABLE 3.1 Records of Philippine fungi from aquatic habitats.—cont'd

Species	Substrate/host	Habitat	Locality	References
Neosartorya Malloch & Cain	Sponge-associated fungi, in mangroves habitat	Brackish	New Washington, Aklan, Panay	Calabon et al. (2019)
Neptunella longirostris (Cribb & J.W. Cribb) K.L. Pang & E.B.G. Jones	Saprophytic	Brackish, Mangroves swamp	Del Carmen, Siargao Island	Besitulo et al. (2010)
Neta Shearer & J.L. Crane	Saprophytic	Freshwater	Liput River, Alegria, Negros Occidental	Cai et al. (2003)
Nia vibrissa R.T. Moore & Meyers	Saprophytic	Marine	Pagbilao Pangasinan	Alias, Jones, and Torres (1999)
Nigrospora Zimm.	Fungal endophytes on mangroves	Brackish	Zambales, Cavite, Batangas, Quezon	Ramirez et al. (2020)
Nodulisporium Preuss	Mangrove (*Rhizophora stylosa*) associated endophytes	Brackish	Camarines Sur	Moron et al. (2018)
Nowakowskiella elegans (Nowak.) J. Schrot.	Saprophytes, on cellulosic substrates (e.g., grass leaves, cotton fibers)	Freshwater	Widespread and common	Dogma (1986)
Nowakowskiella haemisphaerospora Shanor	Saprophytes, on cellulosic materials (e.g., grass leaves, filter papers)	Freshwater	Del Rosario, Sorsogon; Los Baños, Laguna	Dogma (1986)
Oedogoniomyces lymnaeae Kobayashi & M. Ookubo	Saprophytic on fruits materials (e.g., banana) in stagnant water	Freshwater	Los Baños, Laguna	Dogma (1986)
Olpidium allomycetos Karling	Endoparasitic on *Allomyces* (*A. arbuscular*, *A. javanicus*) and invertebrates (rotifers eggs, mayflies, nematodes)	Freshwater	Batac, Ilocos Norte; Mariano Marcos, Sultan Kudarat	Dogma (1975), Dogma (1986)
Olpidium decipiens (Braun) Petersen	Endoparasitic on the filamentous green algae *Oedogonium* sp.	Freshwater	La Trinidad Valley, Benguet	Dogma (1986)
Olpidium gregarium (Nowak.) J. Schrot.	Endoparasitic on eggs of rotifers (e.g., *Brachionus, Lecane*)	Freshwater	Montalban, Rizal	Dogma (1986)

(*Continued*)

I. Fungi in aquatic and terrestrial habitats

54　　　3. Fungi and fungus-like microorganisms in Philippine marine ecosystems

TABLE 3.1　Records of Philippine fungi from aquatic habitats.—cont'd

Species	Substrate/host	Habitat	Locality	References
Olpidium pendulum Zopf	Saprophytic, occurs on pollen grains (e.g., *Pinus*, *Zea*)	Freshwater	Cagayan; Nueva Ecija; Nueva Vizcaya; Pangasinan Bataan; Pampanga; Bulacan Laguna; Batangas; Quezon; Leyte; North Cotabato; Cagayan de Oro; Bukidnon; Zamboanga del Sur; Davao del Norte	Dogma (1986)
Olpidium (A. Braun) J. Schrot.	Endoparasitic on eggs within dead mayflies	Freshwater	Los Baños, Laguna	Dogma (1986)
Olpidium sparrowii Dogma	Endoparasitic on rotifer eggs	Freshwater	Sorsogon	Dogma (1977), Dogma (1986)
Ophioceras dolichostomum (Berk. & M.A. Curtis) Sacc.	Saprophytic	Freshwater	Liput River, Alegria, Negros Occidental	Cai et al. (2003)
Ophioceras Sacc.	Saprophytic	Brackish, Mangroves swamp	Del Carmen, Siargao Island	Besitulo et al. (2010)
Oxydothis nypicola K.D. Hyde	Saprophytic	Brackish, Mangroves swamp	Del Carmen, Siargao Island	Besitulo et al. (2010)
Paecilomyces lilacinus (Thom) Samson	Sponge-associated fungi, in mangroves habitat	Brackish	New Washington, Aklan, Panay	Calabon et al. (2019)
Paecilomyces persicinus Nicot	Sponge-associated fungi, in mangroves habitat	Brackish	New Washington, Aklan, Panay	Calabon et al. (2019)
Paecilomyces roseolus G. Sm.	Sponge-associated fungi, in mangroves habitat	Brackish	New Washington, Aklan, Panay	Calabon et al. (2019)
Paecilomyces formosus Sakag., May. Inoue & Tada ex Houbraken & Samson	Mangrove (*Lumnitzera littorea*) associated endophytes	Brackish	Zambales	Moron et al. (2018)
Paecilomyces javanicus (Friedrichs & Bally) A.H.S. Br. & G. Sm.	Sponge-associated fungi, in mangroves habitat	Brackish	New Washington, Aklan, Panay	Calabon et al. (2019)

I. Fungi in aquatic and terrestrial habitats

2. Marine yeasts, mangrove-derived fungi, and marine fungal endophytes

TABLE 3.1   Records of Philippine fungi from aquatic habitats.—cont'd

Species	Substrate/host	Habitat	Locality	References
Paecilomyces Bainier	Saprophytic, with dye decolorization bioactivities; Sponge-associated fungi, in mangroves habitat	Marine, Brackish	Calatagan, Batangas; New Washington, Aklan, Panay	Torres et al. (2011), Calabon et al. (2019)
Paecilomyces victoriae (Szilvinyi) A.H.S. Br. & G. Sm.	Sponge-associated fungi, in mangroves habitat	Brackish	New Washington, Aklan, Panay	Calabon et al. (2019)
Papulospora Preuss	Saprophytic	Freshwater	Liput River, Alegria, Negros Occidental	Cai et al. (2003)
Passeriniella savoryellopsis K.D. Hyde & Mouzouras	Saprophytic	Brackish, Mangroves swamp	Del Carmen, Siargao Island	Besitulo et al. (2010)
Penicillium canescens Sopp	Saprophytic	Freshwater	Marilao River, Bulacan	Maini et al. (2019)
Penicillium citreonigrum Tsunoda	Sponge-associated fungi, in mangroves habitat	Brackish	New Washington, Aklan, Panay	Calabon et al. (2019)
Penicillium citrinum Thom	Sponge-associated fungi, in mangroves habitat	Brackish	New Washington, Aklan, Panay	Calabon et al. (2019)
Penicillium janthinellum Biourge	Sponge-associated fungi, in mangroves habitat	Brackish	New Washington, Aklan, Panay	Calabon et al. (2019)
Penicillium chrysogenum Thom	Sponge-associated fungi, in mangroves habitat	Brackish	New Washington, Aklan, Panay	Calabon et al. (2019)
Penicillium citrinum Thom	Mangrove (*Bruguiera gymnorrhiza*) associated endophytes	Brackish	Pangasinan	Moron et al. (2018)
Penicillium purpurascens (Sopp) Biourge	Sponge-associated fungi, in mangroves habitat	Brackish	New Washington, Aklan, Panay	Calabon et al. (2019)
Penicillium purpurogenum Stoll	Saprophytic	Marine	Calatagan, Batangas	Solis et al. (2010)
Penicillium rubrum Stoll	Sponge-associated fungi, in mangroves habitat	Brackish	New Washington, Aklan, Panay	Calabon et al. (2019)
Penicillium rugulosum Thom	Sponge-associated fungi, in mangroves habitat	Brackish	New Washington, Aklan, Panay	Calabon et al. (2019)

(Continued)

I. Fungi in aquatic and terrestrial habitats

56
3. Fungi and fungus-like microorganisms in Philippine marine ecosystems

TABLE 3.1 Records of Philippine fungi from aquatic habitats.—cont'd

Species	Substrate/host	Habitat	Locality	References
Penicillium Link	Saprophytic; likely hemibiotrophic casing *ice-ice* disease on its host macroalgae (*Kappaphycus*); also with dye decolorization bioactivities; Sponge and mangrove-associated fungi, in mangroves habitat	Marine, Freshwater	Calatagan, Batangas; Marilao River, Bulacan; Manila Bay; Nasugbu, Batangas; New Washington, Aklan, Panay	Solis et al. (2010), Torres et al. (2011), Sia Su et al. (2014), Maini et al. (2019), Rafael and Calumpong (2019), Calabon et al. (2019)
Penicillium spinulosum Thom	Sponge-associated fungi, in mangroves habitat	Brackish	New Washington, Aklan, Panay	Calabon et al. (2019)
Periconia prolifica Anast.	Saprophytic	Marine, Brackish	Taklong, Guimaras; Boracay, Aklan; Pagbilao, Pangasinan	Alias, Jones, and Torres (1999)
Pestalotiopsis Steyaert	Saprophytic	Marine	Boracay, Aklan; Pagbilao Pangasinan	Alias, Jones, and Torres (1999)
Pestalotiopsis adusta (Ellis & Everh.) Steyaert	Fungal endophytes, on mangroves (*Rhizophora mucronata*)	Brackish, Marine	Marabut, Samar	Apurillo et al. (2019)
Pestalotiopsis cocculi (Guba) G.C. Zhao & N. Li	Fungal endophytes on mangroves	Mangroves	Albay; Catanduanes; Sorsogon; Masbate; Camarines Sur; Camarines Norte	Guerrero et al. (2018)
Pestalotiopsis microspora (Speg.) Bat. & Peres	Mangrove (*Rhizophora apiculata*) associated endophytes	Brackish	Pangasinan	Moron et al. (2018)
Pestalotiopsis Steyaert	Sponge-associated fungi, in mangroves habitat; Mangrove (*Excoecaria agallocha, Rhizophora apiculata*) associated endophytes	Brackish	Camarines Sur; New Washington, Aklan, Panay	Moron et al. (2018), Calabon et al. (2019)
Phaeoisaria clematidis (Fuckel) S. Hughes	Saprophytic	Freshwater	Liput River, Alegria, Negros Occidental	Cai et al. (2003)
Phaeosphaeria I. Miyake	Saprophytic	Freshwater	Liput River, Alegria, Negros Occidental	Cai et al. (2003)
Phaeosphaeriopsis musae M. Arzanlou & Crous	Fungal endophytes, on mangroves (*Rhizophora mucronata*)	Brackish, Marine	Marabut, Samar	Apurillo et al. (2019)

I. Fungi in aquatic and terrestrial habitats

2. Marine yeasts, mangrove-derived fungi, and marine fungal endophytes

57

TABLE 3.1 Records of Philippine fungi from aquatic habitats.—cont'd

Species	Substrate/host	Habitat	Locality	References
Phialophora verrucosa Medlar	Saprophytic, with dye decolorization bioactivities; Fungal endophytes on mangroves	Marine, Brackish	Manila Bay; Zambales; Cavite; Batangas, Quezon	Torres et al. (2011), Ramirez et al. (2020)
Phialophorophoma litoralis Linder	Saprophytic	Brackish, Mangroves swamp	Del Carmen, Siargao Island	Besitulo et al. (2010)
Phlyctidium anatropum (A. Braun) Sparrow	Ectoparasitic on filaments of green alga *Oedogonium*	Freshwater	Los Baños, Laguna	Dogma (1986)
Phlyctochytrium planicorne G.F. Atk.	Weak pathogens of filamentous green algae (e.g., *Cladophora, Mougeotia, Oedogonium*)	Freshwater	Mt. Makiling, Laguna	Dogma (1986)
Phlyctorhiza endogena A.M. Hanson	Saprophyte of chitinous substrata (e.g., insect exuviae)	Freshwater	Atimonan, Quezon; Mt. Makiling, Laguna	Dogma (1986)
Phoma lingam (Tode) Desm.	Saprophytic; likely hemibiotrophic casing *ice-ice* disease on its host macroalgae (*Kappaphycus*)	Marine	Calatagan, Batangas	Solis et al. (2010)
Phoma nebulosa (Pers.) Berk.	Saprophytic	Marine	Calatagan, Batangas	Solis et al. (2010)
Phoma Sacc.	Saprophytic; likely hemibiotrophic casing *ice-ice* disease on its host macroalgae (*Kappaphycus*); also with dye decolorization bioactivities	Marine, Brackish, Mangroves swamp, Freshwater	Del Carmen, Siargao Island; Calatagan, Batangas; Calampit, Ternate, Cavite	Solis et al. (2010), Besitulo et al. (2010), Torres et al. (2011)
Phomatospora berkeleyi Sacc.	Saprophytic	Freshwater	Liput River, Alegria, Negros Occidental	Cai et al. (2003)
Phomopsis mangrovei K.D. Hyde	Saprophytic	Brackish, Mangroves swamp	Del Carmen, Siargao Island	Besitulo et al. (2010)
Phomopsis pittospori (Cooke & Harkn.) Grove	Fungal endophytes, on mangroves (*Rhizophora mucronata*)	Brackish, Marine	Marabut, Samar	Apurillo et al. (2019)
Phomopsis Sacc.	Mangrove (*Avicennia lanata, Camptostemon philippinense*) associated endophytes	Brackish	Zambales	Moron et al. (2018)

(Continued)

I. Fungi in aquatic and terrestrial habitats

58   3. Fungi and fungus-like microorganisms in Philippine marine ecosystems

TABLE 3.1   Records of Philippine fungi from aquatic habitats.—cont'd

Species	Substrate/host	Habitat	Locality	References
Pichia angusta (Teunisson, H.H. Hall & Wick.) Kurtzman	Sponge-associated fungi, in mangroves habitat	Brackish	New Washington, Aklan, Panay	Calabon et al. (2019)
Pleurophragmium bitunicatum Matsush.	Saprophytic	Freshwater	Liput River, Alegria, Negros Occidental	Cai et al. (2003)
Pseudocersosporella	Mangrove associated	Brackish	Bais, Alcantara; Pangangan Island	Rafael and Calumpong (2019)
Pseudohalonectria longirostrum Shearer	Saprophytic	Freshwater	Liput River, Alegria, Negros Occidental	Cai et al. (2003)
Pseudospiropes cubensis Hol.-Jech.	Saprophytic	Freshwater	Liput River, Alegria, Negros Occidental	Cai et al. (2003)
Quintaria lignatilis (Kohlm.) Kohlm. & Volkm.-Kohlm.	Saprophytic	Marine	Boracay, Aklan	Alias, Jones, and Torres (1999)
Ramichloridium Stahel ex de Hoog	Sponge-associated fungi, in mangroves habitat	Brackish	New Washington, Aklan, Panay	Calabon et al. (2019)
Rhizoclosmatium globosum H.E. Petersen	Saprophyte of chitinous substrata	Freshwater	Quezon National Park, Atimonan, Quezon	Dogma (1986)
Rhizoclosmatium H.E. Petersen	Saprophyte of chitinous substrata	Freshwater	Mt. Makiling, Puting Lupa, Calamba, Laguna	Dogma (1986)
Rhizophila marina K.D. Hyde & E.B.G. Jones	Saprophytic	Brackish, Mangroves swamp	Del Carmen, Siargao Island	Besitulo et al. (2010)
Rhizophydium carpophilum Zopf	The obligate endobiotic pathogen in oogonium of *Achlya, Dictyuchus, Brevilegnia, Aphanomyces*	Freshwater	Abra; Cagayan; Cavite; Batangas; Sorsogon; Laguna; Camarines Sur; Davao City; North Cotabato; South Cotabato; Bukidnon; Lanao del Sur	Dogma (1986)
Rhizophydium pollinis-pini (A. Braun) Zopf	Saprophytic, occurs on pollen grains	Freshwater	Laguna de Bay, Mayondon, Laguna	Dogma (1986)
Rhizopus Ehrenb.	Mangrove associated	Brackish	Bais, Alcantara; Pangangan Island	Rafael and Calumpong (2019)

I. Fungi in aquatic and terrestrial habitats

2. Marine yeasts, mangrove-derived fungi, and marine fungal endophytes

59

TABLE 3.1 Records of Philippine fungi from aquatic habitats.—cont'd

Species	Substrate/host	Habitat	Locality	References
Rhizosiphon multiporum Dogma	Endoparasitic on *Anaebaena*, specifically on akinetes	Freshwater	La Trinidad Valley, Benguet	Dogma (1986)
Rhizopus microsporus Teigh	Fungal endophytes on mangroves	Brackish	Albay; Catanduanes; Sorsogon; Masbate; Camarines Sur; Camarines Norte	Guerrero et al. (2018)
Roussoëlla minutella (Penz. & Sacc.) Aptroot	Saprophytic	Freshwater	Liput River, Alegria, Negros Occidental	Cai et al. (2003)
Rozella allomycis Foust	Obligate endoparasite of *Allomyces* (e.g., *A. arbuscula, A. javanicus, A. macrogynus, A. moniliformis, A. neo-moniliformis*)	Freshwater	Batac, Ilocos Norte; Villa Verde, Nueva Vizcaya; Guimaras Islands; Malungon, South Cotabato	Dogma (1986)
Rozella cladochytrii Karling	Obligate endoparasite of fungi	Freshwater	Sta. Cruz, Camarines Sur; Putiao, Sorsogon; Surallah, South Cotabato; Malabang, Lanao del Sur	Dogma (1986)
Rozella myzoctii Dogma	Hyperparasitic on *Myzoctium megastomum*	Freshwater	Montalban, Rizal	Dogma (1986)
Saagaromyces ratnagiriensis (S.D. Patil & Borse) K.L. Pang & E.B.G. Jones	Saprophytic	Brackish, Mangroves swamp	Del Carmen, Siargao Island	Besitulo et al. (2010)
Saccardoella minuta L. Cai & K.D. Hyde	Saprophytic	Freshwater	Liput River, Alegria, Negros Occidental	Cai et al. (2003)
Salsuginea ramicola K.D. Hyde	Saprophytic	Marine	Boracay, Aklan	Alias, Jones, and Torres (1999)
Salsuginea ramicola K.D. Hyde	Saprophytic	Brackish, Mangroves swamp	Del Carmen, Siargao Island	Besitulo et al. (2010)
Sarvoryella paucispora (Cribb and Cribb) Koch	Saprophytic	Marine	Taklong, Guimaras; Boracay, Aklan; Pagbilao, Pangasinan	Alias, Jones, and Torres (1999)
Savoryella aquatica K.D. Hyde	Saprophytic	Freshwater	Liput River, Alegria, Negros Occidental	Cai et al. (2003)

(*Continued*)

I. Fungi in aquatic and terrestrial habitats

60   3. Fungi and fungus-like microorganisms in Philippine marine ecosystems

TABLE 3.1   Records of Philippine fungi from aquatic habitats.—cont'd

Species	Substrate/host	Habitat	Locality	References
Savoryella lignicola E.B.G. Jones & R.A. Eaton	Saprophytic	Marine, Brackish	Taklong, Guimaras; Boracay, Aklan; Pagbilao, Pangasinan; Boracay, Aklan	Alias, Jones, and Torres (1999)
Savoryella longispora E.B.G. Jones & K.D. Hyde	Saprophytic	Brackish, Mangroves swamp	Del Carmen, Siargao Island	Besitulo et al. (2010)
Scedosporium aurantiacum Gilgado, Cano, Gené & Guarro	Sponge-associated fungi, in mangroves habitat	Brackish	New Washington, Aklan, Panay	Calabon et al. (2019)
Scopulariopsis brumptii Salv.-Duval	Saprophytic	Marine	Calatagan, Batangas	Solis et al. (2010)
Siphonaria variabilis H.E. Petersen	Saprophyte of chitinous substrata	Freshwater	Quezon National Park, Atimonan, Quezon	Dogma (1986)
Spiropes caaguazuense (Speg.) M.B. Ellis	Saprophytic	Freshwater	Liput River, Alegria, Negros Occidental	Cai et al. (2003)
Sporidesmiella hyalosperma (Corda) P.M. Kirk	Saprophytic	Freshwater	Liput River, Alegria, Negros Occidental	Cai et al. (2003)
Sporidesmium paludosum M.B. Ellis	Saprophytic	Freshwater	Liput River, Alegria, Negros Occidental	Cai et al. (2003)
Sporoschisma juvenile Boud.	Saprophytic	Freshwater	Liput River, Alegria, Negros Occidental	Cai et al. (2003)
Stachybotrys Corda	Sponge-associated fungi, in mangroves habitat	Brackish	New Washington, Aklan, Panay	Calabon et al. (2019)
Swampomyces Kohlm. & Volkm.-Kohlm.	Saprophytic	Brackish, Mangroves swamp	Del Carmen, Siargao Island	Besitulo et al. (2010)
Swampomyces triseptatus K.D. Hyde & Nakagiri	Saprophytic	Brackish, Mangroves swamp	Del Carmen, Siargao Island	Besitulo et al. (2010)
Talaromyces macrosporus (Stolk & Samson) Frisvad, Samson & Stolk	Saprophytic	Freshwater	Marilao River, Bulacan	Maini et al. (2019)

I. Fungi in aquatic and terrestrial habitats

2. Marine yeasts, mangrove-derived fungi, and marine fungal endophytes

61

TABLE 3.1 Records of Philippine fungi from aquatic habitats.—cont'd

Species	Substrate/host	Habitat	Locality	References
Talaromyces sp.	Saprophytic	Freshwater	Marilao River, Bulacan	Maini et al. (2019)
Tetraploa aristata Berk. & Broome	Saprophytic	Marine, Brackish	Boracay, Aklan; Pagbilao Pangasinan	Alias, Jones, and Torres (1999)
Thalassogena sphaerica Kohlm. & Volkm.-Kohlm.	Saprophytic	Brackish, Mangroves swamp	Del Carmen, Siargao Island	Besitulo et al. (2010)
Tiarosporella paludosa (Sacc. & Fiori) Höhn	Saprophytic	Freshwater	Liput River, Alegria, Negros Occidental	Cai et al. (2003)
Tirisporella beccariana (Ces.) E.B.G. Jones, K.D. Hyde & Alias	Saprophytes, associated with *Nypa fruticans*	Marine, Brackish	Boracay, Aklan; San Esteban, Pampanga	Alias, Jones, and Torres (1999); Jones et al. (1996)
Torpedospora radiata Meyers	Saprophytic	Marine	Taklong, Guimaras; Pagbilao, Pangasinan	Alias, Jones, and Torres (1999)
Trametes cubensis (Mont.) Sacc.	Fungal endophytes on mangroves	Mangroves	Albay; Catanduanes; Sorsogon; Masbate; Camarines Sur; Camarines Norte	Guerrero et al. (2018)
Trematosphaeria mangrovis Kohlm.	Saprophytic	Marine, Brackish, Mangroves swamp	Boracay, Aklan; Del Carmen, Siargao Island	Alias, Jones, and Torres (1999), Besitulo et al. (2010)
Trichocladium achrasporum (Meyers & R.T. Moore) M. Dixon	Saprophytic	Marine, Brackish	Taklong, Guimaras; Boracay, Aklan; Pagbilao, Pangasinan	Alias, Jones, and Torres (1999)
Trichocladium alopallonellum (Meyers & R.T. Moore) Kohlm. & Volkm.-Kohlm.	Saprophytic	Marine, Brackish, Mangroves swamp	Taklong, Guimaras; Boracay, Aklan; Pagbilao, Pangasinan; Del Carmen, Siargao Island	Alias, Jones, and Torres (1999), Besitulo et al. (2010)
Trichoderma aureoviride Rifai	Sponge-associated fungi, in mangroves habitat	Brackish	New Washington, Aklan, Panay	Calabon et al. (2019)
Trichoderma polysporum (Link) Rifai	Sponge-associated fungi, in mangroves habitat	Brackish	New Washington, Aklan, Panay	Calabon et al. (2019)
Trichoderma Pers.	Saprophytic; Mangrove (*Avicennia lanata*, *Avicennia officinalis*) associated endophytes	Marine, Brackish	Manila Bay, Zambales	Sia Su et al. (2014); Moron et al. (2018)

(Continued)

I. Fungi in aquatic and terrestrial habitats

62   3. Fungi and fungus-like microorganisms in Philippine marine ecosystems

TABLE 3.1   Records of Philippine fungi from aquatic habitats.—cont'd

Species	Substrate/host	Habitat	Locality	References
Tricladium lunderi Crane and Shearer	Saprophytic	Marine	Taklong, Guimaras; Pagbilao, Pangasinan	Alias, Jones, and Torres (1999)
Tritirachium oryzae (Vincens) de Hoog	Sponge-associated fungi, in mangroves habitat	Brackish	New Washington, Aklan, Panay	Calabon et al. (2019)
Tylopocladium terricola Medlar	Saprophytic, with dye decolorization bioactivities	Marine	Tamalan, Concepcion, Cavite	Torres et al. (2011)
Valsa brevispora G.C. Adams & Jol. Roux	Fungal endophytes, on mangroves (*Sonneratia alba*)	Brackish, Marine	Babatngon, Leyte	Apurillo et al. (2019)
Vanakripa gigaspora Bhat, W.B. Kendr. & Nag Raj	Saprophytic	Freshwater	Liput River, Alegria, Negros Occidental	Cai et al. (2003)
Vargamyces Tot	Saprophytic	Freshwater	Liput River, Alegria, Negros Occidental	Cai et al. (2003)
Verruculina enalia (Kohlm.) Kohlm. & Volkm.-Kohlm.	Saprophytic	Marine, Brackish, Mangroves swamp	Boracay, Aklan; Del Carmen, Siargao Island	Alias, Jones, and Torres (1999), Besitulo et al. (2010)
Verticillium nigrescens Pethybr.	Fungal endophytes, on mangroves (*Verticillium nigrescens*)	Brackish, Marine	Marabut, Samar	Apurillo et al. (2019)
Verticillium Nees	Saprophytic	Freshwater	Liput River, Alegria, Negros Occidental	Cai et al. (2003)
Vibrissea nypicola K.D. Hyde et Alias	Saprophytic	Brackish, Mangroves swamp	Del Carmen, Siargao Island	Besitulo et al. (2010)
Xylaria cubensis (Mont.) Fr.	Fungal endophytes, on mangroves (*Rhizophora mucronata*)	Brackish, Marine	Marabut, Samar	Apurillo et al. (2019)
Xylaria Hill ex Schrank	Mangrove (*Avicennia* sp.) associated endophytes	Brackish	Camarines Sur	Moron et al. (2018)
Xylomyces Goos, R.D. Brooks & Lamore	Saprophytic	Marine, Brackish	Taklong, Guimaras; Boracay, Aklan; Pagbilao, Pangasinan	Alias, Jones, and Torres (1999)
Zalerion maritimum (Linder) Anast.	Saprophytic	Marine, Brackish	Boracay, Aklan; Pagbilao Pangasinan	Alias, Jones, and Torres (1999)
Zalerion varia Anast.	Saprophytic	Marine	Boracay, Aklan; Pagbilao Pangasinan	Alias, Jones, and Torres (1999)

I. Fungi in aquatic and terrestrial habitats

2. Marine yeasts, mangrove-derived fungi, and marine fungal endophytes

TABLE 3.1 Records of Philippine fungi from aquatic habitats.—cont'd

Species	Substrate/host	Habitat	Locality	References
Undescribed (anamorphic taxa)	Saprophytic	Brackish, Mangroves swamp	Del Carmen, Siargao Island	Besitulo et al. (2010)
Undescribed (ascomycete)	Saprophytic	Brackish, Mangroves swamp	Del Carmen, Siargao Island	Besitulo et al. (2010)
Undescribed (basidiomycetes)	Saprophytic	Brackish, Mangroves swamp	Del Carmen, Siargao Island	Besitulo et al. (2010)

disease (BRS), White Leaf Spot disease (WLS), and Black Leaf Spot disease (BLS). From the infected leaf samples, four genera that belonged to Ascomycota were identified: *Aspergillus, Penicillium, Pseudocercosporella,* and *Rhizopus.* Apurillo et al. (2019) isolated *Verticillium nigrescens* Pethybr. (1919) that causes Verticillium wilt in alfalfa (Hu et al., 2011) and *Marasmiellus palmivorus* (Sharples) Desjardin, (Wilson & Desjardin, 2005), which causes disease in coconuts (Almaliky et al., 2013). Despite their pathogenicity, a type of host-endophyte relationship may occur where a balanced antagonism is achieved. Here, the fungal endophyte invades the host defense without causing an infection. In return, the host plant may produce metabolites that keep the endophytes in check (Schulz et al., 2008).

On the other hand, the diversity of fungi in mangrove-associated sponges in the Philippines was first reported by Calabon et al. (2019). The following genera were commonly identified from the sponges collected from the roots: *Aspergillus, Penicillium, Paecilomyces, Cladosporium,* and *Acremonium.* They also isolated strains that failed to sporulate and were designated as *mycelia sterilia.* Mycologists suggest that the role of fungi in mangrove sponges is involved in nutrient transfer, chemical defense, and degradation of dissolved organic matter (Bugni & Ireland, 2004; Ding et al., 2011; Hunting et al., 2010; Taylor et al., 2007). Fungal species are abundant in the mangrove ecosystem, colonizing specific plant parts and invertebrates. Based on the findings conducted in the Philippines, some fungal species found in the driftwood and intertidal wood are involved in the decomposition. Other species may inhabit or colonize internal plant tissues without causing any harm to the host. While most researchers have identified saprophytic fungi in senescent leaves or decaying plant materials, others have discussed their ability to cause an infection.

2.3 Fungi from macroalgae and seagrasses

Marine fungi are organisms that are recovered from marine habitats. They grow and possibly sporulate in the marine environment, form a symbiotic relationship with other marine organisms, and are metabolically active in the marine environment (Pang et al., 2016). In marine communities, fungi colonize various organic substrates, including seagrasses and algae. Notarte et al. (2018) reported sixteen (16) morphospecies of fungi belonging to the genera *Aspergillus*, *Fusarium*, *Paecilomyces*, *Penicillium*, *Sclerotinia* Fuckel, *Thamnidium* Link, and *Trichoderma*, including five mycelia sterilia isolated from the different substrates of macroalgae and seagrasses collected from Piapi Beach, Dumaguete City, Negros Oriental, Philippines, and were also found to exhibit biological activities (Notarte et al., 2017, 2018). Results showed that *Aspergillus* has the highest frequency among the genera and was widely distributed in the macroalgae and seagrasses. In their study, including fungi derived from the mangrove ecosystem (Table 3.1), *Aspergillus* and some genera are also found in the terrestrial substrates. Jones (1994) reasoned that some may have originated from the terrestrial or freshwater habitat. Their introduction into the marine ecosystem caused them to evolve because of selective selection pressure. Further, they may have undergone physiological adaptations to survive in the marine environment (Kohlmeyer & Kohlmeyer, 1979). Other studies on fungi associated with algae or algicolous fungi were Lavadia et al. (2017) and Solis et al. (2010).

2.4 Biotechnological applications of marine and marine-derived fungi

An intensive study using fungi in search for antimicrobials, novel bioactive compounds, biocontrol properties, and enzymes has been reported in the literature (Bungihan et al., 2010, 2011, 2013a; Dagamac et al., 2008; dela Cruz et al., 2020; de Mesa et al., 2020; Ghadin et al., 2008; Maria et al., 2005; Schulz et al., 2008; Torres & dela Cruz, 2013). In the Philippines, the potential of fungi (specifically molds) from the marine environment has been reported in various studies. For instance, the ability of mangrove fungi (*Aspergillus* sp. PAQ-H, *Aureobasidium* sp. Viala and Boyer (1891) 2LIPA-M, *Colletotrichum* sp. WABA-L, *Fusarium* sp. KAWIT-A, *Paecilomyces* sp. FDCAB-7, *Guignardia* sp. 2SANQ-F, *Penicillium* sp. LIABA-L, and *Phomopsis* sp. MACA-J), from leaf litters to produce xylanase was investigated (Torres & dela Cruz, 2013). Xylanases degrade the polysaccharide xylan, facilitating hydrolysis of one of the major components of plant cell walls called hemicellulose (Collins et al., 2005). Fungi are known to secrete the enzyme extracellularly and play a major role in the depolymerizing of highly robust and recalcitrant lignocellulosic cell walls (Kulkarni et al., 1999). In a similar study, Bacal and Yu (2017) reported the high endoglucanase and xylanase activity of endophytic fungi *Aspergillus tubingensis* Mosseray and *Fomitopsis* sp. P.Karst. Thus, the cellulolytic activities of these organisms are good candidates for industrial applications in the country. Marine-derived fungi could also be directly used for the bioremediation of environmental pollutants (De Padua & dela Cruz, 2021; Mendoza et al., 2010), as well as fungi from freshwater habitats (Carascal et al., 2017).

Mangrove-derived fungi are good sources of bioactive compounds. The active secondary metabolites (tyrosol C, cytosporone B, dothiorelone A, and dothiorelone C) from the endophytic and nonpathogenic *Phyllosticta* sp. Pers. isolated from the mangrove leaf was studied for its active secondary metabolites (Tan et al., 2015). Previous reports have shown that these chemical constituents were proven to have antibacterial, anticancer, and antioxidant properties (Ahn et al., 2008; Bungihan et al., 2013a). In Moron et al. (2018) study, crude culture extracts of *Pestalotiopsis microspora* (Speg.) Bat. & Peres showed very active inhibitory activity against Gram-positive bacteria, while *Fusarium oxysporum* Schltdl. and *Lasiodiplodia theobromae* (Pat.) Griffon & Maubl. are highly inhibitory against yeasts. Notarte et al. (2018) reported that culture crude extracts of *Aspergillus tubingensis* showed cytotoxic activity against P388 and HeLa cancer cells, while *A. fumigatus* Fresen. revealed high antiparasitic activity against *Trypanosoma congolense* Broden (1904). In a similar study, Apurillo et al. (2019) also investigated fungal endophytes' cytotoxic and antibacterial activities. The culture extracts of *Pestalotiopsis adusta* (Ellis & Everth.) Steyaert were most potent against *Pseudomonas aeruginosa* (Schroter) Migula, (1900). Yao et al. (2009) also demonstrated the antimicrobial activities of marine fungi derived from seawater and marine sediments. Collectively, these findings suggest that fungi isolated from the mangroves or marine environment in the Philippines are good sources of enzymes and bioactive compounds that can be exploited for further research.

3. Marine oomycetes

3.1 Oomycetes in mangrove ecosystems

Mangroves are home to a diverse number of microorganisms, many of which are fungal species categorized into saprophytic, parasitic, and symbiotic organisms. Their wide distribution and abundance in the mangrove environment have contributed to their important role in decomposing organic matter, thereby providing smaller animals with their foods. In the aquatic ecosystem, one of the ecologically important groups of microorganisms that thrive on mangrove leaves, barks, soil, and water are the Oomycetes G. Winter (1880), also known as the straminipilan fungi or water molds. Oomycetes were first classified by plant pathologists as lower fungi based on their morphological characteristics such as their filamentous growth habit, reproduction by means of zoospores, and nutrient acquisition. However, based on the evolutionary relationships and molecular phylogenetics, these oomycetes were seen to be not closely related to true fungi but were closely associated with algae and plants (Fry & Grünwald, 2010). For several years, the taxonomical classification of these groups of organisms had intriguing unresolved questions. In the beginning, they were originally grouped with fungi due to similarities in morphology and lifestyle. Later on, molecular and phylogenetic studies revealed their distinct morphological characteristics, and so they were placed in the kingdom Chromalveolata by Cavalier-Smith in 1981 (phylum Heterokontophyta, the 'stramenopiles') with brown and golden algae and diatoms (Cavalier-Smith, 1986, 1999).

Of the roughly 2000 species of oomycetes described so far, only about 60 species have been reported from the marine environment. The highest oomycete diversity seems to be in terrestrial plant pathogens (Thines, 2009; Thines & Kamoun, 2010). Plant pathogenic species, notably those of the genus *Phytophthora* de Bary and *Pythium* Pringsh. are the best-studied oomycetes.

Oomycetes, along with a diverse group of organisms (macroscopic and microscopic), occur in the mangrove ecosystem relatively due to its accommodating environment and its rich source of food. Leaf litters or decomposing leaves are one of the most abundant sources of organic matter in the aquatic system and serve as the primary food source in the food chain. One of the organisms that benefit from the organic materials produced by the mangrove leaves is the filamentous oomycetes and thraustochytrids, which are always associated with the early to late stages of decaying leaves in the salt-containing aquatic environment. They are reported to be more abundant than fungi as they rapidly colonize the leaf materials by producing abundant zoospores, which are chemotactically attracted to decaying mangrove leaves and thus reside within particulate detritus. As a result, they would attach to a suitable substratum where they can encyst, reproduce, and germinate as well (Leaño et al., 2001).

One group of marine oomycetes is the *Halophytophthora* H.H. Ho & S.C. Jong, which inhabit marine and brackish water environments, especially at mangroves, and colonize submerged leaf litter. Among the initial colonizers of fallen mangrove leaves, the straminipilous *Halophytophthora* spp. are among the most frequent species (Nakagiri et al., 1989, pp. 297–301; Newell et al., 1987; Newell & Fell, 1992; Tan & Pek, 1997). They play essential roles in decomposing fallen mangrove leaves as the first colonizer and enriching the leaves' nutrients with their mycelial biomass. This process changes the plant debris into nutritious food useable to consumers (crabs, shrimps, zooplankton, etc.) in the mangrove ecosystems (Fell & Master, 1975, 1980; Nakagiri et al., 1989, pp. 297–301; Newell, 1996). Their abundance is attributed to the efficient production of zoospores (Nakagiri et al., 1989, pp. 297–301) and the ability to compete against higher fungi on colonizing fallen mangrove leaves (Newell & Fell, 1987).

Labyrinthulomycetes Arx, comprised of Thraustochytrids and Labyrinthulids, are also abundant in the estuarine environment. These are marine osmoheterotrophic protists that also thrive in brackish waters and mangrove-rich areas. Thraustochytrids are organisms that can be found on fallen mangrove leaves (Leaño, 2001) and which grow outside of a substratum by absorbing nutrients from within through the ectoplasmic nets that they produced and by breaking down several complex organic substrates like the cellulose cell wall (Bremer, 1995; Raghukumar et al., 1992).

On the other hand, Labyrinthulids are a large group of organisms that are primarily photosynthetic, though not related to plants or even algae. They obtain their nutrition from the surrounding bodies of water or soil and, in some instances, on the tissues of other organisms. Most species behave as parasites on other organisms. An example of which is the infamous *Labyrinthula zosterae* D. Porter & Muehlst., which caused lesions on sea grasses known as the rapid blight disease (Armstrong et al., 2000; Sullivan et al., 2013). Several studies (Perveen et al., 2006; Raghukumar et al., 1992; Yokochi et al., 1998) have noted that labyrinthulids are prevalent on decaying mangrove leaves and very rarely on healthy leaf tissues.

Oomycetes species are generally characterized based on their colonial morphology, sporangial structure, the proliferating pattern of zoosporangiophores, and zoospore release. Specifically, thraustochytrid species are identified through their several distinct morphological characteristics such as the presence of their ectoplasmic net (EN) elements and zoospores that possess heterokont flagella (Lee Chang et al., 2012). Identification of thraustochytrid species is based mostly on their thallic stage morphology, specifically on their spore release and differences in sporogenesis (Arafiles et al., 2011). While the colonies and growth patterns of labyrinthulids are not visible under the naked eye. The EN of this organism can be visually observed by bringing up the agar plate against a light source where its root-like structure made up of colorless aggregated cells is shown (Dee et al., 2019).

Some isolated oomycetes such as the halophytophthoras are identified through phenotypic characters. Colony patterns are observed by using different agar media such as the Vegetable Juice Seawater Agar (VJSA), Potato Carrot Agar (PCA), Peptone Yeast-extract Glucose Seawater Agar (PYGSA), and Potato Sucrose Agar (PSA) (Fig. 3.1). Observation of colony radial growth at different temperatures is usually performed by measuring the mean colony extension rates (MCER). Mature sporangia are observed under a light microscope through their proliferation type, shape, size, papilla, basal plug, and vacuoles. Monitoring of zoospore development and release, presence of discharge or dehiscence tube, vesicle and operculum, and presence of sexual structures and chlamydospores are likewise performed for characterization.

Taxonomic keys are also used to further identify the oomycetes species, such as the taxonomic keys reported by Nakagiri et al. (2000), Nham Tran et al. (2020), and Bongiorni et al. (2005) for thraustochytrids and halophytophthoras. Molecular techniques are used further to identify the oomycete species.

Despite the country's tropical and archipelagic nature where species richness and endemism are supposed to be high, reports on Philippine mangrove oomycetes are still very limited. The first report of mangrove oomycetes in the Philippines was described by Leaño (2001) on Panay Island, and he noted that *Halophytophthora* spp. were found from seven species of mangroves, namely, *Avicennia lanata* Ridl., *A. officinalis, Ceriops decandra, Rhizophora apiculata, Sonneratia* sp., *Xylocarpus granatum, X. moluccensis* (Lam.) M. Roem. These marine oomycetes were identified as *Halophytophthora vesicula* (Anastasiou & Churchl.) H.H. Ho & S.C. Jong, *H. kandeliae* (syn. *Phytopythium kandeliae*) H.H. Ho, H.S. Chang & S.Y. Hsieh, *H. bahamensis* (Fell & Master) H.H. Ho & S.C. Jong, *H. epistomium* (Fell & Master) H.H. Ho & S.C. Jong, *H. spinosa* var. *lobata* (Fell & Master) H.H. Ho & S.C. Jong, and *Halophytophthora* sp.

Most of the recent reports on the occurrence of mangrove oomycetes in the Philippines were published by Bennett and Thines (2017), Bennett, Dedeles, and Thines (2017), Bennett, de Cock, et al. (2017), Bennett, Nam, et al. (2018), Bennett, Devanadera, et al. (2018) and his coworkers (Table 3.2) enlisting two known terrestrial and freshwater pathogen species, *Phytophthora insolita* Ann & W.H. Ko and *Phytophthora elongata* A. Rea, M. Stukely & T. Jung, and three novel species *Phytopythium leanoi* R. Bennett & Thines, *Phytopythium dogmae* R. Bennett & Thines, and *Salispina hoi* R. Bennett & Thines.

Other reported mangrove oomycetes such as *Halophytophthora vesicula* AK1YB2 (Aklan), *H. vesicula* PQ1YB (Quezon), and *Salispina spinosa* ST1YB3 (Davao del Norte) by Caguimbal et al. (2019) and *Halophytophthora* S13005YL1-3.1 isolated from Samal Island, Davao del Norte (Say et al., 2017) were studied for their diverse production of polyunsaturated fatty acids such as eicosapentaenoic acid (EPA) and arachidonic acid (ARA). In consequence, a species of *Salisapilia tartarea* (Nakagiri & S.Y. Newell) Hulvey, Nigrelli, Telle, Lamour and Thines, (Bennett & Thines, 2019) also isolated from Davao del Norte was explored by Marcelo et al. (2018) for its antioxidant (anthraquinones, anthrones, flavonoids, phenols, triterpenes) and α-glucosidase activities using the secondary metabolites. The broth ethyl acetate extract was examined for its anticancer potential and was found to be toxic to HepG2 cells but with low toxicity on normal human fibroblasts. Following this, Devanadera et al. (2019) explored further the cytotoxic and apoptotic potentials of crude fatty acids from three marine oomycetes, *Halophytophthora* spp. (T12GP1 and T12YBP2) and *Salispina hoi* (USTCMS 1611) against human

FIGURE 3.1 (*Top Image*) *Halophytophthora* sp. USTCMS 4123; (*Middle Image*) *Halophytophthora* sp. USTCMS 4158; (*Bottom Image*). *Halophytophthora* sp. USTCMS 4162. Colony pattern on (A) Vegetable juice seawater agar (B) Potato carrot agar (C) Peptone yeast glucose seawater agar and (D) Potato sucrose agar (PSA) after 5 days of incubation in the dark at 25°C; scale bars = 30 mm. (E) Sporangia; scale bar = 30 μm.

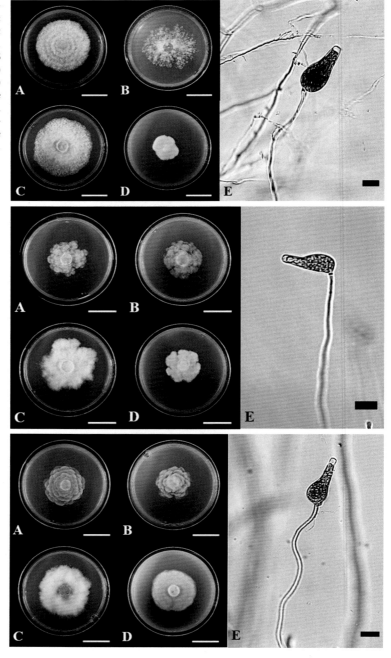

I. Fungi in aquatic and terrestrial habitats

3. Marine oomycetes

69

TABLE 3.2 Philippine oomycetes are recorded in the literature.

Species	Isolation source/host/substrate	Family	References
Lagenidium oophilum Sparrow	Parasite, nematode eggs		
Myzocytium megastomum De Wild.	Obligate endoparasite, *Cladophora* Kutz, *Pithophora* Wittrock *Closterium* Nitzsch ex Ralfs		Dogma (1986), Dogma (1975)
Myzocytium proliferum (Schenk) Schenk	*Spirogyra* Nees		Dogma (1975)
Albugo candida (Pers. ex J.F. Gmel.) Roussel	The obligate pathogen, crucifers	Albuginaceae	Dogma (1986)
Albugo ipomoeae-panduratae (Schwein.) Swingle	The obligate pathogen, *Ipomoea* L.	Albuginaceae	Dogma (1986)
Haliphthoros milfordensis Vishniac	The pathogen, *Scylla serrata* Forssk.	Haliphthoraceae	Leaño (2002)
Haliphthoros philippinensis Hatai, Bian, Batic. & Egusa	Pathogen, *Penaeus monodon* Fabricius *Scylla serrata*	Haliphthoraceae	Hatai et al. (1980), Leaño (2002)
Lagenidium giganteum Couch	Saprotroph, a parasite of mosquito larvae	Lagenidiaceae	Dogma (1986)
Lagenidium humanum Karling	Saprotroph, soil	Lagenidiaceae	Dogma (1986)
Lagenidium pygmaeum Zopf	Saprotroph, soil, and freshwater	Lagenidiaceae	Dogma (1986)
Leptolegniella keratinophilum Huneycutt	Soil (baited using snake skin, hair)	Leptolegniellaceae	Dogma (1986)
Apodachlya minima Coker & J. Leitn.	soil, saprotroph	Leptomitaceae	Dogma (1986)
Leptomitus lacteus C. Agardh	Saprotroph, foul and polluted waters	Leptomitaceae	Dogma (1986)
Olpidiopsis karlingiae Karling	Parasite, *Karlingia rosea* (syn. *Chytridium rosea* de Bary & Woronin	Olpidiopsidaceae	Dogma (1986)
Olpidiopsis luxurians Barrett	Parasite, *Aphanomyces laevis* de Bary	Olpidiopsidaceae	Dogma (1986)
Olpidiopsis pythii (E.J. Butler) Karling	Parasite, *Pythium intermedium* de Bary	Olpidiopsidaceae	Dogma (1975)
Bremia lactucae Regel	The obligate pathogen, lettuce	Peronosporaceae	Dogma (1986)

(Continued)

I. Fungi in aquatic and terrestrial habitats

70 3. Fungi and fungus-like microorganisms in Philippine marine ecosystems

TABLE 3.2 Philippine oomycetes are recorded in the literature.—cont'd

Species	Isolation source/host/substrate	Family	References
Halophytophthora vesicula (≡ Phytophthora vesicula)	Saprotroph, Avicennia lanata, A. officinalis, Ceriops decandra, R. apiculata, Sonneratia sp., X. granatum, X. moluccensis	Peronosporaceae	Leaño (2001), Caguimbal et al. (2019)
Pe. philippinensis (≡ S. philippinensis) (Weston) C. G. Shaw	Obligate pathogen, corn Saccharum L., Sorghum Moench, Euchlaena Schrad.	Peronosporaceae	Weston (1920), Dogma (1986)
Pe. sacchari (≡ S. sacchari) (T. Miyake) Shirai & Hara	The obligate pathogen, sugarcane, Euchlaena, Tripsacum L., Sorghum	Peronosporaceae	Dogma (1986)
Pe. sorghi (≡ S. sorghi) (W. Weston & Uppal) C.G. Shaw	The obligate pathogen, sorghum, Euchlaena, Heteropogon Pers., (1807), Panicum L., Pennisetum Rich., Saccharum, Dichantium Willemet	Peronosporaceae	Dogma (1986)
Pe. spontanea (≡ S. spontanea) (W. Weston) C.G. Shaw	The obligate pathogen, corn, Saccharum	Peronosporaceae	Dogma (1986)
Peronosclerospora miscanthi (≡ Sclerophthora miscanthi) (T. Miyake) C. G. Shaw	The obligate pathogen, Miscanthus Andersoon, Saccharum, Sorghum, Zea mays L.	Peronosporaceae	Dogma (1986)
Phytophthora cactorum (Lebert & Cohn) J. Schröt.	The pathogen, Theobroma cacao L.	Peronosporaceae	Mendiola and Espino (1916)
Phytophthora capsica Leonian	The pathogen, Piper nigrum L.	Peronosporaceae	Tsao et al. (1994)
Phytophthora citrophthora (R.E. Sm. & E.H. Sm.) Leonian	The pathogen, Citrus L. Nephelium L., Sandoricum Cav.	Peronosporaceae	Del Rosario (1968), Tsao et al. (1994)
Phytophthora colocasiae Racib.	The pathogen, Colocasia esculenta L.	Peronosporaceae	Mendiola and Espino (1916)
Phytophthora elongata	Decaying mangrove leaves	Peronosporaceae	Bennett, Dedeles, and Thines (2017)
Phytophthora heveae A.W. Thomps.	Pathogen, Sandoricum koetjape (Burm.f.) Merr.	Peronosporaceae	Tsao et al. (1994)
Phytophthora infestans (Mont.) de Bary	The pathogen, Solanum tuberosum L. tomato, potato	Peronosporaceae	Lee (1921), Dogma (1986)
Phytophthora insolita	Decaying mangrove leaves	Peronosporaceae	Bennett and Thines (2017)
Phytophthora meadii McRae	Pathogen, Hevea brasiliensis (Wild. Ex A. Juss.) Mull.Arg.	Peronosporaceae	Teodoro (1926)

I. Fungi in aquatic and terrestrial habitats

3. Marine oomycetes

71

TABLE 3.2 Philippine oomycetes are recorded in the literature.—cont'd

Species	Isolation source/host/substrate	Family	References
Phytophthora nicotianae Breda de Haan	Pathogen, *Piper nigrum* *Citrus* *Solanum melongena* L. *Carica papaya* L. *Ananas comosus* (L.) Merr. *Citrullus lanatus* (Thunb.) Matsum. & Nakai	Peronosporaceae	Tsao et al. (1994), Lee (1921), Ocfemia (1925), Quimio and Quimio (1974), Quebral et al. (1962)
Phytophthora palmivora (E.J. Butler) E.J. Butler	Pathogen, omnivorous	Peronosporaceae	Reinking (1919), Teodoro (1926), Celino (1933), Ela (1968), Del Rosario (1968), Dogma (1986), Tsao et al. (1994), Borines et al. (2014)
Phytophthora parasitica (synonym of *P. nicotianae*) Dastur	The pathogen, *Cocos* L., *Theobroma* L., *Gossypium* L., *Lycopersicum* Hill, *Solanum* L., *Hibiscus* L., *Musa* L., *Citrus* L., *Rheum* L., *Lilium* L., *Capsicum* L., *Ananas* Mill., *Nicotiana* L., *Ricinus* L., *Grammatophyllum* Blume, *Pollia* Thunb., *Catharanthus* G. Don, *Manihot* Miller, *Vigna* Savi, *Bryophyllum* Salisb.	Peronosporaceae	Dogma (1986)
Phytophthora phaseoli Thaxt.	Pathogen, *Sandoricum koetjape*	Peronosporaceae	Clara (1928)
Phytopythium dogmae	Decaying mangrove leaves	Peronosporaceae	Bennett, de Cock, et al. (2017)
Phytopythium kandeliae (≡ *Halophytophthora kandeliae*)	Saprotroph, *Avicennia lanata*, *R. apiculata*, *Sonneratia* sp.	Peronosporaceae	Leaño (2001)
Phytopythium leanoi	Decaying mangrove leaves	Peronosporaceae	Bennett, de Cock, et al. (2017)
Plasmopara parasitica	The obligate pathogen, crucifers	Peronosporaceae	Dogma (1986)
Plasmopara viticola (Berk. & M.A. Curtis) Berl. & De Toni	The obligate pathogen, grapes	Peronosporaceae	Dogma (1986)
Pseudoperonospora cubensis (Berk. & M.A. Curtis) Rostovzev	The obligate pathogen, cucurbits	Peronosporaceae	Dogma (1986)
Pythium aphanidermatum Paulitz & M. Mazzola	The pathogen, sugarcane, corn, sorghum, radish, tobacco, *Carica*, cucumber, *Solanum*	Pythiaceae	Dogma (1986)
Pythium arrhenomanes Drechsler	Pathogen, corn, wheat, sugarcane	Pythiaceae	Dogma (1986)
Pythium debaryanum R. Hesse	Pathogen, omnivorous species	Pythiaceae	Dogma (1986)

(*Continued*)

I. Fungi in aquatic and terrestrial habitats

72

3. Fungi and fungus-like microorganisms in Philippine marine ecosystems

TABLE 3.2 Philippine oomycetes are recorded in the literature.—cont'd

Species	Isolation source/host/substrate	Family	References
Pythium echinulatum V.D. Matthews	Saprotroph, normally can be pathogenic similar to *Py. debaryanum*, *Py. arrhenomanes*, and *Py. aphanidermatum*	Pythiaceae	Dogma (1986)
Pythium monospermum Pringsh.	Saprotroph, soil, and freshwater	Pythiaceae	Dogma (1986)
Pythium proliferum Schen	Saprotroph, or pathogen of strawberry	Pythiaceae	Dogma (1986)
Pythium torulosum Coker & P. Pratt.	Saprotroph, freshwater, and very moist soils	Pythiaceae	Dogma (1986)
Salisapilia bahamensis (≡ *Phytophthora bahamensis*) (Fell & Master) R. Benn. & Thines	Saprotroph, *Avicennia lanata*	Salisapiliaceae	Leaño (2001)
Salisapilia elongata (≡ *Halophytophthora elongata*) (Fell & Master) R. Benn. & Thines	Decaying mangrove leaves	Salisapiliaceae	Bennett and Thines (2019)
Salisapilia epistomium (≡ *Phytophthora epistomium*) (Fell & Master) R. Benn. & Thines	Saprotroph, *Rhizophora apiculata*, *Sonneratia* sp., *Xylocarpus granatum*	Salisapiliaceae	Leaño (2001)
Salispina hoi	Decaying mangrove leaves	Salispinaceae	Bennett, Nam, et al. (2018), Bennett, Devanadera, et al. (2018)
Salispina lobata (≡ *Phytophthora spinosa* var. *lobata*) (Fell & Master) A.L. Jesus, Marano & Pires-Zottar.	Saprotroph, *X. moluccensis*	Salispinaceae	Leaño (2001)
Salispina spinosa (≡ *Phytophthora spinosa* var. *spinosa*) (Fell & Master) A.L. Jesus, Marano & Pires-Zottar.	Decaying mangrove leaves	Salispinaceae	Caguimbal et al. (2019)
Achlya ambisexualis J.R. Raper	Freshwater, saprotroph	Saprolegniaceae	Dogma (1986)
Achlya proliferoides Coker	Freshwater, saprotroph	Saprolegniaceae	Dogma (1986)
Achyla americana Humphrey	Pathogen, Rice grain rot	Saprolegniaceae	Dogma (1986)

I. Fungi in aquatic and terrestrial habitats

3. Marine oomycetes

73

TABLE 3.2 Philippine oomycetes are recorded in the literature.—cont'd

Species	Isolation source/host/substrate	Family	References
Achyla bisexualis Coker & Couch	Freshwater and moist soils	Saprolegniaceae	Dogma (1986)
Achyla flagellata Cocker	Pathogen, Rice grain rot	Saprolegniaceae	Dogma (1986)
Dictyuchus anomalus Nagai	Freshwater, soil, saprotroph	Saprolegniaceae	Dogma (1986)
Saprolegnia diclina Humphrey	Pathogen of fishes, extremely rare in the Philippines	Saprolegniaceae	Dogma (1986)
Thraustotheca clavata (de Bary) Humphrey	Saprotroph, highland soils	Saprolegniaceae	Dogma (1986)
Aphanomyces cladogamus Drechsler	Parasite, tomato, spinach, eggplant, lettuce, sugar beets	Verrucalvaceae	Dogma (1986)
Aphanomyces helicoides Minden	Freshwater, saprotroph	Verrucalvaceae	Dogma (1986)
Aphanomyces keratinophilus (M. Ôkubo & Kobayasi) R.L. Seym. & T.W. Johnson	Freshwater, saprotroph	Verrucalvaceae	Dogma (1986)
Aphanomyces laevis de Bary	Freshwater, saprotroph; or as a parasite of some desmids, crayfish, and Philippine milkfish (*Chanos* sp.), catfish, carps	Verrucalvaceae	Dogma (1986)
Plectospira gemmifera Drechsler	Pathogen, sugarcane	Verrucalvaceae	Dogma (1986)

Information was taken from the published report of Bennett, R. M., & Thines, M. (2020). An overview of Philippine estuarine oomycetes. *Philippine Journal of Systematic Biology, 14*(1), 1—14. https://doi.org/10.26757/pjsb2020a14007, with copyright permission from the Philippine Journal of Systematic Biology.

adenocarcinoma/breast cancer cells (MCF7) and normal human dermal fibroblasts (HDFn). These studies have shown interesting biological properties of oomycetes, which contribute to potential and beneficial uses for human consumption, biological control, and drug development.

Likewise, a few species of Labyrinthulomycetes, such as thraustochytrids and labyrinthulids, were reported in brackish waters and mangrove-rich areas in the country. Thraustochytrids are known to have high amounts of lipid accumulation, which include essential fatty acids or polyunsaturated fatty acids (PUFAs) such as docosahexaenoic acid (DHA) and eicosapentaenoic acid (EPA) that can reach up to 70% of their biomass. In 2001, Leaño started to document species of thraustochytrids isolated from fallen mangrove leaves on Panay Island. Two genera: *Schizochytrium* S. Goldst & Belsky ex Raghuk. and *Thraustochytrium* Sparrow was reported in his study, which was associated with different mangrove species in Ibajay and Dumaguit Aklan, and Banate Iloilo. In addition, *Aurantiochytrium mangrovei* (Raghuk.)

I. Fungi in aquatic and terrestrial habitats

R. Yokoy & D. Honda (reported as *S. mangrovei* Raghuk.) in the study of Leaño et al. (2003) have explored its fatty acid production. To date, still very few species of thraustochytrids such as *Schizochytrium* sp (SB11) and *Thraustochytrium* sp (SB04) isolated from a mangrove in Subic Bay were reported (Arafiles et al., 2011; Atienza et al., 2012) to contain PUFAs and which have promising potentials as a feed supplement for fish, shrimps, and in the food industry due to their carotenoid content. Nevertheless, the aquaculture sector has motivated some researchers to explore the viability of fatty-rich thraustochytrids' biomass as a supplement to animal feeds. Estudillo-del Castillo et al. (2009) utilized *A. mangrovei* IAo-1 to enrich *Brachionus plicatilis* Muller, a rotifer commercially grown on larval fish feed. Ludevese-Pascual et al. (2013) have isolated *Schizochytrium* sp. LEY7 from Baybay, Leyte Island, and evaluated its enrichment potential for larval snapper feed (*Lutjanus* sp. Bloch). Dela Peña et al. (2016) utilized the biomass from *Schizochytrium* sp. as a supplement for the juveniles of *Haliotis asinine* L. In addition, *Thraustochytrium* sp CR01 isolated from a senescent mangrove leaf in Cavite was explored by Uba et al. (2016) using different sugar products such as liquid sugar, molasses, and corn syrup as glucose substitutes for its carbon source in the culture medium. This study showed that thraustochytrids could utilize other sugar sources for biomass and fatty acid production as sources of DHA and MUFAs.

Additionally, the occurrence of some species of thraustochytrids in Pagbilao Mangrove Park in Quezon Province was reported by the group of Perez et al. (2020) and was preliminarily assumed to belong to either *Aurantiochytrium* R. Yokoyama & D. Honda, *Hondea*, or *Monorhizochytrium* K. Doi & D. Honda because of their similarity in morphological characteristics. The isolates have circular-shaped cells, which are slightly greenish due to their oleaginous nature. Partitioning of their cells is likewise observed with the presence of inconspicuous ectoplasmic nets.

Considerably less studied is the Labyrinthulids, which are the poorly characterized oomycetes and are common in marine and terrestrial environments, although some species were said to be highly virulent. Only three labyrinthulid species (AK1) isolated from Kalibo, Aklan, and two species (DV1, DV2) isolated from Davao del Sur have been studied so far. Owing to their complicated isolation, cultivation, and long-term cell viability, Dee et al. (2019) tried to determine which cultural and optimal growth conditions and cell viability can be effectively applied to these organisms. All three labyrinthulid species were generally characterized by their spindle-shaped cells, which develop into colonies and are surrounded by a common extracellular matrix. These organisms are mostly marine and are characterized by ectoplasmic nets (EN). According to Dee and her coworkers, a broth medium supplemented with carabao grass effectively prolonged their cell viability for up to 4 months, while the agar block method is best for their purification. So far, no sufficient reports about this organism have been explored in the Philippine setting, considering the large bodies of water surrounding our vast archipelago.

3.2 Biotechnological applications of marine oomycetes

The marine ecosystem is considered a rich natural resource of many bioactive compounds such as proteins, sterols, antioxidants, polysaccharides, pigments, and polyunsaturated fatty acids (PUFAs). Studies on marine lipids have gathered significant interest recently, not only due to their bioactive properties but also because it is an important component of aquaculture feed, biofuel production, and energy and biomass transfer between different trophic levels in the global food web.

Studies on Oomycetes, such as *Phytophthora* and *Pythium*, have shown the presence of significant polyunsaturated fatty acids (PUFAs). The most common PUFAs are those with beneficial health effects, such as docosahexaenoic acid (DHA), eicosapentaenoic acid (EPA), and arachidonic acid (ARA). DHA is widely distributed in marine organisms, including deep-sea fish, seaweeds, bacteria, microalgae, and Thraustochytrids; moreover, scale-up production by microalgae and Thraustochytrids has been extensively studied (Kroes et al., 2003; Pang et al., 2015, 2016). Both freshwater and terrestrial species of *Pythium* and *Phytophthora* have also shown production of EPA and ARA (Cheng et al., 1999).

In the study by Pang et al., in 2015, *Halophytophthora* spp. from fallen mangrove leaves in Taiwan were examined and were found to produce polyunsaturated fatty acids such as Eicosapentaenoic acid (EPA) and Arachidonic acid (APA). Thus, the economic importance of these mangrove oomycetes has been more highlighted since several studies have shown the potential of these organisms to produce polyunsaturated fatty acids (Caguimbal et al., 2019; Devanadera et al., 2019; Say et al., 2017).

4. Biosynthesis, isolation, analysis, and applications of fatty acids from Philippine marine oomycetes

4.1 De novo fatty acid biosynthesis

The lipid metabolism, specifically the biosynthetic pathways of fatty acids and TAG, is less tackled and explored in marine fungi than in other aquatic microorganisms and higher organisms. The primary biosynthetic pathways of fatty acid and TAG in lower eukaryotic microorganisms are somehow like those demonstrated in other higher organisms based on their genetic homology, biochemical characteristics in terms of enzymatic reactions, mechanisms of actions, and pathways in their lipid metabolism. Unlike higher organisms, where lipids are synthesized and localized in a specific organelle in the cell or organs, the synthesis of glycerolipids and their sequestration takes place within a single cell (Hu et al., 2008; Xie & Wang, 2015). The accumulation of glycerolipids occurs in a densely packed vesicle called the lipid body located within the cytoplasm of a cell (Fig. 3.2). Both plants and animals cannot synthesize significant amounts of very long-chain PUFAs above 18 carbons. Unlike aquatic fungi, specifically, those marine species can synthesize and store large quantities of very long-chain PUFAs like EPA, DHA, and ARA (Reikhof et al., 2005; Xie & Wang, 2015).

Aside from the Kennedy pathway, an acyl-CoA independent biosynthesis of TAGs is possible in aquatic microorganisms through reactions involving phospholipid diacylglycerol acyltransferase (PDAT) and carnitine palmitoyltransferase (CPT) enzymatic activities, which is like the biosynthesis observed in both plants and yeasts (Cagliari et al., 2011) (Fig. 3.3). This acyl-CoA independent pathway plays a vital role in regulating the membrane lipid composition in response to various stressed environmental and growth conditions. Under stress conditions, aquatic microorganisms rapidly degrade the membranes by accumulating cytosolic lipid bodies. The fatty acid composition of their plasma membrane includes medium-chain (C10–C14), long-chain (C16-18), and very-long-chain (>C20) of fatty acids (Fig. 3.3).

I. Fungi in aquatic and terrestrial habitats

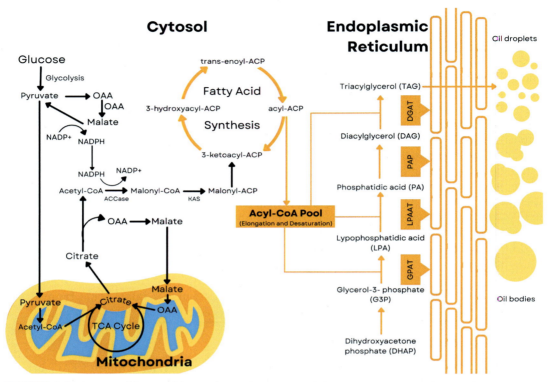

FIGURE 3.2 Process of fatty acid biosynthesis, elongation, and desaturation in marine lower eukaryotic organisms followed by triacylglycerol biosynthesis via Kennedy pathway and the formation of oil droplets and oil bodies thru smooth endoplasmic reticulum.

The lipid composition present in the cells of aquatic microorganisms also depends on their growth phase, wherein lower yield is current in the logarithmic stage, increasing in the late logarithmic phase, and stable at the stationary phase (Hu et al., 2008; Xu et al., 2008). In the exponential growth of the microorganisms, the biosynthesized fatty acids are used to form glycerol-based polar membrane lipids for cell structure. In contrast, the appearance of neutral lipids like TAGs, which do not have a structural function but act as storage, are enhanced during unfavorable or stress conditions (Hu et al., 2008). TAGs are usually used when the cells need energy during metabolism and cellular division (Hu et al., 2008; Huerlimann et al., 2010).

4.2 Fatty acid production and profiling

Extraction of fatty acids from the mycelial mass of oomycetes was based on the methods of Arafiles et al. (2011), Pang et al. (2015), Caguimbal et al. (2019), Devanadera et al. (2019), and Su et al. (2021). Frozen (from the −80°C freezer) samples of mycelial mass were freeze-dried, and then these were homogenized and soaked in ethyl acetate/methanol solution of

4. Biosynthesis, isolation, analysis, and applications of fatty acids from Philippine marine oomycetes

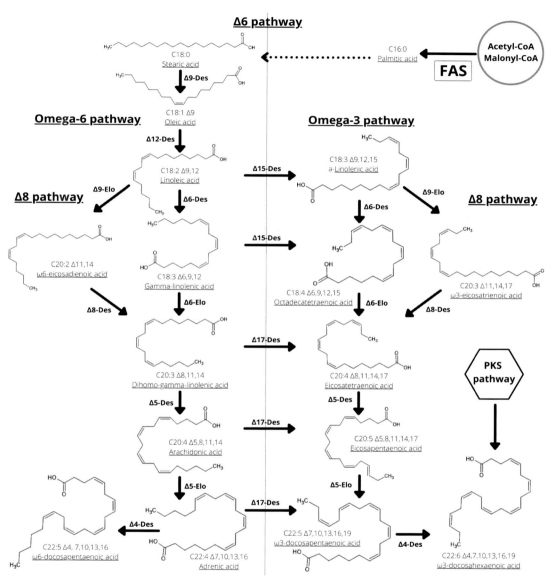

FIGURE 3.3 Monounsaturated and polyunsaturated fatty acid biosynthesis from palmitic and stearic acid thru various pathways (Δ6 and Δ8 pathways, ω3 and ω6 pathways, and PKS pathway) with the help of the enzymes elongases (elo) and desaturases (des).
Note: Data are in terms of percent (%) composition of fatty acids from the total fatty acids content.

chloroform/methanol (2:1 v/v) for 24 h. In the method of Say et al. (2017), fatty acids were extracted by treating the mycelial with t-butyl methyl ester, methanol, and distilled water.

The suspension was collected and concentrated using a rotary evaporator at 40°C. The concentrated sample was re-dissolved using ethyl acetate/methanol solution and was transferred into a glass container, dried, and esterified.

Esterification of fatty acids based on the study of Arafiles et al. (2011), Pang et al. (2015), and Devanadera et al. (2019) used a standard procedure by adding 0.5N methanolic sodium hydroxide solution and heating at 90°C for 15 min. After heating, the solution was cooled at room temperature and mixed with 0.7N HCl-methanol solution and 14% boron trifluoride in methanol solution followed by heating at 90°C for 15 min and cooling down at room temperature. After which, saturated NaCl solution and hexane were added and thoroughly mixed. The upper liquid layer was collected, transferred, and dried in an amber bottle. In the method of Caguimbal et al. (2019) and Say et al. (2017), the MTBE-methanol upper layer from the extraction was transferred into another tube with 10% HCl in methanol and heated at 60°C for 2 h. The dried fatty acid methyl esters (FAMEs) were stored at −80°C for further experiment and analysis.

FAME samples were analyzed by using gas chromatography (GC) according to the protocol of Arafiles et al. (2011), Tilay and Annapure (2012), Pote and Bhadekar (2014), Pang et al. (2015), Deelai et al. (2015), Say et al. (2017), Caguimbal et al. (2019), Devanadera et al. (2019), and Su et al. (2021). Samples were filtered into a sample vial using Whatman No. 1 and were placed on an autosampler injector. The injector was set at 250−270°C and the interface temperature at 270°C, while the column was set at 50−150°C with 15 °C/min temperature increments and 150−250°C with 3°C/min temperature increments. The gas pump was noted at 38.0 cm/min. Sulpeco 37 component FAME Mix (Sigma−Aldrich) was used as the FAME standard for GC analysis of the samples. Some procedures used an internal standard of C19:0 or C20:0 for the quantification of the peak area. The presence of the fatty acids was based on the peaks of samples related to the peaks of standards in the gas chromatography. In contrast, the quantitative data were based on the peak area relative to the peak of each sample and were expressed as the percent fatty acid content of the sample.

In the past years, studies on the fatty acid composition and profiling from fungal-like protists isolated from the Philippine mangroves (Table 3.3) emerged to seek its potential as a new source of essential lipids applied to different industries. As presented in Table 3.3, the dominant fatty acid composition among the marine oomycetes and thraustochytrids are palmitic acid (C16:0) for saturated fatty acids, linoleic acid (C18:1n9) for monounsaturated fatty acids, and linoleic acid (C18:2n-6) for polyunsaturated fatty acids. Traces of long-chain polyunsaturated fatty acids, which are essential, are also found like γ-linolenic acid (C18:3n-6), α-linolenic acid (C18:3n-3), arachidonic acid (C20:4n-6), eicosapentaenoic acid (C20:5n-3), and docosahexaenoic acid (C22:6n-3).

With the widespread use of fuel due to higher demand in today's industrial revolution, scarcity in the supply of fuel happens, and disadvantage in the use of fossil fuel arises since it is not renewable. As part of the solution, alternative fuel sourced from plant and oleaginous microorganisms with renewable biomass has been a significant interest in recent years (Katre et al., 2012). Advantages of using biofuels include higher flash points, reduced emissions, miscibility, compatibility with existing petroleum, lubricity, and biodegradability (Knothe, 2008).

Usage of marine oleaginous microorganisms as a feedstock for biodiesel production shows an advantage in a shorter life cycle, easier to propagate, requires less labor and resources, and

TABLE 3.3 Fatty acid profiles of oomycetes isolated from Philippine mangroves.

Taxon	Microorganisms	Saturated fatty acids				Monounsaturated fatty acids			Polyunsaturated fatty acids					
		C14:0	C15:0	C16:0	C18:0	C16:1n-7	C18:1n-9	C18:1n-7	C18:2n-6	C18:3n-6	C18:3n-3	C20:4n-6	C20:5n-3	C22:6n-3
Peronosporaceae	*Halophytophthora vesicula* AK1YB2[a]	5.9	—	25.2	2.5	3.6	14.9	—	24.7	—	—	20.8	—	—
Peronosporaceae	*H. vesicula* AK1YB2[b]	0.7	—	9.5	1.9	1.3	8.0	—	14.4	—	—	3.0	—	—
Peronosporaceae	*H. vesicula* PQ1YB3[a]	3.5	—	9.5	0.7	3.9	8.5	—	15.5	—	—	16.1	—	—
Peronosporaceae	*H. vesicula* PQ1YB3[b]	0.4	—	14.4	1.9	1.5	8.2	—	12.6	—	—	1.8	—	—
Peronosporaceae	*Halophytophthora* sp. T12GP1[c]	8.6	0.8	32.6	6.8	2.7	17.4	—	13.9	1.1	—	—	1.4	—
Peronosporaceae	*Halophytophthora* sp. T12YBP2[c]	5.9	0.6	30.0	8.1	2.5	15.6	—	16.1	2.0	0.9	—	2.1	—
Salispinaceae	*S. hot*[c]	1.3	—	38.6	4.9	0.9	32.1	—	18.6	—	1.4	—	—	—
Salispinaceae	*Salispina spinosa* ST1YB3[a]	2.7	—	28.3	7.5	3.1	14.9	—	30.7	—	—	20.0	—	—
Salispinaceae	*S. spinosa* ST1YB3[b]	0.8	—	13.4	2.6	0.9	10.3	—	21.6	—	—	4.1	—	—
Thraustochytriaceae	*Schizochytrium* sp.[d]	3.2	—	61.8	1.6	—	—	—	—	0.5	—	—	—	22.6
Thraustochytriaceae	*Thraustochytrium* sp.[d]	—	—	17.7	—	—	70.6	—	11.7	—	—	—	—	—
Thraustochytriaceae	*Thraustochytrium* sp. CR01[e]	4.9	10.6	57.0	3.7	—	3.7	—	—	—	—	—	0.9	4.7
Thraustochytriaceae	*Thraustochytrium* sp. CR01[f]	4.9	13.8	53.9	2.7	—	3.3	—	1.8	—	—	—	1.1	7.3
Thraustochytriaceae	*Thraustochytrium* sp. CR01[g]	—	2.8	24.4	1.4	—	2.9	—	—	—	—	—	—	1.7

[a]*Caguimbal et al. (2019) (Media: 10 g/L glucose. 4 g/L yeast extract, 4 g/L peptone, 50% sterile sea water).*
[b]*Caguimbal et al. (2019) (Media: 20% (v/v) V8 juice, 3 g/L calcium carbonate, 50% sterile sea water).*
[c]*Devanadera et al. (2019) (Media: 5% clarified V8 juice, 4 g/L glucose, 4 g/L yeast extract, 4 g/L peptone, 10 g/L marine salt).*
[d]*Arafiles et al. (2011) (Medium: 10 g/L glucose, 10 g/L yeast extract, 1.0 g/L peptone, and 50% seawater).*
[e]*Uba et al. (2016) (Media: 7.5% (w/v) corn syrup, 1.25 g/L yeast extract. 1.25 g/L peptone, and 50% (v/v) natural seawater).*
[f]*Uba et al. (2016) (Media: 3% (w/v) molasses, 1.25 g/L yeast extract, 1.25 g/L peptone, and 50% (v/v) natural seawater).*
[g]*Uba et al. (2016) (Media: 7.5% (w/v) liquid sugar, 1.25 g/L yeast extract, 1.25 g/L peptone, and 50% (v/v) natural seawater.).*

easy adaptation to environmental stress. These organisms are known to accumulate lipids in triacylglycerols as lipid bodies. The production of these lipid bodies caught the attention for its usage as a potential feedstock for biofuel production. Lipid composition and content of lipid bodies vary among microorganisms with different metabolic processes in response to environmental factors and nutrients. Even though marine oleaginous microorganisms produce lipids, not all species can be used as feedstock for biofuel production. To consider marine oleaginous microorganisms as a suitable feedstock for biofuel production, they should have more than 20% total lipid content. It must produce a long chain of saturated and monounsaturated fatty acids.

Among neutral lipids, triacylglycerol is the most suitable lipid to be converted into biofuel. The methyl esters of MUFAs are liquid at room temperature, which helps in the applicable flow property of biofuel. Unsaturated fatty acids significantly affect the stability of the biofuel products against oxidation during prolonged storage and increase the product's viscosity, s is undesirable in biofuel. Therefore, a high percentage of PUFA is omitted. Thus, the ideal composition of biofuel is composed of methyl esters of saturated and monounsaturated fatty acids with significantly low to no amount of PUFAs (Katre et al., 2012; Ramos et al., 2009).

Lipids from marine oleaginous microorganisms are isolated and traditionally used as a dietary source of essential fatty acids for metabolism and are used as a component for enhancing animal growth in aquaculture (Huerlimann et al., 2010). PUFAs are essential in the growth and development of aquatic animals, and they reduce the mortality rates of larvae and juveniles.

Fish and its oil products are the primary sources of essential fatty acids (MUFAs and PUFAs) required in the diet of humans. But due to the sustainability of marine resources, climate change, and pollutants, the supply of those essential lipids is slowly deteriorating; thus, an alternative source of those essential lipids was identified and isolated in the marine oleaginous microorganisms (Ye et al., 2015). On the same note, PUFAs are also used as nutraceuticals that help in the improvement of human health by improving the immune system and preventing cardiovascular diseases through reduction of blood pressure, lowering of blood viscosity, affecting low-density lipoprotein count, preventing platelet aggregation, preventing cardiac arrhythmia, and improves plasma triglyceride levels (Das, 2008). It is also vital due to its positive effect on arthritis, depression, diabetes mellitus, tumorigenesis, and some cancer; it is also necessary to develop and maintain the brain and retina (Innis, 2008; Ye et al., 2015).

5. Concluding remarks

The Philippines is a tropical country, described by ecologists and conservationists as a biodiversity hotspot. However, of the different biomes with reports on microbial diversity, the Funga in the mangrove environment is left with a considerable knowledge gap. This chapter attempted to bridge this gap and presented that the mangrove biome offers a home to a diverse of less explored and cryptic microorganisms, not only those with ecological significance but also with potential applications in biotechnology and pharmaceutics.

Acknowledgments

We would like to thank the numerous authors whose papers are cited in this paper and Dr. Anthony Buaya for his valuable insights on fungal-like organisms in aquatic habitats. The Research Center for the Natural and Applied Sciences, University of Santo Tomas, Manila, and the National Research Council of the Philippines, Department of Science and Technology, supported most of our research projects on Oomycetes.

References

Ahn, E. Y., Jiang, Y., Zhang, Y., Son, E. I., You, S., Kang, S., Park, J., Jung, J., Lee, B., & Kim, D. (2008). Cytotoxicity of p-tyrosol and its derivatives may correlate with the inhibition of DNA replication initiation. *Oncology Reports, 19*, 527–534.

Alias, S. A., & Jones, E. B. G. (2000). Vertical distribution of marine fungi in *Rhizophora apiculata* at Morib mangrove, Selangor, Malaysia. *Mycoscience, 41*, 431–436.

Alias, S. A., & Jones, E. B. G. (2009). Fungi from mangroves of Malaysia. *Institute of Ocean and Earth Sciences, 8*, 1–109.

Alias, S. A., Jones, E. B. G., & James, T. (1999). Intertidal fungi from the Philippines, with a description of *Acrocordiopsis sphaerica* sp. nov. (Ascomycota). *Fungal Divers, 2*, 35–41.

Almaliky, B. S. A., Abidin, M. A. Z., Kader, J., & Wong, M. Y. (2013). First report of *Marasmiellus palmivorus* causing post-emergence damping-off on coconut seedlings in Malaysia. *Plant Disease, 97*(1), 143.

Apurillo, C. C. S., Cai, L., & dela Cruz, T. E. E. (2019). Diversity and bioactivities of mangrove fungal endophytes from Leyte and Samar, Philippines. *Philippine Science Letters, 12*, 33–48.

Arafiles, K. V., Alcantara, J. O., Batoon, J. L., Galura, F. S., Cordero, P. F., Leaño, F. M., & Dedeles, G. R. (2011). Cultural optimization of thraustochytrids for biomass and fatty acid production. *Mycosphere, 2*, 521–531.

Armstrong, E., & Rogerson, A. (2000). Utilization of seaweed carbon by three surface heterotrophic protists: *Stereomyxa ramose, Nitzschia alba*, and *labyrinthula* sp. *Aquatic Microbial Ecology, 21*, 49–57.

Atienza, G. M. A., Arafiles, K. H. V., Carmona, M. C. M., Garcia, J. P. C., Macabago, A. M. B., Peñacerrada, B. J. D. C., Cordero, P. R. F., Bennett, R. M., & Dedeles, G. R. (2012). Carotenoid analysis of locally isolated thraustochytrids and their potential as an alternative fish feed for *Oreochromis niloticus* (Nile tilapia). *Mycosphere, 3*(4), 420–428.

Bacal, C. J. O., & Yu, E. T. (2017). Cellulolytic activities of a novel *Fomitopsis* sp. and *Aspergillus tubingensis* isolated from Philippine mangroves. *Philippine Journal of Science, 146*(4), 403–410.

Bennett, R. M., de Cock, A. W. A. M., Lévesque, C. A., & Thines, M. (2017). *Calycofera* gen. nov., an estuarine sister taxon to *Phytopythium*, Peronosporaceae. *Mycological Progress, 16*(10), 947–954.

Bennett, R. M., Dedeles, G. R., & Thines, M. (2017). *Phytophthora elongata* (Perosnoporaceae) is present as an estuarine species in Philippine mangroves. *Mycosphere, 8*(7), 959–967.

Bennett, R. M., Devanadera, M. K., Dedeles, G. R., & Thines, M. (2018b). A revision of Salispina, its placement in a new family, Salisapinaceae (Rhipidiales), and description of fourth species, S. hoi sp. nov. *IMA Fungus, 9*(2), 259–269.

Bennett, R. M., Nam, B., Dedeles, G. R., & Thines, M. (2018a). *Phytopythium leanoi* sp. nov. and *Phytopythium dogmae* sp. nov., *Phytopythium* species associated with mangrove leaf litter from the Philippines. *Acta Mycologica, 52*(2), 1–13.

Bennett, R. M., & Thines, M. (2017a). Confirmation that *Phytophthora insolita* (Peronosporaceae) is present as a marine saprotroph on mangrove leaves and first report of the species for the Philippines. *Nova Hedwigia, 105*(1–2), 185–196.

Bennett, R. M., & Thines, M. (2019). Revisiting Salisapiliaceae. *Fungal Systematics and Evolution, 3*, 171–184.

Bennett, R. M., & Thines, M. (2020). An overview of Philippine estuarine oomycetes. *Philippine Journal of Systematic Biology, 14*(1), 1–14.

I. Fungi in aquatic and terrestrial habitats

82   3. Fungi and fungus-like microorganisms in Philippine marine ecosystems

Berkhout, C. M. (1923). *De schimmelgeslachten Monilia, Oidium, Oospora en Torula*. Netherlands: Utrecht University (Thesis 1923).

Besitulo, A., Moslem, M. A., & Hyde, K. D. (2010). Occurrence and distribution of fungi in a mangrove forest on Siargao Island, Philippines. *Botanica Marina, 53*(6), 535−543.

Bongiorni, L., Jain, R., Raghukumar, S., & Aggarwal, R. K. (2005). *Thraustochytrium gaertnerium* sp. nov.: A new thraustochytrid stramenopilan protist from mangroves of Goa, India. *Protist, 156*(3), 303−315.

Borines, L. M., Palermo, V. G., Guadalquiver, G. A., Dwyer, C., Drenth, A., Daniel, R., & Guest, D. I. (2014). Jackfruit decline caused by Phytophthora palmivora (Butler). *Australasian Plant Pathology, 43*, 123−129.

Bremer, G. B. (1995). Lower marine fungi (Labyrinthulomycetes) and the decay of mangrove leaf Litter. *Hydroblologia, 295*, 89−95.

Broden, A. (1904). *Les infections à trypanosomes au Congo chez l'homme et les animaux* (communication préliminaire). *Bulletin de la Societe d'Etudes Coloniales, 11*, 116−139.

Bucher, V. V. C., Hyde, K. D., Pointing, S. B., & Reddy, C. A. (2004). Production of wood decay enzymes, mass loss and lignin solubilization in wood by marine ascomycetes and their anamorphs. *Fungal Divers, 15*, 1−14.

Bugni, T. S., & Ireland, C. M. (2004). Marine derived fungi: A chemically and biologically diverse group of microorganisms. *Natural Product Reports, 21*, 143−163.

Bungihan, M. E., Tan, M. A., Kitajima, M., Kogure, N., dela Cruz, T. E., Takayama, H., & Nonato, M. G. (2010). A new isocoumarin compound from an endophytic fungus *Guignardia* sp. isolated from *Pandanus amaryllifolius* Roxb. *ACGC Chemical Research Communications, 24*, 13−16.

Bungihan, M. E., Tan, M. A., Kitajima, M., Kogure, N., Franzblau, S. G., dela Cruz, T. E. E., Takayama, H., & Nonato, M. G. (2011). Bioactive metabolites of *Diaporthe* sp. P133, an endophytic fungus isolated from *Pandanus amaryllifolius*. *Journal of Natural Medicines, 65*, 606−609.

Bungihan, M. E., Tan, M. A., Takayama, H., dela Cruz, T. E., & Nonato, M. G. (2013a). A new macrolide isolated from the endophytic fungus *Colletotrichum* sp. *Philippine Science Letters, 6*, 57−73.

Cagliari, A., Margis, R., Maraschin, F., Turchetto-Zolet, d., Loss, G., A., & Margis-Pinheiro, M. (2011). Biosynthesis of triacylglycerols (TAGs) in plants and algae. *International Journal of Plant Biology, 2*(e10), 40−52.

Caguimbal, N. A. L. E., Devanadera, M. K. P., Bennett, R. M., Arafiles, K. H. V., Watanabe, K., Aki, T., & Dedeles, G. R. (2019). Growth and fatty acid profiles of *Halophytophthora vesicula* and *Salispina spinosa* from Philippine mangrove leaves. *Letters in Applied Microbiology, 69*, 221−228.

Cai, L., Zhang, K., Mckenzie, E. H., & Hyde, K. D. (2003). Freshwater fungi from bamboo and wood submerged in the Liput River in the Philippines. *Fungal Divers, 13*, 1−12.

Calabon, M. S., Sadaba, R. B., & Campos, W. L. (2019). Fungal diversity of mangrove-associated sponges from New Washington, Aklan, Philippines. *Mycology, 10*(1), 6−21.

Carascal, M. B., del Rosario, M. J. G., Notarte, K. I. R., Huyop, F., Yaguchi, T., & dela Cruz, T. E. E. (2017). Butachlor biodegradation potential of fungi isolated from submerged wood and surface water collected in Taal Lake, Philippines. *Philippine Science Letters, 10*(2), 81−88.

Carrol, G. C., & Petrini, O. (1983). Patterns of substrate utilization by some fungal endophytes from coniferous foliage. *Mycologia, 75*, 53−63.

Cavalier-Smith, T. (1986). The Kingdom Chromista: Origin and systematics. *Progress in Phycological Research, 4*, 309−347.

Cavalier-Smith, T. (1999). Principles of protein and lipid targeting in secondary symbiogenesis: Euglencid, dinoflagellate, and sporozoan plastid origins and the eukaryote family tree. *Journal of Eukaryotic Microbiology, 46*, 347−366.

Celino, M. S. (1933). Blight of cinchona seedlings. *Philippine Agricultural, 23*, 111−123.

Cheng, M. H., Walker, T. H., Hulbert, G. J., & Raman, D. R. (1999). Fungal production of eicosapentaenoic acid and arachidonic acids from industrial waste streams and crude soybean oil. *Bioresource Technology, 67*, 101−110.

Chi, Z., Yan, K., Gao, L., Li, J., Wang, X., & Wang, L. (2008). Diversity of marine yeasts with high protein content and evaluation of their nutritive compositions. *Journal of the Marine Biological Association of the United Kingdom, 88*(7), 1347−1352.

Clara, F. M. (1928). A *Phytophthora* disease of santol seedlings. *Philippine Journal of Science, 35*, 411−425.

Collins, T., Gerday, C., & Feller, G. (2005). Xylanases, xylanase families and thermophilic xylanases. *FEMS Microbiology Reviews, 29*, 3−23.

Dagamac, N. H. A., Sogono, P. G., Cabalfin, R. C. B., Adducul, A. C. Y., & dela Cruz, T. E. E. (2008). Fungal root endophytes from *Musa* spp. as biological control agents against the plant pathogen *Fusarium oxysporum*. *Acta Manilana, 56*, 27−35.

I. Fungi in aquatic and terrestrial habitats

References

Das, U. (2008). Essential fatty acids and their metabolites could function as endogenous HMG-CoA reductase and ACE enzyme inhibitors, anti-arrhythmic, anti-hypertensive, andti-atherosclerotic, anti-inflammatory, cytoprotective, and cardioprotective molecules. *Lipids in Health and Disease, 7*, 37.

De la Peña, M. R., Teruel, M. B., Oclari, J. M., Amar, M. J. A., & Ledesma, E. G. T. (2016). Use of thraustochytrid *Schizochytrium* sp as source of lipid and fatty acid in a formulated diet for abalone *Haliotis asinine* (Linnaeus) juveniles. *Aquaculture International, 24*, 1103–1118.

De Jesus Milanez, G., Ferrer Lirio, M. R., Hagosojos, B. M., & Masangkay, F. R. (2020). First report of *Candida tropicalis* in edible freshwater fish in the Philippines. *Asian Journal of Biological and Life Sciences, 8*(3), 117–119.

De Mesa, R. B. C., Espinosa, I. R., Agcaoili, M. C. R. R., Calderon, M. A. T., Pangilinan, M. V. B., De Padua, J. C., & dela Cruz, T. E. E. (2020). Antagonistic activities of needle-leaf fungal endophytes against *Fusarium* spp. *MycoAsia,* 2020/06.

De Padua, J. C., & dela Cruz, T. E. E. (2021). Isolation and characterization of nickel-tolerant *Trichoderma* strains from marine and terrestrial environments. *Journal of Fungi, 7*(8), 591.

Dee, J. A. S., Ferrer, A. A. V., Pastor, J. A. S., Gacelo, M. C. P., & Dedeles, G. R. (2019). Optimization of cultural conditions for labyrinthula species isolated from mangrove leaves. *Philippine Journal of Systematic Biology, 13*(2), 78–84.

Deelai, S., Suetrong, S., Damrianant, S., Unagul, P., & Sakkayawong, N. (2015). Isolation and identification of native lower fungi for polyunsaturated fatty acid (PUFA) production in Thailand, and the effect of carbon and nitrogen sources on the growth and production. *African Journal of Biotechnology, 14*, 1449–1460.

Del Rosario, M. S. (1968). A handbook of citrus diseases in the Philippines. *UPCA Tech Bulletin, 31*, 1–33.

Dela Cruz, T. E. E., Notarte, K. I. R., Apurillo, C. C. S., Tarman, K., & Bungihan, M. E. (2020). Biomining fungal endophytes from tropical plants and seaweeds for drug discovery. In M. Ozturk, D. Egamberdieva, & M. Pesic (Eds.), *Biodiversity and biomedicine our Future* (pp. 51–62). USA: Elsevier Academic Press.

Devanadera, M. K. P., Bennett, R. M., Watanabe, K., Santiago, M. R., Ramos, M. C., Aki, T., & Dedeles, G. R. (2019). Marine oomycetes (*halophytophthora* and *Salispina*): A potential source of fatty acids with cytotoxic activity against breast adenocarcinoma cells (MCF7). *Journal of Oleo Science, 68*, 1163–1174.

Ding, B., Yin, Y., Zhang, F., & Li, Z. (2011). Recovery and phylogenetic diversity of culturable fungi associated with marine sponges *Clathrina luteoculcitella* and *Holoxea* sp. in the South China Sea. *Marine Biotechnology, 13*, 713–721.

Dogma, I. J. (1974). Studies on chitinophylic *Siphonaria, Diplophlyctis* and *Rizoclosmatium*, Chytridiales. III. *Nephochytrium complicatus* Wiloughby: another *Diplophlctis* with a sexual phase. *Nova Hedwigia, 25*(1-2), 143–159.

Dogma, I. J. (1975). Of Philippine mycology and lower fungi. *Kalikasan, Philippine Journal of Biology, 4*, 69–105.

Dogma, I. J. (1976). *Diplophlyctis asteroidea*, a new species with asexual resting spores. *Transactions of the British Mycological Society, 67*(2), 255–264.

Dogma, I. J. (1977). Philippine zoosporic fungi: *Olpidium sparrowii*, a new chytridiomycete parasite of rotifer eggs. *Kalikasan, 6*(1), 9–20.

Dogma, I. J. (1986). Zoosporic fungi. In , *Vol. 1. Guide to Philippine Flora and Fauna* (pp. 1–199). Quezon City: Natural Resources Management Center, Ministry of Natural Resources and University of the Philippines.

Doty, M. S. (1988). Prodromus Ad Systematica Eucheumatoideorum: A Tribe of Commercial Seaweeds Related to *Eucheuma* (Solieriaceae, Gigartinales). In I. A. Abbott (Ed.), *Taxonomy of Economic Seaweeds with reference to some Pacific and Caribbean species Volume II* (pp. 159–207). California: California Sea Grant, La Jolla.

Ela, V. M. (1968). Notes on diseases of orchids in the Philippines. *Philippine Agricultural, 4*, 531–537.

Estudillo-del Castillo, C., Gapasin, R. S., & Leaño, E. M. (2009). Enrichment potential of HUFA-rich thraustochytrid *Schizochytrium mangrovei* for the rotifer *Brachionus plicatilis*. *Aquaculture, 293*, 57–61.

Fell, J. W., & Master, I. M. (1975). Phycomycetes (*Phytophthora* spp. nov. and *Pythium* sp. nov.) associated with degrading mangrove (*Rhizopora mangle*) leaves. *Canadian Journal of Botany, 53*(24), 2908–2922.

Fell, J. W., & Master, I. M. (1980). The association and potential role of fungi in mangrove detrital systems. *Botanica Marina, 23*, 257–263.

Francis, M. M., Webb, V., & Zuccarello, G. C. (2016). Marine yeast biodiversity on seaweeds in New Zealand waters. *New Zealand Journal of Botany, 54*(1), 30–47.

Fry, W. E., & Grünwald, N. J. (2010). Introduction to oomycetes. *The Plant Health Instructor*. https://doi.org/10.1094/PHI-I-2010-1207-01. http://www.apsnet.org/edcenter/intropp/Pathogen Groups/Pages/IntroOomycetes.aspx

Gao, L., Chi, Z., Sheng, J., Ni, X., & Wang, L. (2007). Single-cell protein production from Jerusalem artichoke extract by a recently isolated marine yeast *Cryptococcus aureus* G7a and its nutritive analysis. *Applied Microbiology and Biotechnology, 77*(4), 825–832.

I. Fungi in aquatic and terrestrial habitats

Ghadin, N., Zin, N. M., Sabaratnam, V., Badya, N., Basri, D. F., Lian, H. H., & Sidik, N. M. (2008). Isolation and characterization of a novel endophytic *Streptomyces* SUK 06 with antimicrobial activity from Malaysian plant. *Asian Journal of Plant Sciences, 7*(2), 189—194.

Guerrero, J. J. G., General, M. A., & Serrano, J. E. (2018). Culturable foliar fungal endophytes of mangrove species in Bicol Region, Philippines. *Philippine Journal of Science, 147*(4), 563—574.

Hatai, K., Bian, Z. B., Baticados, M. C. L., & Egusa, S. (1980). Studies on fungal diseases in crustaceans: II. *Haliphthoros philippinensis* sp. nov. Isolated from cultivated larvae of the jumbo tiger prawn (*Penaeus monodon*). *Transactions of the Mycological Society of Japan, 21*, 47—55.

Holguin, G., Vazquez, P., & Bashan, Y. (2001). The role of sediment microorganisms in the productivity, conservation, and rehabilitation of mangrove ecosystems: An overview. *Biology and Fertility of Soils, 33*(4), 265—278.

Huerlimann, R., de Nys, R., & Heimann, K. (2010). Growth, lipid content, productivity, and fatty acid composition of tropical microalgae for scale-up production. *Biotechnology and Bioengineering, 107*, 245—257.

Hunting, E. R., de Goeij, J. M., Asselman, M., van Soest, R. W. M., & van der Geest, H. G. (2010). Degradation of mangrove-derived organic matter in mangrove-associated sponges. *Bulletin of Marine Science, 86*, 871—877.

Hu, Q., Sommerfeld, M., Jarvis, E., Ghirardi, M., Posewitz, M., Seibert, M., & Darzins, A. (2008). Microalgal triacylglycerols as feedstocks for biofuel production: Perspectives and advances. *The Plant Journal, 54*(4), 621—639.

Hu, X. P., Wang, M. X., Hu, D. F., & Yang, J. R. (2011). First report of wilt on alfalfa in China caused by *Verticillium nigrescens*. *Plant Disease, 95*(12), 1591.

Hyde, K. D., & Jones, E. B. G. (1988). Marine mangrove fungi. P.S.Z.N.I. *Marine Ecology, 9*, 15—33.

Hyde, K. D., & Lee, S. Y. (1995). Ecology of mangrove fungi and their role in nutrient cycling: What gaps occur in our knowledge? *Hydrobiologia, 295*, 107—118.

Innis, S. M. (2008). Dietary omega 3 fatty acids and the developing brain. *Brain Research, 1237*, 35—43.

Johnson, T. W. J., & Sparrow, F. K. J. (1961). *Fungi in oceans and estuaries*. Weinheim: J. Cramer.

Jones, E. B. G. (1994). Ultrastructure and taxonomy of the aquatic ascomycetes order Halosphaeriales. *Canadian Journal of Botany, 73*, 790—801.

Jones, E. B. G., Hyde, K. D., Read, S. T., Moss, S. T., & Alias, S. A. (1996). *Tirisporella* gen. nov., an ascomycete from the mangrove palm *Nypa fruticans*. *Canadian Journal of Botany, 74*(9), 1487—1495.

Katre, G., Joshi, C., Khot, M., Zinjarde, S., & RaviKumar, A. (2012). Evaluation of single cell oil (SCO) from a tropical marine yeast *Yarrowia lipolytica* BCIM 3589 as a potential feedstock for biodiesel. *AMB Express, 2*, 36.

Knothe, G. (2008). "Designer" Biodiesel: Optimizing fatty ester composition to improve fuel properties. *Energy Fuel, 22*, 1358—1364.

Kohlmeyer, J., & Kohlmeyer, E. (1979). *Marine mycology*. New York: Academic Press.

Kulkarni, N., Shendye, A., & Rao, M. (1999). Molecular and biotechnological aspect of xylanases. *FEMS Microbiology Reviews, 23*, 411—456.

Kroes, R., Schaefer, E. J., Squire, R. A., & Williams, G. M. (2003). A review of the safety of DHA 45—oil. *Food and Chemical Toxicology, 41*, 1433—1446.

Kurtzman, C. P., & Fell, J. W. (1998). *The yeasts, A taxonomic study* (4th ed.). Amsterdam: Elsevier.

Kutty, S. N., & Philip, R. (2008). Marine yeasts—a review. *Yeast, 25*, 465—483.

Lavadia, M. G. B., Dagamac, N. H. A., & dela Cruz, T. E. E. (2017). Diversity and biofilm inhibition activities of algicolous fungi collected from two remote islands of the Philippine Archipelago. *Current Research in Environmental & Applied Mycology, 7*(4), 309—321.

Leaño, E. M. (2001). Straminipilous organisms from fallen mangrove leave from Panay, Island, Philippines. *Fungal Divers, 6*, 75—81.

Leaño, E. M. (2002). *Haliphthoros* spp. from spawned eggs of captive mud crab, *Scylla serrata*, broodstocks. *Fungal Divers, 9*, 93—103.

Leaño, E. M., Gapasin, R. S. J., Polohan, B., & Vrijmoed, L. L. (2003). Growth and fatty acid production of thraustochytrids from Panay mangroves, Philippines. *Fungal Divers, 12*, 111—112.

Lee, A. H. (1921). Observations on previously unreported or noteworthy plant diseases in the Philippines. *The Philippine Agricultural Review, 14*, 422—434.

Lee Chang, K. J. L., Dunstan, G. A., Abell, G. C., Clementson, L. A., Blackburn, S. I., Nichols, P. D., & Koutoulis, A. (2012). Biodiscovery of new australian thraustochytrids for production of biodiesel and long-chain omega-3 oils. *Applied Microbiology and Biotechnology, 93*(5), 2215—2231.

Liu, Z. C. T., & Chi, Z. (2012). Occurrence and diversity of yeasts in the mangrove ecosystems in Fujian, Guangdong and Hainan Provinces of China. *Indian Journal of Microbiology, 52*(3), 346—353.

Ludevese-Pascual, G., De la Peña, M., Reyes, O., & Tomalejo, J. (2013). Low-cost production of the marine thraustochytrid isolate, *Schizochytrium* sp. LEY7 as larval live feed enrichment for the mangrove snapper, Lutjanus sp. In

I. Fungi in aquatic and terrestrial habitats

References

C. I. Hendry (Ed.), *Larvi'13—fish & Shellfish Larviculture symposium* (pp. 253—256). Belgium: Ghent University, Laboratory of Aquaculture & Artemia Reference Center.

Maini, Z. A. N., Aribal, K. M. J., Narag, R. M. A., Melad, J. K. L. T., Frejas, J. A. D., Arriola, L. A. M., & Lopez, C. M. (2019). Lead (II) tolerance and uptake capacities of fungi isolated from a polluted tributary in the Philippines. *Applied Environmental Biotechnology, 4*(1), 18—29.

Marcelo, A., Geronimo, R. M., Vicente, C. J. B., Callanta, R. B. P., Bennett, R. M., Ysrael, M. C., & Dedeles, G. R. (2018). TLC screening profile of secondary metabolites and biological activities of *Salisapilia tartarea* SY1P1 isolated from Philippine mangroves. *Journal of Oleo Science, 67*(12), 1585—1595.

Maria, G. I., Sridhar, K. R., & Raviraja, N. S. (2005). Antimicrobial and enzyme activity of mangrove endophytic fungi of southwest coast of India. *International Journal of Agricultural Technology, 1*(1), 67—80.

Mendiola, N., & Espino, R. B. (1916). Some phycomycetous diseases of cultivated plants in the Philippines. *The Philippine Agriculturist and Forester, 5*, 65—72.

Mendoza, R. A. J., Estanislao, K., Aninipot, J. F., Dahonog, R. A., de Guzman, J., Torres, J. M. O., & dela Cruz, T. E. E. (2010). Biosorption of mercury by the marine fungus *Dendryphiella salina*. *Acta Manilana, 58*, 25—29.

Migula, W. (1900). *System der Bakterien* (Volume 2). Germany: Gustav Fischer, Jena.

Moron, L. S., Lim, Y. W., & dela Cruz, T. E. E. (2018). Antimicrobial activities of crude culture extracts from mangrove fungal endophytes collected in Luzon Island, Philippines. *Philippine Science Letters, 11*, 28—36.

Mouzouras, R., Jones, E. B. G., Venkatasamy, R., & Moss, S. T. (1986). Decay of wood by microorganisms in aquatic habitats. *Record of the Annual Convention of the BWPA, 1986*, 1—18.

Nakagiri, A. (2000). Ecology and biodiversity of Halopyhtophthora species. In K. D. Hyde, W. H. Ho, & S. B. Pointing (Eds.), *Aquatic mycology across the millennium* (pp. 153—164). Chiang Mai, Thailand: Fungal Diversity Press.

Nakagiri, A., Tokumasu, S., Araki, H., Koreada, S., & Tubaki, K. (1989). Succession of fungi in decomposing mangrove leaves in Japan. In T. Hattori, Y. Ishida, Y. Maruyama, R. Y. Morita, & A. Uchida (Eds.), *Recent advances in microbial ecology. Proceedings of the 5th international symposium on microbial ecology*. Tokyo, Japan: ISMES Japan Scientific Societies Press.

Nasr, S., Mohammadimehr, M., Vaghei, M. G., Amoozegar, M. A., Abolhassan, S., Fazeli, S., & Yurkov, A. (2017). *Jaminaea pallidilutea* sp. nov. (Microstromatales), a basidiomycetous yeast isolated from plant material of mangrove forests in Iran. *International Journal of Systematic and Evolutionary Microbiology, 67*, 4405—4408.

Newell, S. Y. (1996). Established and potential impacts of eukaryotic mycelial decomposers in marine/terrestrial ecotypes. *Journal of Experimental Marine Biology and Ecology, 200*, 187—206.

Newell, S. Y., & Fell, J. W. (1992). Distribution and experimental responses to substrate of marine oomycetes (*Halophytophthora* spp.) in mangrove ecosystems. *Mycological Research, 96*(10), 851—856.

Newell, S. Y., Miller, J. D., & Fell, J. W. (1987). Rapid and pervasive occupation of fallen mangrove leaves by a marine zoosporic fungus. *Applied and Environmental Microbiology, 53*(10), 2464—2469.

Nham Tran, T. L., Miranda, A. F., Gupta, A., Puri, M., Ball, A. S., Adhikari, B., & Mouradov, A. (2020). The nutritional and pharmacological potential of new Australian thraustochytrids isolated from mangrove sediments. *Marine Drugs, 18*(3), 151.

Notarte, K. I., Nakao, Y., Yaguchi, T., Bungihan, M., Suganuma, K., & dela Cruz, T. E. (2017). Trypanocidal activity, cytotoxicity and histone modifications induced by malformin A1 isolated from the marine-derived fungus *Aspergillus tubingensis* IFM 63452. *Mycosphere, 8*(1), 111—120.

Notarte, K. I., Yaguchi, T., Suganuma, K., & dela Cruz, T. E. (2018). Antibacterial, cytotoxic and trypanocidal activities of marine-derived fungi isolated from Philippine macroalgae and seagrasses. *Acta Botanica Croatica, 77*(2), 141—151.

Ocfemia, G. O. (1925). The *Phytophthora* disease of eggplant in the Philippine islands. *Philippine Agricultural, 14*, 317—328.

Osono, T. (2008). Endophytic and epiphytic phyllosphere fungi of *Camellia japonica*: Seasonal and leaf age dependent variations. *Mycologia, 100*(3), 387—391.

Pang, K.-L., Lin, H.-J., Lin, H.-Y., Huang, Y.-F., & Chen, Y.-M. (2015). Production of arachidonic and eicosapentaenoic acids by the marine oomycetes Halophytophthora. *Marine Biotechnology, 17*, 121—129.

Pang, K. L., Overy, D., Jones, E. B. G., Calado, M. D. L., Burgaud, G., Walker, A. K., Johnson, J. A., Kerr, R. G., Cha, H. J., & Bills, J. F. (2016). "Marine fungi" and "marine-derived fungi" in natural product chemistry research: Toward a new consensual definition. *Fungal Biology Reviews, 30*(4), 163—175.

Paulino, G. V. B., Félix, C. R., Broetto, L., & Landell, M. F. (2017). Diversity of culturable yeasts associated with zoanthids from Brazilian reef and its relation with anthropogenic disturbance. *Marine Pollution Bulletin, 123*(1—2), 253—260.

I. Fungi in aquatic and terrestrial habitats

Perez, J. G. A., Mappala, A. K. A., Icaro, C. K. L., Estrada, A. M. T., Arafiles, K. H. V., & Dedeles, G. R. (2020). Occurrence of thraustochytrids on fallen mangrove leaves from Pagbilao mangrove park, Quezon Province. *Philippine Journal of Systematic Biology, 14*(1), 1–8.

Perveen, Z., Ando, H., Ueno, A., Ito, Y., Yamamoto, Y., Yamada, Y., Takagi, T., Kaneko, T., Kogame, K., & Okuyama, H. (2006). Isolation and characterization of a novel thraustochytrid-like microorganisms that efficiently produces docosahexaenoic acid. *Biotechnology Letters, 28*, 197–202.

Pitt, J. I., & Miller, M. W. (1970). The parasexual cycle in yeasts of the genus Metschnikowia. *Mycologia, 62*(3), 462–473.

Pote, S., & Bhadekar, R. (2014). Statistical approach for production of PUFA from Kocuria sp. BRI 35 isolated from marine water sample. *BioMed Research International, 2014*, 570925.

Primavera, J. H. (2009). *Field guide to Philippine mangroves*. Manila: SEAFDEC.

Quebral, F. C., Pordesimo, A. N., Reyes, T. T., & Tamayo, B. P. (1962). Heart rot of pineapple in the Philippines. *Philippine Agricultural, 46*, 432–450.

Quimio, T. H., & Quimio, A. J. (1974). Compendium of postharvest and common diseases of fruits in the Philippines. *UPCA Tech Bulletin 34.*

Rafael, A., & Calumpong, H. P. (2019). Fungal infections of mangroves in natural forests and reforestation sites from the Philippines. *AACL Bioflux, 12*(6), 2062–2074.

Raghukumar, S. (1992). Bacterivory: A novel dual role for thraustochytrids in the sea. *Marine Biology, 113*, 165–169.

Raghukumar, S. (2008). Marine fungal biotechnology: An ecological perspective. *Fungal Divers, 31*, 19–35.

Ramirez, C. S. P., Go, C. O., Hernandez, S. A. S., Ruiz, H. I., Sabit, M. B., & dela Cruz, T. E. E. (2010). Characterization of marine yeasts isolated from different substrates collected in Calatagan, Batangas. *Philippine Journal of Systematic Biology, 4*(0), 1–11.

Ramirez, C. S. P., Notarte, K. I. R., & dela Cruz, T. E. E. (2020). Antibacterial activities of mangrove leaf endophytic fungi from Luzon Island, Philippines. *Studies in Fungi, 3*(1), 320–331.

Ramos, M. J., Fernández, C. M., Casas, A., Rodríguez, L., & Perez, Á. (2009). Influence of fatty acid composition of raw materials on biodiesel properties. *Bioresource Technology, 100*, 261–268.

Reikhof, W. R., Sears, B. B., & Benning, C. (2005). Annotation of genes involved in glycerolipid biosynthesis in *Chlamydomonas reinhardtii*: Discovery of the betaine lipid synthase BTA1Cr. *Eukaryot Cell, 4*(2), 242–252.

Reinking, O. A. (1919). *Phytophthora faberi* Maubl: The cause of coconut bud rot in the Philippines. *Philippine Journal of Science, 14*, 131–150.

Reynolds, D. R. (1970). Fungi isolated from rice paddy soil at Central Exp. Station, UP Colllege of Agriculture. *Philippine Agricultural, 54*, 55–59.

Royle, J. F. (1839). *Illustrations of the Botany and Other Branches of the Natural History of the Himalayan Mountains and of the Flora of Cashmere* (Volume 1). London: H. Allen & Co.

Say, E. K. P., Yabut, A. T. V., Cinco, N. E. T., Caguimbal, N. A. L. E., Devanadera, M. K. P., Bennett, R. M., Arafiles, K. H. V., Aki, T., & Dedeles, G. R. (2017). Growth and fatty acid production of *Halophytophthora* S13005YL1-3.1 under different salinity and pH levels. *Philippine Agricultural Scientist, 100*, S6–S11.

Schulz, B., Draeger, S., dela Cruz, T. E., Rheinheimer, J., Siems, K., Loesgen, S., Bitzer, J., Schloerke, O., Zeeck, A., Koch, I., Hussain, H., Dai, J., & Krohn, K. (2008). Screening strategies for obtaining novel, biologically active, fungal secondary metabolites from marine habitats. *Botanica Marina, 51*, 219–234.

Shurson, G. C. (2018). Yeast and yeast derivatives in feed additives and ingredients: Sources, characteristics, animal responses, and quantification methods. *Animal Feed Science and Technology, 235*, 60–76.

Sia Su, G., Dueñas, K., Roderno, K., Sison, M., Su, M., Ragragio, E., Guzman, T., & Heralde, F., III (2014). Distribution and diversity of marine fungi in Manila Bay, Philippines. *Annual Research & Review in Biology, 4*(24), 4166–4173.

Solis, M. J., Draeger, S., & dela Cruz, T. E. (2010). Marine-derived fungi from *Kappaphycus alvarezii* and *K. striatum* as potential causative agents of ice-ice disease in farmed seaweeds. *Botanica Marina, 53*, 587–594.

Su, C. J., Ju, W. T., Chen, Y. M., Chiang, M. W., Hsieh, S. Y., Lin, H. J., Jones, E. B., & Gareth Pang, K. L. (2021). Palmitic acid and long-chain polyunsaturated fatty acids dominate in mycelia of mangrove *Halophytophthora* and *Salispina* species in Taiwan. *Botanica Marina*, 1–16.

Sullivan, B. K., Sherman, T. D., Damare, V. S., Lilje, O., & Gleason, F. H. (2013). Potential roles of *Labyrinthula* spp. in global seagrass population declines. *Fungal Ecology, 6*, 328–338.

Tan, M. A., dela Cruz, T. E. E., Apurillo, C. C. S., & Proksch, P. (2015). Chemical constituents from a Philippine mangrove endophytic fungi *Phyllosticta* sp. *Der Pharma Chemica, 7*(2), 43–45.

Tan, T. K., & Pek, C. L. (1997). Tropical mangrove leaf litter fungi in Singapore with an emphasis on Halophytophthora. *Mycological Research, 101*(2), 165–168.

References

Taylor, M. W., Radax, R., Steger, D., & Wagner, M. (2007). Sponge-associated microorganisms: Evolution, ecology, and biotechnological potential. *Microbiology and Molecular Biology Reviews, 71*(2), 295–347.

Teodoro, N. G. (1926). Rubber tree diseases and their control. *The Philippine Agricultural Review, 19*, 63–73.

Thines, M. (2009). Bridging the gulf: Phytophthora and downy mildews are connected by rare grass parasites. *PLoS ONE, 4*(3), e4790.

Thines, M., & Kamoun, S. (2010). Oomycete–plant coevolution: Recent advances and future prospects. *Current Opinion in Plant Biology, 13*, 1–7.

Tilay, A., & Annapure, U. (2012). Novel simplified and rapid method for screening and isolation of polyunsaturated fatty acids producing marine bacteria. *Biotechnology Research International, 2012*, 542721.

Torres, J. M. O., Cardenas, C. V., Moron, L. S., Guzman, A. P. A., & dela Cruz, T. E. E. (2011). Dye decolorization activities of marine-derived fungi isolated from Manila Bay and Calatagan Bay, Philippines. *Philippine Journal of Science, 140*(2), 133–143.

Torres, J. M. O., & dela Cruz, T. E. E. (2013). Production of xylanases by mangrove fungi from the Philippines and their application in enzymatic pretreatment of recycled paper pulps. *World Journal of Microbiology and Biotechnology, 29*(4), 645–655.

Tsao, P. H., Gruber, L. C., Portales, L. A., Gochangco, A. M., Luzaran, P. B., de los Santos, A. B., & Pag, H. (1994). Some new records of *Phytophthora* crown and root rots in the Philippines and in world literature. *Phytopathol, 84*, 871.

Uba, M. O., Duabe, K. C. P., Biene, M. A. C. M., Ortiz, M. K. C. R., Bennett, R. M., & Dedeles, G. R. (2016). Growth and fatty acid profile of *Thraustochytrium* sp. CR01 using different sugar substitutes. *Philippine Journal of Science, 145*(4), 365–371

Vaughan-Martini, A., Kurtman, C. P., Meyer, S. A., & O'Neill, E. B. (2005). Two new species in the *Picia guilloermondii* clade: *Pichia caribbica* sp. nov., the ascosporic state of *Candida fermentati*, and *Candida carpophila* comb. nov. *FEMS Yeast Research, 5*(4-5), 463–469.

Viala, P., & Boyer, G. (1891). *Sur un Basidiomycète inferérieur, parasite des grains de raisins, 112* (pp. 1148–1150). Paris: Comptes Rendues Hebdomaires des Séances de l'Académie de Sciences.

Weston, W. H. (1920). Philippine down mildew of maize. *Journal of Agricultural Research, 19*, 97–122.

Wilson, A. W., & Desjardin, D. E. (2005). Phylogenetic relationships in the gymnopoid and marasmioid fungi (Basidiomyces, eugarics clade). *Mycologie, 97*(3), 667–679.

Xie, Y., & Wang, G. (2015). Mechanisms of fatty acid synthesis in marine fungus-like protists. *Applied Microbiology and Biotechnology, 99*, 8363–8375.

Xu, Z., Yan, X., Pei, L., Luo, Q., & Xu, J. (2008). Changes in fatty acids and sterols during batch growth of *Pavlova viridis* in photobioreactor. *Journal of Applied Phycology, 20*, 237–243.

Yao, M. L. C., Villanueva, J. D. H., Tumana, M. L. S., Calimag, J. G., Bungihan, M. E., & dela Cruz, T. E. (2009). Antimicrobial activities of marine fungi isolated from seawater and marine sediments. *Acta Manilana, 57*, 19–27.

Ye, C., Qiao, W., Yu, X., Ji, X., Huang, H., Collier, J. L., & Liu, L. (2015). Reconstruction and analysis of the genome-scale metabolic model of *Schizochytrium limacinum* SR21 for docosahexaenoic acid production. *BMC Genomics, 16*, 799.

Yokochi, T., Honda, D., Higashihara, T., & Nakahara, T. (1998). Optimization of docosahexaenoic acid production by *Schizochytrium limacinum* SR21. *Applied Microbiology and Biotechnology, 49*, 72–76.

CHAPTER 4

Species and functional diversity of forest fungi for conservation and sustainable landscape in the Philippines

Nelson M. Pampolina[1,2], Edwin R. Tadiosa[3], Jessa P. Ata[1], Janine Kaysee R. Soriano[1], Jason A. Parlucha[4] and Jennifer M. Niem[2]

[1]Department of Forest Biological Science, College of Forestry and Natural Resources, University of the Philippines Los Baños, College, Laguna, Philippines; [2]UPLB Museum of Natural History, University of the Philippines Los Baños, College, Laguna, Philippines; [3]Science Department, College of Sciences, Bulacan State University, Malolos, Bulacan, Philippines; [4]Department of Wood Science and Technology, College of Forestry and Environmental Science, Central Mindanao University, Maramag, Bukidnon, Philippines

1. Introduction

The evolution of the Philippine archipelago million years back was immense that triggered the development of unique structures and landscape resulting in the formation of around 7641islands and islets covering 30,000 km^2. Based on theories, the formation of the Philippine islands was naturally brought by volcanic eruptions that occurred from the Pacific Ocean and not from the break of Pangea (Yumul et al., 2008). Its isolation for a long period of years from the mainland had created unique habitats of various life forms across islands (Brown & Thatje, 2014; Mittermeier et al., 1999). Geographically, the location of the Philippine islands is within the equatorial region in the tropics at 21 and 5°N latitude and 116 and 16°E longitude. Two distinct climatic regimes prevail, i.e., dry (November–May) and wet (June–October) with cooler (25–26°C) and hot (27–29°C) mean temperature range. The

Mycology in the Tropics
https://doi.org/10.1016/B978-0-323-99489-7.00009-3
© 2023 Elsevier Inc. All rights reserved.

physiography of the island varied with nearly half being mountainous, reaching to 2,900 masl and other low-lying areas up to the coastal region. The geomorphic and edaphic characteristics of the island exhibit variable features reflective of volcanic debris (Paris et al., 2014).

These configurations across the Philippine islands from ridge to reef developed forest formations with distinct floristic dominance, architectural features, aesthetic values, and ecological variations favoring diversification of species and habitats of life forms (Culmsee & Leuschner, 2013). These attributes had earned the country as among the 17 declared megadiverse lifeforms worldwide, both in terrestrial and marine ecosystems, containing 70% of the world's biodiversity due to its geographical isolation, evolving habitats, and high endemism (Keong, 2015). Records would further show that the number of endemic species of vascular and non-vascular plants, together with vertebrate and invertebrate animals, were higher than all other countries (Marchese, 2015). Such diversity and endemism would include microorganisms, particularly fungi, that are known to thrive in a broad range of habitats above and below ground surface, even under the most adverse conditions.

This chapter highlights the Philippine forest fungi in different forest formations across the archipelago living as saprophytes on forest debris, parasites, or symbionts on diverse plant groups. The sections within this chapter provide information on the diverse fungal habitats within forest ecosystems and the wide spectrum of interactions between fungi and forest trees. The taxonomic groups of fungi across landcape of Philippine forests, curation systems, and conservation measures to sustain their environmental functions and ecosystem services are also discussed. The chapter ends with the challenges and opportunities in the conservation of fungal communities in the Philippines.

2. Forest formations in the Philippines

Based on the dominant tree species, six distinct vegetation types, that is, mangrove, beach, molave, dipterocarp, pine, and mossy forests were first described by Whitford (1911). Following this classification as recognized in Southeast Asia, 12 different forest formations were characterized in detail based on floristic composition, structure, and physiognomy by Fernando et al. (2008) and Whitmore (1984). The forest formations include the following: tropical lowland evergreen, tropical lower montane, tropical upper montane, tropical subalpine, tropical semi-evergreen, tropical moist deciduous, forest over limestone or karst, forest over ultramafic rocks, beach forest, mangrove forest, peat swamp forest, and freshwater swamp. The ecological dynamics within forest formations serve as unique habitats to fungal communities and food chain that regulate essential processes of nutrient, energy cycling and carbon source from various kinds of forest plants, debris, and/or litter (Osono, 2020). Fungal associations with plants significantly contribute to the overall productivity and sustainability of forest ecosystems (Read & Perez-Moreno, 2003).

The tropical lowland evergreen forest, sometimes known as neotropical rainforest, is structurally tall with emergent trees reaching 45 m with 100 cm diameter (Ashton & Zhu, 2020). Underneath the forest canopy are dominant and codominant individuals often with woody climbers. The vertical profile would show less undergrowth but abundant forest litters that serve as substrate to forest fungi. This forest formation that is geographically in eastern part of the archipelago usually favors abundant rain year-round of more than 200 mm suitable for dipterocarp trees and other big indigenous trees growing at elevation of lowland to 900 and sometimes a 1000 m. This forest type accounts for having the highest number of species that are highly dense and diverse floral composition where fruiting bodies of fungi coexist.

The tropical lower montane lies in between the lowland evergreen and upper montane rainforest at elevation gradient of 800—1300 m altitude for larger mountain ranges. Trees are structurally less emergent with canopies lower than those found in the lowland evergreen but with generally abundant epiphytes. The dipterocarps (Dipterocarpaceae) remain dominant, represented by the lauan-apitong type with inclusion of non dipterocarps like lauraceous and myrtaceous groups of trees (Hamann et al., 1999; Ingle, 2003; Malabrigo, 2013). The undergrowth plants are naturally abundant as exemplified by araceous, begonias, and ground orchids, however organic matter from decomposing litter remains favorable for fruiting of fungi. In the northern part of the country, pine (*Pinus kesiya* Royle ex Gordon) trees represent the tropical lower montane rainforest where understory has abundant mycorrhizal fungi that are their natural habitat while associated with the roots of pines.

The tropical upper montane is referred to as mossy forest in reference to the abundance of mosses, liverworts and frequent vascular and non-vascular epiphytes that grow on bole of studded trees, branches, and forest floor at elevation above 1000 m (Banwa, 2011; Hipol et al., 2007). Beyond this forest formation is the tropical subalpine up to 2400 m where structure and physiognomy of forest plants are nearly similarly stunted with heath vegetation (Amoroso et al., 2011). The climatic conditions here are basically moist brought by normal cloudy formation suitable for coniferous trees and dwarf scrub where organic matter accumulates on forest floor for heterotrophic life forms like fungi.

The tropical semi-evergreen rainforest is common in regions with distinct dry seasons that exhibit both evergreen and deciduous trees as physiological adaptation to water stress (Zhu et al., 2021). Structurally there are patches of emerging trees reaching 40 m tall. Their canopies provide space for intermediate growth forming dense undergrowth layers that is more diverse due to abundant light. This condition provides space for fungi that inhabit soil and forest litter favoring growth of dipterocarps and non dipterocarp wildlings. Similarly, the tropical moist deciduous forest formation also occurs in drier regions where water is periodically limited and located in coastal hilly land. Structurally trees do not form buttresses having lower branches and develop uneven canopies with open ground. Macrofungi inhabit soil and few forest litter on the forest floor and on dead plant parts.

Forest over limestone or karst forest is an outcrop of sedimentary rocks that is calcium-rich lifted million years by tectonic force forming topography of cliffs or tower of karst that are found in lower areas or mountainous slopes in some parts of the country (Tang et al., 2011). Succession of forest grows on surface with limited minerals suitable for endemic species that are naturally dispersed by wild animals or wind (Alviola et al., 2015; Sedlock et al., 2014). Vegetation is typically shorter than those growing on normal soil and often shedding their leaves. Some tree species under molave type including leguminous trees favor this kind of habitat particularly in dry regions, though few dipterocarp species were recorded (Aureo et al., 2020; Tolentino et al., 2020). Limited organic matter is accumulated on ground but some beneficial endomycorrhizae form association with nitrogen fixing plants to provide limited minerals. By contrast, ultramafic forest typically evolved from weathered material on igneous or metamorphic soil where heavy metals are generally rich. Plants grow slowly and short, exhibiting sclerophyllous characters and showing potentials as phytoremediators (Lillo et al., 2019). Usually, endomycorrhizal fungi are most associated with fine roots of higher plants in this vegetation.

The beach forest represents vegetation along the coastal region extending along mangrove and limestone or mangrove forest in the country. Forest plants varied from herbaceous creepers of low stature to patches of woody trees on sandy shore represented by monocotyledonous pandan and coconut to leguminous and nonleguminous trees (Co & Tan, 1992; Primavera & Sadaba, 2012). Most of these trees form synergistic association with nitrogen-fixing bacteria, endomycorrhizal fungi, and actinomycetes (Cabutaje et al., 2020). Stretching from the beach is the mangrove forest that connects along the river mouth and intertidal zone. In this area, the water is brackish or saline that favors stilt and pneumatophores-forming mangrove trees. The ground layer of mangrove forest is dominated with wildlings of their kinds and saplings growing on muddy soil that is a carbon sink (Gevania & Pampolina, 2009). Presence of macrofungi is usually found on decomposing litter and boles with butt or center rotting wood parts (Jones et al., 1988; Torres & dela Cruz, 2013).

The peat swamp forest refers to lowland areas inundated with water that allows accumulation of forest litters forming peat on waterlogged soil (Jauhiainen et al., 2005). The soil pH is generally acidic to slight and showed vegetation zonation in reference to varying sizes of trees from tall pole to intermediate in size and becoming pygmy where peat increases beneath (Aribal & Fernando, 2014, 2018). Macrofungi are usually observed on forest litter that help in slow decomposition and above ground on dead stems and branches. Occasionally, this forest formation is connected to freshwater swamp from mineral-rich rivers and streams where pH is neutral to basic. Forest plants are represented by species of pandan and palms with patches of trees that produce stilt roots and buttress.

3. Ecological functions and substrates of forest fungi

Ecologically, forest fungi can be categorized as saprotrophs, pathogens, or mutualists. Saprotrophic fungi obtain food by assimilation of nutrients from organic matter on forest litter, soil, and wood debris through decomposition process that facilitates nutrient cycling (Asiegbu & Kovalchuk, 2021). Litter trapped on forest canopies is also a potential substrate of saprotrophic fungi and is eventually washed as throughfall (Dighton, 2007). The decomposition of forest organic matter by specific groups of fungi is through enzymes, mineralizing lignocellulosic components into simple compounds absorbed through extensive hyphae (Kowalczyk et al., 2014; Kubartová et al., 2015; Osono, 2020). Fungal hyphae of saprotrophic fungi (e.g., *Marasmiellus, Collybia, Mycena* and *Marasmius*) act as binder of forest litters in soil preventing nutrient loss after soil erosion and leaching in tropical forest (Dighton, 2007; Lodge & Asbury, 1988). Saprophytic basidiomycetes are most common Philippine Forest fungi, dominated by Polyporaceae found on many decaying woody forest substrates (Angeles et al., 2016; Arenas et al., 2015; Parlucha et al., 2021). In agroforestry ecosystem, forest fungi are exemplified by *Auricularia* under rain tree (*Samanea saman* Jacq. Merr.) coconut (*Cocos nucifera* L.), ipil-ipil (*Leucaena leucocephala* Lam. de Wit), mahogany

(*Swietenia macrophylla* King), mango (*Mangifera indica* L.), and rubber tree (*Hevea brasiliensis* (HBK) Muell.-Arg (Musngi et al., 2005).

Pathogenic fungi cause diseases to trees as host from natural or plantation forests. This happens when air or water-borne fungal spores colonize wound on stumps, trunk, and roots, developing mycelial growth (Gonthier & Thor, 2013). For example, *Armillaria* spp. that initially behave as saprophytes of decaying debris or stump could threaten living trees by attacking their roots (Guillaumin & Legrand, 2013).

Polyporaceae are fungal families that cause brown rots in gymnosperms by degrading cellulose and hemicelluloses leaving lignin that darkens wood (Schwarze et al., 2013). White rot fungi are more apparent on angiosperm caused by ascomycetes and other basidiomycetes (Schwarze et al., 2013). These fungi consume wood lignin that slightly degrades cellulose and hemicellulose, leaving a whitish and physically weakened wood. Soft rot is caused by ascomycetes and deuteromycetes attacking wood superficially when exposed on moist condition (Goodell et al., 2008). Other orders of pathogenic fungi-like Aphyllophorales, Agaricales, Tremellales, Lycoperdales, and Nidulariales were found as wood-rotting fungi in molave forest of La Union, Philippines (Tadiosa & Arsenio, 2014). More pathogenic fungi were reported from protected areas in provinces of Quezon and Laguna exemplified by species of *Trametes, Stereum, Biscogniauxia, Polyporus,* and *Fomes* (Parlucha et al., 2021).

The mutualist fungi are referred here as ectomycorrhizas and endomycorrhizas that associate beneficially through extensive fungal hyphae with fine roots of higher plants (Smith & Read, 2010). Basidiomycete and ascomycete represent ectomycorrhizas that link on forest trees like dipterocarps and non-dipterocarps by colonizing epidermal and cortical cell walls, forming Hartig networks while developing fungal sheath on the root surface (Brundrett, et al., 1996, pp. 20–158; Pampolina et al., 2002). Glomeromycetes are endomycorrhizas that commonly produce vesicles and arbuscles within cells at epidermal and cortical layers which are predominant in land plants (Asiegbu & Kovalchuk, 2021; Wang & Qiu, 2006).

The presence of endomycorrhizal fungi in plant roots does not manifest infections or cause any symptoms of disease throughout the plant life cycle (Wilson, 1995). The hyphae of endomycorrhizas spread through the root axis enveloping the cortical cells and penetrating toward the direction of endodermis (Kottke & Oberwinkler, 1986). This orientation eases the nutrient transport between the endodermis and hyphae.

Aside from mycorrhizas and pathogens, there are fungal species known as endophytes that lives inside host tissues without causing any harm or adverse effects. The ecological functions of most endophytes are still unknown. Some endophytes act as a counterdefense mechanism from invading pathogens through synthesis of bioactive compounds (Terhonen et al., 2019), induction of plant defenses, stimulation of secondary metabolites, or promotion of plant growth (Gao et al., 2010). The diversity of endophytes in mangroves and a few native tree species were reported in local literature (Guerrero et al., 2018; Moron & Lim, 2018; Solis et al., 2016).

Another important interaction of fungi within forests is among arthropods which help facilitate effective deterioration process, natural selection, and maintenance of species diversity (Araújo & Hughes, 2016; Gessner et al., 2010; Kambach et al., 2021). Fungal associations with insects could be as entomopathogens and as dispersal agents (Vogel et al., 2017).

Entomopathogens or fungi parasitizing insects were reported in the Philippines on plant crops, suggesting their potential in biological control of diseases (Inglis & Tigano, 2006; Mongkolsamrit et al., 2020). Aside from transmission by wind, fungal spores are contained in the exoskeleton of beetles (Order: Coleoptora) which exhibit how inoculum is being transmitted and initiate wood decay in the forests (Jacobsen et al., 2017). In natural stands, wood decay of the Philippine teak (*Tectona philippinensis* Benth. & Hook) by polypore fungi were found to coexist with various insect taxa such as beetles, moths, ants, and termites (Briones et al., 2017).

4. Fungal growth requirements

The high diversity of fungal species in the forest is a result of the viable biophysical conditions of their essential growth requirements. Diversity of plants, precipitation, and properties of soil are three of the most important regulating factors affecting fungal species in the forest (Li et al., 2020). Their survival and productivity increases with tree species diversity, canopy cover, and tree basal area. Unmanaged conditions of natural forest with high deadwood matter in the forest floor promotes the diversity of mycorrhizal and wood degrading fungi (Tomao et al., 2020). Other than diversity of host plants, the tree physiology, competition and influence of understory also affects ectomycorrhizal diversity (Collado et al., 2018).

Precipitation strongly influences growth and productivity of fungal species where fungal diversity positively correlated to mean annual rainfall though these are less understood (Angeles et al., 2016; McGuire et al., 2015). Some fungal species are diverse during dry months (Hawkes et al., 2011; Peñuelas et al., 2012) but found abundant during wet periods in Makiling forest (Lapitan et al., 2010). Collado et al. (2018) suggested that precipitation does not regulate diversity of fungi for saprophytes provided temperature and moisture requirements are favorable. These contrasting results should be explored in support of conservation strategies of highly vulnerable fungal species at community level.

Soil properties such as nutrient content and pH also affect fungal growth. Mycorrhizal fungi were responsive to different soil properties as a result of increasing forest secondary succession (Li et al., 2020). Nitrogen is a significant limiting factor in both lowland and montane tropical forests (Nottingham et al., 2018). Soil pH was found to influence alpha diversity compared to beta diversity of soil fungal community (Wang et al., 2015). Other factors such as wind speed, temperature, and biological agents of spore dispersal also contribute to fungal diversity (Angeles et al., 2016).

5. Fungal diversity and mycological collection across forest formations and landscape

A diverse group of organisms in the different Philippine protected and nonprotected areas belong to the Kingdom Fungi. Recognized as distinct from flora and fauna, the fungi are a large group of eukaryotic, spore-bearing, and achlorophyllous organisms that constitute an abundant element of terrestrial biota in the Philippines. About 72,000 species have been described and recorded worldwide out of the approximately 1.5 million fungal species estimated to occur on Earth (Hawksworth, 2004). In the Philippines, about 4968 species have been described (Tadiosa, 2012), which mainly belong to the groups: egg fungi, zygospore forming fungi, sac fungi, club fungi, and imperfect fungi, and could potentially have diverse ecological functions. These are spread out across diverse forest formations in Philippine Forest reserves, protected landscapes and seascapes, and watersheds that have been surveyed so far (Fig. 4.1). The list of species in each respective site is shown in Appendix Table 4.1.

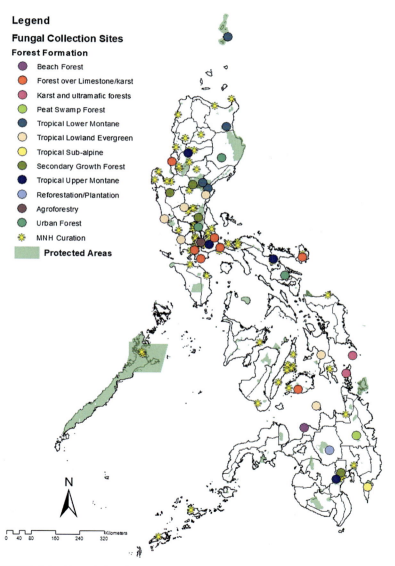

FIGURE 4.1 Fungal distribution across different forest formation in the Philippines.

I. Fungi in aquatic and terrestrial habitats

96 4. Species and functional diversity of forest fungi for conservation and sustainable landscape in the Philippines

APPENDIX TABLE 4.1 Forest fungi documentation in the Philippines.

Species	Documented sites
Agaricaceae	
Agaricus bolohizus	1
Agaricus campestris Linn.	1,4,5,8
Agaricus diminutivus	1
Agaricus goosensiae	1
Agaricus hemilasius	1
Agaricus luzoniensis	1
Agaricus merrillii	1
Agaricus perfuscus	1
Agaricus augustus Fr.	1,5
Agaricus pocillator	1
Agaricus rhoadsii	1
Agaricus sylvaticus	1
Agaricus moelleri	26
Agaricus spp.	1,2,4,6,9,11, 12,13,15,25,17
Agaricus arvensis Schaeff.	5, 31
Calvatia spp.	1, 4, 27, 33
Calvatia cythiformis	31
Coprinus atramentarius (Bull.) Fr.	1,4,5
Coprinus disseminatus (Pers.) Gray	4,7,8,16,20,21,22
Coprinus niveus (Pers.) Fr.	1,2,4
Coprinus plicatilis (Curt.) Fr.	4,8
Coprinus fibrillosus	1
Coprinus micaceus	1
Coprinus radiatus	1
Coprinus comatus	5
Coprinus stercoreus (Fr., Epicrisis)	2
Coprinus cinereus (Schaeff.) Gray	32
Coprinus spp.	1,4,5,6,9,10,17,20,31,33
Chlorophyllum molybdites	1

I. Fungi in aquatic and terrestrial habitats

APPENDIX TABLE 4.1 Forest fungi documentation in the Philippines.—cont'd

Species	Documented sites
Chlorophyllum spp.	1,5,27
Lepiota aspera (Pers.) Quel.	4,7
Lepiota cristata (Bolt.) Kumm.	1,4,5,8,11
Lepiota aluviina	1
Lepiota americana	1
Lepiota besseyi	1
Lepiota citrophylla	1
Lepiota clypeolaria	1
Lepiota flagellata	1
Lepiota phytaenodes	1
Lepiota spp.	1,4,33
Lycoperdon pyriforme Schmach,	2,29
Lycoperdon echinatum Pers.	4
Lycoperdon microspermum	1
Lycoperdon spp.	1,4,16,20,22,28
Macrolepiota rhacodes (Vittadini.) Singer	4,7
Macrolepiota procera (Scop.) Gray	5
Macrolepiota spp.	4
Leucocoprinus spp.	1,26
Leucocoprinus luteus	1
Leucocoprinus fragilissimus (Berk. & M.A. Curtis) Pat.	2
Leucocoprinus cepaestipes (Sowerby)	29
Tulostoma mommosu	1
Amanitaceae	
Amanita fulva (Schaeff.) Fr.	4
Amanita sp.	1,4
Amanita hemibapha	1
Amanitopsis spp.	1
Amanitopsis vaginata	1
Amanita angustilamellata	1

(Continued)

98 4. Species and functional diversity of forest fungi for conservation and sustainable landscape in the Philippines

APPENDIX TABLE 4.1 Forest fungi documentation in the Philippines.—cont'd

Species	Documented sites
Aphelariaceae	
Aphelaria dendroides	1
Aphelaria flagelliformis	1
Aphelaria incarnata	6
Aphelaria tasmanica	6
Auriculariaceae	
Auricularia auricula-judae	1,2,3,4,5,6,7,8,9,10,11,12,13,14,16,17,19,20,23,24,26,31,32,33
Auricularia cornea (Ehrenb.) Endl.	1,3,14
Auricularia auricula (L.) Underw.	8,24
Auricularia delicata (Fr.) Henn.	1,3,4,6,8,24,26
Auricularia fusco-succinea (Mont.) Henn.	1,3,7
Auricularia mesenterica (Dicks.) Pers.	2,3,5,7,8,9,12,14
Auricularia polytricha (Mont.) Sacc.	1,2,3,4,6,7,8,10,11,12,13,14,16,17,20,21,22,26,31,32
Auricularia ornata	1
Auricularia peltata	1
Auricularia spp.	1,2,3,4,6,9,10,13,15,17,20,26
Auriscalpiaceae	
Clavicorona spp.	1,25,26
Clavicorona candelabrum	6,25
Clavicorona javanica	6,25
Bolbitiaceae	
Conocybe spp.	1,4,5,33
Conocybe tenera (Schaeff.) Fayod	1,2,5
Panaeolus spp.	1,4,32,33
Panaeolus campanulatus	12
Panaeolus semi-ovatus (Sowerby) Fr.	8
Panaeolus papilionaceus (Bull. ex Fries) Quelet	1,2
Panaeolus foenisecii	1
Boletaceae	
Strobilomyces strobilaceus (Stop.) Berk.	4
Boletus spp.	1,5,6,32

I. Fungi in aquatic and terrestrial habitats

5. Fungal diversity and mycological collection across forest formations and landscape

APPENDIX TABLE 4.1 Forest fungi documentation in the Philippines.—cont'd

Species	Documented sites
Boletinellaceae	
Boletinellus spp.	1,8
Bondarzewiaceae	
Heterobasidion annosum	1
Cantharellaceae	
Cantharellus cibarius	2,5
Cantharellus aureus	1
Cantharellus infundibuliformis (Scop.) Fr.	1,3,5,7,8,13,14
Cantharellus spp.	1,4,5,6,9,10,16,20, 21,24,26,30
Ceratostomataceae	
Melanospora spp.	1,2,26
Clavariaceae	
Clavaria vermicularis	2
Clavulinopsis gracillima	12
Clavulinopsis corallinorosacea (Cleland) Corner,	29
Clavaria spp.	1,11,33
Clavaria straminea	1
Clavariales	1
Clavulinaceae	
Clavulina gracilis	1
Clavulina leveillei	1
Clavulina sp.	1
Clavulina banahaoensis	6
Clavulina rugosa	6
Clavulina cristata	31
Coniophoraceae	
Coniophora puteana (Schum.) Karst.	2,7,8
Corticiaceae	
Corticium caeruleum (Schrad.) Fr.	3
Corticium salmonicolor Berk. & Br.	8,14
Corticium confluens (Fr.) Fr.	5,7

(Continued)

I. Fungi in aquatic and terrestrial habitats

APPENDIX TABLE 4.1 Forest fungi documentation in the Philippines.—cont'd

Species	Documented sites
Corticium evolvens Fr.	3
Corticium roseum Pers.	3
Corticium spp.	1,2,4,5,6,8,9,10,17,26
Pulcherricium caeruleum (Lam.) Parm.	8
Cortinariaceae	
Cortinarius spp.	1,4,5,12,13
Cortinarius callisteus (Fr.) Fr.	1,5,7,8
Crepidotaceae	
Crepidotus spp.	1,2,13,15,30,32
Crepidotus herbarum (Peck) Sacc.	1,5
Crepidotus mollis (Schaeff.) Quel.	2,8,12
Crepidotus variabilis (Pers.) P. Kumm.	2
Dacrymycetaceae	
Dacrymyces palmatus (Schwein.) Bres.	2,7,13
Dacrymyces chrysospermus Berk. & M.A. Curtis,	29
Dacryopinax spathularia (Schwein.) Martin	1,2,4,5,6,8,11,12,14,16,20,22,23,24
Dacrymyces cupularis	12
Dacryopinax sp.	1
Calocera spp.	13,26
Dacryomyces sp.	6
Diatrypaceae	
Diatrype sp.	7
Diplocystaceae	
Astraeus hygrometricus	2
Entolomataceae	
Enteloma serrulatum (Pers.) Hesl.	4
Entoloma lividum (Bull.) Quelet.	1,5,8
Entoloma spp.	1,4,5
Entoloma cetratum (F.) M. M. Moser	2
Claudopus sp.	1
Clitopilus spp.	1

I. Fungi in aquatic and terrestrial habitats

5. Fungal diversity and mycological collection across forest formations and landscape

APPENDIX TABLE 4.1 Forest fungi documentation in the Philippines.—cont'd

Species	Documented sites
Exidiaceae	
Guepinia spathularia	1
Guepinia sp.	28
Exidia sp.	4
Exidia saccharina (Alb. & Schwein.) Fr.	7
Exidia thuretiana (Lev.) Fr.	24
Fistulinaceae	
Fistulina sp.	1
Fomitopsidaceae	
Daedalea ambigua Berk.	1,2,3,5,7,8,14,16,20,30
Daedalea quercina	12
Daedalea amanitoides Beauv.	3,8,14
Daedalea flava Lev.	1,3,8,14
Daedalea hobsoni Berk.	3,14
Daedalea palisoti Fr.	4,14
Daedalea repanda (Pers.) Gaud.	3
Daedaleopsis confragosa (Bolt.) J. Schrot.	8
Daedalea dickinsii	31
Daedalea spp.	1,2,4,5,8,11,14,15, 20,23,24
Fomitopsis pinicola	2
Fomitopsis sp.	1
Laetiporus sulphurous (Bull.) Murrill	11
Ganodermataceae	
Amauroderma rogusom (Bl. & Nees.) Bres.	2,4,8,11,12,14,16,20,24
Amauroderma auriscalpium Lloyd.	14
Amauroderma rude (Berk.) Torrend,	29
Amauroderma spp.	4,6,20,23,26
Ganoderma applanatum (Pers.) Pat.	1,2,3,4,5,6,7,8,9,10,11,12,13,14,15,16,17,18,19,20,22,23,24,30,31,33
Ganoderma lucidum (Leys.) Karst.	1,2,4,5,6,8,10,11,12,13,14,15,17,20,31,32
Ganoderma lobatum (Schwein.) Atk.	4

(Continued)

I. Fungi in aquatic and terrestrial habitats

4. Species and functional diversity of forest fungi for conservation and sustainable landscape in the Philippines

APPENDIX TABLE 4.1 Forest fungi documentation in the Philippines.—cont'd

Species	Documented sites
Ganoderma tsugae Murrill	8,32
Ganoderma sinense	1
Ganoderma mangiferae (Lev.) Pat.	14
Ganoderma japonicum	31
Ganoderma adspersum (Schulzer) Donk	31
Ganoderma australe (Fr.) Pat	32
Ganoderma spp.	1,2,4,6,8,9,11,14,15,20,23,24,26,27,28, 30,32,33
Geastraceae	
Geastrum hydrometricus	1
Geastrum triplex Jungh.	5,14,31
Geastrum spp.	1,13,25,26
Geastrum saccatum Fr.	8,13,16,20,22
Geastrum fimbriatum	31
Gomphaceae	
Ramaria stricta	2
Ramaria spp.	1,11,13
Ramaria zippelli	1
Helotiaceae	
Hymenoscyphus herbarum (Pers.) Dennis	4
Hymenoscyphus spp.	1,25
Helotium sp.	6
Hericiaceae	
Hericium spp.	1,26,27,28
Hydnaceae	
Hydnum sp.	1,5,6,8
Hydnangiaceae	
Laccaria ochropurpurea	1
Laccaria laccata	1,12
Laccaria spp.	1,5
Hydnodontaceae	

I. Fungi in aquatic and terrestrial habitats

5. Fungal diversity and mycological collection across forest formations and landscape

APPENDIX TABLE 4.1 Forest fungi documentation in the Philippines.—cont'd

Species	Documented sites
Scytinopogon angulisporus	1
Scytinopogon spp.	1,2,25
Hymenochaetaceae	
Phellinus gilvus (Schw.) Fr.	1,2,7,12,14,31
Phellinus nigrolimitatus (Romell) Bourd.	3
Phellinus punctatus (Karst.) Pilat	2,3
Phellinus igniarius (L.) Quel.	2,31
Phellinus pini	28
Phellinus cinereus	12
Phellinus linteus	12,31
Phellinus spp.	13
Phellinus robustus (Karst.) Bourd.	7
Phellinus torulosus (Pers.) Bourdot & Galzin	7
Phellinus spp.	1,2,3,4,5,6,8,9,17,20,26,28,30,33
Polystictus spp.	1,2,6,25,26,27
Polystictus flabelliformis	1
Polystictus affinis	1,14
Polystictus connexus (Liv.) Cooke	14
Polystictus incomptus Fr.	14
Polystictus occidentalis (Klotz.) Fr.	14
Polystictus touteopia	6,14
Hymenochaete tenuissima (Berk.) Berk	1,32
Hymenochaete rubiginosa (Dicks.) Lev.	4,5,7,8,12
Hymenochaete mougeotii	1
Hymenochaete spp.	4,5,6,26
Coltricia cinnamomea (Jacq.) Murrill,	29
Coltricia spp.	1,11,28
Inonotus sp.	1
Fuscoporia torulosa	14
Hymenogastraceae	

(*Continued*)

I. Fungi in aquatic and terrestrial habitats

APPENDIX TABLE 4.1 Forest fungi documentation in the Philippines.—cont'd

Species	Documented sites
Hebeloma spp.	1,4,8
Flammula sapinea	1
Flammula spumosa	1
Flammula spp.	1
Hebeloma mesophaeum (Pers.) Quel.	2
Hypocreaceae	
Hypocrea spp.	1,6
Hygrophoraceae	
Hygrocybe coccinea (Schaeff.)Kumm.	4,12
Hygrocybe chlorophana	12
Hygrocybe intermedia	12
Hygrocybe miniata (Fr.) Kumm.	1,4,5,33
Hygrophorus pratensis Fr.	1,5,8
Hygrocybe spp.	4,5,8,11,28
Hygrophorus spp.	1,5
Hygrophoropsis aurantiaca (Fr.) Maire	8
Ischnodermataceae	
Ischnoderma resinosum (Schrader) P. Karsten	26
Inocybaceae	
Inocybe spp.	4,33
Inocybe peckii	1
Incrustoporiaceae	
Tyromyces chioneus (Fries) P. Karsten	26
Laetiporaceae	
Phaeolus sp.	1
Lyophyllaceae	
Termitomyces albuminosus	1,5
Termitomyces cartilaginous	1
Termitomyces cartilaguis	1
Termitomyces clypeatus	1,5
Termitomyces eurhizus (Berk.)Heinm	1,5

I. Fungi in aquatic and terrestrial habitats

5. Fungal diversity and mycological collection across forest formations and landscape

APPENDIX TABLE 4.1 Forest fungi documentation in the Philippines.—cont'd

Species	Documented sites
Termitomyces microcarpus	1
Termitomyces striatus (Beeli) R. Heim.	8,31
Termitomyces robustus	1
Termitomyces spp.	1,8,26
Nidulariaceae	
Cyathus striatus Willd.	2,5,8,13,33
Cyathus rudis Pat.	14
Cyathus spp.	2,26
Marasmiaceae	
Marasmius rotula (Scop.) Fr.	1,2,4,5,7,8,13,14,16,20,33
Marasmius haematocephala	2
Marasmius scorodonius	1,12
Marasmius siccus (Schweinitz) Fries	12,33
Marasmius spp.	1,2,4,5,7,11,12,13,15,24,26,28,31,32,33
Marasmius arborescens (Henn.) Beeli	29
Marasmius foetidis	1
Marasmius androsaceus (Linn.) Fr.	1,4,16,20
Marasmius ramealis (Bull.) Fr.	1,4,5,16,20
Marasmius patanilorde	1
Marasmius pilopus	1
Marasmius plicatus	1,33
Marasmius oreades (Bolton) Fr.	33
Marasmius haematocephalus (Mont.) Fr.	2
Marasmiellus spp.	1,26,32
Marasmiellus ramealis (Bull.) Singer	33
Omphalotus spp.	4,5,11
Trogia sp.	11
Trogia infundibuliformis Berk & Broome	2
Crinipellis scabella (Alb. & Schwein.)	2
Meruliaceae	

(Continued)

I. Fungi in aquatic and terrestrial habitats

APPENDIX TABLE 4.1 Forest fungi documentation in the Philippines.—cont'd

Species	Documented sites
Cymatoderma elegans Jungh.	2,5,11,12,29
Cymatoderma sp.	1,2,9,13,20,23,33
Cymatoderma africanum Boidin	4
Podoscypha bolleana (Mont.) Boidin	4
Podoscypha petalodes (Berk.) Pat.	24,33
Podoscypha subaffinis (Berk.& Curt.) Pat.	4
Podoscypha sp.	26,31
Merulius incarnatus Schw.	8
Merulius sp.	6,28
Aquascypha hydrophora (Berk.) D.A. Reid	24
Bjerkandera adusta (Willd.) P. Karst.	24
Phlebia sp.	28
Meripilaceae	
Rigidoporus microporus (Sw.) Overeem	24,32
Mycenaceae	
Mycena spp.	2
Mycena galericulata (Scop.) Gray	2,4
Mycena acicula (Schaeff.) Fr.	2,8
Mycena galopus (Pers. ex Fr.) Kummer	2,4,7
Mycena pura (Pers.) Kumm.	2,4,8,12
Mycena vulgaris	1,2,12
Mycena alcalina (Pers.) Gillet	1,2
Mycena cinerella (P. Karst.) P.Karst.	2
Mycena clavularis (Batsch) Sacc.	2
Mycena fibula (Bull. Kühner)	1,2
Mycena spp.	1,4,5,6,7,9,11,22,13,26,32
Favolaschia pustulosa (Jungh.) Kuntze	2
Peniophoraceae	
Peniophora sp.	33
Pluteaceae	

I. Fungi in aquatic and terrestrial habitats

5. Fungal diversity and mycological collection across forest formations and landscape

APPENDIX TABLE 4.1 Forest fungi documentation in the Philippines.—cont'd

Species	Documented sites
Pluteus salicinus (Pers.) P. Kumm.	24
Pluteus cervinus (Schaeff.) P. Kumm.,	1,29
Pluteus nanus	1
Pluteus pulverulentus	1
Pluteus spp.	1,4,5,8,9,33
Pluteus umbrosus (Pers.) Kumm.	4,7
Volvariella volvacea (Bull.) Singer	1,4,8
Volvariella spp.	1,25,27
Pleurotaceae	
Pleurotus pulmonarius (Fr.) Quel.	1,2,7
Pleurotus ostreatus (Jacq. ex Fr.) Kummer	1,5,7,8,12
Pleurotus porrigens (Pers.) P. Kumm.	8,31
Pleurotus flabellatus	1
Pleurotus limpidus	1
Pleurotus dryinus (Pers.) P. Kumm.	32
Pleurotus opuntiae	1,14
Pleurotus ostreatus (Jacq. ex Fr.) Kummer	1,11
Pleurotus cornucopiae	1
Pleurotus cystidiosus	1
Pleurotus sapidus	1
Pleurotus ulmarius	1
Pluteus cervinus	1
Pleurotus spp.	1,5,6,9,13,15,17,20,25,28
Phallaceae	
Aseroe rubra Labill.	4
Dictyophora merrillii	1
Dictyophora duplicata (Bosc.) E. Fisch.	5,29
Dictyophora spp.	1
Mutinus sp.	1
Mutinus bambusinus (Zoll.) E. Fisch.	29
Phallus indusiatus Vent.	11
Physalacriaceae	

(Continued)

I. Fungi in aquatic and terrestrial habitats

4. Species and functional diversity of forest fungi for conservation and sustainable landscape in the Philippines

APPENDIX TABLE 4.1 Forest fungi documentation in the Philippines.—cont'd

Species	Documented sites
Armillaria sp.	1,4,6,8,11,13
Oudemansiella canarii (Jungh.) Hohn.	1,4,11
Oudemansiella radicata (Relh.) Singer	4,11
Oudemansiella sp.	1,4,8,13
Phyllotopsidaceae	
Pleurocybella porrigens	31
Platygloeaceae	
Jola sp.	7
Polyporaceae	
Microporus xanthopus (Fr.) Kuntze	1 to 9, 11,12,13,14, 15,16,17,19,20,21,22,23,24,26,29,30,32
Microporus affinis (Blume & T.Nees.) Kuntze	2,3,4,7,8,11,12,13,15,24,32
Microporus flabelliformis	13
Microporus picipes	16,18,20
Microporus subaffinis (Lloyd) Imazeki	32
Microporus spp.	1,6,7,9,13,15,26, 30,33
Microporus vernicipes (Berk.) Kuntze	4,6,11,12,24,30
Coriolus versicolor (Lev.) Pat.	5,7,14
Coriolus sp.	8
Lenzites acuta Berk.	3
Lenzites striata (Swartz.) Fr.	1,3,5
Lenzites repanda (Pers.) Fr.	8
Lenzites elegans (Spreng.) Pat.	7,12,24
Lenzites betulina (L.) Fr.	14
Lentinus auracariae	1
Lentinus strigosus (Schwein.) Fr.	8,11
Lentinus glabratus	1
Lentinus sajor-caju	1,31
Lentinus squarollus	1
Lentinus tigrinus (Bull.) Fr.	1,2,7,24,31,32
Lentinus velutinus Fr.	2

I. Fungi in aquatic and terrestrial habitats

5. Fungal diversity and mycological collection across forest formations and landscape

APPENDIX TABLE 4.1   Forest fungi documentation in the Philippines.—cont'd

Species	Documented sites
Lentinus crinipillis	1
Lentinus spp.	2,4,11,13,15,16, 18,20,23,25,28,32,33
Lenzites spp.	1,2,4,17,20,26,27,30
Hexagonia apiaria (Pers.) Fr.	2,3,5,8,14,
Hexagonia tenuis (Hook.) Fr.	1,2,3,4,5,6,7,8,9,11,12,13,14,15,16,17,19,20,21,24,30,32
Hexagonia nitida Durieu & Mont.	2,7,12
Hexagonia glaber (Beauv.) Ryvarden	2,4,7
Hexagonia spp.	1,2,3,13,17,20,33
Trametes corrugata (Pers.) Bres.	1,2,3,5,6,7,8,9,11,16,17,19,20,21,23,30
Trametes versicolor (L.) Lloyd.	1,2,5,7,8,30,31
Trametes gibbosa	2,11,13
Trametes aspera Jungh.	1,8,14
Trametes hirsuta (Wulf.) Pilat	7,11,12,24,26,31
Trametes spp.	1,2,3,4,6,9,10,17,20,26,27,32,33
Trametes aspera Jungh.	3
Trametes membranacea (Sw.) Kreisel	33
Trametes pubescens	1,31
Trametes elegans (Spreng.) Fr.	31,32
Trametes ochracea (Pers.) Gilb & Ryvarden	2,26
Panus rudis Fr.	1,3,4,5,8
Panus relutinus	1
Panus strigosus	1
Panus stugasus	1
Panus spp.	1,6,9
Poria straminea Bres.	1,3,7
Poria tricolor Bres.	3
Poria latemarginata (Fr.) Karst.	14
Poria spp.	1,2,3,4,5,6,8,9,25,26,27,28,33
Favolus alveolaris (DC.) Quel.	1,2,3,11
Favolus tenuiculus P. Beauv.	24

(Continued)

I. Fungi in aquatic and terrestrial habitats

4. Species and functional diversity of forest fungi for conservation and sustainable landscape in the Philippines

APPENDIX TABLE 4.1 Forest fungi documentation in the Philippines.—cont'd

Species	Documented sites
Favolus reniformis Murr.) Sacc & Trotter	7,8,11
Favolus spp.	2,3,4,6,26,28
Fomes australis Fr.	3,16,20
Fomes albo-marginatus (Lev.) Cke.	3
Fomes pachyphloeus Pat.	1,2,3,7,12,14
Fomes caryophylli (Rac.) Bres.	3,5,7,8,14
Fomes cinereus (Berk.) Sacc.	3
Fomes fomentarius (Linn.) Kickx.	3,4,12
Fomes gibbosus Nees.	3,16
Fomes auberianus (Mont.) Murr.	16,20
Fomes calignosus Berk.	16,20
Fomes carneus (Bl. and Nees) Cke.	16
Fomes subresinosus Murr.	16,20
Fomes subungulatus Murr.	20
Fomes linteus (Berk.) & Curt.) Cke.	3,14
Fomes roseus	2,16,20,21
Fomes senex (Nees. & Mont.) Cooke	1,3,5,7,14
Fomes gilvus (Shwein.) Lloyd.	2,3,5,8,14
Fomes spp.	1,2,3,4,6,9,10,11,15,17,20,23,25,26,27, 28,30,32
Polyporus arcularius (Batsch) Fr.	2,3,11,13
Polyporus adustus (Willd.) Fr.	3
Polyporus beccarianus Ces.	3
Polyporus bicolor Jungh.	3
Polyporus cumingii Berk.	3
Polyporus cuticularis (Bull.) Fr.	3,14
Polyporus durus Jungh.	3,7,8,
Polyporus fissus Berk.	3
Polyporus gilvus (Schw.) Fr.	3,7,8
Polyporus grammocephalus Berk.	3,5,7,11,12,14
Polyporus hirsutus (Wulf.) Fr.	1,2,3,4,5,6,7,8,9,11,12,14,15,17,18,19,20,23,30

I. Fungi in aquatic and terrestrial habitats

APPENDIX TABLE 4.1   Forest fungi documentation in the Philippines.—cont'd

Species	Documented sites
Polyporus pinsitus	2,5,9,12,14
Polyporus picipes Fr.	2,3,4,5,7,11,13,17,20,22,32
Polyporus semilaccatus	1,3
Polyporus sanguineus	1,2,26,28,31
Polyporus roseus (Alb. & Schw.) Fr.	3,8
Polyporus lignosus	1
Polyporus luteus	1
Polyporus marianus	1
Polyporus meyenii	1
Polyporus meyeri	1
Polyporus occidentalis	1
Polyporus ostreiformis	1
Polyporus rhodophaeus	1
Polyporus rubiculis	1
Polyporus sanuineneus	1
Polyporus affinis Blume & T. Nees	1,12
Polyporus alveolaris	12,13
Polyporus badius (Pers.) Schwein.	2,13,24
Polyporus melanopus	12
Polyporus versicolor	12
Polyporus capucinus Mont.	16
Polyporus carbonaceus Berk.	16
Polyporus caryophylli Racib.	16
Polyporus velutinus Fr.	16,20
Polyporus vernicipes Berk.	16,20
Polyporus versiformis Berk.	16,20
Polyporus spp.	1,2,4,5,6,7,8,9,10,11,12,13,14,15,17,20,23,25,26,27,28,30,32,33
Polyporus trigonis	26
Polyporus squamosus (Huds.) Fr.	7,11
Polyporus varius (Pers.) Fr.	2,7,9,13,17,20,26

(*Continued*)

APPENDIX TABLE 4.1 Forest fungi documentation in the Philippines.—cont'd

Species	Documented sites
Pycnoporus sanguineus (L.) Murill.	3,4,5,7,11,12,14,15,16,18,19,20,21,22,23,24,26,30,32
Pycnoporus coccineus (Fr.) Bondartsev. & Singer	7
Pycnoporus cinnabarinus (Jacq.) P. Karst.	24
Trichaptum sp.	1
Trichaptum abietinum (Dicks.) Ryvarden	8
Earliella scabrosa (Pers.) Gilbn. & Ryv.	2,11,12,13,16,18,20,23,24
Nigroporus vinosus (Berk.) Murrill	24
Cryptoporus volvatus	28
Lignosus rhinocerus (Cooke) Ryvarden	29
Pterulaceae	
Deflexula fascicularis	1,2,6,25
Deflexula subsimplex	1
Pterula intermedia	1,6
Pterula subulata Fr.	1
Pterula sp.	1
Pterula taxiformis	1
Pterula verticellata	1
Pterulicium sp.	1
Pterulicium xylogenium	1
Psathyrellaceae	
Psathyrella piluliformis	2
Psathyrella delineata	12
Psathyrella spp.	7,8,13,24
Psathyra spp.	1
Psathyrella disseminata	1
Psathyrella candolleana	31
Coprinellus disseminatus (Pers.) J.E. Lange	11
Coprinellus micaceus (Bull.) Fr.	2,11
Coprinopsis atramentaria (Bull.) Redhead	32

I. Fungi in aquatic and terrestrial habitats

APPENDIX TABLE 4.1 Forest fungi documentation in the Philippines.—cont'd

Species	Documented sites
Coprinopsis lagopus (Fr.) Redhead, Vilgalys & Moncalvo	32
Coprinopsis sp.	26
Coprinellus sp.	33
Parasola plicatilis (Curtis) Redhead	33
Russulaceae	
Lactarius spp.	1,4,5,11,12,27,32,33
Lactarius hygropheroides	1
Lactarius glaucescens	12
Lactarius piperatus (L.) Pers.	1,2,8
Russula albida	1
Russula spp.	1,5,8,16,20,23
Russula emetica Fr.	1,5
Russula sanguinea (Bull.) Fr.	7
Russula virescens	1,12
Lactifluus piperatus (L.) Kuntze,	1
Schizophyllaceae	
Schizophyllum commune	1,2,3,4,5,6,7,8,9,10,11,12,13,14,15,16,17,18,19,20,22,23,24,26,30,31,32,33
Schizophyllum spp.	1,28
Sclerodermataceae	
Scleroderma cepa	1
Scleroderma lycoperdonoides	1
Scleroderma spp.	1,16,20,22
Septobasidiaceae	
Septobasidium spp.	6
Serpulaceae	
Serpula lachrymans (Wulf.) Schroter.	11
Sprassidaceae	
Sparassis sp.	27
Stereaceae	
Stereum spp.	1,2,4,5,6,7,8,13,15,17,20,25,26,27,28
Stereum complicatum (Fr.) Fr.	2,7,8,

(Continued)

114 4. Species and functional diversity of forest fungi for conservation and sustainable landscape in the Philippines

APPENDIX TABLE 4.1 Forest fungi documentation in the Philippines.—cont'd

Species	Documented sites
Stereum ostrea (Bl.& Nees.) Fr.	1,2,5,8,11,12,14,32
Stereum sanguinolentum (Alb.& Schwein) Fr.	7,8,11
Stereum rugosum (Pers.) Fr.	8,26
Stereum pustulatum	6
Stereum striatum	28
Stereum lobatum (Kunze ex Fr.) Fr.	32
Stereum subtomentosum Pouzar	26
Stereum hirsutum (Willd.) S.F. Gray	1,3,12,13,32
Aleurodiscus aurantius (Pers.) J. Schrot.	24
Aleurodiscus wakefieldiae Boidin & Beller	24
Aleurodiscus canidus	28
Strophariaceae	
Pholiota spp.	1,4,8
Hypholoma fasciculare (Huds.) Kummer	8
Gymnopilus spp.	1
Gymnopilus sapineus (Fr.) Maire	24
Psilocybe spp.	1,5,8
Psilocybe banderillensis	12
Agrocybe spp.	1,5,8
Stropharia rugoso-annulata (Farlow) Murril	5,8,12
Stropharia semiglobata (Batsch) Quel	2
Nematoloma fasciculare (Huds.) P. Karst.	1
Steccherinaceae	
Irpex lacteus Fr.	3
Irpex flavus (Fr.) Fr.	1,7,14
Irpex spp.	1,6,9,25,26,27
Junghuhnia collabens (Fr.) Ryvarden	24
Suillaceae	
Suillus sp.	1
Suillus granulatus	1

I. Fungi in aquatic and terrestrial habitats

5. Fungal diversity and mycological collection across forest formations and landscape

APPENDIX TABLE 4.1 Forest fungi documentation in the Philippines.—cont'd

Species	Documented sites
Thelephoraceae	
Thelephora spp.	1,6,11,13,26
Thelephora terrestris (Ehrenb.) Fr.	7,8,14
Tremellaceae	
Tremella fuciformis Berk.	2,5,7,8,11,12,20,23,24
Tremella foliacea	2
Tremella mesenterica (Schaeff.) Retz.	24
Tremella spp.	1,6,26,33
Tremella calocera	6
Tricholomataceae	
Collybia albuminosa	2
Collybia dryophila	1,12,15
Collybia maculata (Alb.& Schw.) Kummer	7,8,11
Collybia fuscopurpurea	1
Collybia reineckeana	1
Clitocybe dealbata (Sowerby) P. Kumm.	1,2
Clitocybe gibba (Pers.) P. Kumm.	1,2
Clitocybe spp.	1,4,5,7,8,32
Collybia spp.	1,4
Omphalina spp.	1,4
Tricholomopsis spp.	1,4,33
Tricholoma malalfum	1
Tricholoma saponaceum (Fr.) P.Kumm.	5
Tricholoma melaleucum	1
Tricholoma lascivum (Fries) Gillet	26
Tricholoma spp.	1,5,8,13
Trypetheliaceae	
Porothelium sp.	6
Proto-Merulius sp.	6

(Continued)

I. Fungi in aquatic and terrestrial habitats

APPENDIX TABLE 4.1 Forest fungi documentation in the Philippines.—cont'd

Species	Documented sites
Tuberaceae	
Tuber sp.	4
Wrightoporiaceae	
Stecchericium sp.	28
Ascomycota	
Sarcosomataceae	
Galiella spp.	1,2,4,26,27
Galiella rufa (Shwein.) Nannf.& Korf.	5,11,29
Sarcoscyphaceae	
Cookeina tricholoma (Mont.) Kuntze	1,2,4,5,6,8,11,13,16,20,21,23,24,30
Cookeina sulcipes (Berk.) Kuntze	1,4,8,11,29
Cookeina sp.	1,26,31
Cookeina speciosa	1
Philllipsia domingensis Berk.	4,12,13
Phillipsia spp.	1,13
Xylariaceae	
Xylaria allantodea	2,4
Xylaria hypoxylon (Linn.) Grev.	4,12
Xylaria polymorpha (Pers.) Grev.	2,3,5,7,8,11,12,13,26,30
Xylaria digitata (L.) Grev.	11
Xylaria spp.	1,2,4,6,9,10,11,13,17,20,26,27,33
Kretzschmaria sp.	3
Xylaria filiformis (Alb. & Schwein) Fr.	3,12
Xylaria fissilis Ces.	7
Xylaria ridleyi Mass.	3,5
Xylaria longipes Nitschke	24
Xylaria allantoides (Berk.) Fr.,	29
Xylaria longiana Rehm	26
Pyronemataceae	
Aleuria aurantia (Pers.) Fuckel.	4
Aleuria sp.	33

I. Fungi in aquatic and terrestrial habitats

5. Fungal diversity and mycological collection across forest formations and landscape

APPENDIX TABLE 4.1   Forest fungi documentation in the Philippines.—cont'd

Species	Documented sites
Otidea sp.	4
Otidea sp.	13
Patella paludosa	1
Patella sp.	6
Octospora humosa (Fr.) Dennis	4,5,7,8
Octospora spp.	4
Octospora spp	5
Scuttelinia scutellata (Linn.) Lamb.	4
Tarzetta spp.	7,13
Hyaloscyphaceae	
Dasyscypha spp.	1
Nectriaceae	
Nectria cinnabarina (Tode) Fr.	1,6,24
Nectria spp.	1,25,26,27
Pleonectria sp.	1
Cordycipitaceae	
Cordyceps sp.	1
Cordyceps militaris	26
Diaporthaceae	
Diaporthe sp.	1
Lachnaceae	
Dasyscyphus apalus (Berk.& Broome) Dennis	4
Pezizaceae	
Peziza repanda Pers.	2,4,5,8,11
Peziza spp.	1,25,26
Sarcosphaera sp.	1
Plicariella scabrosa (Cooke) Spooner	26
Hypoxylaceae	
Daldinia concentrica (Bolt.) Ces.& de Not.	2,3,4,5,7,8,11,13,16,17,199,20,23,24,26,30
Daldinia fissa Lloyd	3

(Continued)

I. Fungi in aquatic and terrestrial habitats

4. Species and functional diversity of forest fungi for conservation and sustainable landscape in the Philippines

APPENDIX TABLE 4.1 Forest fungi documentation in the Philippines.—cont'd

Species	Documented sites
Daldinia polyporum (Lv.) Sacc.	3
Daldinia vernicosa (Schw.) Ces.	3
Daldinia spp.	1,3,6,17,20, 25,26,28
Hypoxylon spp.	1,13,24
Hypoxylon multiforme	12
Hypoxylon fragiforme (Persoon) J. Kickx f.	26
Annulohypoxylon sp.	13
Hysteriaceae	
Hysterium spp.	26
Hysterium angustatum Pers.	24

1 Mt. Makiling Forest Reserve (Quimio, 2001; Nacua et al., 2018, & Authors collections).
2 Quezon Protected Landscape (Brazas et al., 2020 and Authors collection).
3 Phil Teak Fungi—Lobo Batangas (Authors collection).
4 Bazal-Baubo Watershed, Aurora (Authors collection).
5 Mt. Maculot (Authors collection).
6 Mt. Banahaw (Authors collection).
7 Mt. Iraya Batanes Protected Landscape and Seascape (Authors collection).
8 Mt. Palaypalay-Mataas na Gulod Protected Landscape (Arenas et al., 2015; Angeles et al., 2016, Authors collection).
9 Taal Volcano Protected Landscape (Tadiosa & Briones, 2013).
10 Hinulugang Taktak Protected Landscape (Antipolo City, Rizal) (Authors collection).
11 Catanduanes Watershed Forest Reserve (Viga, Panganiban, Bagamanoc) (Authors collection).
12 Mt. Timpoong -Hibokhibok Natural Monument (Camiguin) (Authors collection).
13 Rajah Sikatuna Protected Landscape (Bilar, Bohol) (Authors collection).
14 San Fernando City Forest, La Union Province (Authors collection).
15 Nueva Vizcaya Watershed (Authors collection).
16 Sierra Madre Mountain Range Cagayan Side (Authors collection).
17 Initao-Buenavista Protected Landscape and Seascape (Authors collection).
18 Wetlands and Peatlands Collection (Authors collection).
19 Subic Watershed Forest Reserve (Authors collection).
20 Mt. Arayat National Park (Authors collection).
21 Malagos Watershed Reservation (Authors collection).
22 Mt. Hamiguitan Range Wildlife Sanctuary (Authors collection).
23 Dinagat Island Forest (Authors collection).
24 Cavinti Underground River and Cave Complex (Niem & Baldovino, 2015).
25 Mt.Data, Benguet side, Mt.Province (MNH collection).
26 Mt.Isarog, Camarines (Paguirigan et al., 2020, MNH collection).
27 Mt.Pangasugan, Leyte (MNH collection).
28 Mt.Pulag, Kabayan, Benguet (MNH collection).
29 Mt Cleopatra Needle Forest Reserve (Kim et al., 2021).
30 Angat Watershed Forest Reserve, Norzagaray Bulacan (Liwanag et al., 2017).
31 Isabela State Univ. Forest Reserve (Jacob et al., 2017).
32 Mt. Umubi Situated at Upper Casecnan Area of Alfonso Castañeda, Nueva Vizcaya (Torres et al., 2020).
33 Kalikasan Forest Park Bicol University, Legazpi City, Bicol, Eastern Philippines (Guerrero et al., 2020)

I. Fungi in aquatic and terrestrial habitats

5.1 Forest and watershed reservation

5.1.1 San Fernando Forest

The San Fernando Forest in San Fernando, La Union is an example of a forest over limestone. The most dominant tree species in the forests is molave (*Vitex parviflora* Juss.) that serves as host or substrate of fungi. This mountainous forest located in the eastern boundaries of the city is being considered as one of the most important forest ecosystems in Ilocos Region thereby indicating the need for conservation and protection. This forest is also home to 56 morphospecies, where 40 of which were polypores and 16 were agarics. Some of these species, being the most destructive fungi, are largely responsible for the decay of living trees. The site was assessed to have high macrofungal diversity attributed to high species richness and evenness as indicated by Simpson's diversity indices (1\D) of 0.87202 and 0.85971, and Shannon (H') indices of 2.21849 for agarics and 2.69332 for polypores. Interestingly, the fungal species richness decreased when the condition of the area was relatively dry. Among the most notable was the polypore species *Fuscoporia torulosa* MFSLP-12 (Hymenochaetaceae), which was determined to have antiglycemeric potential.

5.1.2 Sierra Madre Mountain Range

The Sierra Madre Mountain Range in Cagayan, Northern Luzon is represented by the tropical lowland evergreen, lower montane, upper montane and limestone forest formations. The most dominant tree species in these forests were dipterocarps that serve as host or fungal substrates while molave (*V. parviflora* Juss) groups dominated limestone forest. The groups of fungi recorded from these habitats were responsible for reproductive spoilage and wooddecay as these infect and destroy living and dead trees. Bracket-like fruiting bodies are good examples of these groups represented by 38 taxa, 21 genera and 13 families. Polyporaceae had the highest diversity in terms of the number of species composed of 14 different kinds. These species characteristically produce basidiocarps varying from leathery to woody with deep pores or tubes.

Other families included Coriolaceae which consisted of six species bearing basidiocarps that are usually thin tough caps and have zoned surface. There were four species of Xylariaceae whose ascocarps are usually black, cylindrical, club-shaped, often compressed, and leathery. Ganodermataceae, consisted of three species where basidiocarps varied from corky to woody, mostly semiflattened, and characterized with minute pores. This wood-decaying fungal family produces the largest basidiocarps with a diameter of around 28—34 cm.

Families under Schizophyllaceae, Dacrymycetaceae, Hymenochaetaceae, Corticiaceae, Sarcoscyphaceae, Thelephoraceae, and Nidulariaceae had only one species each. Their basidiocarps varied in consistency from fleshy to leathery. Fleshy *Auricularia* consists of three species with gelatinous basidiocarp that are mostly ear-shaped, abundant during the rainy season, and much smaller than Polyporaceae and Ganodermataceae. Frequently occurring fungi were *Pycnoporus sanguineus* (L.) Murrill followed by *Schizophyllum commune* Fr., *Ganoderma applanatum* (Pers.) Pat, *Auricularia auricula-judae* (Fr.) Quel., and *Daedalea ambigua* Berk. In general, there were relatively high fungal species diversity in dipterocarp forest compared with the molave forest and other forest ecosystems.

5.1.3 Palali -Mamparang Mountain Range (PMMR)

The mountain range of Palali-Mamparang where Didipio is located is bordered by province of Nueva Vizcaya and Quirino province. The climatic type of Didipio falls under Type 4 (Corona classification) with mean annual rainfall ranging from 2289 to 2917 mm. The natural forests represent tropical lowland evergreen rainforest and tropical lower montane rainforest, with elevation ranges of 700—950 masl. Valuable and threatened Philippine endemic tree species at PMMR includes *Shorea polysperma* (Blanco) Merr., *S. contorta* Vidal, and *S. palosapis* (Blanco) Merr.

The fungal sampling at Didipio was conducted using transect and quadrat techniques (Malabrigo et al., 2017). A total of 21 orders, 41 families, 60 genera, and 72 species of fungi were recorded from nine transects with low to high diversity (H' = 1.717—3.454) . In general, the dense forest formations have relatively higher fungal diversity. In all orders, wood decay fungi (e.g., *Polyporus*, *Stereum*, *Microporus*) were most abundant with few beneficial mycorrhizas (e.g., *Hygrocybe*, *Amanita*, *Russula*) that indicates level of disturbance. Other fungal species are potential food source and have medicinal uses such as species of *Pleurotus* and *Ganoderma*, respectively. Most fungal substrate are decaying wood and leaf litters suggesting that fungi primarily break wood debris to facilitate natural pruning and decomposition process.

5.1.4 Nueva Vizcaya Watershed

Nueva Vizcaya Watershed is located in Alfonso, Castaneda which is represented by tropical lower montane forest formation. A total of 45 macrofungi belonging to six orders, 15 families and 25 genera was collected, identified and described in the taxonomic checklist. Polyporaceae was recorded as the most abundant macrofungi family present in the area. Out of all macrofungi, 25 species were used by indigenous Bugkalots as either food or medicine. The forest formation sorrounding the Bugkalot tribal community is a habitat for the different macrofungal species. Further studies on seasonal and spatial patterns are essential to determine the species richness and distribution of macrofungi in the community. The cultural values of macrofungi are interesting to explore for promising bioactivity potentials.

5.1.5 Subic Watershed Forest Reserve

The Subic Watershed Forest Reserve is represented by tropical lowland forest and tropical lower montane forest formations. The dominant tree species in these forests that serve as host or fungal substrates were *Shorea negrosensis* Foxw., *Parkia timoriana* (DC) Merr., *Anisoptera thurifera* (Blanco) Merr., *Dipterocarpus grandifloras* Blanco, *Shorea guiso* (Blanco) Blume, *Pterospermum diversifolium* Blume, *Pterocarpus indicus* Wild forma *indicus*, *Albizzia saman* (Jacq) F. Muell., *Macaranga tanarius* (L.) Muell-Arg, *Koorsiodendron pinnatum* (Blanco) Merr. A total of 15 macrofungal taxa belonging to eight families and five genera was collected, identified, and described. Polyporaceae was recorded as the most abundant macrofungi family present in the area. The new recorded species in this area were *Microporus xanthopus* (Fr.) Kuntz, *Daldinia concentrica* (Bolt.) Ces. & de Not, *Polyporus hirsutus* (Wulf.) Fr., *Schizophyllum commune* Fr., *Ganoderma applanatum* (Pers.) Pat., and *Pycnoporus sanguineus* (L.) Murrill.

5.1.6 La Mesa Watershed Reservation (LMWR)

The La Mesa Watershed Reservation is a protected area that is the source of potable water in Metro Manila. The LMWR is characterized with a moderate rolling terrain with varying elevations from 42 to 259 masl and gentle slopes. Climatic conditions are wet during May to December and dry in January to April.

The reservation has a long history of rehabilitation programs using native and exotic trees. The macrofungi associated with plantations of narra, mahogany, and molave were assessed to determine possible wood-decay taxa in support of plantation management programs in the LMWR. A total of 21 taxa, from eight orders, 11 families, 19 genera and 10 species were recorded. Basidiomycetes dominated all sampling plots compared to Ascomycetes. Most taxa identified was under the family of bracket fungus, Polyporaceae, followed by Stereaceae. The recorded species were as follows: *Daldinia concentrica* (Bolton) Ces. & De Not., *Coprinellus disseminatus* (Pers.) J.E. Lange, *Schizophyllum commune* Fr., *Auricularia polytrichta* (Mont.) Sacc., *Pycnoporus sanguineus* (L.) Murrill, *Daedaleopsis* sp., *Ganoderma applanatum* (Pers.) Pat, *Hexagonia tenuis* (Fr.) Fr., *Lentinus crinitus* (L.) Fr., *Microporus xanthopus* (Fr.) Kuntze. Most of these species were saprobic fungi degrading most woody substrates such as wood, branches, and twigs, except for *C. disseminatus* thriving on soil substrate. These species represent dry season macrofungi thriving in LMWR.

5.1.7 Mount Makiling Forest Reserve

The ASEAN heritage Mount Makiling Forest Reserve (MMFR) in Los Baños, Laguna is composed of tropical lowland rainforest and tropical lower montane forest formations covering four subwatersheds (i.e., Molawin-Dampalit, Tigbi, Cambantoc, and Greater Sipit subwatersheds) (Lapitan et al., 2018). The site has a climatic condition of pronounced wet (May to December) and dry (January to April) with annual precipitation of 1645—2299 mm (Pagaduan & Afuang, 2012). The macrofungi in all subwatersheds ranged from 13 to 36 genera comprising of 45 species (Lapitan et al., 2010, 2018). Another study conducted in Molawin-Dampalit revealed a total of 27 taxa belonging to Basidiomycota (26) and Ascomycota (1), with dominance of Polyporaceae (Parlucha et al., 2021). In this study, a total of 11 species were possibly new records that were undocumented in previous studies conducted by Quimio (2001) and Nacua et al. (2018). These fungal species include *Microporus xanthopus* (Fr.) Kuntze, *Daedalea ambigua* Berk., *Pycnoporus sanguineus* (L.) Murrill, *Fomes pachyphloeus* Pat., *Hexagonia tenuis* (Hook.) Fr., *Phellinus gilvus* (Schw.) Fr., *Ganoderma lucidum* (Leys.) Karst, *Poria straminea* Pres., *Lenzites striata* (Swartz.) Fr, *Hymenochaete tenuissima* Berk, and *Trametes corrugata* (Pers.) Bres. Most of these are saprophytes which grow only on dead woody plants. However, some have evolved to acquire the ability to invade living trees and eventually infect the heartwood. They enter through wounds of trees and cause decay as they establish and grow unrestricted within the heartwood (Eusebio, 1998).

Fungal flora across the four subwatersheds were also distinct due to the varied lower montane vegetation brought about by conservation management systems. The Molawin-Dampalit, Tigbi, Cambantoc, and Greater Sipit subwatersheds equally share diverse fungal flora. Results of the transect sampling method recorded 39 genera in all subwatersheds, where majority of the taxa were wood decay fungi. Species diversity indicated low to moderate diversity with Shannon index (H′) ranging from 0.8798 to 1.343. *Corticium* had the highest frequency in Cambantoc, Tigbi, and Greater Sipit. Other common fungal genera were wood-inhabiting fungi such as *Polyporus, Polystictus* and *Xylaria*. Sporocarps of ectomycorrhizas such as *Amanita* and *Scleroderma* were also found in the subwatersheds of MMFR (Lapitan et al. 2010, 2018).

5.1.8 Catanduanes Watershed Forest Reserve

Catanduanes Watershed Forest Reserve is located in the municipalities of Viga, Panganiban, and Bagamanoc with a lowland montane forest formation. Using transect line (TL) method stretching 2,000 m, 6 TLs and 60 quadrats measuring 150 m^2 at 200 m interval were established in the different areas of the three municipalities. Fungal species encountered along the TL were collected, recorded and photo-documented. Field sampling documented 66 species of fungi that belong to 42 genera and 28 families, with a total of 532 individuals, of which 58 were basidiomycetes and eight ascomycetes considered as new island records. The most common species documented includes *Cookeina tricholoma* (Mont.) Kuntze, *Cymatoderma elegans* Jungh., *Dacryopinax spathularia* (Schwein.) G.W. Martin, *Earliella scabrosa* (Pers.) Gilb. & Ryvarden, *Galiella rufa* (Schwein.) Nannf., *Ganoderma applanatum* (Pers.) Pat., *Hexagonia tenuis* (Hook.) Fr., *Microporus xanthopus* (Fr.) Kuntze, *Phallus indusiatus* Vent., *Pycnoporus sanguineus* (Fr.) Murr., *Schizophyllum commune* Fr., and *Trametes gibbosa* (Pers) Fr. Further field surveys of the Watershed are anticipated to uncover a rich and diverse fungal flora including the unexplored area. Fungal diversity research efforts are encouraged to evaluate the effects of some human disruptions on the ecology of the watershed.

5.1.9 Malagos Watershed Reserve

Malagos Watershed Reservation in Davao City is represented by Tropical Lower montane forest formation. The most dominant tree species in this forest were *Trema orientalis* (L.) Blume, *Dillenia philippinensis* Rolfe, *Colona serratifolia* Cav., *Pterocarpus indicus* Wild. *forma* indicus, *Milletia pinnata* (L.) Panigrahi, *Lagerstroemia speciosa* (L.) Pers., and *Buchanania arborescens* (Blume) Blume that serve as host or fungal substrates. A total of 16 macrofungi belonging to five families and seven genera was collected, identified and described in the taxonomic checklist. Polyporaceae was recorded as the most abundant macrofungi family present in the area. The newly recorded species in the area were *Fomes albomarginatus* (Lev.) Cke., *Fomes roseus* (Alb. & Schw.) Cke., *Auricularia polytricha* (Mont.) Sacc., *Schizophyllum commune* Fr., *Daldinia concentrica* (Bolt.) Ces. & de Not, *Daedalea ambigua* Berk., *Daedalea repanda* (Pers.) Gaud., *Ganoderma applanatum* (Pers.) Pat., *Hexagonia tenuis* (Hook.) Fr., *Microporus xanthopus* (Fr.) Kuntze, *Pycnoporus sanguineus* (L.) Murrill, *Trametes corrugata* (Pers.) Bres., *Auricularia auricula-judae* (Fr.) Quel, *Lepiota cristata* (Bolt.) Kumm., *Coprinus disseminatus* (Pers.) J.E. Lange, and *Dacryopinax spathularia* (Schwein.) Martin.

5.1.10 Dinagat Island Watershed

Dinagat Island Watershed is located in the municipalities of Loreto, Tubajon, Libjo, and Cagdianao which is represented by Tropical Evergreen Forest, Tropical Upper Montane, Forest over limestone, Tropical Lower montane, and ultramafic forest formations. The dominant plant species were *Xanthostemon verdugonianus* Naves, including highly endemic, *Gomphandra dinagatensis* (Merr.) Merr. that serve as host or fungal substrates. A total of 34 macrofungi belonging to 14 families and 18 genera were collected, identified, and described in the taxonomic checklist. Polyporaceae was recorded as the most abundant macrofungi family present in the area. The new recorded species in this area were *Auricularia polytricha* (Mont.) Sacc., *Coprinus plicatilis* (Curt.) Fr., *Hexagonia tenuis* (Hook.) Fr., *Schizophyllum commune* Fr., *Amauroderma rugosum* (Berk.) Torrend., *Ganoderma lucidum* (Leys.) Karst., *Hygrocybe miniata* (Fr.) Kumm., *Marasmius rotula* (Scop.) Fr., *Cymatoderma elegans* Jungh., *Fomes fomentarius* (L.) Kickx., *Earliella scabrosa* (Pers.) Gilb. & Ryvarden, *Tremella fuciformis* Berk., *Microporus xanthopus* (Fr.) Kuntze, *Panus rudis* Fr., *Daldinia concentrica* (Bolt.) Ces. & de Not., and *Xylaria allantodea* (Berk.) Fr.

5.2 Protected Landscape and Seascape

5.2.1 *Batanes Protected Landscape and Seascape*

The Batanes Protected Landscape and Seascape, particularly Mt. Iraya and Mt. Matarem, is composed of tropical lowland and tropical lower montane forest formations. The most dominant tree species in these forests that serve as host or fungal substrates were *Pterocarpus indicus* Wild forma *indicus, Pterospermum niveum* S. Vidal, *Wendlandia luzonensis* DC, *Pongamia pinnata* (L.) Pierre, and *Neonauclea reticulata* (Havil) Merr. The group of fungi recorded from these fungal habitats were characteristically responsible for wood-decay. These include bracket-like fruiting bodies represented by 68 species, 27 genera, and 18 families. Polyporaceae had the highest species diversity that is composed of 20 different species. Sixty-eight (68) species documented and collected represent 15% of the total species reported in the province. Of these, 12 species of Philippine fungi were newly recorded, which includes *Auricularia polytricha* (Mont.) Sacc., *Cantharellus infundibuliformis* (Scop.) Fr., *Ganoderma applanatum* (Pers.) Pat., *Hexagonia tenuis* (Hook.) Fr., *Microporus xanthopus* (Fr.) Kuntze, *Polyporus picipes* Fr., *Pycnoporus sanguineus* (L.) Murrill, *Schizophyllum commune* Fr., *Trametes corrugata* (Pers.) Bres., *Tremella fuciformis* Berk., *Xylaria polymorpha* (Pers.) Grev., and *Tarzetta cupularis* (L.) Srvcek.

5.2.2 *Mt. Palaypalay Mataas na Gulod Protected Landscape & Mt. Maculot*

The tropical lowland evergreen forest dominated both Mt. Palaypalay-Mataas na Gulod Protected Landscape (MPMGPL) in Cavite and Mt Maculot in Batangas. The MPMGPL was categorized by the National Integrated Protected Areas System as one of the priority areas for biodiversity conservation.

There were 434 fungal specimens collected under 24 families, 37 genera, and 41 species in MPMGPL. Common anthropogenic disturbances in the landscape are due to gathering of minor forest products, slash-and-burn farming, and quarrying that necessitates evaluation of human influence to fungal ecology of the landscape.

The sampled plots in the lower montane forest formation of Mount Maculot had 304 individuals under 34 families, 63 genera, and 100 species. The dominant fungal species were: *Ganoderma* (Pers.) Pat., *Auricularia auricula-judae* (Fr.) Quel., *Hexagonia tenuis* (Hook.) Fr., *Polyporus picipes* Fr., *Microporus xanthopus* (Fr.) Kuntze, *Fomes gilvus* (Schw.) Fr., *Schizophyllum commune* Fr., *Daldinia concentrica* (Bolt.) Ces. & de Not., *Dacryopinax spathularia* (Schwein.) Martin, *Cantharelllus infundibuliformis* (Scop.) Fr., and *Termitomyces eurhizus* (Berk.) Heim. The type of fungal habitat appeared to influence the number of macroscopic fungi in Mt. Maculot which is considered one of the most diverse forest ecosystems in the Philippines.

5.2.3 Taal Volcano Protected Landscape (TVPL)

Taal Volcano Protected Landscape in Talisay, Batangas is classified as forest over limestone formation. This protected area is considered as one of the most diverse forest ecosystems in CALABARZON (Cavite, Laguna, Batangas, Rizal, and Quezon) region that need conservation and protection. Two climatic types are evident, dry (November until April) and wet (May until October). Field sampling of fungi within the 200—600 masl elevation resulted in the collection and identification of 75 species belonging to 36 genera and 23 families. Accordingly, these collections are first fungal documentations at TVPL (Tadicsa and Briones 2013). The reported species in TVPL were *Cookeina sulcipes* (Berk.) Kuntze, *Galiella rufa* (Schwein.) Nannf. & Korf, *Dictyophora duplicate* (Bosc) E.Fisch., *Cymatoderma elegans* Jungh., *Microporus vernicipes* (Berk.) Kuntze and *Xylaria longipes* Nitschke.

5.2.4 Hinulugang Taktak Protected Landscape

Hinulugang Taktak Protected Landscape was declared as one of the protected area classified as urban landscape in Antipolo, Rizal. The dominant tree species were *Pterocarpus indicus* Willd forma *indicus*, *Delonix regia* (Boj.ex Hook) Raf., *Albizzia saman* (Jacq) F. Muell., *Cananga odorata* (Lamk.) Hook f & Thoms, and *Mangifera indica* (L.). Collections of fungi in this landcape comprised of 24 species belonging to 12 families and 10 genera where group of Polyporaceae remained abundant family. The new recorded species in this area were *Ganoderma applanatum* (Pers.) Pat., *Ganoderma lucidum* (Leys.) Karst., *Cantharellus* sp., *Auricularia auricula-judae* (Fr.) Quel., *Auricularia polytricha* (Mont.) Sacc., *Schizophyllum commune* Fr., and *Trametes* sp.

5.2.5 Mount Banahaw and San Cristobal Protected Landscape (MBSCPL)

Mt. Banahaw is an important watershed with forest formation classified as lowland to submontane forest having annual precipitation of 2300—4500 mm. Latest assessment recorded 35 species of fungi representing 19 families. Polyporaceae species were found dominant such as *Fomes* sp., *Microporous affinis* Pat. and *Trametes* sp. (Parlucha et al., 2021). Mycenaceae species were also recorded to include *Mycena* sp. and *Favolaschia* sp.. Ascomycetes were also found such as *Bisporella sulfurina* (Quel.) S.E. Carp. of Helotiaceae and *Xylaria* sp. of Xylariaceae.

Tadiosa et al. (2015) also conducted extensive survey of macroscopic fungi in MBSCPL accounting to 62 species, 30 families, and 49 genera. Among the species includes *Favolaschia sp.* and *Xeromphalina sp.* (Mycenaceae), *Oudemansiella sp.* (Physalacriaceae), *Pholiota lignicola* Peck Jacobsson and *Pholiota sp. (Strophariaceae)*, *Pisolithus sp.* (Sclerodermataceae), *Cantharellus sp.* (Cantharellaceae), *Cymatoderma sp..* (Meruliaceae), *Trametes sp.* (Polyporaceae), *Russula sp.* (Russulaceae), *Thelephora sp.* (Thelephoraceae).

5.2.6 Quezon Protected Landscape (QPL)

Quezon protected landscape is characterized with lowland rainforest and karst covering 983 ha. of closed forest located in municipalities of Pagbilao, Padre Burgos, and Atimonan, Quezon Province. The area has pronounced rainfall period during November to January. Brazas et al. (2020) recorded 53 species of macrofungi representing 19 different families and 31 genera. Most recorded species were Polyporaceae with *Hexagonia* and *Polyporous* genera as most dominant. A follow-up study of Parlucha et al. (2021) conducted during dry period recorded 22 species representing 15 families. Polyporaceae was consistent as the highest recorded species. Only *Biscogniauxia nummularia* (Bull.) Kuntze of Xylariaceae family belongs to Ascomycota. Both studies revealed high diversity values (H' = 4.14 and 3.4) and hypothesized that there could be more species in the site.

5.2.7 Buenavista Protected Landscape (BPL)

The Buenavista Protected Landscape has forest over limestone formation situated in the municipality of Mulanay in Quezon Province. The protected landscape has a total aggregate size of 284.27 ha. which consists of three parcel areas.

Preliminary assessment of macrofungi composition followed quadrat and opportunistic sampling in BPL since it was proclaimed as a protected landscape in 2000. A total of 127 macrofungi were described consisting of 47 genera, and 24 families dominated by basidiomycetes. The most abundant macrofungus recorded was *Favolus tenuiculus* P. Beauv, which is an edible fungus.

The largest ecological group were the saprophytic fungi, accounting to more than 90% of the total fungal diversity while the rest were mycorrhizal and plant pathogenic. Mycorrhizal fungi include *Volvariella volvacea* (Bull.) Singer, *Agaricus* sp., *Boletinellus* sp. and *Cuphophyllus* sp. with *Septobasidium sp., Phellinus sp., Laetiporus* sp. and *Hexagonia sp.*as pathogenic species. A number of edible macrofungi include *Auricularia auricula auricula* (Fr.) Quel., *Auricularia polytricha* (Mont.) Sacc., *Lentinus* sp., *Volvariella volvacea* (Bull.) Singer, *Schizophyllum commune* Fr., *Laetiporus., Favolus tenuiculus* P. Beauv., and *Pleurotus* sp. Their potential as an alternative food source can be explored by the locals particularly those who dwell in the forests where resources are scarce. Some macrofungi exhibit medicinal properties important in pharmacology such as *Ganoderma lucidum* (Leys.) Karst., *G. applanatum* Pers.P at., *Pycnoporus sanguineus* (L.) Murrill, *Auricularia auricula-judae* (Fr.) Quel., *A. polytricha* (Mont.) Sacc., *Microporus xanthopus* (Fr.) Kuntze, *Schizophyllum commune* Fr., *Lentinus* sp., and *Trametes* sp.

5.2.7.1 Initao-Buenavista Protected Landscape and Seascape

The Initao-BPL and Seascape in Misamis Oriental is represented by forest over limestone formation. The topography varies from coastal plains, to rolling hills across terrestrial and marine ecosytems with unique biological features of regional and national significance. Forest vegetation is abuounds with *Pterocarpus, Barringtonia, Koorsiodendron, Pandanus,* and *Nauclea*. A total of 40 fungal taxa were collected along the trail, on substrates like soil, rotten trunks and roots. The dominant species were *Auricularia auricula-judae* (Fr.) Quel., *Auricularia polytricha* (Mont.) Sacc., *Daldinia concentrica* (Bolt.) Ces. & de Not., *Hexagonia tenuis* (Hook.) Fr., *Microporus xanthopus* (Fr.) Kuntze, *Polyporus hirsutus* (Wulf.) Fr., *Trametes corrugata* (Pers.) Bres., and *Polyporus picipes* Fr.

5.3 National Parks and Wildlife Sanctuary

5.3.1 Mt. Arayat National Park

The Mount Arayat National Park is located in Magalang, Pampanga which is represented by tropical lowland evergreen forest formation. The most dominant tree species in these forests were *Pterocarpus indicus* WIlld forma *indicus, Parkia timoriana, Sterculia foetida* L., *Ziziphus*

126 4. Species and functional diversity of forest fungi for conservation and sustainable landscape in the Philippines

talanai (Blanco) Merr., *Premna odorata* Blanco, *Toona calantas* Merr. & Rolf., *Dracontomelon dao* (Blanco) Merr., *Mallotus philippensis* (Lamk.) Muell-Arg. And *Macaranga bicolor* Muell-Arg. that serve as host or fungal substrates. A total of 45 macrofungal taxa belonging to 15 families and 10 genera was collected, identified and described. Polyporaceae was recorded as the most abundant macrofungi family present in the area. Newly recorded species in this landscape were *Schizophyllum commune* Fr., *Auricularia polytricha* (Mont.) Sacc., *Amauroderma rogusom* (Berk.) Torrend., *Cookeina tricholoma* (Mont.) Kuntze, *Hexagonia tenuis* (Hook.) Fr., *Trametes corrugata* (Pers.) Bres., *Coprinellus disseminatus* (Pers.) J.E. Lange, *Dacryopinax spathularia* (Schwein.) Martin, *Microporus xanthopus* (Fr.) Kuntze, *Earliella scabrosa* (Pers.) Gilb. & Ryvarden, *Polyporus picipes* Fr., and *Fomes roseus* (Alb. & Schw.) Cke.

5.3.2 Cavinti Underground River and Cave Complex

Situated within the Sierra Madre Mountain Range, the Cavinti Underground River and Cave Complex (CURCC) in Brgy. Paowin, Cavinti, Laguna is a natural forested area surrounded by agroforests, lush vegetation, and a number of cave complexes important for ecotourism.

The climatic condition in this landscape is characterized by pronounced dry months from March to May and wet throughout the rest of the year. Exploratory and opportunistic sampling from seven sites revealed 507 collections of macrofungi, belonging to 41 species, 34 genera and 20 families (Niem & Baldovino, 2015). The family containing the greatest number of species was Polyporaceae which accounted for 29.07% of the total relative fungal density. The most common genus was *Microporus* with three species while highest relative density was recorded for *Daldinia concentrica* (Bolt.) Ces. & de Not., followed by *Rigidoporus microporus* (Swartz.) Overeem, *Auricularia auricula*(Fr.) Quel., *Lentinus tigrinus* (Bull.) Fr., *Schizophyllum commune* Fr., and *Microporus xanthopus*(Fr.) Kuntze. Those less occurring fungal species (<0.20%) were *Nectria cinnabarina* (Tode) Fr., *Cookeina tricholoma* (Mont.) Kuntze, *Xylaria longipes* Nitschke, *Pluteus salicinus* (Pers.) P. Kumm., *Gymnopilus sapineus* (Fr.) Murrill, *Auricularia delicata* (Fr.) Henn., *Dacryopinax spathularia* (Schwein.) Martin, *Daedalea* sp., *Amauroderma rugosum* (Berk.) Torrend., *Ganoderma applanatum* (Pers.) Pat., *Aquascypha hydrophora* (Berk.) D.A. Reid., *Hexagonia tenuis* (Hook.) Fr., *Microporus affinis* Pat., *Exidia thuretiana* (Lev.) Fr., *Tremella fuciformis* Berk. In addition, one basidiomycetous fungus, *Multiclavula vernalis* (Schwein.) R.H. Petersen (Petersen, 1967), previously not recorded in the Philippines was identified from CURCC.

The surveyed fungal communities were found to have diverse ecological functions. Saprophytic fungi comprised the largest (60%) ecological group recorded, as represented by *Daldinia concentrica*. Saprophytes are important in nutrient cycling via biodegradation and biodeterioration of organic materials. These organisms maintain forest health by returning organic material to the soil, making the carbon source available to plants and other organisms (Senn-Irlet et al., 2007). Other taxa of macrofungi are parasitic or plant pathogenic and as such, they infect the tissues of the living plant by extracting organic substances from their hosts (Leonard, 2010). If left untreated, this may reduce the fitness of the host plant and may even lead to its eventual death. *Nectria, Rigidoporus, Podoscypha, Lenzites, Pycnoporus* and *Trametes* are the genera of plant pathogenic macrofungi present in CURCC.

Notably, certain macrofungi in CURCC such as *Cookeina, Daedalea, Microporus,* and *Trametes* may also have bioremediation potential which can degrade various environmental

I. Fungi in aquatic and terrestrial habitats

contaminants. For instance, wood-rotting fungi have been used for the decolorization of dyes and industrial effluents (Gadd, 2001). There are also lignolytic fungi, such as those causing white rot, that can degrade straw, saw dust, or corn cobs.

5.3.3 Mt. Hamiguitan Range Wildlife Sanctuary

The Mt. Hamiguitan Range Wildlife Sanctuary (MHRWS) located in Davao Oriental was declared as a UNESCO World Heritage Site. The MHRWS is represented by tropical sub-alpine forest formation dominanted by tree species *Trema orientalis* (L.) Blume., *Dillenia philippinensis* Rolfe., *Colona serratifolia* Cav., *Pterocarpus indicus* Willd form *indicus*, *Milletia pinnata* (L.) Panigrahi, *Lagerstroemia speciosa* (L.) Pers. and *Buchanania arborescens* (Blume) Blume, that serve as host or fungal substrates. A total of 27 macrofungal taxa belonging to 12 families and 16 genera was collected, identified and described. Polyporaceae was recorded as the most abundant macrofungi family present in the sanctuary. The new record species were *Pycnoporus sanguineus* (L.) Murrill, *Schizophyllum commune* Fr., *Microporus xanthopus* (Fr.) Kuntze, *Ganoderma applanatum* (Pers.) Pat., *Auricularia polytricha* (Mont.) Sacc., *Coprinellus disseminatus* (Pers.) J.E Lange, and *Cookeina tricholoma* (Mont.) Kuntze.

5.4 Wetlands and peatlands

The Agusan Marsh and Wildlife Sanctuary in San Francisco, Agusan del Sur which is represented by peat swamp forest. Fungal collections have a total of 25 taxa belonging to 15 families and 10 genera. Polyporaceae was recorded as the most abundant. The new record species are *Earliella scabrosa* (Pers.) Gilb. & Ryvarden, *Schizophyllum commune* Fr., *Russula emetica* (Schaffer) Persoon, *Ganoderma applanatum* (Pers.) Pat., *Polyporus hirsutus* (Wulf.) Fr., *Pycnoporus sanguineus* (L.) Murrill, and *Polyporus picipes* Fr.

5.5 Mycological collections

Most of the first forest macrofungi collection were deposited at the Mycological Herbarium of the Museum of Natural History in University of the Philippines Los Baños (UPLB). These collections were part of the fungal expeditions led by Don Reynolds, an American Peace Corps volunteer and a mycologist. In 1965, Reynolds and his troupe of students from the Department of Plant Pathology at UPLB traversed Batanes to Sulu and amassed a sizable fungal assemblage that included but not limited to forest macrofungi. From then on, the collection continued to build up through the years from research of local mycologists and from systematics mycology students.

At present, the UPLB MNH Mycological Herbarium serves as a repository of fungal collection which houses around 4000 macrofungal specimens. From among these, more than 3000 are from the Phylum Basidiomycota, comprising almost 80% of the total collection while only 6% of the collection belong to Phylum Ascomycota.

The forest formations were well-represented in the fungal collections of the UPLB MNH Mycological Herbarium. Under the tropical upper montane rainforest formation, about 155 fungal taxa were documented from high altitude mountain ecosystems in Mt. Pulag (Benguet), Mt. Data (Mountain. Province), Mt. Banahaw (Quezon), Mt. Isarog (Camarines Sur),

I. Fungi in aquatic and terrestrial habitats

128   4. Species and functional diversity of forest fungi for conservation and sustainable landscape in the Philippines

and Mt. Apo (Davao region). Tropical lower montane fungi comprise the greatest number of collections which totals to 553 taxa from Mt. Polis (Mountain Province), Mt. Pangasugan (Leyte), Mt. Posuey (Abra), and Mt. Makiling (Laguna). This estimate excludes the fungal collections (424 taxa) labeled as College, Laguna that could possibly add up to this number. Mt. Manungal (Cebu) and the Quezon National Park (Quezon), representing the tropical lowland evergreen rainforest, have 16 fungal taxa. Representing the beach forest and limestone formation includes the island ecosystems in Palaui (Cagayan Valley), Alabat (Quezon), Anda (Pangasinan), Lipata (Quezon), Catanduanes (Bicol), Jolo and Tawi-tawi (Sulu) which totals to 126 fungal taxa.

6. Challenges in mycology for conservation and sustainable landscape

The Philippines faces rapid environmental changes, particularly the conversion of forest land use and changes in climate, that threaten biodiversity including forest fungi. Forest ecosystem disturbances have been demonstrated to alter fungal composition and function (McGuire et al., 2015; Robinson et al., 2020), with the possibility of dysbiosis that favors pathogenic fungi due to the loss of saprophytic competitors (Shi et al., 2019). While efforts to improve forest cover had been implemented through policies and reforestation programs, satellite observations continue to track forest loss in the country (Perez et al., 2020), where observed deforestation hotspots brought about by unsustainable practices often overlap with primary forests (Araza et al., 2021). This direct reduction and/or fragmentation of forest habitats has been further exacerbated by the vulnerability to climate change of dominant and structurally important Philippine dipterocarp forests (Pang et al., 2021), that are hosts to diverse fungal residents. Nonetheless, forest revegetation has been shown to improve fungal diversity and function (Liu et al., 2019; McGuire et al., 2015). Thus, besides strengthening forest protected areas, a holistic approach that integrates the sustainable use of forest resources, restoration of degraded forests, and climate change mitigation in the Philippines is important for the conservation of forest fungal resources.

An in-depth knowledge on the distribution and biological and ecological knowledge of forest macro- and microfungi is important to inform forest managers and policy makers in formulating plans for priority forest areas that are critical fungal niches and identifying priority fungal species for conservation. While studies on forest fungi in the Philippines remain to be fragmented and underfunded, different avenues for forest mycological research can provide new discoveries crucial in fungal conservation. For example, research on population dynamics, colonization requirements, trophic guilds, and reproductive biology across diverse forest fungal species will help assess conservation status, target conservation efforts on critically endangered fungal groups in the country, and establish effective fungal recruitment strategies (Dahlberg et al., 2010; Lonsdale et al., 2008; Molina et al., 2011). Various modern molecular approaches can also refine taxonomy of morphologically similar fungal species endemic to the country (Rodriguez et al., 2004), and capture the diversity of unaccounted cryptic fungal species reportedly abundant in the tropics (Arnold et al., 2000; Fröhlich & Hyde, 1999; Tomita, 2003). Further, as data for Philippine Forest fungi are currently patchy, a comprehensive and updated checklist coupled with an accessible database of Philippine

I. Fungi in aquatic and terrestrial habitats

forest fungi that is fulfilled through regular surveys, mycomonitoring, and active collection would be useful tools for additional insights on their diversity and ecology (Halme et al., 2012; Sette et al., 2013).

Understanding sociopolitical dimensions is key to successful biodiversity conservation. Citizen science, that is engaging the public to report the occurrence of either rare or common forest fungi in their localities, would be useful to boost fungal diversity assessment and management (May et al., 2019) and a means to improve social awareness (Irga et al., 2020). However, while others argue for a far more committed community requiring social capital coupled with policy reforms to achieve positive biodiversity outcomes (Pretty & Smith, 2004), the Philippines still has a lot to learn from its past experiences in forest policies and strategies involving local participation (Utting, 2000). Nonetheless, current policies that aim to conserve and/or protect Philippine forest ecosystems could help maintain unique fungal assemblages that have yet been explored. Lastly, regional and global collaborations with mycologists and fungi enthusiasts are needed to put forward the agenda of fungal resources conservation to policy makers, help establish an accessible mycological database in the country, and contribute to the global accounting of fungal diversity.

Acknowledgments

The authors acknowledge the numerous authors whose papers are cited in this paper.

References

Alviola, P. A., Macasaet, J. P. A., Afuang, L. E., Cosico, E. A., & Eres, E. G. (2015). Cave-dwelling bats of Marinduque island, Philippines. *Museum Publications in Natural History, 4*, 1—17.

Amoroso, V. B., Laraga, S. H., & Calzada, B. V. (2011). Diversity and assessment of plants in Mt. Kitanglad range natural Park, Bukidnon, southern Philippines. *Gard Bull Singapore, 63*, 219—236.

Angeles, L. P., Arma, E. J. M., Basaca, C. W., Biscocho, H. E. H., Castro, A. E., Cruzate, S. M., Garcia, R. J. G., Maghari, L. M. E., Pagadora, R. S., & Tadiosa, E. R. (2016). Basidiomycetous fungi in Mt. Palay-palay protected landscape, Luzon island, Philippines. *Asian Journal of Biodiversity, 7*(1), 79—94. https://doi.org/10.7828/ajob.v7i1.867

Araújo, J. P. M., & Hughes, D. P. (2016). Diversity of entomopathogenic fungi. Which groups conquered the insect Body? *Advances in Genetics, 94*, 1—39. https://doi.org/10.1016/bs.adgen.2016.01.001

Araza, A. B., Castillo, G. B., Buduan, E. D., Hein, L., Herold, M., Reiche, J., Gou, Y., Villaluz, M. G. Q., & Razal, R. A. (2021). Intra-annual identification of local deforestation hotspots in the Philippines using earth observation products. *Forests, 12*(8), 1—16. https://doi.org/10.3390/f12081008

Arenas, M. C., Tadiosa, E. R., Alejandro, G. J. D., & Reyes, R. G. (2015). Macroscopic fungal flora of Mts. Palaypalay — Mataas na Gulod protected landscape, southern Luzon, Philippines. *Asian Journal of Biodiversity, 6*(1), 1—22. https://doi.org/10.7828/ajob.v6i1.693

Aribal, L. G., & Fernando, E. S. (2014). Vascular plants of the peat swamp forest in Caimpugan, Agusan del Sur province on Mindanao Island, Philippines. *Asian Journal of Biodiversity, 5*, 1—17. https://doi.org/10.7828/ajob.v5i1.478

Aribal, L. G., & Fernando, E. S. (2018). Plant diversity and structure of the Caimpugan peat swamp forest on Mindanao Island, Philippines. *Mires & Peat, 22*, 1—16. https://doi.org/10.19189/MaP.2017.OMB.309

Arnold, A. E., Maynard, Z., Gilbert, G. S., Coley, P. D., & Kursar, T. A. (2000). Are tropical fungal endophytes hyperdiverse? *Ecology Letters, 3*(4), 267—274.

Ashton, P., & Zhu, H. (2020). The tropical-subtropical evergreen forest transition in East Asia: An exploration. *Plant Diversity, 42*(4), 255—280.

Asiegbu, F. O., & Kovalchuk, A. (2021). An introduction to forest biome and associated microorganisms. In *Forest microbiology* (pp. 3—16). Academic Press. https://doi.org/10.1016/b978-0-12-822542-4.00009-7

Aureo, W. A., Reyes, T. D., Mutia, F. C. U., Jose, R. P., & Sarnowski, M. B. (2020). Diversity and composition of plant species in the forest over limestone of Rajah Sikatuna protected landscape, Bohol, Philippines. *Biodiversity Data Journal, 8*. https://doi.org/10.3897/BDJ.8.e55790

Banwa, T. P. (2011). Diversity and endemism in mossy/montane forests of central cordillera region, northern Philippines. *Biodiversity, 12*(4), 212–222.

Brazas, F. P., Taglinao, L. P., Revilla, A. G. M., Javier, R. F., & Tadiosa, E. R. (2020). Diversity and taxonomy of basidiomycetous fungi at the northeastern side of Quezon protected landscape, southern Luzon, Philippines *Journal of Agricultural Science and Technology A, 10*, 1–11.

Briones, R. U., Tadiosa, E. R., & Manila, A. C. (2017). Threats on the natural stand of philippine teak along verde island passage marine corridor (VIPMC), Southern Luzon, Philippines. *Journal of Environmental Science and Management, 20*(2), 54–67.

Brown, A., & Thatje, S. (2014). Explaining bathymetric diversity patterns in marine benthic invertebrates and demersal fishes: Physiological contributions to adaptation of life at depth. *Biological Reviews, 89*(2), 406–426.

Brundrett, M. (1996). *Working with mycorrhizas in forestry and agriculture: International mycorrhizal workshop in Kaping, China in 1994*. Australian Centre for International Agricultural Research.

Cabutaje, E. M., Pecundo, M. H., & dela Cruz, T. E. E. (2020). Diversity of myxomycetes in typhoon-prone areas: A case study in beach and inland forests of Aurora and Quezon province, Philippines. *Sydowia, 73*, 113–132.

Collado, E., Camarero, J. J., Martínez de Aragón, J., Pemán, J., Bonet, J. A., & de-Miguel, S. (2018). Linking fungal dynamics, tree growth and forest management in a Mediterranean pine ecosystem. *Forest Ecology and Management, 422*(April), 223–232. https://doi.org/10.1016/j.foreco.2018.04.025

Co, L. L., & Tan, B. C. (1992). Botanical exploration in Palanan wilderness, Isabela province, the Philippines: First report. *Flora Malesiana Bulletin, 11*, 49–53.

Culmsee, H., & Leuschner, C. (2013). Consistent patterns of elevational change in tree taxonomic and phylogenetic diversity across Malesian mountain forests. *Journal of Biogeography, 40*(10), 1997–2010.

Dahlberg, A., Genney, D. R., & Heilmann-Clausen, J. (2010). Developing a comprehensive strategy for fungal conservation in Europe: Current status and future needs. *Fungal Ecology, 3*(2), 50–64. https://doi.org/10.1016/j.funeco.2009.10.004. Elsevier Ltd.

Dighton, J. (2007). 16 Nutrient cycling by saprotrophic fungi in terrestrial habitats. *Environmental and Microbial Relationships, 4*, 287. Springer.

Eusebio, M. A. (1998). *Pathology in forestry*. Ecosystems Research and Development Bureau and Department of Environment and Natural Resources.

Fernando, E. S., Suh, M. N., Lee, J., & Lee, D. K. (2008). Forest formation of the Philippines. In *ASEAN – Korea environmental cooperation unit (AKECU)*. GeoBook Publishing Co.

Fröhlich, J., & Hyde, K. D. (1999). Biodiversity of palm fungi in the tropics: Are global fungal diversity estimates realistic? *Biodiversity and Conservation, 8*(7), 977–1004. https://doi.org/10.1023/A:1008895913857

Gadd, G. M. (Ed.). (2001). *Fungi in bioremediation (No. 23)*. Cambridge University Press.

Gao, F. K., Dai, C. C., & Liu, X. Z. (2010). Mechanisms of fungal endophytes in plant protection against pathogens. *African Journal of Microbiology Research, 4*(13), 1346–1351.

Gessner, M. O., Swan, C. M., Dang, C. K., McKie, B. G., Bardgett, R. D., Wall, D. H., & Hättenschwiler, S. (2010). Diversity meets decomposition. *Trends in Ecology and Evolution, 25*(6), 372–380. https://doi.org/10.1016/j.tree.2010.01.010

Gevania, D. T., & Pampolina, N. M. (2009). Plant diversity and carbon storage of a *Rhizophora* stand in verde passage, san juan, Batangas, Philippines. *Journal of Environmental Science and Management, 12*, 1–10.

Gonthier, P., & Thor, M. (2013). Annosus root and butt rots. In *Infectious forest diseases* (pp. 128–158). CAB International Wallingford (UK. https://doi.org/10.1079/9781780640402.0128

Goodell, B., Qian, Y., & Jellison, J. (2008). Fungal decay of wood: Soft rot-brown rot-white rot. *ACS Symposium Series, 982*, 9–31. https://doi.org/10.1021/bk-2008-0982.ch002. ACS Publications.

Guerrero, J. J. G., General, M. A., & Serrano, J. E. (2018). Culturable foliar fungal endophytes of mangrove species in Bicol Region, Philippines. *Philippine Journal of Science, 147*(4), 563–574.

Guerrero, J. J., Banares, E. N., General, M. A., & Imperial, J. T. (2020). Rapid survey of macro-fungi within an urban forest fragment in Bicol, eastern Philippines. *Austrian Journal of Mycology, 28*, 37–43.

Guillaumin, J. J., & Legrand, P. (2013). Armillaria root rots. In *Infectious forest diseases* (pp. 159–177). CAB International Wallingford (UK. https://doi.org/10.1079/9781780640402.0159

References

Halme, P., Heilmann-Clausen, J., Rämä, T., Kosonen, T., & Kunttu, P. (2012). Monitoring fungal biodiversity — towards an integrated approach. *Fungal Ecology, 5*(6), 750—758. https://doi.org/10.1016/j.funeco.2012.05.005

Hamann, A., Barbon, E. B., Curio, E., & Madulid, D. A. (1999). A botanical inventory of a submontane tropical rainforest on Negros Island, Philippines. *Biodiversity & Conservation, 8*, 1017—1031.

Hawkes, C. V., Kivlin, S. N., Rocca, J. D., Huguet, V., Thomsen, M. A., & Suttle, K. B. (2011). Fungal community responses to precipitation. *Global Change Biology, 17*(4), 1637—1645. https://doi.org/10.1111/j.1365-2486.2010.02327.x

Hawksworth, D. L. (2004). Fungal diversity and its implications for genetic resource collections. *Studies in Mycology, 50*(1), 9—17.

Hipol, R. M., Tolentino, D. B., Fernando, E. S., & Dadiz, N. M. (2007). Life strategies of mosses in Mt. Pulag, Benguet province, Philippines. *Philippine Journal of Science, 136*, 11—18.

Ingle, N. R. (2003). Seed dispersal by wind, birds, and bats between Philippine montane rainforest and successional vegetation. *Oecologia, 134*, 251—261.

Inglis, P. W., & Tigano, M. S. (2006). Identification and taxonomy of some entomopathogenic Paecilomyces spp. (Ascomycota) isolates using rDNA-ITS sequences. *Genetics and Molecular Biology, 29*(1), 132—136. https://doi.org/10.1590/S1415-47572006000100025

Irga, P. J., Dominici, L., & Torpy, F. R. (2020). The mycological social network a way forward for conservation of fungal biodiversity. *Environmental Conservation, 47*(4), 243—250. https://doi.org/10.1017/S0376892920000363

Jacob, J. K. S., Romorosa, E. S., & Kalaw, S. P. (2017). Species listing of macroscopic fungi in Isabela State University, Isabela as baseline information. *International Journal of Agricultural Technology, 13*(7.1), 1199—1203.

Jacobsen, R. M., Kauserud, H., Sverdrup-Thygeson, A., Bjorbækmo, M. M., & Birkemoe, T. (2017). Wood-inhabiting insects can function as targeted vectors for decomposer fungi. *Fungal Ecology, 29*, 76—84. https://doi.org/10.1016/j.funeco.2017.06.006

Jauhiainen, J., Takahashi, H., Heikkinen, J. E., Martikainen, P. J., & Vasander, H. (2005). Carbon fluxes from a tropical peat swamp forest floor. *Global Change Biology, 11*, 1788—1797.

Jones, E. G., Uyenco, F. R., & Follosco, M. P. (1988). Fungi on driftwood collected in the intertidal zone from the Philippines. *Asian Marine Biology, 5*, 103—106.

Kambach, S., Sadlowski, C., Peršoh, D., Guerreiro, M. A., Auge, H., Röhl, O., & Bruelheide, H. (2021). Foliar fungal endophytes in a tree diversity experiment are Driven by the identity but not the diversity of tree species. *Life, 11*(10), 1081. https://doi.org/10.3390/life11101081

Keong, C. Y. (2015). Sustainable resource management and ecological conservation of mega-biodiversity: The Southeast Asian big-3 reality. *International Journal of Environmental Science and Development, 6*(11), 876.

Kim, D. H., Ha, N. M., Manalo, M. M. Q., Baldovino, M., & Lee, J. K. (2021). Checklist of mushrooms of Mt. Cleopatra needle forest reserve in Palawan island, Philippines. *Journal of Korean Society of Forest Science, 110*, 289—294. https://doi.org/10.14578/jkfs.2021.110.2.289

Kottke, I., & Oberwinkler, F. (1986). M y c o r r h i z a o f forest trees — structure and f u n c t i o n *. *Trees Structure and Function*, 1—24.

Kowalczyk, J. E., Benoit, I., & De Vries, R. P. (2014). Regulation of plant Biomass utilization in Aspergillus. *Advances in Applied Microbiology, 88*, 31—56. https://doi.org/10.1016/B978-0-12-800260-5.00002-4

Kubartová, A., Ottosson, E., & Stenlid, J. (2015). Linking fungal communities to wood density loss after 12 years of log decay. *FEMS Microbiology Ecology, 91*(5), 1—11. https://doi.org/10.1093/femsec/fiv032

Lapitan, P. G., Castillo, M. L., Dolom, P. C., Villanueva, T. R., Peralta, E. O., Eleazar, P. M., Pulhin, J. M., Balahadia, N. M., Balatibat, J. B., Calimag, C. A., Bantayan, N. C., Pampolina, N. M., & Mora, A. M. (2018). *Science-based management and upland community development in the Philippines*. UP Press. Diliman Quezon city Philippines.

Lapitan, P. G., Fernando, E., Suh, M. H., Fuentes, R. U., Shin, Y. K., Pampolina, N. M., Castillo, M. L., Cereno, R. P., Lee, J. H., Han, S., Choi, T. B., & Lee, D. K. (2010). *Biodiversity and natural resources conservation in protected areas of Korea and the Philippines*. AKECU. GeoBook.

Leonard, P. L. (2010). *A guide to collecting and preserving fungal specimens for the Queensland herbarium*. Queensland Herbarium, Department of Environment and Resource Management.

Li, S., Huang, X., Shen, J., Xu, F., & Su, J. (2020). Effects of plant diversity and soil properties on soil fungal community structure with secondary succession in the Pinus yunnanensis forest. *Geoderma, 379*(August), 114646. https://doi.org/10.1016/j.geoderma.2020.114646

I. Fungi in aquatic and terrestrial habitats

Lillo, E. P., Fernando, E. S., & Lillo, M. J. R. (2019). Plant diversity and structure of forest habitat types on Dinagat Island, Philippines. *Journal of Asia-Pacific Biodiversity, 12*, 83–105.

Liu, G. Y., Chen, L. L., Shi, X. R., Yuan, Z. Y., Yuan, L. Y., Lock, T. R., & Kallenbach, R. L. (2019). Changes in rhizosphere bacterial and fungal community composition with vegetation restoration in planted forests. *Land Degradation and Development, 30*(10), 1147–1157. https://doi.org/10.1002/ldr.3275

Liwanag, J. M. G., Santos, E. E., Flores, F. R., Clemente, R. F., & Dulay, R. M. R. (2017). Species listing of macrofungi in Angat watershed reservation, Bulacan province, Luzon island, Philippines. *International Journal of Biology, Pharmacy and Allied Sciences, 6*(5), 1060–1068.

Lodge, D. J., & Asbury, C. E. (1988). Basidiomycetes reduce export of organic matter from forest slopes. *Mycologia, 80*(6), 888–890. https://doi.org/10.1080/00275514.1988.12025745

Lonsdale, D., Pautasso, M., & Holdenrieder, O. (2008). Wood-decaying fungi in the forest: Conservation needs and management options. *European Journal of Forest Research, 127*(1), 1–22. https://doi.org/10.1007/s10342-007-0182-6

Malabrigo, P. L. (2013). Vascular flora of the tropical montane forests in Balbalasang-Balbalan national Park, Kalinga province, northern Luzon, Philippines. *Asian Journal of Biodiversity, 4*, 1–22.

Malabrigo, P. L., Jr., Pampolina, N. M., Balatibat, J. B., Tinio, C. E., Aguilon, D. J., Tingzon, K., Labatos, B. V., Umali, A. G., & Tobias, A. B. (2017). Ecological assessment and monitoring of biodiversity in terrestrial and Aqua-International Journal of Biology, Pharmacy and Allied Sciencestic ecosystems in Didipio gold copper Project. In *Poster presented at: 19th International botanical congress; 2017 July 23–29; Shenzhen, China.*

Marchese, C. (2015). Biodiversity hotspots: A shortcut for a more complicated concept. *Global Ecology and Conservation, 3*, 297–309.

May, T. W., Cooper, J. A., Dahlberg, A., Furci, G., Minter, D. W., Mueller, G. M., Pouliot, A., & Yang, Z. (2019). Recognition of the discipline of conservation mycology. *Conservation Biology, 33*(3), 733–736. https://doi.org/10.1111/cobi.13228

McGuire, K. L., D'Angelo, H., Brearley, F. Q., Gedallovich, S. M., Babar, N., Yang, N., Gillikin, C. M., Gradoville, R., Bateman, C., Turner, B. L., Mansor, P., Leff, J. W., & Fierer, N. (2015). Responses of soil fungi to logging and oil palm Agriculture in Southeast Asian tropical forests. *Microbial Ecology, 69*(4), 733–747. https://doi.org/10.1007/s00248-014-0468-4

Mittermeier, R. A., Myers, N., Mittermeier, C. G., & Robles Gil, P. (1999). *Hotspots: Earth's biologically richest and most endangered terrestrial ecoregions.* CEMEX, SA, Agrupación Sierra Madre, SC.

Molina, R., Horton, T. R., Trappe, J. M., & Marcot, B. G. (2011). Addressing uncertainty: How to conserve and manage rare or little-known fungi. *Fungal Ecology, 4*(2), 134–146. https://doi.org/10.1016/j.funeco.2010.06.003

Mongkolsamrit, S., Khonsanit, A., Thanakitpipattana, D., Tasanathai, K., Noisripoom, W., Lamlertthon, S., Himaman, W., Houbraken, J., Samson, R. A., & Luangsa-ard, J. (2020). Revisiting Metarhizium and the description of new species from Thailand. *Studies in Mycology, 95*(May), 171–251. https://doi.org/10.1016/j.simyco.2020.04.001

Moron, L. S., & Lim, Y. (2018). Antimicrobial activities of crude culture extracts from mangrove fungal endophytes collected in Luzon Island, Philippines. *Philippine Science Letters, 11*, 28–36.

Musngi, R. B., Abella, E. a, & Lalap, A. L. (2005). Four species of wild Auricularia in Central Luzon , Philippines as sources of cell lines for researchers and mushroom growers. *Journal of Agricultural Technology,* 279–300.

Nacua, A. E., Pacis, H. Y. M., Manalo, J. R., Soriano, C. J. M., Tosoc, N. R. N., Padirogao, R., Clementa, K. J., & Deocaris, C. C. (2018). Macrofungal diversity in Mt. Makiling forest reserve, Laguna, Philippines: With floristic update on roadside samples in Makiling Botanic gardens (MBG). *Biodiversitas Journal of Biological Diversity, 19*(4), 1579–1585.

Niem, J. M., & Baldovino, M. M. (2015). Initial checklist of macrofungi in the karst area of Cavinti, Laguna. *Museum Publications in Natural History, 4*, 55–61.

Nottingham, A. T., Hicks, L. C., Ccahuana, A. J. Q., Salinas, N., Bååth, E., & Meir, P. (2018). Nutrient limitations to bacterial and fungal growth during cellulose decomposition in tropical forest soils. *Biology and Fertility of Soils, 54*(2), 219–228. https://doi.org/10.1007/s00374-017-1247-4

Osono, T. (2020). Functional diversity of ligninolytic fungi associated with leaf litter decomposition. *Ecological Research, 35*(1), 30–43. https://doi.org/10.1111/1440-1703.12063

Pagaduan, D. C., & Afuang, L. E. (2012). Understorey bird species diversity along elevational gradients on the northeastern slope of Mt. Makiling, Luzon, Philippines. *Asia Life Sciences Journal, 21*(2), 585–607.

Paguirigan, J. A. G., David, B. A. P., Elsisura, R. N. M. S., Gamboa, A. J. R., Gardaya, R. F. P., Ilagan, J. P. N., Mendiola, J. P. L., Pineda, P. B., Samelin, R. N., & Pangilinan, M. V. (2020). Species listing and distribution of macrofungi in Consocep mountain Resort, Tigaon and mount Isarog national Park, Goa, Camarines Sur. *Philippine Journal of Systematic Biology, 14*, 1–9. https://doi.org/10.26757/pjsb2020a14005

Pampolina, N. M., Dell, B., & Malajczuk, N. (2002). Dynamics of ectomycorrhizal fungi in a Eucalyptus globulus plantation: Effect of phosphorus fertilization. *Forest Ecology and Management, 158*, 291–304.

Pang, S. E. H., De Alban, J. D. T., & Webb, E. L. (2021). Effects of climate change and land cover on the distributions of a critical tree family in the Philippines. *Scientific Reports, 11*(1), 1–13. https://doi.org/10.1038/s41598-020-79491-9

Paris, R., Switzer, A. D., Belousova, M., Belousov, A., Ontowirjo, B., Whelley, P. L., & Ulvrova, M. (2014). Volcanic tsunami: A review of source mechanisms, past events and hazards in Southeast Asia (Indonesia, Philippines, Papua New Guinea). *Natural Hazards, 70*(1), 447–470.

Parlucha, J. A., Soriano, J. K. R., Yabes, M. D., Pampolina, N. M., & Tadiosa, E. R. (2021). Species and functional diversity of macrofungi from protected areas in mountain forest ecosystems of Southern Luzon, Philippines. *Tropical Ecology, 62*(3), 359–367. https://doi.org/10.1007/s42965-021-00152-7

Peñuelas, J., Rico, L., Ogaya, R., Jump, A. S., & Terradas, J. (2012). Summer season and long-term drought increase the richness of bacteria and fungi in the foliar phyllosphere of Quercus ilex in a mixed Mediterranean forest. *Plant Biology, 14*(4), 565–575. https://doi.org/10.1111/j.1438-8677.2011.00532.x

Perez, G. J., Comiso, J. C., Aragones, L. V., Merida, H. C., & Ong, P. S. (2020). Reforestation and deforestation in Northern Luzon, Philippines: Critical issues as observed from space. *Forests, 11*(10), 1–20. https://doi.org/10.3390/f11101071

Petersen, R. H. (1967). Notes on clavarioid fungi. VII. Redefinition of the *Clavaria vernalis-C. mucida* complex. *The American Midland Naturalist, 77*, 205–221.

Pretty, J., & Smith, D. (2004). Social capital in biodiversity conservation and management. *Conservation Biology, 18*(3), 631–638. https://doi.org/10.1111/j.1523-1739.2004.00126.x

Primavera, J., & Sadaba, R. B. (2012). *Beach forest species and mangrove associates in the Philippines*. Aquaculture Department, Southeast Asian Fisheries Development Center.

Quimio, T. H. (2001). *Common mushroom in Mt Makiling*. Museum of Natural History.

Read, D. J., & Perez-Moreno, J. (2003). Mycorrhizas and nutrient cycling in ecosystems - a journey towards relevance? *New Phytologist, 157*(3), 475–492. https://doi.org/10.1046/j.1469-8137.2003.00704.x

Robinson, S. J. B., Elias, D., Johnson, D., Both, S., Riutta, T., Goodall, T., Majalap, N., Mcnamara, N. P., Griffiths, R., & Ostle, N. (2020). Soil fungal community characteristics and mycelial production across a disturbance gradient in lowland dipterocarp rainforest in Borneo. *Frontiers in Forests and Global Change, 3*(June). https://doi.org/10.3389/ffgc.2020.00064

Rodriguez, R. J., Cullen, D., Kurtzman, C. P., Khachatourians, G. G., & Hegedus, D. D. (2004). Molecular methods for discriminating taxa, monitoring species, and assessing fungal diversity. *Biodiversity of Fungi: Inventory and Monitoring Methods*, 77–102.

Schwarze, F. W. M. R., Engels, J., & Mattheck, C. (2013). *Fungal strategies of wood decay in trees*. Springer Science & Business Media.

Sedlock, J. L., Jose, R. P., Vogt, J. M., Paguntalan, L. M. J., & Cariño, A. B. (2014). A survey of bats in a karst landscape in the central Philippines. *Acta Chiropterologica, 16*, 197–211.

Senn-Irlet, B., Heilmann-Clausen, J., Genney, D., & Dahlberg, A. (2007). *Guidance for conservation of macrofungi in Europe* (pp. 1–39). *European Council for Conservation of Fungi* (ECCF).

Sette, L. D., Pagnocca, F. C., & Rodrigues, A. (2013). Microbial culture collections as pillars for promoting fungal diversity, conservation and exploitation. *Fungal Genetics and Biology, 60*(July), 2–8. https://doi.org/10.1016/j.fgb.2013.07.004

Shi, L., Dossa, G. G. O., Paudel, E., Zang, H., Xu, J., & Harrison, R. D. (2019). Changes in fungal communities across a forest disturbance gradient. *Applied and Environmental Microbiology, 85*(12). https://doi.org/10.1128/AEM.00080-19

Smith, S. E., & Read, D. J. (2010). *Mycorrhizal symbiosis*. Elsevier Academic press.

Solis, M. J. L., Dela Cruz, T. E., Schnittler, M., & Unterseher, M. (2016). The diverse community of leaf-inhabiting fungal endophytes from Philippine natural forests reflects phylogenetic patterns of their host plant species Ficus benjamina, F. elastica and F. religiosa. *Mycoscience, 57*(2), 96–106. https://doi.org/10.1016/j.myc.2015.10.002

Tadiosa, E. R. (2012). The growth and development of mycology in the Philippines. *Fungal Conservation, 2*, 18–22.

Tadiosa, E. R., Arenas, M. C., & Reyes, R. G. (2015). Macroscopic fungi of Mts. Banahaw-san cristobal protected landscape northwestern side, with a description of nidula banahawensis sp. nov.(Basidiomycota). *Asian Journal of Biodiversity, 6*(2).

Tadiosa, E. R., & Arsenio, J. S. (2014). A taxonomic study of wood-rotting basidiomycetes at the molave forest of san fernando city, La union province, Philippines. *Asian Journal of Biodiversity, 5*(1), 92–108. https://doi.org/10.7828/ajob.v5i1.483

Tadiosa, E. R., & Briones, R. U. (2013). Fungi of Taal volcano protected landscape, southern Luzon, Philippines. *Asian Journal of Biodiversity, 4*(1).

Tang, J. W., Lü, X. T., Yin, J. X., & Qi, J. F. (2011). Diversity, composition and physical structure of tropical forest over limestone in Xishuangbanna, south-west China. *Journal of Tropical Forest Science, 23*, 425–433.

Terhonen, E., Blumenstein, K., Kovalchuk, A., & Asiegbu, F. O. (2019). Forest tree microbiomes and associated fungal endophytes: Functional roles and impact on forest health. *Forests, 10*(1), 1–32. https://doi.org/10.3390/f10010042

Tolentino, P. J., Navidad, J. R. L., Angeles, M. D., Fernandez, D. A. P., Villanueva, E. L. C., Obeña, R. D. R., & BUOT JR, I. E. (2020). Biodiversity of forests over limestone in Southeast Asia with emphasis on the Philippines. *Biodiversitas Journal of Biological Diversity, 21*, 1597–1613.

Tomao, A., Antonio Bonet, J., Castaño, C., & de-Miguel, S. (2020). How does forest management affect fungal diversity and community composition? Current knowledge and future perspectives for the conservation of forest fungi. *Forest Ecology and Management, 457*(July 2019), 117678. https://doi.org/10.1016/j.foreco.2019.117678

Tomita, F. (2003). Endophytes in Southeast Asia and Japan: Their taxonomic diversity and potential applications. *Fungal Diversity, 14*, 187–204.

Torres, J. M. O., & dela Cruz, T. E. E. (2013). Production of xylanases by mangrove fungi from the Philippines and their application in enzymatic pretreatment of recycled paper pulps. *World Journal of Microbiology and Biotechnology, 29*, 645–655.

Torres, M. L., Tadiosa, E. R., & Reyes, R. G. (2020). Species listing of macrofungi on the Bugkalot tribal community in Alfonso Castañeda, Nueva Vizcaya, Philippines. *Current Research in Environmental & Applied Mycology (Journal of Fungal Biology), 10*(1), 475–493. https://doi.org/10.5943/cream/10/1/37

Utting, P. (2000). *Towards participatory conservation: An introduction.* Ateneo de Manila University Press.

Vogel, H., Schmidtberg, H., & Vilcinskas, A. (2017). Comparative transcriptomics in three ladybird species supports a role for immunity in invasion biology. *Developmental and Comparative Immunology, 67*, 452–456. https://doi.org/10.1016/j.dci.2016.09.015

Wang, B., & Qiu, Y. L. (2006). Phylogenetic distribution and evolution of mycorrhizas in land plants. *Mycorrhiza, 16*(5), 299–363. https://doi.org/10.1007/s00572-005-0033-6

Wang, J. T., Zheng, Y. M., Hu, H. W., Zhang, L. M., Li, J., & He, J. Z. (2015). Soil pH determines the alpha diversity but not beta diversity of soil fungal community along altitude in a typical Tibetan forest ecosystem. *Journal of Soils and Sediments, 15*(5), 1224–1232. https://doi.org/10.1007/s11368-015-1070-1

Whitford, H. N. (1911). *The forests of the Philippines.* Issue 10. Bureau of Printing.

Whitmore, T. C. (1984). *Tropical rain forests of the par East.* Clarendon Press.

Wilson, D. (1995). Endophyte: The evolution of a term, and clarification of its use and Definition. *Oikos, 73*(2), 274. https://doi.org/10.2307/3545919

Yumul, G. P., Dimalanta, C. B., Maglambayan, V. B., & Marquez, E. J. (2008). Tectonic setting of a composite terrane: A review of the philippine island arc system. *Geosciences Journal, 12*(1), 7.

Zhu, H., Ashton, P., Gu, B., Zhou, S., & Tan, Y. (2021). Tropical deciduous forest in Yunnan, southwestern China: Implications for geological and climatic histories from a little-known forest formation. *Plant Diversity, 43*, 444–451.

Further reading

Kristanti, R. A., Hadibarata, T., Toyama, T., Tanaka, Y., & Mori, K. (2011). Bioremediation of crude oil by white rot fungi Polyporus sp. S133. *Journal of Microbiology and Biotechnology, 21*(9), 995–1000. https://doi.org/10.4014/jmb.1105.05047

Martin, F., Kohler, A., Murat, C., Veneault-Fourrey, C., & Hibbett, D. S. (2016). Unearthing the roots of ectomycorrhizal symbioses. *Nature Reviews Microbiology, 14*(12), 760–773. https://doi.org/10.1038/nrmicro.2016.149

Matos, A. J. F. S., Bezerra, R. M. F., & Dias, A. A. (2007). Screening of fungal isolates and properties of Ganoderma applanatum intended for olive mill wastewater decolourization and dephenolization. *Letters in Applied Microbiology, 45*(3), 270–275. https://doi.org/10.1111/j.1472-765X.2007.02181.x

Reyes, R. G., Lou, L., Lopez, M. A., Kumakura, K., & Kalaw, S. P. (2009). Coprinus comatus , a newly domesticated wild nutriceutical mushroom in the Philippines. *Journal of Agricultural Technology, 5*(2), 299–316.

Rillig, M. C., Wright, S. F., Nichols, K. A., Schmidt, W. F., & Torn, M. S. (2001). Large contribution of arbuscular mycorrhizal fungi to soil carbon pools in tropical forest soils. *Plant and Soil, 233*(2), 167–177. https://doi.org/10.1023/A:1010364221169

Romero-Silva, R., Sánchez-Reyes, A., Díaz-Rodríguez, Y., Batista-García, R. A., Hernández-Hernández, D., & Tabullo de Robles, J. (2019). Bioremediation of soils contaminated with petroleum solid wastes and drill cuttings by Pleurotus sp. under different treatment scales. *SN Applied Sciences, 1*(10), 1–9. https://doi.org/10.1007/s42452-019-1236-3

Tadiosa, E. R. (1998). Some noteworthy species of wood-rotting fungi found in the forested hills of La union province, northern Luzon, Philippines. *UST Journal of Graduate Research, 25*(2), 55–58.

Tadiosa, E. R., Arsenio, J. J., & Marasigan, M. C. (2007). Macroscopic fungal diversity of Mt. Maculot, Cuenca, Batangas, Philippines. *Journal of Nature Studies, 6*(1 & 2).

Treseder, K. K., & Allen, M. F. (2000). Mycorrhizal fungi have a potential role in soil carbon storage under elevated CO_2 and nitrogen deposition. *New Phytologist, 147*(1), 189–200. https://doi.org/10.1046/j.1469-8137.2000.00690.x

Yu, T. E., Egger, K. N., & Peterson, L. R. (2001). Ectendomycorrhizal associations - characteristics and functions. *Mycorrhiza, 11*(4), 167–177. https://doi.org/10.1007/s005720100110

Bioluminescent mushrooms of the Philippines

Carlo Oliver M. Olayta[1,2] *and*
Thomas Edison E. dela Cruz[1,3,4]

[1]The Graduate School, University of Santo Tomas, Manila, Philippines; [2]Laboratory Equipment and Supplies Office, University of Santo Tomas, Manila, Philippines; [3]Department of Biological Sciences, College of Science, University of Santo Tomas, Manila, Philippines; [4]Fungal Biodiversity, Ecogenomics and Systematics (FBeS) Group, Research Center for the Natural and Applied Sciences, University of Santo Tomas, Manila, Philippines

1. Introduction

Bioluminescence is a natural phenomenon among living organisms where they produce lights of different spectra, colors, and patterns (Deheyn & Latz, 2007). In contrast to fluorescence and phosphorescence where light production is mediated by a light-absorbing pigment (Kahlke & Umbers, 2016), bioluminescence is an enzyme-driven, biochemical reaction involving luciferase which reacts with molecular oxygen and acts on luciferin as the substrate (Weitz, 2004). Bioluminescence has been observed from a wide array of organisms, from bacteria and fungi to arthropods, mollusks, and other animals (Fig. 5.1). There are about 10,000 species from 800 genera across different kingdoms of the natural world that display natural bioluminescence (Haddock et al., 2010), but its function in many of these organisms is still quite unknown, albeit some studies suggest that it may be a form of visual communication, for attraction of preys or mates, or as a defense mechanism (Herring, 1994). Intensity of light produced by bioluminescence ranges from 400 to 720 nm and may appear violet to nearly infrared (Kahlke & Umbers, 2016). In this paper, we talk about fungal bioluminescence and listed species of bioluminescent fungi with particular emphasis on those found in Southeast Asia and particularly, the Philippines.

FIGURE 5.1 Some bioluminescent organisms. (A) Bacteria. Liquid culture of *Vibrio fischeri*, (B) Fungi. Dark photo of *Mycena lucentipes* and (C) Planktons. *Noctiluca scintillans* on waters of New Zealand. Credits: (A) Miller, S.D., Haddock, S.H., Elvidge, C.D., & Lee, T.F. (2007). Milky Seas: A new science Frontier for Nighttime Visible-Band Satellite Remote sensing. Joint EUMETSAT meteorological satellite conference and 15th satellite meteorology & oceanography conference of the American meteorological society, Amsterdam, The Netherlands, September 24–28. (B) Cassius V. Stevani, Institute of Chemistry, University of São Paulo, Brazil. (C) Richard Paul J. Yulo.

2. Bioluminescence in fungi

Early reports of bioluminescent fungi were mentioned in the papers of Desjardin et al. (2008) and Vladimir et al. (2012). They reported that the great philosopher, Aristotle (384–322 BCE), during the ancient Greek period, observed light emitted from a rotting wood, which he called then as "shining wood" or "foxfire." They also mentioned that Pliny the Elder (Gaius Plinus Secundus, 23–79 AD) described a bioluminescent white fungus found on rotting trees with a sweet taste and which possess pharmacological properties as noted in his *Historia Naturalis*. In their papers, they also mentioned that G.E. Rumph (1637–1706), a Dutch Physician, wrote in his *Herbarium Amboiense* that the native people of the Moluccas in Indonesia used bioluminescent fungi as a source of light to guide them in trekking the dark forests. The people of Micronesia used these fungi as head ornamentations during rituals and smear them on their face also to scare their enemies.

Fungal bioluminescence occurs in the mycelia, the fruiting bodies or both (Weitz, 2004). Emission of light can also occur in some stages of the fungal life cycle. The light intensity emitted by the fruiting bodies or the mycelia is said to be brighter when young unlike the aged or matured ones, although some species still emit luminescence even at maturity (Shimomura, 2012). Oftentimes, these fungi are generally saprobes, while some are pathogenic to plants (Vladimir et al., 2012). Intensity of light emitted by these fungi ranges from 520 to 530 nm (Desjardin et al., 2008). Interestingly, it was reported that fungal bioluminescence is controlled by a circadian clock (Oliveira et al., 2015).

Almost all bioluminescent systems including those in fungi are produced by a reaction when the substrate luciferin (or 3-hydroxyhispidin) is oxidized by the enzyme luciferase in the presence of molecular oxygen (Bubyrev et al., 2019; Kaskova et al., 2017). The type of luciferin-luciferase system appears to be shared by the different lineages of bioluminescent fungi (Oliveira et al., 2012). However, the function of bioluminescence in fungi is still vague, although it has been hypothesized that these fungi use bioluminescence to attract arthropods and other invertebrates for spore dispersal (Sivinski, 1981, 1998). This hypothesis is supported by the observation that some species produce bioluminescence on the lamellae or gills such as in *Panellus stipticus* (Bull.) P. Karst. and in spores of *Mycena rorida* var. *lamprospora* (= *Mycena lamprospora* (Corner) E. Horak) (Bermudes et al., 1992). The functions of bioluminescence in other stages of the fungal life cycle, for example, mycelia, would be a promising topic for research.

Bioluminescent fungi are found almost everywhere. In temperate Europe, about 40 species from 9 genera of Basidiomycetes are bioluminescent. Twenty-six of these are identified from the genus *Mycena*. We listed in Table 5.1 an early species list of bioluminescent fungi by Wassink (1978, 1979) with its updated taxonomic names. A more comprehensive list of bioluminescent fungi can be found in the paper "Fungi bioluminescence revisited" by Desjardin et al. (2008). Most of the bioluminescent fungi are found attached on dead trees or growing on decaying leaves or trunks (Deheyn & Latz, 2007; Herring, 1994; Vladimir et al., 2012). Recently, Desjardin et al. (2008) noted about 64 species of luminescent fungi. Chew, Desjardin et al. (2014) revised the number to 81 species. Vydryakova et al. (2009) also recognized over 80 species of bioluminescent fungi belonging to nine genera, all of which were described as members of the Phylum Basidiomycota. Five luminescent species of *Armillaria* were recently discovered in North America (Mihail, 2015) along with two new species from Brazil (Desjardin et al., 2016). In Japan, eight new species of bioluminescent fungi were identified by Terashima et al. (2016). Three new species from the genus *Mycena* were also discovered in Taiwan, that is, *Mycena jingyinga* C.C. Chang, C.Y. Chen, W.W. Lin & H.W. Kao, *Mycena luguensis* C.C. Chang, C.Y. Chen, W.W. Lin & H.W. Kao, and *Mycena venus* C.C. Chang, C.Y. Chen, W.W. Lin & H.W. Kao, but their luminescence appear in their mycelia rather than in their fruiting bodies (Chang et al., 2020). With these recent additions, the total number of known bioluminescent fungal species worldwide is 108 (Chang et al., 2020).

3. Bioluminescent fungi in Southeast Asia

In Southeast Asia, one of the early records of bioluminescent fungi was that of *Mycena illuminans*, which was collected by Volkens in April 1902 in the island of Java in Indonesia and was found growing on the trunk and branches of the palm *Calamus* sp. (Hennings, 1903). Malaysia had the most numerous records of bioluminescent fungi. In the inventory conducted by Corner (1954, 1994), *Mycena pruinoso-viscida*, a newly described species, and *Mycena*

140 5. Bioluminescent mushrooms of the Philippines

TABLE 5.1 Early identified luminescent fungi from the listing of Wassink (1978, 1979).

Genus	Species[a]
Armillaria	*Armillaria mellea* (Vahl) P. Kumm. *Armillaria fuscipes* Petch.
Pleurotus (syn. *Omphalotus*)	*Pleurotus olearius* (DC.) Gillet [current name: *Omphalus olearius* (DC.) Singer] *Pleurotus japonicus* Kawam. [current name: *Omphalotus japonicus* (Kawam.) Kircnm. & O.K. Mill. *Pleurotus noctilucens* (Lév.) Sacc. [current name: *Nothopanus noctilucens* (Lév.) Singer]
Panellus (syn. *Panus*)	*Panellus stipticus* (Bull.) P. Karst.
Mycena	*Mycena polygramma* (Bull.) Gray *Mycena tintinnabulum* (Paulet) Quél. *Mycena galopus* (Pers.) P. Kumm. *Mycena epipterygia* (Scop.) Gray
	Mycena sanguinolenta (Alb. & Schwein.) P. Kumm. *Mycena dilatata* (Fr.) Gillet [current name: *Mycena stylobates* (Pers.) P. Kumm.] *Mycena stylobates* (Pers.) P. Kumm. *Mycena zephirus* (Fr.) P. Kumm.
	Mycena parabolica (Fr.) Quél. *Mycena galericulata* (Scop.) Gray
	Mycena avenacea (Fr.) Quél. [current name: *Mycena olivaceomarginata* (Massee) Massee] *Mycena illuminans* Henn. *Mycena chlorophos* (Berk. & M.A. Curtis) Sacc. *Mycena lux-coeli* Corner
	Mycena noctilucens Corner *Mycena pruinosoviscida* Corner [current name: *Roridomyces pruinosoviscidus* (Ccrner) Blanco-Dios] *Mycena sublucens* Corner [current name: *Roridomyces sublucens* (Corner) Blanco-Dios] *Mycena rorida* (Fr.) Quél. [current name: *Roridomyces roridus* (Fr.) Rexer]
	Mycena manipularis (Berk.) Sacc. [current name: *Favolus manipularis* Berk.] *Mycena pseudostylobates* Kobayasi *Mycena daisyogunensis* Kobayasi
	Mycena photogena Kominami* *Mycena microillumina* Kawam.* *Mycena yapensis* Kawam.* *Mycena citronella* (Pers.) P. Kumm. [current name: *Agaricus citrinellus* Pers.]
Omphalia	*Omphalia flavida* Maubl. & Rangel [current name: *Mycena citricolor* (Berk. & M.A. Curtis)]
Polyporus	*Polyporus rhipidium* Berk. [current name: *Panellus pusillus* (Pers. ex Lév.)]
Dictyopanus	*Dictyopanus pusillus* (Pers. ex Lév.) Singer [current name: *Panellus pusillus* (Pers. ex Lév.)] *Dictyopanus luminescens* Corner
	Dictyopanus gloeocystidiatus Corner [current name: *Panellus pussillus* (Pers. ex Lév.)] *Dictyopanus foliicola* Kobayasi

I. Fungi in aquatic and terrestrial habitats

TABLE 5.1 Early identified luminescent fungi from the listing of Wassink (1978, 1979).—cont'd

Genus	Species[a]
Marasmius	*Marasmius phosphorus* Kawam.*
Locellina	*Locellina noctilucens* Henn.
	Locellina illuminans Henn.

[a]*Names with asterisk (*) are considered invalid as reported by Desjardin et al. (2007) and Chang and Ju (2017). These scientific names are also not listed in MycoBank (https://www.mycobank.org/) and Index Fungorum (http://www.indexfungorum.org/names/Names.asp).*

pruinoso-viscida var. *rabaulensis*, were reported in this country. In Peninsular Malaysia, three bioluminescent fungal species were also later recorded, that is, *Neonothopanus nambi* (Speg.) R.H. Petersen & Krisai, *Pleurotus decipiens* Corner, and *Pleurotus eugrammus* var. *radicicolus* Corner (1981). In addition, other reports of Corner (1950, 1954, 1986, 1994) and later by Chew et al. (2013) identified *Filoboletus manipularis* (Berk.) Singer., *Panellus luminescens* Corner, *Panellus pusillus* (Pers. ex Lév.) Burds. & O.K. Mill., *Mycena chloropos*, *Mycena lamprospora* (Corner) E. Horak., and *Mycena illuminans* Henn. Desjardin et al. (2010) described seven species of *Mycena* as bioluminescent from specimens collected in different countries which included Malaysia, Japan, Puerto Rico, Brazil, Belize, Dominican Republic, and Jamaica. Four are new species, *Mycena luxaeterna* Desjardin, B.A. Perry & Stevani, *M. luxarboricola* Desjardin, B.A. Perry & Stevani, *M. luxperpetua* Desjardin, B.A. Perry & Lodge, and *M. silvaelucens* B.A. Perry & Desjardin while three represented new reports of luminescence in previously described species, that is, *Mycena* aff. *abieticola*, *M. aspratilis* Maas Geest. & de Meijer, and *M. margarita* (Murrill) Murrill. Of particular interest is *Mycena silvaelucens* B.A. Perry & Desjardin which abundantly grows on barks of standing dipterocarpous trees in lowland dipterocarp forests in Borneo, Malaysia. Chew, Desjardin et al. (2014) also described 15 species within a period of 3 years from 25 forests in Peninsular Malaysia. Bioluminescent fungi were found in 13 localities, four were novel: *Panellus luxfilamentus* A.L.C. Chew & Desjardin, *Mycena coralliformis* A.L.C. Chew & Desjardin, *Mycena nocticaelum* A.L.C. Chew & Desjardin, and *Mycena gombakensis* A.L.C. Chew & Desjardin. They later described four new species of bioluminescent *Mycena*, also collected from Peninsular Malaysia. These were *Mycena cahaya* A.L.C. Chew & Desjardin, *Mycena sinar* A.L.C. Chew & Desjardin, *Mycena sinar* var. *tangkaisinar* A.L.C. Chew & Desjardin, and *Mycena seminau* A.L.C. Chew & Desjardin (Chew, Tan et al., 2014).

In Singapore, Corner (1950) also reported a bioluminescent fungus, *Dictyopanus luminescens* Corner, which was found on dead leaves of palm *(Rhapis flabelliformis* L'Hér, syn. = *Rhapis excelsa* (Thunb.) A. Henry.) in Singapore Botanical Garden. *Mycena rorida* var. *lamprospora* Corner was recorded on dead leaves and twigs, but its luminescence was noted only on its fresh damp spores seen around the base of the stalk. *Mycena illuminans* was also recorded in Singapore growing on dead palm leaves and trunks (Corner, 1954). Other species of bioluminescent fungi from Singapore were also reported in social media. For example, a Facebook group in Singapore posted photographs of bioluminescent mushrooms found along the trails of Bukit Batok Nature Park (Fig. 5.2). Citizen scientists are important collaborators or partners for documenting our biodiversity, including the elusive bioluminescent mushrooms.

FIGURE 5.2 Bioluminescent mushrooms from Singapore as reported by citizen scientists. *Omphalotus nidiformis* (Berk.) O.K. Mill., with light (A) and in darkness (B) (photo credit: Elmer Gono) and *Mycena manipularis*, with light (C) and in darkness (D) (photo credit: John Wong). Both species were photographed in Bukit Batok National Park, Singapore.

In Indonesia, one of the early records of bioluminescent fungi was that of G. E. Rumph who reported that the people of Amboine (Moluccas) used luminescent fruiting bodies as a lamp to guide their way in dark forests (Desjardin et al., 2008). It was also used as decorations for their ritual dances (Vladimir et al., 2012). Corner (1954) also described a novel species of *Mycena*, *Mycena sublucens*, in Ambon Island, with luminous stipe and lamellae. Other species recorded by Corner (1954) included *Mycena illuminans* which was found on dead palm leaves and trunks or on wood in the island of Java, and *Mycena manipularis* which grew on sticks, fallen trunks and stumps in the islands of Sumatra, Borneo, and in the volcanic island of Krakatoa.

Among the countries in Southeast Asia, Thailand holds the most numerous records of fungal species (Hyde et al., 2018). However, of these fungi, none were described to be bioluminescent. In Vietnam, *Mycena chloropohos* was recorded in Cattien National Park in Dongnai Province, Southern Vietnam (Tham, 2014) while *Neonothopanus nambi* (Speg.) R.H. Petersen & Krisai was collected in a rainforest in Southern Vietnam which was eventually used for the study on modulation of fungal bioluminescence (Kaskova et al., 2017). For other countries in Southeast Asia, that is, Brunei, Cambodia, Laos, and Myanmar including Timor Leste, there are no records of bioluminescent fungi. However, it is possible that these countries also hold a vast variety of bioluminescent fungi as seen with other countries in the Indochina region.

4. Bioluminescent fungi in the Philippines

For the Philippines, only one species of bioluminescent fungi was so far recorded. In the study of Corner (1954), *Mycena manipularis* was noted in the Philippine Island, albeit no indication of the exact locality. The second author also personally observed the presence of bioluminescent mushrooms along the trails of Mt. Makiling in Laguna. This observation was also confirmed by two professors from UST, Dr. Irineo Dogma Jr and Dr. Gina Dedeles, during their visits to Mt. Makiling. There were also sightings of bioluminescent fungi in Masungi Georeserve in Rizal Province as attested by the park management. Records of *Mycena manipularis* from a farm in Agusan del Sur and Batangas were noted from posts in social media. It is very clear that the full records of bioluminescent fungi in the Philippines are nonexistent. This scenario certainly merits the need for taxonomic expeditions to document fungal species including bioluminescent mushrooms in many of our forest habitats. One of the primary motivations for writing this book chapter is the need to raise awareness on the many groups of fungi that remained undocumented in the country despite their reported presence in social media and/or by personal observations. While macrofungi are constantly being documented in the country, particularly those utilized by our indigenous communities (de Leon et al., 2012, 2013, 2018, 2019), many fungal groups remained understudied, and certainly requires more attention from Filipino mycologists.

5. Uses and applications of bioluminescent fungi

The function of bioluminescence in fungi had not been clearly elucidated unlike the bioluminescence in other organisms such as arthropods, chordates, and bacteria. However, it was hypothesized that bioluminescence in fungi is a by-product of a biochemical reaction such as the degradation of lignin (lignolysis) where the bioluminescent fungus *Panellus* detoxifies itself from peroxides produced during this process (Bermudes et al., 1992; Lingle, 1993). It has also been shown that the production of light by fungi is a way for the fungus to release energy in the form of light instead of heat as a by-product of an enzyme-mediated biochemical reaction (Fox, 2000; Herring, 1994). There are also assumptions about their ecological significance. For example, bioluminescence is thought to be as a photo-signal to some arthropods to attract them to the fruiting bodies, thereby dispersing the fungal spores in the environment like the function of the foul-smell produced by the stink horn fungus (*Phallus impudicus* L.) which attracts flies. In the experiment of Sivinski (1981), arthropods were more numerous on test tubes containing the bioluminescent mycelia of *Mycena* in forest floor litter. Similarly, Oliviera et al. (2015) also demonstrated the attraction of insects to acrylic models of the mushroom *Neonothopanus gardneri* (Berk. ex Gardner) Capelari, Desjardin, Perry, Asai & Stevani which emitted 530 nm LED light. The result of their study showed that more insects were

144   5. Bioluminescent mushrooms of the Philippines

trapped on the glue-coated acrylic model of the mushroom when the light is on. Fungi-feeding insects are said to be phototactic to light emissions ranging from 300 to 650 nm wavelengths (Jess & Bingham, 2004). With these studies, bioluminescent fungi had more advantages in terms of spore dispersal since insects, particularly those that are capable of flight, can carry their spores across distance (Desjardin et al., 2008; Weinstein et al., 2016). In contrast, Sivinski (1981) stated that bioluminescence could be a defense mechanism of the fungi against nocturnal insects that feed on their fruiting bodies or mycelia. Further studies are needed to support this hypothesis.

Bioluminescence has also its societal use. For example, bioluminescent bacteria have been used to evaluate toxicity of pollutants (Fernández-Piñas et al., 2014; Girotti et al., 2008). Similarly, bioluminescent fungi have also been used in toxicity testing. For example, the naturally bioluminescent fungi, *Armillaria mellea* and *Mycena citricolor*, were used as a novel bioluminescence-based bioassay for toxicity testing of 3,5-dichlorophenol (3,5-DCP), pentachlorophenol (PCP), copper and zinc (Weitz et al., 2002). The bioluminescent basidiomycete fungus *Gerronema viridilucens* Desjardin, Capelari & Stevani was also used to assess the acute toxicity of metal cations, including Na^+, K^+, Li^+, Ca^{2+}, Mg^{2+}, Co^{2+}, Zn^{2+}, Ni^{2+}, Mn^{2+}, Cd^{2+}, and Cu^{2+} (Mendes & Stevani, 2010). The bioluminescent fungi, *Panellus stypticus* and *Omphalotus olearius*, were also antagonized by *Trichoderma harzianum*, resulting in reduced light production (Bermudes et al., 1991). While the assay looked at the ability of *Trichoderma* to inhibit growth of the bioluminescent fungi, it can also be used as potential sensor for detection of bioactivities. If, however, bioluminescent fungi will be used for biomonitoring of toxic pollutants, it is important to consider the effect of culture conditions on its growth and luminescence (Weitz et al., 2001).

6. Concluding remarks

Knowledge on the taxonomy, diversity, and distribution of bioluminescent fungi in the Philippines is generally lacking. This is a sad scenario given reports and observations of the presence of these organisms in the country. If bioluminescent fungi will be observed and documented in many of our forest habitats, these fungi can later be explored for their beneficial properties, for example, as sources of bioactive metabolites for drug discovery or for the development of biosensors against environmental pollutants.

Acknowledgments

The authors acknowledge the numerous authors whose papers are cited in this paper, and to those who gave permission to use their photos: Mr. Elmer Gono and Mr. John Wong for the bioluminescent mushrooms from Singapore and to Richard Paul J. Yulo for the bioluminescent dinoflagellates from the waters of New Zealand.

References

Bermudes, D., Boraas, M. E., & Nealson, K. N. (1991). *In vitro* antagonism of bioluminescent fungi by *Trichoderma harzianum. Mycopathologia, 115,* 19−29.

Bermudes, D., Petersen, R. H., & Nealson, K. H. (1992). Low-level bioluminescence detected in *Mycena haematopus* basidiocarps. *Mycologia, 84*(5), 799−802.

References

Bubyrev, A. I., Tsarkova, A. S., & Kaskova, Z. M. (2019). Optimization of fungal luciferin synthesis. *Russian Journal of Bioorganic Chemistry, 45*(2), 183–185.

Chang, C. C., Chen, C. Y., Lin, W. W., & Kao, H. W. (2020). *Mycena jingyinga, Mycena luguensis,* and *Mycena venus*: Three new species of bioluminescent fungi from Taiwan. *Taiwania, 65*(3), 396–406.

Chang, Y. Y., & Ju, Y. M. (2017). Small agarics in Taiwan: *Mycena albopilosa* sp. nov. and *Gloiocephala epiphylla. Botanical Studies, 58*(1), 19. https://doi.org/10.1186/s40529-017-0173-y

Chew, A. L. C., Desjardin, D. E., Tan, Y., Musa, M. Y., & Sabaratnam, V. (2014). Bioluminescent fungi from Peninsular Malaysia—a taxonomic and phylogenetic overview. *Fungal Diversity, 70,* 149–187.

Chew, A. L. C., Tan, Y. S., Desjardin, D. E., Musa, M. Y., & Sabaratnam, V. (2013). Taxonomic and phylogenetic re-evaluation of *Mycena illuminans. Mycologia, 105,* 1325–1335.

Chew, A. L. C., Tan, Y.-S., Desjardin, D. E., Musa, M. Y., & Sabaratnam, V. (2014). Four new bioluminescent taxa of *Mycena* sect. Calodontes from Peninsular Malaysia. *Mycologia, 106*(5), 976–988.

Corner, E. J. H. (1950). Descriptions of two luminous tropical agarics (*Dictyopanus* and *Mycena*). *Mycologia, 42*(3), 423–431.

Corner, E. J. H. (1954). Further descriptions of luminous agarics. *Transactions of the British Mycological Society, 37*(3), 256–271.

Corner, E. J. H. (1981). The Agaric Genera Lentinus, Panus, and Pleurotus *with particular reference to Malaysian species* (p. 169). Nova Hedwigia.

Corner, E. J. H. (1986). The agaric genus *Panellus* Karst. (including *Dictyopanus* Pat.) in Malaysia. *Gard Bull Singapore, 39,* 103–147.

Corner, E. J. H. (1994). Agarics in Malesia I. Tricholomatoid II. Mycenoid. *Beih* Nova Hedwigia, *109,* 1–271.

De Leon, A. M., Cruz, A. S., Evangelista, A. B. B., Miguel, C. M., Pagoso, E. J. A., dela Cruz, T. E. E., Nelsen, D. J., & Stephenson, S. L. (2019). Species listing of macrofungi found in the Ifugao indigenous community in Ifugao Province, Philippines. *The Philippine Agricultural Scientist, 102*(2), 118–131.

De Leon, A. M., Fermin, S. M. C., Rigor, R. P. T., Kalaw, S. P., dela Cruz, T. E. E., & Stephenson, S. L. (2018). Ethnomycological report on the macrofungi utilized by the indigenous community in the Ifugao Province, Philippines. *The Philippine Agricultural Scientist, 101*(2), 194–205.

De Leon, A. M., Luangsa-ard, J. J. D., Karunarathna, S. C., Hyde, K. D., Reyes, R. G., & dela Cruz, T. E. E. (2013). Species listing, distribution, and molecular identification of macrofungi in six Aeta tribal communities in Central Luzon, Philippines. *Mycosphere, 4*(3), 478–494.

De Leon, A. M., Reyes, R. G., & dela Cruz, T. E. E. (2012). An ethnomycological survey of the macrofungi utilized by the Aeta communities in Central Luzon, Philippines. *Mycosphere, 3*(2), 251–259.

Deheyn, D. D., & Latz, M. I. (2007). Bioluminescence characteristics of a tropical terrestrial fungus (Basidiomycetes). *Luminescence, 22,* 462–467.

Desjardin, D. E., Capelari, M., & Stevani, C. (2007). Bioluminescent *Mycena* species from São Paulo, Brazil. *Mycologia, 99*(2), 317–331.

Desjardin, D. E., Oliveira, A. G., & Stevani, C. V. (2008). Fungi bioluminescence revisited. *Photochemical and Photobiological Sciences, 7*(2), 170–182.

Desjardin, D. E., Perry, B. A., Lodge, D. J., Stevani, C. V., & Nagasawa, E. (2010). Luminescent *Mycena*: New and noteworthy species. *Mycologia, 102*(2), 459–477.

Desjardin, D. E., Perry, B. A., & Stevani, C. V. (2016). New luminescent mycenoid fungi (Basidiomycota, Agaricales) from São Paulo state, Brazil. *Mycologia, 108,* 1165–1174.

Fernández-Piñas, F., Rodea-Palomares, I., Leganés, F., González-Pleiter, M., & Angeles Muñoz-Martín, M. (2014). Evaluation of the ecotoxicity of pollutants with bioluminescent microorganisms. *Advances in Biochemical Engineering/Biotechnology, 145,* 65–135.

Fox, R. T. V. (2000). Biology and life cycle. In *Armillaria Root Rot: Biology and control of honey fungus* (pp. 3–43). Intercept Ltd.

Girotti, S., Ferri, E. N., Fumo, M. G., & Maiolini, E. (2008). Monitoring of environmental pollutants by bioluminescent bacteria. *Analytica Chimica Acta, 608*(1), 2–29.

Haddock, S. H. D., Moline, M. A., & Case, J. F. (2010). Bioluminescence in the sea. *Annual Review of Marine Science, 2,* 443–493.

Hennings, P. (1903). Ein Stark phphoreszierender javanischer Agaricus (*Mycena illuminans* P. Henn. n. sp.). *Nova Hedwigia, 42,* 309–310.

I. Fungi in aquatic and terrestrial habitats

Herring, P. J. (1994). Luminous fungi. *Topics in Catalysis, 8*(4), 181–183.

Hyde, K., Norphanphoun, C., Chen, J., Dissanayake, A., Doilom, M., Hongsanan, S., Jayawardena, R., Jeewon, R., Perera, R. H., Thongbai, B., Wanasinghe, D. N., Wisitrassameewong, K., Tibpromma, S., & Stadler, M. (2018). Thailand's amazing diversity: Up to 96% of fungi in northern Thailand may be novel. *Fungal Diversity, 93*, 215–239.

Jess, S., & Bingham, J. (2004). The spectral specific responses of *Lycoriella ingenua* and *Megaselia halterata* during mushroom cultivation. *The Journal of Agricultural Science, 142*, 421–430.

Kahlke, T., & Umbers, K. D. L. (2016). Bioluminescence. *Current Biology, 26*(8), 313–314.

Kaskova, Z. M., Dörr, F. A., Petushkov, V. N., Purtov, K. V., Tsarkova, A. S., Rodionova, N. S., Mineev, K. S., Guglya, E. B., Kotlobay, A., Baleeva, N. S., Baranov, M. S., Arseniev, A. S., Gitelson, J. I., Lukyanov, S., Suzuki, Y., Kanie, S., Pinto, E., Di Mascio, P., Waldenmaier, H. E., Pereira, T. A., Carvalho, R. P., Oliveira, A. G., Oba, Y., Bastos, E. L., Stevani, C. V., & Yampolsky, I. V. (2017). Mechanism and color modulation of fungal bioluminescence. *Science Advances, 3*(4), Article e1602847. https://doi.org/10.1126/sciadv.1602847

Lingle, W. L. (1993). Bioluminescence and ligninolysis during secondary metabolism in the fungus *Panellus Journal of Bioluminescence and Chemiluminescence, 8*, 100.

Mendes, L. F., & Stevani, C. V. (2010). Evaluation of metal toxicity by a modified method based on the fungus *Gerronema viridilucens* bioluminescence in agar medium. *Environmental Toxicology and Chemistry, 29*(2), 320–326.

Mihail, J. D. (2015). Bioluminescene patterns among North America Armillaris species. *Fungal Biology, 119*(6), 528–537.

Miller, S. D., Haddock, S. H., Elvidge, C. D., & Lee, T. F. (2007). Milky Seas: A new science Frontier for Nighttime Visible-Band Satellite Remote sensing. In *Joint EUMETSAT meteorological satellite conference and 15th satellite meteorology & oceanography conference of the American meteorological society, Amsterdam, The Netherlands, September 24–28*.

Oliveira, A. G., Desjardin, D. E., Perry, B. A., & Stevani, C. V. (2012). Evidence that a single bioluminescent system is shared by all known bioluminescent fungal lineages. *Photochemical and Photobiological Sciences, 11*(5), 848–852.

Oliveira, A. G., Stevani, C. V., Waldenmaier, H. E., Viviani, V., Emerson, J. M., Loros, J. J., & Dunlap, J. C. (2015). Circadian control sheds light on fungal bioluminescence. *Current Biology, 25*(7), 964–968.

Shimomura, O. (2012). Chapter 10: Luminous fungi. In *Bioluminescence: Chemical principles and methods (revised edition)*. World Scientific.

Sivinski, J. M. (1981). Arthropods attracted to luminous fungi. *Psyche, 88*, 383–390.

Sivinski, J. M. (1998). Phototropism, bioluminescence, and the diptera. *Florida Entomologist, 81*, 282–292.

Terashima, Y., Takahashi, H., & Taneyama, Y. (2016). *The fungal flora in southwestern Japan: Agarics and boletes*. Tokai University Press.

Tham, L. (2014). One new luminous mushroom species for the macrofungi flora of Vietnam *Mycena chlorophos* (Berk.: Curt.) Sacc. *Academia Journal of Biology, 29*(1), 32–36.

Vladimir, S.,B., Osamu, S., & Josef, I. G. (2012). Luminescence of higher mushrooms. *Journal of Siberian Federal University. Biology, 5*(4), 331–351.

Vydryakova, G. A., Psurtseva, N. V., Belova, N. V., Pashenova, N. V., & Gitelson, J. I. (2009). Luminous mushrooms and prospects of their use. *Mikologiya I Fitopatologiya, 43*, 369–376.

Wassink, E. C. (1978). Luminescence in fungi. In P. J. Herring (Ed.), *Bioluminescence in action* (p. 171). Academic Press.

Wassink, E. C. (1979). On fungus luminescence. *Mededelingen Landbouwhogeschool Wageningen, 79*(5).

Weinstein, P., Delean, S., Wood, T., & Austin, A. D. (2016). Bioluminescence in the ghost fungus *Omphalotus nidiformis* does not attract potential spore dispersing insects. *IMA Fungus, 7*(2), 229–234.

Weitz, H. J. (2004). Naturally bioluminescent fungi. *Mycologist, 18*(1), 4–5.

Weitz, H. J., Ballard, A. L., Campbell, C. D., & Killham, K. (2001). The effect of culture conditions on the mycelial growth and luminescence of naturally bioluminescent fungi. *FEMS Microbiology Letters, 202*(2), 165–170.

Weitz, H. J., Campbell, C. D., & Killham, K. (2002). Development of a novel, bioluminescence-based, fungal bioassay for toxicity testing. *Environmental Microbiology, 4*(7), 422–429.

CHAPTER

6

Lichens in the Philippines: diversity and applications in natural product research

Thomas Edison E. dela Cruz[1,2], Jaycee Augusto G. Paguirigan[1,3] and Krystle Angelique A. Santiago[4]

[1]Department of Biological Sciences, College of Science, University of Santo Tomas, Manila, Philippines; [2]Fungal Biodiversity, Ecogenomics and Systematics (FBeS) Group, Research Center for the Natural and Applied Sciences, University of Santo Tomas, Manila, Philippines; [3]Korean Lichen Research Institute, Sunchon National University, Suncheon, Korea; [4]School of Science, Monash University Malaysia, Bandar Sunway, Selangor Darul Ehsan, Malaysia

1. Introduction

Lichen as a dual association of a fungus and an alga has been first introduced by Schwendener in 1867, as mentioned by Hawksworth and Grube (2020). This classic definition of lichens identifies the heterotrophic fungus component, the mycobiont, as the "shelter," thereby providing protection to its partner alga while the autotrophic alga, also known as the photobiont, is shown as the "food producer," and nourishes the symbiosis (Grimm et al., 2021). However, with the continuous discovery and increasing evidence of the presence of a plethora of associated fungal and bacterial components inhabiting this symbiotic organism (Muggia & Grube, 2018), the classical definition of lichens as an association between two organisms is being revised. Lichens are now re-defined as a miniature and complex ecosystem (Fig. 6.1), with an indefinite number of participants other than its major symbionts (Hawksworth & Grube, 2020; Honegger et al., 2013). Polyphasic approaches such as metagenomic, metabarcoding and/or gene sequence analysis and the traditional culture-based methods and morphological descriptions revealed this high diversity of associated bacterial

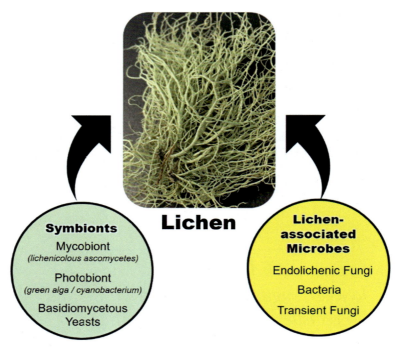

FIGURE 6.1 Lichens as an association of multiple microbial partners.

and fungal microbiota that were suggested to play important roles for the survival of lichens (Honegger et al., 2013; Noh et al., 2021; Zhang et al., 2015, 2016).

The complexity of the lichen association resulted in various studies exploring their ecological functions and screening for their economic importance. Lichens are conceivably identified as one of the early colonizers of terrestrial habitats and play a key role in soil formation (Seaward, 2008). Lichens colonizing rock surfaces enhance the chemical weathering of rocks through the production of lichen acids as they grow, thereby hastening the formation of soil (Jackson, 2015). Furthermore, lichens accumulate various elements such as nitrogen, phosphorus, and sulfur, rendering these elements available for other colonizing organisms, for example, plants (Seaward, 2008). The ability of the lichen thalli to absorb elements from the atmosphere including air pollutants led to their application as biomonitors and bioindicators of air pollution and ecological disturbances (Conti et al., 2012; Root et al., 2021; Sancho et al., 2007; Shrestha et al., 2012). Biomonitoring of air quality by lichens have been effectively done in different ecoregions of the world from temperate regions (Carreras & Pignata, 2002; Loppi et al., 2004; Zvěřina et al., 2014) to tropical countries (Bajpai et al., 2013; Fernández et al., 2011; Ng et al., 2006).

Another remarkable importance of lichens is their production of bioactive secondary metabolites. Extensive studies on the assessment, characterization and identification of valuable lichen metabolites have been done for over a decade now. In fact, an estimate of 1050 secondary metabolites is now known from lichens (Verma & Behera, 2019), which also exhibited a myriad of biological activities including antimicrobial (Jha et al., 2017; Santiago et al., 2010, 2013), antiviral (Odimegwu et al., 2019), antiinflammatory (Studzińska-sroka & Dubino 2018), cytotoxic (Hawrył et al., 2020; Tram et al., 2020), antioxidant (de Jesus et al., 2016), and herbicidal (Gazo et al., 2019) activities, to name a few.

Lichens cover approximately 6%—8% of Earth's land surface, inhabiting a vast spectrum of habitats and growing under various climatic conditions including environments that are considered extreme for many organisms such as the polar regions in the Arctic (Kristinsson et al., 2010) and the extremely dry Mojave Desert in USA (Proulx et al., 2016). Examples of ecosystems where lichens are also reported include mangrove ecosystems (Lucban & Paguirigan, 2019), tropical and temperate forests, drylands, and tundras (Asplund & Wardle, 2016; Nash, 2008). The lichen thalli in these ecosystems colonize a variety of natural substrata, including tree barks, wood, rock, soil, leaf surfaces, and shells, and other man-made substrates like walls, mortar, asbestos, glass, iron poles, plastics, and many more (Fig. 6.2). About 18,500 lichen species are recorded worldwide, with nearly 20% of known fungal species are associated with the lichen thalli (Nash, 2008). With this number, lichens contribute to our global biodiversity.

FIGURE 6.2 Lichens grow on natural habitats such as leaves (A), tree bark (B), and rock (C), and on artificial substrates including concrete statues (D), aluminum strip (E), iron fence (F), and wooden beams (G).

2. Early studies on Philippine lichens

The first reported publication on lichens in the Philippines was that of the Finnish scientist E. A. Wainio in 1909. He published *Lichenes Insularum Philippinarum* on the third issue of the Philippine Journal of Science where he listed 44 species belonging to genera *Cetraria*, *Eumitria*, *Oropogon*, *Parmelia*, *Ramalina*, *Stereocaulon*, and *Usnea*. They were all reported as new records in the Philippines. Eighteen highlighted species were new to science. The 44 species were as follows: ***Eumitria endorhodina*** Wain., ***E. endochroa*** var. ***papillata*** Wain., ***E. endochroa*** var. ***farinosa*** Wain., *Usnea florida* var. *subcomosa* (L.) Wain. (Addit. Lich. Antill.), *U. perpexlans* (Stirt.) Wain., *U. australis* Fr., *U. subinermis* Wain., ***U. pycnoclada*** Wain., ***U. philippina*** var. ***primaria*** Wain., ***U. philippina*** var. ***mearnsii*** Wain., *U. trichodea* Ach., *U. trichodea* var. *rubiginosa* Hepp., ***U. squarrosa*** Wain., *U. longissima* Ach. var. *typica* Wain., ***U. longissima*** var. ***misamensis*** (Ach.) Wain., ***U. furcata*** var. ***communis*** Wain., ***U. furcata*** var. ***marivelensis*** Wain., *Oropogon loxensis* (Fée) Th. Fr., *Ramalina pollinaria* var. *insularis* (Westr.) Ach. Wain., *R. gracilenta* var. *torulosa* (Ach.) Nyl., *R. linearis* (Sw.) Müll. Arg., *R. subfraxinea* Nyl., *R. vittata* Nyl., ***Cetraria straminea*** Wain., *Parmelia perlata* Kraempelh., *P. zollingeri* Hepp., *P. nilgherrensis* Nyl., *P. corniculans* Nyl. Fl., ***P. merrillii*** Wain. (now known as *P. merrillii* (Vain.) Hale), *P. coralloidea* (Mey. & Flot.), *P. latissima* Fée var. *cristifera* (Tayl.) Hue., ***P. clandelii*** var. ***clemensae*** (Harm.) Wain. (now known as *P. sieberi* (C.W. Dodge) A.A. Spielm & Marcelli), *P.* (sect. *irregularis*) *cetrata* Ach., ***P. manilensis*** Wain., *P. hookeri* Tayl., *P.* (sect. *sublinearis*) *americana* (Mey. & Flot.) Mont., *P. sorocheila* Wain. Lich. Nov. Rar. I, Hedwigia, *P. subdissecta* Nyl. Fl., ***P.* (div. *endoxantha*) *biformis*** Wain., ***P.* (div. *endoxantha*) *biformis*** f. ***pauaiensis*** Wain. (now known as *P. biformis* f. *pauaiensis* Vain.), ***P.* (div. *endoxantha*) *biformis*** f. ***dataensis*** Wain. (now known as *P. biformis* f. *dataensis* Vain.), *Stereocaulon nesaeum* Nyl. var. *zeorina* Wain., *Stereocaulon nesaeum* Nyl. var. *lecideoides* Wain., *S. graminosum* Schaer., and *S. arbuscula* Nyl. Prodr. Fl. Nov. Gran. This early study on Philippine lichens was followed by the several works of Herre (1946, 1950, 1951, 1963 and 1957) and Hale (1972), which reported a total of 76 lichens in the country. Gruezo (1979) in his Compendium of Philippine Lichens also mentioned some of the early lichen collections by F. J. F Meyen, a German botanist, which included nine species, five new varieties, one form, and four new records. These lichen collections are now preserved at the Botanisches Institut of the University of Kiel in Germany.

3. Taxonomic diversity of Philippine lichens

Lichens in the Philippines are equally diverse with those reported from other tropical countries in Southeast Asia like Thailand (554 lichen species) as cited by Wolseley et al. (2002), and in the Neotropics like Mexico (2722 lichen species) as records were compiled by Herrera-Campos et al. (2014). In Gruezo's Compendium of Philippine lichens (1979), a total of 789 species were reported in the country. However, recently Paguirigan et al. (2020) updated the list and recorded a total of 1234 validated species names distributed into 65 families and 229 genera. They also identified the lichen families Graphidacea and Parmeliaceae as the most species rich followed by Pilocarpaceae, Gomphillaceae, Ramalinaceae, Trypetheliaceae, Caliciaceae, Physciaceae, and Roccellaceae (Fig. 6.3). Graphidaceae are usually epiphytic

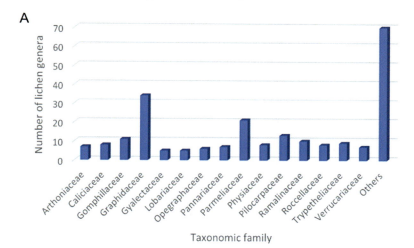

FIGURE 6.3 Distribution of lichen genera per taxonomic families. *Source: Paguirigan, J.A.G., dela Cruz, T.E.E., Santiago, K.A.A., Gerlach, A., & Aptroot, A. (2020). A checklist of lichens known from the Philippines. CREAM, 10(1): 319–376.*

organisms on trees and are widely distributed in tropical countries. For example, the genus *Graphis* alone had a total of 109 species reported in the Philippines. The study of Tabaquero et al. (2013) contributed significantly to the increase in the number of species in this genus. Other lichen genera with high number of reported species in the country belong to *Usnea* (71) *Porina* (39), *Leptogium* (38), *Parmotrema* (36), *Pyrenula* (31), *Lecidea* (26), *Lobaria* (26), and *Ocellularia* (26) (Table 6.1).

Collections of lichens in the different provinces of the country have been noted since 1909. Recent reports on lichen collections in the Philippines are studies conducted in the provinces of Bukidnon (Azuelo & Puno, 2018; Cababan et al., 2020; Timberza et al., 2017), Batangas, Laguna, and Quezon (de Jesus et al., 2016; Lucban & Paguirigan, 2019), Davao (Medina & Carreon, 2018), Kalinga (Galinato et al., 2017), Negros Oriental (Paguirigan et al., 2019), and Palawan (Sipman et al., 2013). Despite these continuous efforts to collect and identify lichens from different provinces of the country, only 32 (40%) of the 81 provinces in the country have lichen reports in the past decade (Fig. 6.4). The most notable includes provinces of Bukidnon, Ilocos Norte, Nueva Vizcaya, Palawan, and Pangasinan. Some of the recent novel species of lichens described from the Philippines are *Myriotrema subviride* Rivas Plata, Sipman & Lücking, *Ocellularia gigantospora* Rivas Plata, Sipman & Lücking, *O. leucocavata*, *O. sublaeviusculoides*, and *Thelotrema philippinum* Rivas Plata, Sipman & Lücking belonging under Thelotrematoid Graphidaceae (Rivas Plata et al., 2014).

4. Philippine lichens in natural product research

Owing to the diversity of thalli morphologies, associated microorganisms, and habitat types, it is not surprising that lichens are screened and tapped for drug discovery and other

6. Lichens in the Philippines: diversity and applications in natural product research

TABLE 6.1 List of top 27 lichen genera with corresponding number of identified species.

Lichen genera	Number of identified species
Arthonia	24
Bacidia	20
Cladonia	25
Coenogonium	23
Graphis	109
Heterodermia	15
Hypotrachyna	15
Lecanora	14
Lecidea	26
Leptogium	38
Lobaria	26
Ocellularia	26
Opegraphina	22
Pannaria	19
Parmotrema	36
Phaeographis	15
Physcia	15
Porina	39
Pseudocyphellaria	18
Pyrenula	31
Pyxine	14
Relicina	17
Sarcographa	13
Sticta	22
Strigula	19
Thelotrema	21
Usnea	71

Source: Paguirigan, J.A.G., dela Cruz, T.E.E., Santiago, K.A.A., Gerlach, A., & Aptroot, A. (2020). A checklist of lichens known from the Philippines. CREAM, 10(1): 319–376.

biopharmaceutical applications. In the Philippines, the fruticose lichen *Usnea philippina* was first reported to treat stomach pain (Quisumbing, 1951). Other studies explored Philippine lichens for antibiotic properties and other natural products. Listed in Table 6.2 are biological activities reported from Philippine lichens in the last 70 years.

I. Fungi in aquatic and terrestrial habitats

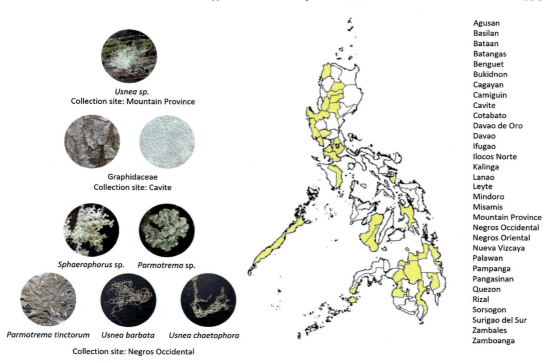

FIGURE 6.4 List of Philippine provinces with reported lichen collections from 1951 to 2020. Representative lichens taken from selected provinces were also included.

4.1 Antimicrobial activities

The crude extracts of three unidentified fruticose lichens belonging to the genera *Ramalina*, *Usnea*, and *Stereocaulon* were tested against *Streptococcus pyogenes*, the penicillin-resistant *Micrococcus pyogenes* var. *aureus*, *Bacillus subtilis*, and the acid-fast bacilli *Mycobacterium tuberculosis* 607 by Santos and colleagues (1964). Sevilla-Santos (1980) also reported the chemical constituents and potential antibiotic use of lichen *Usnea montagnei*. Other more recent studies on the antimicrobial properties of lichens are the studies of Santiago et al. (2010, 2013). Crude extracts of the lichen genera *Usnea*, *Ramalina*, *Stereocaulon*, and *Cladonia*, collected from various sites within Luzon Island displayed strong antibacterial activities (>19 mm zone of inhibition, ZOI) against common bacterial pathogens (Santiago et al., 2010). From these crude extracts, different bioactive lichen acids were identified: barbatic, stictic, diffractaic, galbinic, norstictic, salazinic, and usnic acids. Santiago et al. (2013) further studied the antimicrobial activities of the lichen *Usnea* collected from Benguet, Cavite, Ifugao, Mountain Province and Nueva Vizcaya against common bacterial pathogens including the acid-fast bacilli *Mycobacterium phlei* and *Nocardia asteroides*. Results also revealed that most of the crude extracts were active (>19 mm ZOI) against *Staphylococcus aureus*, *Bacillus subtilis*, *Mycobacterium phlei*, and *Nocardia asteroides* with the lichen metabolites usnic, norstictic, and salazinic acids as the bioactive compounds. De Jesus et al. (2016) worked on three corticolous lichens, *Parmotrema*

TABLE 6.2 Biological activities of Philippine lichens from 1951 to 2021.

Lichen taxa	Sampling site	Bioactivities	Reference
Usnea philippina	Unknown	Antibacterial	Quisumbing (1951)
Usnea sp.	Unknown	Antibacterial	Santos et al. (1964)
Ramalina sp.			
Stereocaulon sp.			
Usnea montagnei	Unknown	Antimicrobial	Sevilla-Santos (1980)
Usnea baileyi	Bataan, Batangas, Benguet	Antibacterial	Santiago et al. (2013)
Ramalina dendriscoides	Batangas, Cavite, Laguna, Quezon		
Stereocaulon massartianum	Bataan		
Cladonia gracilis			
Usnea ceratina	Ifugao	Antibacterial	Santiago et al. (2013)
U. lapponica			
U. subscabrosa	Benguet		
U. fragilescens			
Canoparmelia aptata	Quezon	Antibacterial, antioxidant, cytotoxic	de Jesus et al. (2016)
Pannaria sp.			
Parmotrema gardneri			
Usnea filipendula	Bukidnon	Antibacterial	Timbreza et al. (2017)
U. fragilescens			
Ramalina farinacea	Batangas		
Ramalina farinacea	Guimaras Island	Antimicrobial, herbicidal	Gazo et al. (2019)
R. roesleri			
R. nervulosa			
Usnea baileyi	Mountain province	Antimicrobial, antioxidant	Santiago, dela Cruz et al. (2021), Santiago, Edrada-Ebel et al. (2021)
U. bismolliuscula			
U. pectinata			
Ramalina peruviana	Cavite	Antioxidant	Galinato et al. (2021)

I. Fungi in aquatic and terrestrial habitats

gardneri, Pannaria sp., and *Canoparmelia aptata*, and were found active against *Pseudomonas aeuruginosa* and *Staphylococcus aureus*. Furthermore, the lichen *Ramalina* and *Usnea*, collected from Mt. Banoi, Batangas in Luzon and Dahilayan, Bukidnon in Mindanao, also significantly inhibited methicillin-resistant *Staphylococcus aureus* (18 mm ZOI), multi-drug resistant *Streptococcus pneumoniae* (13−19 mm ZOI) and other Gram-positive bacteria such as *S. aureus* and *B. subtilis* (10−18 mm ZOI) (Timbreza et al., 2017). The lichen *Ramalina* from Guimaras Island in Visayas were tested by Gazo et al. (2019) against bacteria and common plant pathogenic fungi such as *Fusarium oxysporum, F. solani, F. verticillioides, Colletotrichum capsica*, and *C. gleosporioides*. Strong antibacterial activities were observed against Gram-positive bacteria; however, none were observed against Gram-negative bacteria and fungi. Three species of *Usnea* collected from Sagada, Mountain Province were found active against the Gram-positive *Staphylococcus aureus* (10−18 mm ZOI) and the yeast *Candida albicans* (7−24 mm ZOI) (Santiago, dela Cruz et al., 2021; Santiago, Edrada-Ebel et al., 2021).

4.2 Cytotoxic activities

The lichen *Parmotrema gardneri* is the first and so far the only lichen in the country to be studied for its anticancer potential (de Jesus et al., 2016). In their study, de Jesus et al. (2016) showed a high inhibition against human gastric adenocarcinoma (AGS) and human lung carcinoma (A549) cell lines with IC_{50} values of 12.29 μg/mL and 20.24 μg/mL, respectively. The lichen metabolite atranorin, which was detected in this study, may have contributed to the inhibition of human cancer cell lines.

4.3 Antioxidant activities

The accumulation of reactive oxygen species, which damages cells and often causes chronic and degenerative diseases such as cancer, has been the focus of several researches, particularly the search for new antioxidants. Interestingly, the Philippine lichens were previously tapped for these metabolites, albeit limited studies have so far been done. Three corticolous lichens, *Canoparmelia aptata, Pannaria* sp., and *Parmotrema gardneri*, were among the first lichens in the country to be evaluated for their radical scavenging activity (RSA) and ferric reducing antioxidant power (FRAP) (de Jesus et al., 2016). However, results revealed low antioxidant potential (26.07−36.04% RSA, 21.75−28.29 ppm ferrous sulfate ($FeSO_4$)) in all three lichen samples, suggesting low antioxidant activities of these lichens. In another study, three *Usnea*, identified as *U. baileyi, U. bismolliuscula*, and *U. pectinata*, showed strong antioxidant activities (Santiago, dela Cruz et al., 2021). The total phenolic content (TPC), total flavonoid content (TFC), and DPPH radical scavenging activities were determined, with their TPC ranging from 14.13 to 24.18 mg Gallic Acid Equivalent/g of each extract, and TFC ranging from 1.288 to 3.957 mg Quercetin Equivalent (QE)/g of extract. Strong antioxidant activities were also observed as indicated by their IC_{50} values of 8.543−19.20 mg/mL. Similarly, the lichen *Ramalina peruviana* collected from Tagaytay, Cavite in Luzon also showed antioxidant property, reported as 45.89% RSA (Galinato et al., 2021). These studies demonstrated that different lichen species exhibit varying antioxidant potential. It was suggested, however, that the presence of the associated endolichenic fungi (ELF) residing inside the lichen may have contributed to the antioxidant activities of the lichen host (Galinato et al., 2021).

4.4 Herbicidal activities

The potential of lichens to serve as herbicidal agents was also reported in the country. In the study of Gazo et al. (2019), the crude extracts of three *Ramalina*, reported as *R. farinacea, R. nervulosa,* and *R. roesleri*, which were collected from Guimaras Island in Visayas, were evaluated as biocontrol agents against the rice weeds *Fimbristylis miliacea* and *Leptochloa chinensis*, and weedy rice (*Oryza* sp.). A decrease in shoots (up to 39%) and root lengths (up to 27%), and reduction of total chlorophyll content (up to 44%) of rice weeds and weedy rice were observed, suggesting the potential of *Ramalina* extracts as herbicidal agents for weed management. Such results can provide preliminary information for further herbicidal studies of other lichen genera, which may be tapped by agriculture-based industries for the production of biocides or herbicides.

5. Pioneering studies on endolichenic fungi from the Philippines

Another group of microorganisms inhabiting the lichen association, known as the endolichenic fungi (ELF), has recently gained attention as alternative sources of valuable compounds and as potential candidate for drug discovery. Much of the lifestyle of these nonexpressive symptom fungi resemble that of plant endophytic fungi (Santiago & Ting, 2019), although studies on their role/s in the lichen symbiosis are rather limited. In the Philippines, three lichen genera were recently explored for their associated ELF, namely, *Parmotrema, Ramalina,* and *Usnea,* and their bioactivities (Galinato et al., 2021; Santiago, dela Cruz et al., 2021; Santiago, Edrada-Ebel et al., 2021; Tan et al., 2020). Tan and colleagues (2020) isolated nine ELF inhabiting the lichen *Parmotrema rampoddense* collected from Sagada, Mountain Province, with *Fusarium proliferatum, Nemania primolutea,* and *Daldinia eschsholtzii* exhibiting the strongest antibacterial activities against *E. faecalis* (15 mm ZOI) and *S. aureus* (19 mm ZOI). These bioactivities led to the purification and identification of three secondary metabolites, bis(2-ethylhexyl)terephthalate, acetyl tributyl citrate, and fusarubin, with acetyl tributyl citrate exhibiting moderately strong activity against *Klebsiella pneumoniae, Pseudomonas aeruginosa,* and *Staphylococcus aureus.* Furthermore, Galinato et al. (2021) recently discovered 11 ELF belonging to the genera *Colletotrichum, Daldinia, Hypoxylon, Nemania, Nigrospora,* and *Xylaria* isolated from the fruticose lichen *Ramalina peruviana* collected from Tagaytay City, Cavite. A comparative study on the antioxidant potential between the associated fungi and its lichen host was done. Results revealed that ELF (RSA range = 23%−89%) exhibited greater antioxidant activities than the lichen host (RSA = 45.89%), which indicate the potential of these associated fungi as sources of antioxidants. This was further supported by the works of Santiago, dela Cruz et al. (2021) and Santiago, Edrada-Ebel et al. (2021). Three species of *Usnea, U. baileyi, U. bismolliuscula* and *U. pectinata,* collected from Sagada, Mountain Province, were explored for their ELF. A total of 101 ELF species were isolated and identified belonging to 12 genera under the classes Sordariomycetes and Eurotiomycetes (Santiago, dela Cruz et al., 2021). Crude extracts of selected ELF were found to exhibit moderately strong antimicrobial activities against *Staphylococcus aureus* (10−14 mm ZOI), *Escherichia coli* (7−11 mm ZOI), and *Candida albicans* (4−14 mm ZOI). Furthermore, these ELF crude extracts also exhibited stronger antioxidant activities ($IC_{50} = 0.57-19.63$ mg/mL) than their lichen hosts ($IC_{50} = 8.543-19.20$ mg/mL). Such findings confirmed that ELF may be a significant alternative for the slow-growing lichens as sources of novel drugs. As a continuation of the exploration of ELF from Philippine lichens, a metabolomic approach was performed on both the lichen *Usnea* and their associated ELF to understand their chemical nature (Santiago, Edrada-Ebel et al., 2021). It was revealed that ELF produces secondary metabolites distinct from those of their lichen hosts. In addition, this study led to a significant insight that ELF may have produced metabolites that protect the lichen host against "invading" or harmful microorganisms. These few studies are the pioneering researches in the country that provided initial peek on the biodiversity and metabolic profiles of ELF inhabiting Philippine lichens. With over a 1000 lichens reported from the country, a vast number of ELF species are waiting to be discovered.

6. Concluding remarks and future direction

The number of lichen species recorded for the Philippines was so far recorded as 1234 species due to the concerted effort of Filipino lichenologists in the country. Looking at the sampling localities where lichen research had been conducted, it is possible that several novel species await discovery in many of our unexplored areas in the Philippines. The fact that many novel taxa could be discovered in the country and its status as a biodiversity hotspot, this paper also raises an urgent call to study lichens in the Philippines before many of its lichen flora disappear or get extinct due to habitat loss, global climate change, and other anthropogenic activities. Thus, there is a need to support the trainings of the next generation of Filipino lichenologists and increase appreciation of lichens by the scientific community and the public in the Philippines.

The Philippine lichen flora is one of the untapped sources of bioactive compounds in the country. The medical and industrial benefits of the lichen metabolites, as presented in this chapter, clearly show that lichens are potential candidates in drug discovery, as well as agricultural applications. As such, these studies elucidate a huge opportunity to investigate the remaining unexplored Philippine lichens. In the future, several researches can be done to maximize the production of these beneficial metabolites. These include the cultivation of lichen associated microorganisms, incorporation of metabolic engineering and biotechnological approaches, and application of high-throughput technologies to harness more interesting and valuable compounds without compromising the conservation of lichens. These research improvements may provide an ultimate advantage of continuously producing lichen metabolites, which are of significance in the natural product research and other researches that contribute to the country's sustainability.

Acknowledgments

The authors acknowledge the numerous authors whose papers are cited in this paper. K.A.A.S. gratefully acknowledges the School of Science, Monash University Malaysia for financial support (CNI-000099).

References

Asplund, J., & Wardle, D. A. (2016). How lichens impact on terrestrial community and ecosystem properties. *Biological Reviews, 92*(3), 1720–1738.

Azuelo, A. G., & Puno, G. R. (2018). Moss and lichen diversity in Mt. Kalatungan Range Natural Park, Bukidnon, Philippines. *International Journal of Biosciences, 12*(3), 248–258.

Bajpai, R., Shukla, V., & Upreti, D. K. (2013). Impact assessment of anthropogenic activities on air quality, using lichen *Remototrachyna awasthii* as biomonitor. *International Journal of Environmental Science and Technology, 10*, 1287–1294.

Cababan, M. L., Memoracion, M. M., & Naive, M. A. (2020). Diversity of lichen flora in Mt. Kitanglad Range Natural Park, Kaatuan, Lantapan, Bukidnon. *Pollution, 6*(3), 481–489. https://doi.org/10.22059/poll.2019.276167.584

Carreras, H. A., & Pignata, M. L. (2002). Biomonitoring of heavy metals and air quality in Cordoba City, Argentina, using transplanted lichens. *Environmental Pollution, 117*(1), 77–87.

Conti, M. E., Finoia, M. G., Bocca, B., Mele, G., Alimonti, A., & Pino, A. (2012). Atmospheric background trace elements deposition in Tierra del Fuego region (Patagonia, Argentina), using transplanted *Usnea barbata* lichens. *Environmental Monitoring and Assessment, 184*, 527–538.

de Jesus, E. E., Hur, J. S., Notarte, K. I. R., Santiago, K. A. A., & dela Cruz, T. E. E. (2016). Antibacterial, antioxidant and cytotoxic activities of the corticolous lichens *Canoparmelia aptata*, *Pannaria* sp., and *Parmotrema gardneri* collected from Mt. Banahaw, Quezon, Philippines. *Current Research on Environmental and Applied Mycology, 6*, 173–183.

Fernández, R., Galarraga, F., Benzo, Z., Márquez, G., Fernández, A. J., Requiz, M. G., & Hernández, J. (2011). Lichens as biomonitors for the determination of polycyclic aromatic hydrocarbons (PAHs) in Caracas Valley, Venezuela. *International Journal of Environmental Analytical Chemistry, 91*(3), 230–240.

Galinato, M. G. M., Bungihan, M. E., Santiago, K. A. A., Sangvichien, E., & dela Cruz, T. E. E. (2021). Antioxidant activities of fungi inhabiting *Ramalina peruviana*: Insights on the role of endolichenic fungi in the lichen symbiosis. *Current Research on Environmental and Applied Mycology, 11*, 119–136.

Galinato, M. G. M., Mangubat, C. B., Leonor, D. S., Cababa, G. R. C., Cipriano, B. P. S., & Santiago, K. A. A. (2017). Identification and diversity of the fruticose lichen *Usnea* in Kalinga, Luzon Island, Philippines. *Current Research on Environmental and Applied Mycology, 7*(4), 249–257.

Gazo, S. M. T., Santiago, K. A. A., Tjitrosoedirjo, S. S., & dela Cruz, T. E. E. (2019). Antimicrobial and herbicidal activities of the fruticose lichen *Ramalina* from Guimaras Island, Philippines. *Biotropia, 26*, 23–32.

Grimm, M., Grube, M., Schiefelbein, U., Zühlke, D., Bernhardt, J., & Riedel, K. (2021). The lichen's microbiota, still a mystery? *Frontiers in Microbiology, 12*.

Gruezo, W. (1979). Compendium of Philippine lichens. *Kalikasan, Philippines Journal of Biology, 8*(3), 267–300.

Hale, M. E., Jr. (1972). New species of *Parmelia* (Lichenes) from India and the Philippines. *Bryologist, 75*(1), 97–101.

Hawksworth, D. L., & Grube, M. (2020). Lichens redefined as complex ecosystems. *New Phytologist, 227*, 1281–1283.

Hawryl, A., Hawryl, M., Hajnos-Stolarz, A., Abramek, J., Bogucka-Kocka, A., & Komsta, Ł. (2020). HPLC fingerprint analysis with the antioxidant and cytotoxic activities of selected lichens combined with the chemometric calculations. *Molecules, 25*, 4301.

Herre, A. W. (1946). The lichen flora of the Philippines. *Journal of the Arnold Arboretum, 27*(4), 408–412.

Herre, A. W. (1950). New lichens from California, New Mexico and the Philippines. *Bryologist, 53*(4), 296–299.

Herre, A. W. (1951). New lichens and other additions to the lichen flora of the Philippine Islands. *Bryologist, 54*(4), 283–290.

Herre, A. (1957). New records of Philippine and other tropical pacific with description of five new species. *Philippine Journal of Science, 86*, 13–34.

Herre, A. W. (1963). The lichen genus *Usnea* and its species at present known from the Philippines. *Philippine Journal of Science, 92*(1), 41–76.

Herrera-Campos, M. D. L. A., Lücking, R., Pérez-Pérez, R. E., Miranda-González, R., Sánchez, N., Barcenas-Peña, A., Carrizosa, A., Zambrano, A., Ryan, B. D., & Nash, T. H., III (2014). Biodiversity of lichens in Mexico. *Revista Mexicana de Biodiversidad, 85*, S82–S99.

Honegger, R., Axe, L., & Edwards, D. (2013). Bacterial epibionts and endolichenic actinobacteria and fungi in the Lower Devonian lichen *Chlorolichenomycites salopensis*. *Fungal Biology, 117*, 512–518.

Jackson, T. A. (2015). Weathering, secondary mineral genesis, and soil formation caused by lichens and mosses growing on granitic gneiss in a boreal forest environment. *Geoderma, 251–252*, 78–91.

References

Jha, B. N., Shrestha, M., Pandey, D. P., Bhattarai, T., Bhattarai, H. D., & Paudel, B. (2017). Investigation of antioxidant, antimicrobial and toxicity activities of lichens from high altitude regions of Nepal. *BMC Complementary Medicine and Therapies, 17*, 282.

Kristinsson, H., Zhurbenko, M., & Steen Hansen, E. (2010). *Pan-arctic checklist of lichens and lichenicolous fungi*. Conservation of Arctic Flora and Fauna Technical Report No. 20.

Loppi, S., Frati, L., Paoli, L., Bigagli, V., Rossetti, C., Bruscoli, C., & Corsini, A. (2004). Biodiversity of epiphytic lichens and heavy metal contents of *Flavoparmelia caperata* thalli as indicators of temporal variations of air pollution in the town of Montecatini Terme (central Italy). *Science of the Total Environment, 326*, 113–122.

Lucban, M. C., & Paguirigan, J. A. G. (2019). Occurrence of manglicolous lichens in Calabarzon, Philippines. *Studies in Fungi, 4*(1), 263–273.

Medina, M. N. D., & Carreon, H. G. (2018). Lichens and bryophytes in the University of Mindanao Matina Campus, Davao City, Philippines. *University of Mindanao International Multi Disciplinary Research Journal, 3*(1), 1–7.

Muggia, L., & Grube, M. (2018). Fungal diversity in lichens: From extremotolerance to interactions with algae. *Life, 8*, 15.

Nash, T. H. (2008). *Lichen Biology*. New York, USA: Cambridge University Press.

Ng, O.-H., Tan, B. C., & Obbard, J. P. (2006). Lichens as bioindicators of atmospheric heavy metal pollution in Singapore. *Environmental Monitoring and Assessment, 123*, 63–74.

Noh, H.-J., Park, Y., Hong, S. G., & Lee, Y. M. (2021). Diversity and physiological characteristics of Antarctic-lichens associated bacteria. *Microorganisms, 9*.

Odimegwu, D. C., Ngwoke, K., Ejikeugwu, C., & Esimone, C. O. (2019). Lichen secondary metabolites as possible antiviral agents. In B. Ranković (Ed.), *Lichen secondary metabolites* (pp. 199–214). Springer.

Paguirigan, J. A. G., Bernal, K. D., Dayto, C. G. A., Ramos, M. E. D., Vigo, G. D. S., & dela Cruz, T. E. E. (2019). Foliose lichens along the trails of Casaroro Falls and Pulangbato Falls in Valencia, Negros Oriental, Philippines. *Studies in Fungi, 4*(1), 43–49.

Paguirigan, J. A. G., dela Cruz, T. E. E., Santiago, K. A. A., Gerlach, A., & Aptroot, A. (2020). A checklist of lichens known from the Philippines. *Current Research on Environmental and Applied Mycology, 10*(1), 319–376.

Proulx, M. W., Knudsen, K., & Claire, L., St (2016). A checklist of Mojave Desert lichens, USA. *North American Fungi, 11*(1), 1–49.

Quisumbing, E. (1951). *Medicinal plants of the Philippines*. Bureau of Printing.

Rivas Plata, E., Sipman, H. J. M., & Lücking, R. (2014). Five new thelotremoid Graphidaceae from the Philippines. *Phytotaxa, 189*(1), 282–288.

Root, H. T., Jovan, S., Fenn, M., Amacher, M., Hall, J., & Shaw, J. D. (2021). Lichen bioindicators of nitrogen and sulfur deposition in dry forests of Utah and New Mexico, USA. *Ecological Indicators, 127*, 107727.

Sancho, L. G., Green, T. G. A., & Pintado, A. (2007). Slowest to fastest: Extreme range in lichen growth rates supports their use as an indicator of climate change in Antarctica. *Flora, 202*, 667–673.

Santiago, K. A. A., Borricano, J. N. C., Canal, J. N., Marcelo, D. M. A., Perez, M. C. P., & dela Cruz, T. E. E. (2010). Antibacterial activities of fruticose lichens collected from selected sites in Luzon Island, Philippines. *Philippine Science Letters, 3*, 18–29.

Santiago, K. A. A., dela Cruz, T. E. E., & Ting, A. S. Y. (2021). Diversity and bioactivity of endolichenic fungi in *Usnea* lichens of the Philippines. *Czech Mycology, 73*, 1–19.

Santiago, K. A. A., Edrada-Ebel, R., dela Cruz, T. E. E., Cheow, Y. L., & Ting, A. S. Y. (2021). Biodiscovery of potential antibacterial diagnostic metabolites from the endolichenic fungus *Xylaria venustula* using LC-MS-based metabolomics. *Biology, 10*, 191.

Santiago, K. A. A., Sangvichien, E., Boonpragob, K., & dela Cruz, T. E. E. (2013). Secondary metabolic profiling and antibacterial activities of different species of *Usnea* collected in Northern Philippines. *Mycosphere, 4*, 267–280.

Santiago, K. A. A., & Ting, A. S. Y. (2019). Endolichenic fungi from common lichens as new sources for valuable bioactive compounds. In M. Akhtar, M. Swamy, & U. Sinniah (Eds.), *Natural bio-active compounds* (pp. 105–127). Springer.

Santos, P., Lat, B., & Palo, M. (1964). The antibiotic activities of some Philippine lichens. *Philippine Journal of Science, 93*, 325–335.

Seaward, M. R. D. (2008). Environmental role of lichens. In T. Nash, III (Ed.), *Lichen biology* (pp. 274–298). Cambridge University Press.

Sevilla-Santos, P. (1980). Studies of Philippine lichens III Part II. Chemical constituents and antibiotics. *Acta Manilana Series A, 19*, 130.

I. Fungi in aquatic and terrestrial habitats

Shrestha, G., Petersen, S. L., & Clair, L. L., St (2012). Predicting the distribution of the air pollution sensitive lichen species Usnea hirta. *Lichenologist, 44*, 511–521.

Sipman, H. J. M., Diederich, P., & Aptroot, A. (2013). New lichen records and a catalogue of lichens from Palawan Island, the Philippines. *Philippine Journal of Science, 142*, 199–210.

Studzińska-sroka, E., & Dubino, A. (2018). Lichens as a source of chemical compounds with anti-inflammatory activity. *Herba Polonica, 64*, 56–64.

Tabaquero, A. L., Bawingan, P., & Lücking, R. (2013). Key and checklist of Graphidaceae lichens in Kalahan Forest Reserve, Nueva Vizcaya, Philippines. *Philippine Journal of Systematic Biology, 7*, 22–38.

Tan, M., Castro, S., Oliva, P. M., Yap, R. P., Nakayama, A., Magpantay, H., & dela Cruz, T. E. E. (2020). Biodiscovery of antibacterial constituents from the endolichenic fungi isolated from Parmotrema rampoddense. *3Biotech, 10*, 212.

Timbreza, L. P., delo Reyes, J. L., Flores, C. H., Perez, R. J. L. A., Stockel, M. A., & Santiago, K. A. A. (2017). Antibacterial activities of the lichen *Ramalina* and *Usnea* collected from Mt. Banoi, Batangas and Dahilayan, Bukidnon, against multi-drug resistant (MDR) bacteria. *Austrian Journal of Mycology, 26*, 27–42.

Tram, N. T. T., Anh, D. H., Thuc, H. H., & Tuan, N. T. (2020). Investigation of chemical constituents and cytotoxic activity of the lichen *Usnea undulata*. *Vietnam Journal of Chemistry, 58*, 63–66.

Verma, N., & Behera, B. C. (2019). Future directions in the study of pharmaceutical potential of lichens. In B. Ranković (Ed.), *Lichen secondary metabolites: Bioactive properties and pharmaceutical potential* (pp. 237–260). Springer.

Wolseley, P. A., Aguirre-Hudson, B., & McCarthy, P. M. (2002). Catalogue of the lichens of Thailand. *Bulletin of the Natural History Museum Botany series, 32*(1), 13–59.

Zhang, T., Wei, X.-L., Wei, Y.-Z., Liu, H.-Y., & Yu, L.-Y. (2016). Diversity and distribution of cultured endolichenic fungi in the Ny-Ålesund region, Svalbard (high Arctic). *Extremophiles, 20*, 461–470.

Zhang, T., Wei, X.-L., Zhang, Y.-Q., Liu, H.-Y., & Yu, L.-Y. (2015). Diversity and distribution of lichen-associated fungi in the Ny-Ålesund region (Svalbard, high Arctic) as revealed by 454 pyrosequencing. *Scientific Reports, 5*, 14850.

Zvěřina, O., Láska, K., Červenka, R., Kuta, J., Coufalík, P., & Komárek, J. (2014). Analysis of mercury and other heavy metals accumulated in lichen *Usnea antarctica* from James Ross Island, Antarctica. *Environmental Monitoring and Assessment, 186*, 9089–9100.

PART II

Fungi in agriculture, health, and environment

CHAPTER

7

Plant diseases caused by fungi in the Philippines

Mark Angelo O. Balendres

Plant Pathology Laboratory, Institute of Plant Breeding, College of Agriculture and Food Science, University of the Philippines Los Baños, Laguna, Philippines

1. Introduction

The World Bank considers the Philippines as one of the most dynamic countries in the Southeast Asian region, with sustained economic growth in the last decade, despite the deep contraction the country experienced in 2020 due to the COVID-19 pandemic (The World Bank, 2021, p. 61). Agriculture, specifically crop production, played an important role in this sustained growth. The nearly two-year-long pandemic highlighted the ever-increasing importance of food security in addressing the needs of the growing human population in times of global crisis. Among the staples, rice (*palay*) remains the most produced crop commodity at 19, 294.9 mt, followed by corn (maize) at 8, 118.5 mt and cassava at 2, 607.8 mt (Philippine Statistics Authority, 2021, p. 76). Banana production reached 9, 056.1 mt for the export crop commodities, followed by pineapple (2, 702.6 mt) and mango at 739.2 mt. The vegetable production volume is still small compared to the proportion of staples and export commodities, but vegetable crops remain a vital component in the Filipino diet. For instance, eggplant, onion, and tomato were the most produced vegetables in 2020, with 242.7, 229.5, and 222.0 mt, respectively (Philippine Statistics Authority, 2021, p. 76).

The agriculture sector is not without problems and challenges. Low productivity is a long-standing problem, particularly in crops with high potential for market expansion (Brown et al., 2018). The Philippines, as an archipelago and as a country located in the pacific, also receives frequent heavy rains and typhoons that devastate crops, further limiting yield. Climate change impacts are also considered threats to efforts to increase food production, establish food safety, and achieve food security. Among the many challenges the agriculture industry faces, those caused by biotic factors (pests and diseases) can have a long-term impact on crop production and product quality. If left unmanaged and environmental conditions are

164   7. Plant diseases caused by fungi in the Philippines

favorable, pests and diseases could wipe out an entire farm and have indirect negative impacts, including loss of jobs and livelihood.

Plant diseases are significant constraints in successful and profitable crop production. Plant diseases can reduce crop yield (and subsequently reduce profit), increase the cost of production (e.g., cost of management measures and additional fertilizers), negatively affect product/harvest quality (reduction of commercial harvests), and in some cases can predispose consumers to diseased fruits and vegetables that may contain toxins, which are produced by microbes causing the diseases or pathogens. Growers are constantly faced with mitigating the detrimental impacts of plant diseases. Plant disease-causing microbes or pathogens in the tropics favor warm, humid conditions, and thus, an increase in temperature could be critical in disease development. A disease epidemic or outbreak can result in significant yield reduction and, in some cases, total yield loss. Frequent rain further aids the dissemination and spread of destructive pathogen inoculum, with moisture providing stimulant cues for rapid pathogen germination and activation. These issues are particularly important for the Philippines due to its geographic position where these conducive conditions occur year-round. Crops that host these destructive pathogens are therefore constantly under threat.

The Philippines has a long history of plant diseases with economic importance. For instance, among the earliest scientific record of fungal plant disease in the country was in 1908, when a disease called "smut" was found in sugarcane (Robinson, 1908). The disease was caused by the fungal pathogen *Ustilago scitaminea* Syd (Robinson, 1908) (syn. *Sporisorium scitamineum* Syd.) M. Piepenbr., M. Stoll & Oberw. Then, within the next 113 years, records of plant diseases have increased, with more than 50% of the reports caused by fungal pathogens. Today, more than 180 genera and more than 600 species of fungi-causing plant diseases are reported. These reports have been listed in the *Host Index of Plant Diseases in the Philippines* (Tangonan, 1999) and published papers following the book's publication.

This chapter reviews the status of mycology from the perspective of plant pathology and agriculture. This chapter will focus on true fungi (phyla Zygomycota, Ascomycota, Basidiomycota, Chytridiomycota, Blastocladiomycota, and Mucoromycota) (James, Kauff et al., 2006, James, Letcher et al., 2006; Spatafora et al., 2017), associated with diseases of economically important crops (including mushroom) in the Philippines. The Oomycetes and Plasmodiophorids were previously regarded as part of the Kingdom Fungi. Many pathogens belonging to these two groups have been reported in the country and have known significant negative impacts on their host plants, for example, *Phytophthora palmivora* E.J. Butler in cacao, *Peronosclerospora philippinensis* (W. Weston) C.G. Shaw in corn (Baker, 1916), *Pythium* spp. in various crops (Tangonan, 1999), *Plasmodiophora brassicae* Woronin in cabbage (Elayda, 1938), and *Spongospora subterranea* (Wallr.) Lagerheim in potato (Balendres & Masangcay, 2020). However, they are now known to belong to other groups outside of the true fungi. Here, also highlighted are fungal pathogens that have been reported in the last 2 decades and those reported for the first time (or in a new plant host) in the country.

2. Genera of plant pathogenic fungi in the Philippines

There are 188 genera of plant pathogenic fungi reported in the Philippines belonging to phyla Ascomycota, Basidiomycota, Blastocladiomycota, Chytridiomycota, and

II. Fungi in agriculture, health and environment

Mucoromycota (Fig. 7.1). Ascomycota makes up the greatest number of genera (77.66%) and species (79.80%). *Haematonectria* (Yago & Chung, 2011), *Neoscytalidium* (Taguiam et al., 2020b), and *Paramyrothecium* (Aumentado & Balendres, 2022) are newly reported genera in the country that were associated with diseases in citrus, dragon fruit, and eggplant, respectively. More than 600 fungal species are associated with plant diseases in the country. The genus *Cercospora* holds the greatest number of species (88) reported, followed by the genus *Colletotrichum* (30), *Fusarium* (25), *Helminthosporium* (20), and *Puccinia* (20). These genera are known to cause diseases in the above-ground parts of the plants, with some exceptions in the *Fusarium* group, where some species (e.g., *Fusarium oxysporum*) are responsible for the destructive soil-borne plant diseases.

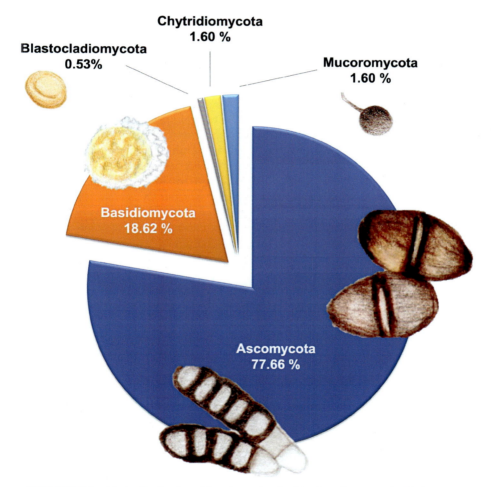

FIGURE 7.1 Phyla distribution of fungi associated with plant disease in the Philippines.

3. Plant diseases caused by fungi in Phylum Ascomycota

Almost 80% of the fungal genera and species associated with plant diseases belong to Phylum Ascomycota (Fig. 7.1, Table 7.1). This group constitutes many dreaded and destructive fungal species that reduce plant yield, affect the livelihood of the farmers, and impact the relevant crop industries. The Fusarium wilt and anthracnose are among the most difficult to control plant diseases due to the complexity and biology of the fungal pathogens causing these diseases. For instance, *Fusarium* species produce chlamydospores that enable the fungus to survive in the soil for up to 30 years in the absence of the host (Leslie & Summerell, 2006), making control measures less effective. On the other hand, *Colletotrichum* species, causative agents of anthracnose, are grouped into species complexes (multiple species per group or clade) and have the ability to enter a "quiescent" or "latent" infection stage (Cannon et al., 2007), resulting in delayed infection.

The genera with the most commonly reported incidence and impact on crop production are *Fusarium* sp., *Colletotrichum* sp. (Fig. 7.2), *Cercospora* sp., *Aspergillus* sp., *Lasiodiplodia* sp., *Alternaria* sp., *Erysiphe* sp., and *Diaporthe* sp. Infection by fungal pathogens (Fig. 7.3) usually results in leaf and fruit spots, rotting, anthracnose, wilting, blight, powdery mildew, canker, surface mold, and scab. Among the economically important plant hosts of fungal pathogens from Phylum Ascomycota in the country are rice, corn, sorghum, soybean, citrus, sugarcane, rubber, mango, peanut, banana, papaya, coffee, mungbean, tomato, pepper, cotton, cassava, eggplant, onion, and avocado. Some of the notable fungal species are *Fusarium oxysporum* f. sp. *cubense* (E.F. Sm.) W.C. Snyder and H.N. Hansen, Tropical race 4 (TR4) causing Panama wilt in banana, multiple *Colletotrichum* species causing anthracnose in vegetables and fruits, *Lasiodiplodia theobromae* (Pat.) Griffon and Maubl. (syn. *Botryodiplodia theobromae* Pat.) causing rots in fruits, *Colletotrichum falcatum* Went causing red rot in sugarcane, *Hemileia vastatrix* Berk. and Broome causing rust in coffee, and *Erysiphe polygoni* DC. causing powdery mildew in vegetables.

Panama wilt, caused by *Fusarium oxysporum* f. sp. *cubense* Tropical Race 4 (Foc TR4), remains the most destructive disease in commercial banana plantations in the country. In 2018, more than 15, 000 ha of Cavendish banana in Mindanao were killed due to Panama wilt (Lo, 2018). Foc TR4 was not found in samples from Luzon, Philippines, but another race that causes wilting in banana has been reported (Aguilar-Hawod et al., 2019). Hence, biosecurity programs to ensure that FocTR4 does not spread from Mindanao to other islands are still in place. Most recently, the weed *Eleusine indica* harbored the Foc TR4 pathogen based on an inoculation study (Catambacan & Cumagun, 2022). However, although colonized by Foc TR4, the inoculated weed plants remained asymptomatic and thus highlighted the importance of weed management in banana plantations (Catambacan & Cumagun, 2022).

Another difficult to control fungal disease caused by an ascomycete is anthracnose. *Colletotrichum* species attack the host plant (or fruit) at pre- and postharvest stages. Although any part or organ of the plant can be infected, anthracnose becomes more important when the infection occurs in fruits, the economically important part of the plant (except for leafy vegetables). A dark, necrotic, sunken lesion commonly occurs in infected tissues (Fig. 7.3) that renders the fruits less appealing to the consumer. In vegetables, an increasing incidence of pepper anthracnose has been observed. Moderate to severe infection was observed, resulting

TABLE 7.1 Genera of phytopathogenic fungi belonging to Phylum Ascomycota associated with plant diseases in the Philippines.

Genus	Some notable species	Plant-host	Disease	Reference
Acremonium	*A. psidii*	*Psidium guajava*	Root rot/wilt	Quimo (1985)
Acrocylindrium	*A. oryzae*	*Oryza sativa*	Seedborne rot	Borja (1980)
Acrosporium	*A. tingitianum*	*Citrus* sp.	Powdery mildew	Kobayashi (1995)
Aithaloderma	*A. longisetum*	*Coffea* spp.	Sooty mold	Reinking (1918)
Alternaria	*A. solani*	*Solanum lycopersicon*	Early blight	Fajardo (1934)
	A. brassicae	*Brassica oleracea* var. *capitata*	Blight	Reinking (1918)
Arthrobotrys	*Arthrobotrys* sp.	*Sorghum bicolor*	Grain mold	Dayan and Dalmacio (1982)
Ascochyta	*A. gossypii*	*Gossypium hirsitum*	Blight	PCAR (1975)
Aspergillus	*A. flavus*	*Zea mays*	Seedborne rot	Yap and Kulshreshtha (1978)
	A. niger	*Glycine max*	Seed rot/decay	Prado (1975)
Asperisorium	*A. caricae*	*Carica papayae*	Black powdery spot	Kobayashi (1995)
Asterinella	*A. atuhlmanni*	*Ananas comosus*	Leaf spot	Baker (1916)
Asteroma	*A. phaseoli*	*Phaseolus vulgaris*	Black pod/spot	Baker (1916)
Asterostomella	*A. horrida*	*Capparis microcantha*	Sooty mold	Kriengyakul and Quimio (1983)
Bakerophoma	*B. sacchari*	*Saccharum officinarum*	Leaf spot	Baker (1916)
Balansia	*B. claviceps*	*Panicum* spp.	Whip tail	Reinking (1919)
Bipolaris	*B. maydis*	*Zea mays*	Leaf spot	Ong Sotto and De Guzman (1987)
Botryodiplodia	*B. theobromae*	*Hevea brasiliensis*	Dieback	Teodoro (1926)
Botryosphaeria	*B. ribis*	*Persea americana*	Fruit rot	Teodoro (1959)
Botrytis	*B. cinerea*	*Fragaria chiloensis*	Gray mold/rot	Quimio and Sumabat (1983)
Calonectria	*C. rigidiuscula*	*Theobromae cacao*	Brown bark	Gandia and Garcia (1980)
Camptomeris	*C. albizziae*	*Paraserianthes falcataria*	Yellow leaf	Dayan (1989)
Capnodium	*C. mangifera*	*Mangifera indica*	Sooty mold	Quimio and Quimio (1974)
Cephalosporium	*C. sacchari*	*Saccharum officinarum*	Wilt	Lee (1923)
Ceratocystis	*C. fimbriata*	*Hevea brasiliensis*	Moldy rot	PCAR (1976)

(Continued)

168

7. Plant diseases caused by fungi in the Philippines

TABLE 7.1 Genera of phytopathogenic fungi belonging to Phylum Ascomycota associated with plant diseases in the Philippines.—cont'd

Genus	Some notable species	Plant-host	Disease	Reference
Cercospora	*C. capsici*	*Capsicum* sp.	Leaf spot	Orillo et al. (1959)
	C. oryzae	*Oryza sativa*	Narrow brown leafspot	Teodoro and Bogayong (1926)
	C. kikuchii	*Glycine max*	Seed rot/purple stain seed	Ilag and Marfil (1980)
Cercosporella	*C. tinosporae*	*Tinospora rumphii*	Leaf spot	Acosta (1992)
Cercosporidium	*C. personatum*	*Arachis hypogaea*	Leaf spot	Paningbatan and Ilag (1981)
Chaetomium	*C. olivaceum*	*Glycine max*	Seed rot/decay	Prado (1975)
Chaetothyrium	*C. anonicola*	*Anona maricata*	Sooty mold	Kriengyakul and Quimio (1983)
Chloridium	*C. musae*	*Musa* sp.	Leaf speckle	San Juan (1980)
Cladosporium	*C. fulvum*	*Solanum lycopersicon*	Leaf spot	Fajardo (1937)
Claviceps	*C. purpurea*	*Saccharum officinarum*	Ergot	Ocfemia (1931)
Colletotrichum	*C. musae*	*Musa* sp.	Postharvest anthracnose	Balendres et al. (2020)
	C. fructicola	*Selenicereus* sp./ *Hylocereus* sp.	Anthracnose	Evallo et al. (2021)
	C. tropicale	*Selenicereus* sp./ *Hylocereus* sp.	Anthracnose	Evallo, Taguiam, Bengca, et al. (2022)
	C. asianum	*Mangifera indica*	Anthracnose	Alvarez et al. (2020)
	C. truncatum	*Capsicum* sp.	Anthracnose	Balendres, Karlovsky, and Cumagun (2019)
	C. scovillei	*Capsicum* sp.	Anthracnose	Cueva et al. (2018)
Conidiocarpus	*Conidiocarpus* sp.	*Chrysophyllum cainito*	Sooty mold	Kriengyakul and Quimio (1983)
Coniothyrium	*C. zingiberi*	*Zingiber officinale*	Leaf spot	Stevens and Atienza (1931)
Cordana	*C. musae*	*Musa* sp.	Leaf spot	Pordesimo (1970)
Corynespora	*C. cassiicola*	*Solanum lycopersicon*	Target spot	Dimayacyac and Balendres (2021)
Cryphonectria	*C. nitschkeri*	*Eucalyptus* spp.	Canker	Kobayashi et al. (1978)
Cryptosporella	*C. viticola*	*Vitis vinifera*	Dead arm	Teodoro (1960)
Curvularia	*C. lunata*	*Citrus* sp.	Brown rot	Tangonan (1999)

II. Fungi in agriculture, health and environment

3. Plant diseases caused by fungi in Phylum Ascomycota

TABLE 7.1 Genera of phytopathogenic fungi belonging to Phylum Ascomycota associated with plant diseases in the Philippines.—cont'd

Genus	Some notable species	Plant-host	Disease	Reference
Cylindrocladium	C. scoparium	Pinus spp.	Damping-off	Tangonan (1999)
Cytospora	C. sacchari	Saccharum officinarum	Sheath rot	Reyes and Quebral (1967)
Daldinia	Daldinia sp.	Albizia falcataria	Wood decay	De Guzman (1974)
Deightoniella	D. toruluda	Musa sp.	Black spotting	Angeles (1980)
Diaporthe	D. citri	Citrus sp.	Melanose	Reyes and Orillo (1957)
Dichotomella	D. areolata	Artocarpus heterophyllus	Leaf disease	Baker (1914)
Didymella	D. caricae	Carica papayae	Fruit rot	Reinking (1918)
Dimeriella	D. sacchari	Saccharum officinarum	Red leafspot	Rivera and Cano (1966)
Diplocarpon	D. rosae	Rosa sp.	Black spot	Sacay (1939)
Diplodia	D. natalensis	Mangifera indica	Stem-end rot	Quimio and Quimio (1974)
Dothiorella	D. mali	Malus sylvestris	Fruit rot	Quimio (1983)
Dreschlera	D. oryzae	Oryza sativa	Leafspot	Mendoza and Molina (1980)
Elsinoe	E. mangiferae	Mangifera indica	Scab	Quimio and Quimio (1974)
Epicoccum	E. sorghinum	Selenicereus sp. Hylocereus sp.	Stem lesion	Taguiam et al. (2020a)
Eriosporella	E. calami	Bambusa spp.	Leafspot	Dayan (1988)
Erysiphe	E. polygoni	Vigna radiata	Powdery mildew	Reinking (1918)
Exserohilum	E. holmi	Zea mays	Leaf spot	Ong Sotto and De Guzman (1987)
Fusarium	F. oxysporum f. sp. cubense	Musa sp.	Fusarium wilt	Lee and Serrano (1920)
	F. moniliforme (F. verticillioides)	Zea mays	Kernel blast	Exconde (1980)
	F. solani	Solanum lycopersicon	Seedling rot	Teferi (1988)
Gaeumannomyces	G. graminis	Oryza sativa	Crown sheath rot	Tangonan (1999)
Gibberella	G. fujikuroi	Oryza sativa	Hear rot	Teodoro and Serrano (1926)
Gliocephalotrichum	G. bulbilium	Nephelium lappaceum	Black rot	Quimio and Abilay (1982)
Gliocladium	C. roseum	Agaricus bisporus	Mushroom rot	Quimio and Abilay (1977)
Gloeocercospora	G. sorghi	Sorghum bicolor	Zonate leaf spot	Karganilla and Elazegui (1970)

(*Continued*)

II. Fungi in agriculture, health and environment

170 7. Plant diseases caused by fungi in the Philippines

TABLE 7.1 Genera of phytopathogenic fungi belonging to Phylum Ascomycota associated with plant diseases in the Philippines.—cont'd

Genus	Some notable species	Plant-host	Disease	Reference
Gloeosporium	*G. canavaliae*	*Canavalia gladiata*	Stem canker	Baker (1914)
Glomerella	*G. musarum*	*Musa textilis*	Anthracnose	Ocfemia (1924)
Golovinomyces	*G. cichoreacearum*	*Zinnia* sp.	Powdery mildew	Balendres et al. (2021)
Guignardia	*G. bidwelli*	*Vitis vinifera*	Black rot	Teodoro (1960)
Haematonectria	*H. haematococca*	*Citrus* sp.	Twig blight	Yago and Chung (2011)
Haplobasidium	*H. musae*	*Musa* sp.	Malayan leaf spot	San Juan (1980)
Helicomina	*H. tricophila*	*Eriobatrya japonica*	Sooty mold	Kriengyakul and Quimio (1983)
Helminthosporium	*H. maydis*	*Zea mays*	Southern leaf blight	Pioneer (1986)
Heterosporium	*H. echinulatum*	*Dianthus caryophyllus*	Fairy ringspot	Teodoro (1961)
Humicola	*Humicola* sp.	*Dipterocarpus grandiflorus*	Wilt	Quinones (1980)
Hypomyces	*H. haematococca*	*Theobromae cacao*	Brown bark rot	Ocfemia and Celino (1933)
Irenopsis	*I. benguetensis*	*Ficus benjamin*	Sooty mold	Kriengyakul and Quimio (1983)
Isariopsis	*I. grisea*	*Phaseolus vulgaris*	Angular leaf spot	Fajardo (1934)
Kabatiella	*K. zeae*	*Zea mays*	Eyespot	Exconde (1980)
Lasiodiplodia	*L. theobromae*	*Musa* sp.	Fruit rot	Mortuza and Ilag (1995)
Leptosphaeria	*L. sacchari*	*Saccharum officinarum*	Ringspot	Reyes et al. (1980)
Leptostroma	*Leptostroma* sp.	*Bambusa* sp.	Leaf spot	Dayan (1988)
Leptothyrium	*L. circumscissum*	*Mangifera indica*	Angular leaf spot	Tangonan (1999)
Leptoxyphium	*Leptoxyphium* sp.	*Sorghum bicolor*	Sooty mold	Kriengyakul and Quimio (1983)
Leveillula	*L. taurica*	*Solanum lycopersicon*	Powdery mildew	Orillo et al. (1959)
Macrophoma	*M. luzonensis*	*Mangifera indica*	Gray leaf spot	Kobayashi (1995)
Macrophomina	*M. phaseoli*	*Vigna radiata*	Seedborne rot	Carreon (1989)
Macrosporium	*M. parasiticum*	*Allium* sp.	Black stalk rot	Teodoro (1923)
Marssonina	*M. rosae*	*Rosa* sp.	Black spot	Divinagracia (1980)
Melanconium	*M. sacchari*	*Saccharum officinarum*	Rind disease	Baker (1916)

II. Fungi in agriculture, health and environment

TABLE 7.1 Genera of phytopathogenic fungi belonging to Phylum Ascomycota associated with plant diseases in the Philippines.—cont'd

Genus	Some notable species	Plant-host	Disease	Reference
Meliola	*M. mangiferae*	*Mangifera indica*	Black mildew	Baker (1914)
Metasphaerella	*M. albescens*	*Oryza sativa*	Leaf scald	Ou (1973)
Monilia	*M. fructigena*	*Helianthis annuus*	Root/Stem rot	Nueda (1978)
Mycogone	*M. perniciosa*	*Agaricus bisporus*	Swollen cap/stem	Liyag (1969)
Mycosphaerella	*M. fijiensis*	*Musa* sp.	Black leaf streak	Hapitan and Reyes (1970)
Myrothecium	*M. roridum*	*Echinochloa* spp.	Seedborne rot	Angelito (1980)
Myxotrichum	*Myxotrichum* sp.	*Pinus* sp.	Damping-off	De Guzman et al. (1991)
Nectriella	*N. zingiberi*	*Zingiber officinale*	Red rhizome rot	Stevens and Atienza (1931)
Nematospora	*Nematospora* sp.	*Citrus* sp.	Dry fruit rot	Lee (1924)
Neoscytalidium	*N. dimidiatum*	*Selenicereus* sp./ *Hylocereus* sp.	Stem canker	Taguiam et al., 2020b
Nigrospora	*E. sphaerica*	*Selenicereus* sp./ *Hylocereus* sp.	Brown spot	Taguiam et al., 2020c
Oidium	*O. mangiferae*	*Mangifera indica*	Powdery mildew	Tangonan (1999)
Ophiobolus	*O. oryzinus*	*Oryza sativa*	Black sheath rot	Baker (1916)
Paramyrothecium	*P. foliicola*	*Solanum melongena*	Crater rot	Aumentado and Balendres (2022)
Penicillium	*P. digitatum*	*Citrus* sp.	Green mold rot	Reinking (1918)
Pestalotia	*P. mangiferae*	*Mangifera indica*	Leaf spot	Quimio and Quimio (1974)
Pestalotiopsis	*P. arachidis*	*Arachis hypogaea*	Leaf disease	Bautista and Natural (1991)
Phaeochaetia	*P. psidii*	*Psidium guajava*	Sooty mold	Kriengyakul and Quimio (1983)
Phaeoisariopsis	*P. anthocephala*	*Anthocephalus* sp.	Brown leaf spot	Kobayashi and De Guzman (1988)
Phoma	*P. lingam*	*Brassica pekinensis*	Black leg	Orillo et al. (1959)
Phomopsis	*P. vexans*	*Solanum melongena*	Blight	Fajardo (1934)
Phyllachora	*P. sorghi*	*Sorghum bicolor*	Black/tar leaf spot	Reinking (1918)
Phyllactinia	*Phyllactinia* sp.	*Rosa* sp.	Powdery mildew	Rodrigo and Seggay (1953)
Phyllosticta	*P. zingiberi*	*Zingiber officinale*	Leaf spot	Chanliongco (1966)
Phymatotrichum	*P. omnivorum*	*Gossypium hirsutum*	Red rot	Teodoro (1958)
Physalospora	*P. rodina*	*Citrullus lanatus*	Stem-end rot	Teodoro (1959)

(Continued)

172

7. Plant diseases caused by fungi in the Philippines

TABLE 7.1 Genera of phytopathogenic fungi belonging to Phylum Ascomycota associated with plant diseases in the Philippines.—cont'd

Genus	Some notable species	Plant-host	Disease	Reference
Pionnotes	*P. capillacea*	*Persea americana*	Dieback	Baker (1916)
Pithomyces	*P. chartarum*	*Glycine max*	Seed rot/decay	Prado (1975)
Pleocyta	*P. sacchari*	*Saccharum officinarum*	Rind disease	Baker (1916)
Podosphaera	*P. leucotricha*	*Malus sylvestris*	Powdery mildew	Ramos et al. (1970)
Pseudocercospora	*P. purpurea*	*Persea americana*	Leaf spot	Kobayashi (1995)
Pyrenochaeta	*P. oryzae*	*Oryza sativa*	Sheath blotch	Shahjahan and Mew (1986)
Pyricularia	*P. oryzae*	*Oryza sativa*	Neck rot	Reyes (1939)
Rhynchosporium	*R. oryzae*	*Oryza sativa*	Leaf scald	Tangonan (1999)
Rosellinia	*R. zingiberi*	*Zingiber officinale*	Black rhizome rot	Stevens and Atienza (1931)
Saccharomyces	*Saccharomyces* sp.	*Zea mays*	Seedborne rot	Aquino (1953)
Sarocladium	*S. oryzae*	*Oryza sativa*	Sheath rot	Tangonan (1999)
Sclerotinia	*S. libertiana*	*Brassica oleracea* var. *capitata*	Drop disease	Elayda (1938)
Septogloeum	*S. arachidis*	*Arachis hypogaea*	Leaf spot	Baker (1914)
Septoria	*S. lycopersici*	*Solanum lycopersicon*	Leaf spot	Erdinal and Celino (1940)
Sphaceloma	*S. fawcetti*	*Citrus* sp.	Fruit spot	Dangan (1987)
Sphaerostilbe	*S. repens*	*Hevea brasiliensis*	Red rot	Teodoro (1926)
Stagonospora	*S. sacchari*	*Saccharum officinarum*	Leaf Scorch	Van Dillewizn and Lopez (1964)
Stemphylium	*S. lycopersici*	*Solanum lycopersicon*	Gray leaf spot	Amadura (1980)
Thielavia	*T. basicola*	*Nicotiana* sp.	Root rot	Lee (1921)
Thielaviopsis	*T. paradoxa*	*Saccharum officinarum*	Pineapple disease	Reinking (1920)
Trabutia	*T. ficum*	*Ficus benjamin*	Black leaf spot	Reinking (1919)
Trichoconis	*Trichoconis* sp.	*Sorghum bicolor*	Grain mold	Dayan and Dalmacio (1982)
Trichoderma	*T. harzianum*	*Glycine max*	Seedling rot	Tandang (1992)
Trichothecium	*T. roseum*	*Malus sylvestris*	Fruit rot	Quimio and Abilay (1982)
Triposporium	*Triposporium* sp.	*Ficus benjamin*	Sooty mold	Kriengyakul and Quimio (1983)
Trotteria	*T. venturiodes*	*Glycine max*	Black mildew	Teodoro (1958)
Uncinula	*U. necator*	*Vitis vinifera*	Powdery mildew	Teodoro (1960)

II. Fungi in agriculture, health and environment

3. Plant diseases caused by fungi in Phylum Ascomycota

TABLE 7.1 Genera of phytopathogenic fungi belonging to Phylum Ascomycota associated with plant diseases in the Philippines.—cont'd

Genus	Some notable species	Plant-host	Disease	Reference
Ustilaginoidea	U. virens	Oryza sativa	Kernel smut	Baker (1914)
Ustulina	U. zonata	Hevea brasiliensis	Coller root tor	Teodoro (1926)
Venturia	V. inaequalis	Malus sylvestris	Scab	Ramos et al. (1970)
Verticillium	V. alboatrum	Gossypium hirsitum	Wilt	PCAR (1975)
Xylaria	Xylaria sp.	Albizia falcataria	Wood decay	De Guzman (1974)

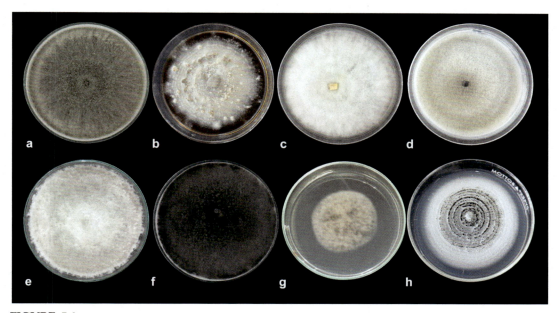

FIGURE 7.2 *Neoscytalidium dimidiatum* (A), *Epicoccum sorghinum* (B), *Colletotrichum tropicale* (C), *Colletotrichum fructicola* (D), *Nigrospora sphaerica* (E), *Lasiodiplodia theobromae* (F), *Corynespora cassiicola* (G), and *Paramyrothecium foliicola* (H) that were recently associated with plant diseases in the country. *Photo credits: John Darby Taguim (A, B, and E), Edzel Evallo (C, D, and F), Deborah Anne Dimayacyac (G), and Herbert Dustin Aumentado (H) of the University of the Philippines Los Baños.*

in the abandonment of the field, reduction of the marketable fruits, and, on one farm, a yield loss of 100% (Dela Cueva & Balendres, 2017).

Paramount to the effective management of anthracnose is the accurate identification of the pathogen. Like many fungi, *Colletotrichum* species were traditionally identified based on their morphology, examined under a microscope, cultural growth in synthetic media, and host

FIGURE 7.3 Canker (A), lesion (B), and anthracnose (D) in dragon fruit stems, crater rot (C) and powdery mildew (E) in eggplant, rust in grapes (F), anthracnose in papaya (G), and mango (H), post-harvest anthracnose in pepper (I), and target spots in *Commelina benghalensis*. Photo credits: *John Darby Taguim (A and B), Edzel Evallo (D), Deborah Anne Dimayacyac (J), Cris Cortaga (H), and Herbert Dustin Aumentado (G) of the University of the Philippines Los Baños and Madora Dacer (I) of the Department of Agriculture -Regional Crop Protection Center.*

pathogenicity. Nevertheless, these methods were found inadequate to precisely identify the *Colletotrichum* species due to the pathogen's overlapping morphological features and, in many cases, host range (Wharton & Diéguez-Uribeondo, 2004). Indeed, several species are grouped into major clades or species complexes (Cannon et al., 2007). Therefore, to identify the *Colletotrichum* species, a polyphasic approach (Cai et al., 2009) is being employed where molecular data or DNA sequences of target gene regions (e.g., internal transcribed spacer, glyceraldehyde-3-phosphate dehydrogenase, B-tubulin, actin, etc.) are used in addition to other traditional methods.

Before molecular data were used in fungal species identification, 21 reported *Colletotrichum* species infecting 63 plant species, of which 22 are vegetable crops, have been recorded (Tangonan, 1999). When organized based on the current *Colletotrichum* systematics (Cannon et al., 2007), these species can be grouped into four main groups (*C. truncatum, C. dematium, C. gloeosoporioides,* and *C. graminicola* species complexes) and with five singleton species (*C. agaves, C. phomoides, C. gossypii, C. heveae,* and *C. lagenarium*). In addition, there were seven unverified and unclassified species in the records (*C. araceae, C. coffeanum, C. euchroum, C. papayae, C. nigrum, C. melongena,* and *C. ficus*), of which the name was based from the host-plant that these fungi have initially been isolated, and reidentifying them would be challenging due to lack of original specimens. Nevertheless, *Colletotrichum* species causing anthracnose in various crops have been identified and their identity validated by molecular data (Table 7.2).

Another frequently reported fungal pathogen is *Lasiodiplodia theobromae*. This fungus was first reported in the Philippines in 1918 (Reinking, 1918). Two new plant species have recently been added to the fungus' host range list, bringing its host range to 43 plant species. The fungus has been associated with dragon fruit stem canker (Evallo, Taguiam, Bengoa, Balendres et al., 2022) and jackfruit brown rot (Taylaran et al., 2021).

Colletotrichum falcatum Went is the red rot-causing fungal pathogen known to be hosted only by *Sorghum* spp. (Matias & Alovera, 1983) and sugarcane, *Saccharum officinarum* (Reinking, 1919) in the country. Red rot is one of the most important sugarcane diseases worldwide that affects cane yield and juice quality. It is the primary constraint in major sugarcane plantations in India (Viswanathan & Rao, 2011). Duttamajumder (2008) has even likened the red rot epidemic in India to other global plant diseases, for example, the potato blight in Ireland, that impacted agriculture and society. The red rot pathogen was first recognized in the Philippines in the late 1920s (Serrano & Marquez, 1926). A few studies were reported until the mid-1970s (Roldan & Tecson, 1935). Two other fungal pathogens (*Fusarium sacchari* (E.J. Butler) W. Gams and *F. proliferatum* (Matsush.) Nirenberg) were identified and were associated with red rot-like symptoms in sugarcane (Dela Cueva et al., 2019). The same findings were obtained in Brazil (Costa et al., 2021). The results from the Philippines and Brazil indicate that *C. falcatum* is not the only fungal pathogen that can induce red rot in sugarcane and thus, accurate identification of the red rot-causing pathogen in effective disease management.

Within this phylum, several fungal species are also toxigenic (Balendres, De Torres, & Dela Cueva, 2019). Toxigenic fungi, aside from causing diseases on plants or contaminating grains, can also produce mycotoxins (e.g., fumonisin and aflatoxin) that have carcinogenic potential on humans and animals. For example, in the Philippines, there are 12 reported fungal species (belonging to *Fusarium* sp., *Aspergillus* sp., and *Penicillium* sp.) isolated from various crop commodities known to produce mycotoxins (Balendres, Karlovsky, & Cumagun, 2019). The most significant and frequently isolated were *Fusarium verticillioides* (Sacc.) Nirenberg and *Aspergillus flavus* Link produce fumonisin and aflatoxin, respectively. In addition, new genotypes of aflatoxigenic fungi have recently been reported in the Philippines (AboDalam et al., 2020). Toxigenic fungi are important in agriculture because the high contamination levels of these fungi in feedstocks can have a detrimental effect on animals and consumers that directly consume these grain products (e.g., rice, corn grits, corn kernels, etc.). Recently, through molecular technique, some isolates of *F. verticillioides* and *F. proliferatum* from sugarcane were also found to contain the fumonisin biosynthesis FUM1 gene (De Torres et al., 2020).

Several new country reports of fungal pathogens have been recorded (Table 7.2). All fungal pathogens are from the Phylum Ascomycota. The stem canker pathogen *Neoscytalidium dimidiatum* (Penz.) Crous and Slippers (Fig. 7.1) was commonly isolated in dragon fruit orchards and could be considered an emerging fungal plant pathogen (Taguiam et al., 2020b, 2021a). Several *Colletotrichum* species have been associated with papaya, persimmon, mango, dragon fruit, and pepper anthracnose (Table 7.2). Other new fungal pathogen records are

7. Plant diseases caused by fungi in the Philippines

TABLE 7.2 *Colletotrichum* species associated with anthracnose and new fungal pathogen records (new occurrence or new plant species) in the Philippines with reported molecular data.

Genera	Species	Plant-host	Disease	Reference
Colletotrichum	*C. asianum*	*Mangifera indica*	Anthracnose	Alvarez et al. (2020)
	C. brevisporum	*Carica papaya*	Anthracnose	Laurel, De Torres, et al. (2021)
	C. fructicola	*Diospyros* sp.[a]	Anthracnose	Evallo, Taguiam, and Balendres (2022)
	C. fructicola	*Mangifera indica*	Anthracnose	Dela Cueva et al. (2021)
	C. fructicola	*Selenicereus* sp./ *Hylocereus* sp.	Anthracnose	Evallo et al. (2021)
	C. musae	*Musa* sp.	Anthracnose	Balendres et al. (2020)
	C. scovillei	*Capsicum annum*	Anthracnose	Cueva et al. (2018)
	C. theobromicola	*Mangifera indica*	Anthracnose	Dela Cueva et al. (2021)
	C. tropicale	*Selenicereus* sp./ *Hylocereus* sp.	Anthracnose	Evallo, Taguiam, Bengoa, Maghirang, et al. (2022)
	C. tropicale	*Mangifera indica*	Anthracnose	Dela Cueva et al. (2021)
	C. truncatum	*Carica papaya*	Anthracnose	Laurel, De Torres, et al. (2021)
	C. truncatum	*Capsicum annum*	Anthracnose	Balendres, Karlovsky, and Cumagun (2019)
Athelia	*A. rolfsii*	*Musa* sp.	Various	Acabal et al. (2019)
Cercospora	*C. basellae-albae*	*Basella alba*	Necrotic spots	Mahamuda Begum and Cumagun (2010)
Corynespora	*C. cassiicola*	*Commelina benghalensis*	Leaf spot	Dimayacyac and Balendres (2021)
Epicoccum	*E. sorghinum*	*Saccharum officinarum*	Leaf spot	Laurel at al. (2021b)
	E. sorghinum	*Selenicereus* sp./ *Hylocereus* sp.	Necrotic spot	Taguiam et al. (2020a)
Fusarium	*F. proliferatum*	*Saccharum officinarum*	Red rot-like	Dela Cueva et al. (2019)
	F. sacchari	*Saccharum officinarum*	Red rot-like	Dela Cueva et al. (2019)
Haematonectria	*H. haematococca*	*Citrus reticulata*	Twig blight	Yago and Chung (2011)
Lasiodiplodia	*L. theobromae*	*Selenicereus* sp./ *Hylocereus* sp.	Stem canker	Evallo, Taguiam, Bengoa, and Balendres (2022)
	L. theobromae	*Artocarpus heterophyllus*	Brown rot	Taylaran et al. (2021)
Neoscytalidium	*N. dimidiatum*	*Selenicereus* sp./ *Hylocereus* sp.	Stem canker	Taguiam et al. (2020b)
Nigrospora	*N. sphaerica*	*Selenicereus* sp./ *Hylocereus* sp.	Brown spot	Taguiam et al. (2020c)
Pseudocercospora	*P. jahnii*	*Tabebuia pallida*	Necrotic spots	Acabal et al. (2014)
Paramyrothecium	*P. foliicola*	*Solanum melongena*	Crater rot	Aumentado and Balendres (2022)

[a]*Fruits imported.*

Cercospora basellae-albae R.K. Srivast., S. Narayan and A.K. Srivast (Begum & Cumagun, 2010), *Epicoccum sorghinum* (Sacc.) Aveskamp, Gruyter and Verkley (Laurel, Magdalita et al., 2021; Taguiam et al., 2020a, 2021b), *Paramyrothecium foliicola* L. Lombard and Crous (Aumentado & Balendres, 2022), *Pseudocercospora jahnii* (Syd.) U. Braun and Crous (Acabal et al., 2014) *and Nigrospora sphaerica* (Sacc.) E.W. Mason (Taguiam et al., 2020c).

4. Plant diseases caused by fungi in Phylum Basidiomycota

There are 35 genera of plant pathogenic fungi under the phylum Basidiomycota (Table 7.3). The symptoms of infected plant hosts are rust, rot, pink disease, blight, smut, and dieback, with rust and smut (Fig. 7.4) being the most common. The rust fungi (order Pucciniales) are among the largest order of fungi (Aime & McTaggart, 2021). In the Philippines, there are 47 accepted species of rust fungi belonging to 12 families. The frequently reported genera are *Puccinia* sp., *Uromyces* sp., *Phakopsora* sp., and *Kuehneola* sp. The fungus *Phakopsora pachyrhizi* Syd. and P. Syd has a wide host range, especially in the Fabaceae family, and was previously considered a major soybean disease (Quebral, 1988). *Puccinia* species are the most commonly reported rust fungi, and these fungi have a broad host range (Tangonan, 1999). In addition, there are three genera associated with the smut of various plants in the country. These are *Ustilago* sp., *Urocystis* sp., and *Entyloma* sp.

Another group within the Phylum Basidiomycota is the wood rotting or decaying fungi. *Fomes* sp. is an ungulate-shaped fungus, and their presence in fruit trees could result in the rotting of the plant part where these fungi are attached. *Fomes* sp. has been associate with rots in coffee, citrus, rambutan, rubber, cassava, and cacao (Tangonan, 1999).

The three other most important fungal genera and species from this phylum are *Rhizoctonia solani* J.G. Kühn, *Athelia rolfsii* (Curzi) C.C. Tu and Kimbr. (syn. *Sclerotium rolfsii*), and *Oncobasidium theobromae* P.H.B. Talbot and Keane (syn. *Ceratobasidium theobromae*). *Rhizoctonia* sp. and *Athelia* sp. are the top two fungal genera with the most number of plant hosts recorded in the country. *Rhizoctonia solani* causes blight in beans, avocado, corn, durian, lima bean, mungbean, peanut, pepper, sorghum (Tangonan, 1999) and necrotic leaf spot on coffee (Priyatmojo et al., 2001). It also causes damping-off in lettuce, papaya, cabbage, cucumber and celery. *Rhizoctonia solani* is also the causative agent of soreshin in tobacco (Reinking, 1918) and cotton (Teodoro, 1958). An important disease caused by *R. solani* is sheath blight in rice (Palo, 1926; Sandoval & Cumagun, 2019) and corn (Pascual et al., 2000; Sales, 1966). Sheath blight in rice can cause up to 80% grain yield reduction (Ou, 1985; Ren et al., 2001). In a recent study, no subdivision was detected among the four populations of *R. solani* in rice from the Philippines (Cumagun et al., 2020).

The genus *Sclerotium* consists of seven fungal species pathogenic to various crops grown in the country. The frequently reported species is *S. rolfsii* (*A. rolfsii*), commonly associated with damping-off and stem, or root, rot. This fungus was first reported in the Philippines in 1918 in various crops, including tomato, sugarcane, and coffee (Reinking, 1918). In 2019, *A. rolfsii* was associated with corn rot, leaf rot, splitting of pseudostem, and yellowing of leaves in bananas (Acabal et al., 2019).

178 7. Plant diseases caused by fungi in the Philippines

TABLE 7.3 Genera of phytopathogenic fungi belonging to Phylum Basidiomycota associated with plant diseases in the Philippines.

Genus	Some notable species	Plant-host	Disease	Reference
Armillaria	*A. mellea*	*Coffea* sp.	Root rot	Teodoro and Gomez (1926)
Coleosporium	*C. plumeriae*	*Plumeria acuminata*	Rust	Tangonan et al. (1997)
Corticium	*C. salmonicolor*	*Cirtus* sp.	Pink disease	Batoon and Macabasco (1968)
Cyphella	*C. heveae*	*Hevea brasiliensis*	White stem blight	Teodoro (1926)
Endophyllum	*E. blumea*	*Blumea balsamifera*	Rust	Divinagracia and Ros (1985)
Entyloma	*E. oryzae*	*Oryza sativa*	Leaf smut	Baker (1914)
Fomes	*F. lignosus*	*Manihot esculenta*	Tip blight	Divinagracia (1974)
Ganoderma	*G. pseudoferreum*	*Hevea brasiliensis*	Red rot	PCAR (1976)
Hemileia	*H. vastatrix*	*Coffea* sp.	Rust	Barret (1912)
Hydnum	*Hydnum* sp.	*Anacardium occidentale*	Branch rot	Rimando (1988)
Kuehneola	*K. gossypii*	*Gossypium hirsutum*	Rust	Baker (1914)
Lenzites	*L. striata*	*Dipterocarpus grandiflorus*	Brown rot	Eusebio (1969)
Marasmiellus	*M. inoderma*	*Musa textilis*	Sheath rot	Tangonan (1999)
Marasmius	*M. sacchari*	*Saccharum officinarum*	Stem rot	Hines (1918)
Melampsora	*M. lini*	*Linum usitassimum*	Rust	Garrido and Torres (1936)
Oncobasidium	*O. theobromae*	*Theobromae cacao*	Vascular streak dieback	San Juan (1983)
Phakopsora	*P. pachyrhizi*	*Phaseolus lunatus*	Rust	Mendoza (1978)
Phellinus	*P. noxious*	*Hevea brasiliensis*	Brown rot	PCAR (1976)
Physopella	*P. fici*	*Ficus benjamin*	Rust	Baker (1914)
Pleurotus	*P. ostreatus*	*Pinus* sp.	Wood rot	Quiniones and Zamora (1987)
Polyporus	*Polyporus* sp.	*Hevea brasiliensis*	Rot	Teodoro (1926)
Poria	*P. hypolateritia*	*Hevea brasiliensis*	Wet rot	Teodora (1926)
Puccinia	*P. polysora*	*Zea mays*	Leaf rust	Reyes (1956)
Rhizoctonia	*R. solani*	*Zea mays*	Leaf sheath blight	Sales (1966)
Rigidoporus	*R. lignosus*	*Hevea brasiliensis*	White rot	Bayaron and Rivera (1995)
Sclerotium/ Athelia	*A. rolfsii*	*Lycopersicon esculentum*	Blight	Reinking (1918)

II. Fungi in agriculture, health and environment

5. Plant diseases caused by fungi in Phylum Blastocladiomycota, Chytridiomycota, and Mucoromycota 179

TABLE 7.3 Genera of phytopathogenic fungi belonging to Phylum Basidiomycota associated with plant diseases in the Philippines.—cont'd

Genus	Some notable species	Plant-host	Disease	Reference
Septobasidium	S. pseudopedicellatum	Citrus sp.	Felt disease	Del Rosario (1968)
Sphacelotheca	S. cruenta	Sorghum bicolor	Kernel smut	Uichanco (1959)
Thelephora	Thelephora sp.	Eucalyptus sp.	Heart rot	Tangonan (1999)
Tilletia	T. barclayana	Oryza sativa	Kernel smut	Reyes (1933)
Trametes	T. personii	Durio zibethinus	Bark disease	Reinking (1919)
Urocystis	U. cepulae	Allium sp.	Smut	Elayda (1935)
Uromyces	U. appendiculatus	Vigna radiata	Rust	Reinking (1918)
Uromycladium	U. tepperianum	Paraserianthes falcataria	Rust	Almonicar (1990)
Ustilago	U. scitaminea	Saccharum officinarum	Smut	Robinson (1908)

Oncobasidium theobromae is a basidiomycetous fungus with economic importance to the Philippines. The fungus was reported in the early 1980s and was associated with vascular streak dieback (VSD) in cacao (San Juan, 1983). It has been associated with vascular streak dieback of guava, jackfruit, mango, mangosteen, chico, lanzones, marang, rambutan, rubber, and citrus (Tangonan, 1999). The fungus's current proposed name is *Rhizoctonia theobromae* (Oberwinkler et al., 2013), but *C. theobromae* (Samuels et al., 2012) is still commonly used. However, VSD-like symptoms can also be incited by other pathogens. For example, the ascomycetous fungus *Lasiodiplodia theobromae* has been reported to cause VSD-like symptoms in cacao seedlings and mature trees (Alvindia & Gallema, 2017).

5. Plant diseases caused by fungi in Phylum Blastocladiomycota, Chytridiomycota, and Mucoromycota

The genera of plant pathogenic fungi in Phylum Blastocladiomycota, Chytridiomycota, and Mucoromycota, combined, are relatively small compared to those in the Phylum Ascomycota and Basidiomycota (Fig. 7.1). *Physoderma maydis* (Miyabe) Miyabe is the only known fungal plant pathogen from the phylum Blastocladiomycota. The fungus was first reported in the country in 1941 and is the causative agent of the brown spot in corn (Reyes, 1941). This fungus has not been detected or known to cause disease in other crops in the country. Within the phylum Chytridiomycota, only three genera (*Olpidium* sp., *Synchytrium* sp., and *Woroninella* sp.), each having a single species, have been recorded in the country. *Olpidium brassicae* (Woronin) P.A. Dang. is the causative agent of damping-off in cabbage (Elayda, 1938), while *Woroninella dolichi* (Cooke) Syd and *Synchytrium psophocarpi* (Racib.) Gäum. induce orange galls in winged bean (Baker, 1916) and hyacinth bean (Reinking, 1918), respectively.

FIGURE 7.4 Diseases (A and C) caused by fungi (C and D) in phylum Basidiomycota. Coffee rust (A, B) and corn smut (C, D). *Photo Credit: Mari Neila Seco (C and D) of the University of the Philippines Los Baños.*

Genera and species belonging to the phylum Mucoromycota are rot-inducing pathogens. *Choanephora* sp., *Mucor* sp., and *Rhizopus* sp. have been associated with fruit rot and soft rot diseases. *Rhizopus* sp. is the most frequently reported of these three genera, with four fungal species recorded (*R. stolonifer*, *R. arrhizus*, *R. artocarpi*, and *R. nigricans*). *Choanephora cucurbitarum* (Berk. & Ravenel) Thaxt. (Quimio et al., 1982) and *C. infundibulifera* (Curr.) Sac. (De Villa, 1982) cause fruitlet rot in jackfruit. *Mucor* sp. has been isolated from citrus (Reinking, 1921), peanut (Quitco, 1989), and sunflower (Quilantang, 1979), showing fruit rot, seedling discoloration, and seedling rot, respectively (Table 7.4).

6. Concluding remarks and future direction

TABLE 7.4 Genera of phytopathogenic fungi belonging to Phylum Blastocladiomycota, Chytridiomycota, and Mucoromycota associated with plant diseases in the Philippines.

Genus	Phylum	Some notable species	Plant-host	Disease	Reference
Physoderma	Blastocladiomycota	*P. maydis*	*Zea mays*	Brown spot	Reyes (1941)
Olpidium	Chytridiomycota	*O. brassicae*	*Brassica oleracea* var. *capitata*	Damping-off	Elayda (1938)
Synchytrium	Chytridiomycota	*S. psophocarpi*	*Psophocarpus tetragonolobus*	Orange gall	Baker (1916)
Woroninella	Chytridiomycota	*W. dolichi*	*Dolichos lablab*	Orange gall	Reinking (1918)
Choanephora	Mucoromycota	*C. cucurbitarum*	*Artocarpus heterophyllus*	Fruitlet rot	Quimio et al. (1982)
Mucor	Mucoromycota	*Mucor* sp.	*Citrus* sp.	Fruit rot	Reinking (1921)
Rhizopus	Mucoromycota	*R. stolonifer*	*Ipomoea batatas*	Soft rot	Reinking (1918)

6. Concluding remarks and future direction

Plant pathogenic fungi are important in agriculture. They cause various diseases in crops, and severe infection could result in crop yield reduction, increased production cost, fruit quality reduction, and even loss of livelihood to farmers. Furthermore, in some cases, contamination of fungi in grain products could predispose consumers to toxins detrimental to health. Hence, these fungal pathogens, and their associated diseases, are an impediment to economic growth in achieving food security and sustaining food safety.

There are over 180 genera and over 600 species of plant pathogenic fungi reported in the country. They cause various diseases in crops, including high-valued ones, that are significant to the country's agriculture sector. Paramount to practical and effective diseases management is accurately identifying the causative fungal pathogen. With singleton species, traditional methods in disease identification would be straightforward, but those considered species complexes would require a polyphasic approach, combining molecular data with conventional identification methods. This has been the case in identifying *Colletotrichum* species associated with anthracnose.

Continuous efforts to control the spread of fungal pathogens and mitigate the impact of fungal diseases in crop production are warranted. Among the actions that would contribute to managing further disease outbreaks and effects are spreading awareness about fungal diseases to various stakeholders (growers, field technicians, extensionist, policy-makers, etc.) and enhancing the investment in research that directly benefits the growers (and consumers). Increasing border security to ward-off unwanted fungal pathogens is also important. Still, these would need a collaborative effort among those who import, those involved in detecting these fungal plant pathogens, and even the consumers. Early detection can aid in the early planning and management of disease incursions. Still, paramount to achieving this is the availability of rapid, sensitive, and cost-effective disease detection protocols and methods. In the field, remain vigilant, and an integrated disease management approach is required, especially when dealing with diseases caused by soil-borne pathogens, which are very challenging. For instance, management of *Fusarium oxysporum* f. sp. *cubense* TR4 in the banana will likely be achieved with the combination of proper field sanitation, deployment of best farm practices, application of cultural methods, use of biological control agents, use of less-susceptible varieties, and, when needed, use of less-toxic chemicals. Planning, monitoring, and coordination on farms remain important in the integrated disease management approach.

Acknowledgments

The author thanks Scientist and Emeritus Professor Naomi Tangonan of the University of Southern Mindanao for providing plant pathologists with a list of fungal diseases of plants from the Philippines. This list, along with new papers published in the last 20 years, was vital in the preparation of this chapter. The author is also grateful to John Joel Tandang for the technical assistance and John Darby Taguiam, Edzel Evallo, Deborah Anne Dimayacyac, Herbert Dustin Aumentado, Cris Cortaga, Madora Abril Dacer, and Mari Neila Seco for providing disease and fungi culture photos. Research on plant diseases by the author is supported by the Department of Agriculture — Bureau of Agriculture Research, the Department of Science and Technology, the DOST — Philippine Council for Agriculture, Aquatic, and Natural Resources Research and Development (PCAARRD), and the Institute of Plant Breeding, College of Agriculture and Food Sciences, University of the Philippines Los Baños. Finally, the author dedicates this work to all Filipino (and foreign) plant pathologists, past and present, who have contributed to increasing the knowledge of fungal diseases affecting important crops in the Philippines.

References

AboDalam, T. H., Amra, H., Sultan, Y., Magan, N., Carlobos-Lopez, A. L., Cumagun, C. J. R., & Yli-Mattila, T. (2020). New genotypes of aflatoxigenic fungi from Egypt and the Philippines. *Current Research in Environmental & Applied Mycology, 10*(1), 142—155.

Acabal, B. D., Dalisay, T. U., Groenewald, J. Z., Crous, P. W., & Cumagun, C. J. R. (2019). *Athelia rolfsii* (= *Sclerotium rolfsii*) infects banana in the Philippines. *Australasian Plant Disease Notes, 14*(1), 1—3.

Acabal, B. D., Groenewald, J. Z., Crous, P. W., & Cumagun, C. J. R. (2014). First report of *Pseudocercospora jahnii* in the Philippines. *Australasian Plant Disease Notes, 9*(1), 1—2.

Acosta, E. A. (1992). *Morphology, pathogenicity, and identity of* Cercosporella *species attacking makabuhay (*Tinospora rumphii *Boerl.).* BS thesis. UPLB.

Aguilar-Hawod, K. G. I., de la Cueva, F. M., & Cumagun, C. J. R. (2019). Genetic diversity of *Fusarium oxysporum* f. sp. *cubense* causing Panama wilt of banana in the Philippines. *Pathogens, 9*(1), 32.

Aime, M. C., & McTaggart, A. R. (2021). A higher-rank classification for rust fungi, with notes on genera. *Fungal Systematics and Evolution, 7*(1), 21—47.

Almonicar, R. S. (1990). Pathogenicity test and toxin bioassay of the needle blight fungus affecting *Pinus caribaea*. *Philipp Phytopathology, 26*, 53. Abstr.

Alvarez, L. V., Hattori, Y., Deocaris, C. C., Mapanao, C. P., Bautista, A. B., Cano, M. J. B., Naito, K., Kitabata, S., Motahashi, K., & Nakashima, C. (2020). *Colletotrichum asianum* causes anthracnose in Philippine mango cv. Carabao. *Australasian Plant Disease Notes, 15*(1), 1—5.

Alvindia, D. G., & Gallema, F. L. M. (2017). *Lasiodiplodia theobromae* causes vascular streak dieback (VSD)—like symptoms of cacao in Davao Region, Philippines. *Australasian Plant Disease Notes, 12*(1), 1—4.

Amadura, F. L. (1980). *Studies on gray leafspot of tomato caused by* Stemphylium lycopersici *(Enjoji) Yamamoto*. BS thesis. UPLB.

References

Angeles, A. I. (1980). *Pathogenicity of* Deightoniella torulosa *(Syd.) Ell. on dwarf Cavendish banana (*Musa cavendishii *Lamb.* BS thesis. UPLB.

Angelito, F. I. (1980). Survey of the seedborne fungi of common weed species in CLSU. *CLSU International Journal of Science & Technology, 5 & 6*, 21.

Aquino, R. V. (1953). Cercospora *leafspot of sweetpotato*. BS thesis. UPCA.

Aumentado, H. D., & Balendres, M. A. (2022). Identification of *Paramyrothecium foliicola* causing crater rot in eggplant and its potential hosts under controlled conditions. *Journal of Phytopathology, 170*, 148–157.

Baker, C. F. (1914). A review of some Philippine plant diseases. *The Philippine Agriculturist and Forester, 3*, 157–166.

Baker, C. F. (1916). Additional notes on Philippine plant diseases. *The Philippine Agriculturist and Forester, 5*, 73–78.

Balendres, M. A., De Torres, R., & Dela Cueva, F. (2019). Culture storage age and fungal re-isolation from host-tissue influence *Colletotrichum* spp. virulence to pepper fruits. *Journal of Phytopathology, 167*(9), 510–515.

Balendres, M. A. O., Karlovsky, P., & Cumagun, C. J. R. (2019). Mycotoxigenic fungi and mycotoxins in agricultural crop commodities in the Philippines: A review. *Foods, 8*(7), 249.

Balendres, M. A., & Masangcay, T. (2020). Confirmed record of *Spongospora subterranea* subsp. *subterranea* in potato cv. Igorota in Northern Philippines and the susceptibility of tomato cv. Yellow plum to *Spongospora* root infection. *Archives of Phytopathology and Plant Protection, 53*(1–2), 37–47.

Balendres, M. A., Mendoza, J., & Dela Cueva, F. (2020). Characteristics of *Colletotrichum musae* PHBN0002 and the susceptibility of popular banana cultivars to postharvest anthracnose. *Indian Phytopathology, 73*(1), 57–64.

Balendres, M. A., Taguiam, J. D., Evallo, E., & Sison, M. L. (2021). Premature defoliation in *Zinnia* sp. and mycoparasitism of *Ampelomyces quisqualis* against the powdery mildew pathogen *Golovinomyces cichoracearum* from the Philippines. *Indian Phytopathology, 74*(1), 283–284.

Barrett, O. W. (1912). Coffee blight in the Philippines. *Philippine Agricultural Review, 5*, 333–334.

Batoon, M. P., & Macabasco, C. B. (1968). Ang chico. *Plant Industry Digest, 30–31*, 37–38.

Bautista, E. A., & Natural, M. P. (1991). Etiology of a new leaf disease of peanut. *Philippines Phytopathology, 27*, 49–50.

Bayaron, T. B., & Rivera, T. S. (1995). Incidence and distribution of white root rot disease of rubber in Mindanao. *Philippines Phytopathology, 31*, 135.

Begum, M. M., & Cumagun, C. J. R. (2010). First record of *Cercospora basellae-albae* from the Philippines. *Australasian Plant Disease Notes, 5*(1), 115–116.

Borja, M. S. (1980). *Seedborne fungi and their relative incidence on seedboard varieties of rice in the Philippines.* BS thesis. UPLB.

Brown, E. O., Decena, F. L. C., & Ebora, R. V. (2018). *The current state, challenges and plans for Philippine agriculture.* FFTC Agriculture Policy Platform (FFTC-AP).

Cai, L., Hyde, K. D., Taylor, P. W. J., Weir, B., Waller, J., Abang, M. M., … Shivas, R. G. (2009). A polyphasic approach for studying *Colletotrichum. Fungal Diversity, 39*(1), 183–204.

Cannon, P. F., Damm, U., Johnston, P. R., & Weir, B. S. (2007). Colletotrichum—current status and future directions. *Studies in Mycology, 59*(1), 129–145.

Carreon, M. C. (1989). *Seed mycoflora of mungbean (*Vigna radiata *L.) Wilczek) and their pathogenic importance.* BS thesis. UPLB.

Catambacan, D. G., & Cumagun, C. J. R. (2022). The weed *Eleusine indica* as an alternative host of *Fusarium oxysporum* f. sp. *cubense* Tropical race 4 causing Fusarium wilt in Cavendish banana. *Journal of Phytopathology.* https://doi.org/10.1111/jph.13095

Chanliongco, R. (1966). Leafspot disease of ginger. *Agriculture et los Banosa, 6*, 16–18.

Costa, M. M., Silva, B. A., Moreira, G. M., & Pfenning, L. H. (2021). *Colletotrichum falcatum* and *Fusarium* species induce symptoms of red rot in sugarcane in Brazil. *Plant Pathology, 70*(8), 1807–1818.

Cueva, F. D., Mendoza, J. S., & Balendres, M. A. (2018). A new Colletotrichum species causing anthracnose of chilli in the Philippines and its pathogenicity to chilli cultivar Django. *Crop Protection, 112*, 264–268.

Cumagun, C. J. R., McDonald, B. A., Arakawa, M., Castroagudín, V. L., Sebbenn, A. M., & Ceresini, P. C. (2020). Population genetic structure of the sheath blight pathogen *Rhizoctonia solani* AG-1 IA from rice fields in China, Japan and the Philippines. *Acta Scientiarum Agronomy, 42*.

Dangan, J. M. (1987). *Morphology and pathogenicity of* Sphaceloma *spp. causing fruit spot anthracnoses.* PhD thesis. UPLB.

Dayan, M. P. (1988). Survey, identification, and pathogenicity of pests and diseases of bamboo in the Philippines. *Sylvatrop: The Philippine Forest Research Journal, 13*(1 & 2), 61–77.

Dayan, M. P. (1989). Moluccan Sau *Albizia falcataria* (L.) Back. *RISE, 1*, 84–97.

II. Fungi in agriculture, health and environment

184   7. Plant diseases caused by fungi in the Philippines

Dayan, M. P., & Dalmacio, S. C. (1982). Sorghum grain molds: Identification, incidence, and pathogenicity. *Philipp Phytopathology, 18*, 68—77.

De Guzman, E. D. (1974). *Study of pathological problems in the nursery, plantations, and second growth stands of Nasipit Lumber Co. Res. Repts* (p. 214). UPLB.

De Guzman, E. D., Militante, E. P., & Lucero, R. (1991). Forestry nursery diseases and insects in the Philippines. In J. R. Sutherland, & S. G. Glover (Eds.), *Proceedings of 1st Mtg IUFRO working party S2.07-09 Info. Rept. BC-X-331Diseases and insects in forest nurseries* (pp. 101—104). Pacific Forestry Centre.

De Torres, R., Dela Cueva, F., & Balendres, M. A. (2020). First report on the detection of fumonisin biosynthetic (FUM1) gene in *Fusarium verticillioides* and *F. proliferatum* associated with sugarcane diseases. *Indian Phytopathology, 73*(3), 555—559.

De villa, E. A. (1982). *Studies on Choanephora fruitlet rot of jackfruit*. BS thesis. UPLB.

Del Rosario, M. S. (1968). A handbook of citrus diseases in the Philippines. *UPCA Technical Bulletin, 24*, 31.

Dela Cueva, F., & Balendres, M. (2017). Elucidating the cause and impact of a chilli anthracnose outbreak in Batangas, Philippines. *Journal of Tropical Plant Pathology, 53*, 37—51.

Dela Cueva, F., De Torres, R., de Castro, A., Mendoza, J., & Balendres, M. A. (2019). Susceptibility of sugarcane to red rot caused by two *Fusarium* species and its impact on stalk sugar level. *Journal of Plant Pathology, 101*(3), 639—646.

Dela Cueva, F. M., Laurel, N. R., Dalisay, T. U., & Sison, M. L. J. (2021). Identification and characterisation of *Colletotrichum fructicola*, *C. tropicale* and *C. theobromicola* causing mango anthracnose in the Philippines. *Archives of Phytopathology and Plant Protection, 54*(19—20), 1989—2006.

Dimayacyac, D. A., & Balendres, M. A. (2021). *Commelina benghalensis* harbors *Corynespora cassiicola*, the tomato target spot pathogen. *International Journal of Pest Management*, 1—7.

Divinagracia, G. G. (1974). *Survey and control of pathogens attacking cassava and sweetpotato 4. Fungal diseases. Res. Repts* (p. 161). UPLB. Abstr.

Divinagracia, G. G. (1980). Diseases of orchids, roses, and other ornamental plants in the Philippines. In *Proceedings Symposium Philippines Phytopathology 1917—1977* (pp. 375—388). PPS, UPLB.

Divinagracia, G. G., & Ros, L. B. (1985). Note: Diseases of selected medicinal plants in the Philippines. *Philippines Agriculturist, 68*, 297—308.

Duttamajumder, S. K. (2008). *Red rot of sugarcane*. Indian Institute of sugarcane research.

Elayda, A. (1935). Bermuda onion (*Allium cepa* L.). *The Philippine Journal of Agriculture, 6*, 175—193.

Elayda, A. (1938). Commercial raising of cabbage (*Brassica oleracea* L.). *Philippine Journal of Agriculture, 9*, 103—117.

Erdinal, D. M., & Celino, M. S. (1940). *Septoria* leafspot of tomato. *Philippines Agriculturist, 29*, 293.

Eusebio, M. A. (1969). Degradation of lignin in wood by white-rot and brown-rot fungi. *Philippines Phytopathology, 5*, 4.

Evallo, E., Taguiam, J. D., & Balendres, M. A. (2022). *Colletotrichum fructicola* associated with fruit anthracnose of persimmon. *Journal of Phytopathology, 170*(3), 194—201.

Evallo, E., Taguiam, J. D., Bengoa, J., & Balendres, M. A. (2022). First report of *Lasiodiplodia theobromae* causing stem canker of *Selenicereus monacanthus* in the Philippines. *Journal of Plant Pathology*, 1—2.

Evallo, E., Taguiam, J. D., Bengoa, J., Maghirang, R., & Balendres, M. A. (2021). First report of *Colletotrichum fructicola*, causing anthracnose of *Hylocereus* plants, in the Philippines. *Czech Mycology, 73*(1).

Evallo, E., Taguiam, J. D., Bengoa, J., Maghirang, R., & Balendres, M. A. (2022). First report of *Colletotrichum tropicale* on dragon fruit and the response of three *Selenicereus* species to anthracnose. *International Journal of Pest Management*, 1—8.

Exconde, O. R. (1980). Corn diseases and their control. In *Proceeding of symposium* Philippines phytopathology *1917—1977. PPS, Inc., December 14—15, 1977* (pp. 98—106).

Fajardo, T. G. (1934). Plant disease problems confronting track farmers in Trinidad Valley and vicinity of Baguio, Mt. Province, Philippine Islands. *Philippine Journal of Science, 53*, 67—95.

Fajardo, T. G. (1937). The tomato leaf mold (*Cladosporium fulvum* Cke.) a new serious disease of tomato in Baguio, Mt. Province. *Philippine Journal of Agriculture, 8*, 163—186.

Gandia, I. M., & Garcia, A. S. (1980). Diseases of coffee and cacao in the Philippines. In *Proceeding of Symposium* Philippines phytopathology *1917—1977, PPS Inc, December 14—15, 1997* (pp. 299—313).

Garrido, T. G., & Torres, J. P. (1936). Flax in the Philippines. *Philippine Journal of Agriculture, 7*, 229—243.

Hapitan, J. C., Jr., & Reyes, T. T. (1970). Black leaf streak of bananas in the Philippines. *Philippine Agriculturist, 54*, 46—54.

II. Fungi in agriculture, health and environment

References

185

Hines, C. W. (1918). Diseases, insects, and plant pests of sugarcane in the Philippines. *Philippine Agricultural Review, 11*, 275—277.

Ilag, L. L., & Marfil, V. E. (1980). Occurrence of mungbean scab in the Philippines. *Philippines Phytopathology, 16*, 5.

James, T. Y., Kauff, F., Schoch, C., Matheny, P. B., Hofstetter, V., Cox, C., Celio, G., Gueidan, C., Fraker, E., Miadlikowska, J., Lumbsch, H. T., Rauhut, A., Reeb, V., Arnold, A. E., Amtoft, A., Stajich, J. E., Hosaka, K., Sung, G.-H., Johnson, D., ... Vilgalys, R. (2006). Reconstructing the early evolution of the fungi using a six gene phylogeny. *Nature, 443*, 818—822.

James, T. Y., Letcher, P. M., Longcore, J. E., Mozley-Standridge, S. E., Porter, D., Powell, M. J., Griffith, G. W., & Vilgalys, R. (2006). A molecular phylogeny of the flagellated Fungi (Chytridiomycota) and a proposal for a new phylum (Blastocladiomycota). *Mycologia, 98*, 860—871.

Karganilla, A. D., & Elazegui, F. A. (1970). Local diseases of sorghum. *Philippines Phytopathology, 6*, 83—88.

Kobayashi, T. (1995). *Diseases of tropical fruit trees* (p. 143). Assoc. Int'l. Coop. Agr. & Forestry.

Kobayashi, T., & De Guzman, E. D. (1988). Monograph of tree diseases in the Philippines with taxonomic notes on their associated microorganisms. *Forestry and Forest Products Research Institute, 351*, 99—200.

Kobayashi, T., Suto, Y., & De Guzman, E. D. (1978). *Cercospora* needle blight of pines in the Philippines. *European Journal of Plant Pathology Band, 9*, 166—175 (1979)/Left 3-4, S.

Kriengyakul, V., & Quimio, T. H. (1983). Additional sooty molds in the Philippines. *Philippines Phytopathology, 19*, 17—23.

Laurel, N. R., De Torres, R. L., Mendoza, J. V. S., Balendres, M. A. O., & Dela Cueva, F. M. (2021). Identification of *Epicoccum sorghinum* and its effect on stalk sugar yield. *Sugar Tech, 23*(6), 1383—1392.

Laurel, N. R., Magdalita, P. M., & Dela Cueva, F. M. (2021). Identification and characterization of *Colletotrichum brevisporum* and *C. truncatum* causing papaya anthracnose in the Philippines. *Journal of Phytopathology, 169*(11—12), 692—700.

Lee, A. H. (1921). Observations on previously unreported or noteworthy plant diseases in the Philippines. *Philippines Agricultural Revolution, 14*, 422—434.

Lee, A. H. (1923). California scaly bark rot of citrus trees in the Philippines. *Philippine Agricultural Review, 16*, 219—226.

Lee, A. H. (1924). Dry rot of citrus fruits caused by *Nematospora* species. *Philippine Journal of Science, 24*, 719—733.

Lee, A. H., & Serrano, F. B. (1920). Banana wilt in the Philippines. *Philippine Agricultural Review, 13*, 128.

Leslie, J. F., & Summerell, B. A. (2006). Fusarium laboratory workshops—a recent history. *Mycotoxin Research, 22*(2), 73.

Liyag, B. S. (1969). *Morphology, pathogenicity, and identity of a fungus attacking the straw mushroom* (Volvariella volvacea *(Bull ex Fr.) Sing.*). BS thesis. UPCA.

Lo, B. (2018). *Farmers in the Philippines fight to save bananas from deadly fungus.* CGTN America. https://america.cgtn.com/2018/05/02/farmers-in-the-philippines-fight-to-save-bananas-from-deadly-fungus.

Matias, R. C., & Alovera, H. C. (1983). Weed hosts of *Colletotrichum falcatum* Went. causing red rot disease of sugarcane. *Abstr. Bibl. Res. Plt. Pathol. USM.*

Mendoza, T. C. (1978). *Host range of* Phakopsora pachyrhizii *Syd., cause of soybean rust in the Philippines.* BS thesis. UPLB.

Mendoza, A. M., & Molina, R. P. (1980). A study on seed-borne fungi associated with rice seed and their effects on rice seedlings [conducted in the Philippines]. *Araneta Research Journal (Philippines).*

Mortuza, M. G., & Ilag, L. L. (1995). Effect of temperature and humidity on the germination and growth of *Lasiodiplodia theobromae*, cause of stem-end rot of mango. *Philippines Phytopathology, 31*, 1—8.

Nueda, G. G. (1978). Preliminary studies on sunflower diseases in Central Luzon. *Philippines Phytopathology, 14*, 13—14.

Oberwinkler, F., Riess, K., Bauer, R., Kirschner, R., & Garnica, S. (2013). Taxonomic re-evaluation of the *Ceratobasidium-Rhizoctonia* complex and *Rhizoctonia butinii*, a new species attacking spruce. *Mycological Progress, 12*(4), 763—776.

Ocfemia, G. O. (1924). Notes on some economic plant diseases new in the Philippine Islands. *Philippines Agriculturist, 13*, 163—165.

Ocfemia, G. O. (1931). Notes on some economic plant diseases new in the Philippines. *Philippines Agriculturist, 19*, 581—589.

Ocfemia, G. O., & Celino, M. S. (1933). A brown bark-rot of cacao trunk. *Philippines Agriculturist, 21*, 665—673.

II. Fungi in agriculture, health and environment

Ong Sotto, R., & De Guzman, R. (1987). Identification of some graminicolous *Helminthosporium* isolates. *Philippines Phytopathology*, *35*, 28.

Orillo, F. T., Schafer, L. A., & Revilla, B. A. (1959). Common diseases of vegetable crops and their control in the Philippines. *Plant Industry Digest*, *4*, 67.

Ou, S. H. (1973). *A handbook of rice diseases in the tropics* (p. 58). IRRI.

Ou, S. H. (1985). *Rice diseases* (2nd ed.). Farnham Royal: Commonwealth Agricultural Bureau.

Palo, M. A. (1926). Rhizoctonia disease of rice. I. A study of the disease and of the influence of certain conditions upon the viability of the sclerotial bodies of the causal fungus. *Philippines Agriculturist*, *15*, 361–375.

Paningbatan, R. A., & Ilag, L. L. (1981). Two leafspots of peanut in the Philippines: Etiology and host response to infection. *Philippines Agriculturist*, *64*, 351–364.

Pascual, C. B., Raymundo, A. D., & Hyakumachi, M. (2000). Efficacy of hypovirulent binucleate *Rhizoctonia* sp. to control banded leaf and sheath blight in corn. *Journal of General Plant Pathology*, *66*(1), 95–102.

PCAR. (1975). *The Philippines recommends for cotton, 1975*. Los Baños, Laguna.

PCAR. (1976). *The Philippines recommends for rubber, 1976*. Los Baños, Laguna.

Philippines Statistics Authority. (2021). *2021 selected statistics on agriculture*. Philippine Statistics Authority.

Pioneer Overseas Corp. (1986). *Atlas of corn and sorghum diseases in the Philippines* (p. 61).

Pordesimo, A. N. (1970). Some diseases of banana. *Agriculture et los Banosa*, *9*, 10–11/14.

Prado, F. V. (1975). *Survey of fungi associated with soybean seeds*. MS thesis. UPLB.

Priyatmojo, A., Escopalao, V. E., Tangonan, N. G., Pascual, C. B., Suga, H., Kageyama, K., & Hyakumachi, M. (2001). Characterization of a new subgroup of *Rhizoctonia solani* anastomosis group 1 (AG-1-ID), causal agent of a necrotic leaf spot on coffee. *Phytopathology*, *91*(11), 1054–1061.

Quebral, F. C. (1988). *What one should know about plant diseases*. College of Agriculture, UPLB, Philippines, pp: 65.

Quilantang, J. R. (1979). Survey of seedborne fungi in sunflower field of CLSU. *CLSU Science*, *5 & 6*, 81 (Abstr).

Quimio, T. H. (1983). Additional rots of imported apples in the Philippines. *Philippines Agriculture*, *66*, 156–159.

Quimio, A. J. (1985). *Survey of diseases and insect pests of some minor fruits and their control. Res. Highlights* (p. 22). PCARRD Network. Abstr.

Quimio, T. H., & Abilay, L. E. (1977). Unreported species of *Cercospora* in the Philippines. *Nova Hedwigia*, *28*, 533–541.

Quimio, T. H., & Abilay, L. E. (1982). Some unreported fungal genera and species in the Philippines. *Philippines Agriculturist*, *65*, 253–258.

Quimio, T. H., & Quimio, A. J. (1974). Compendium of postharvest and common diseases of fruits in the Philippines. *UPCA Techincal Bulletin*, *34*, 44.

Quimio, A. J., & Sumabat, R. S. (1983). Studies on the chemical control of the major diseases of minor fruits in the Philippines. In *PCARRD Res. High-lights, Los Baños, Laguna*.

Quimio TH, Abilay LE, De Villa EA, & Elauria J. 1982. Unreported and/or undescribed fungal species and genera in the Philippines. *Philippines Phytopathology*. *18*:17, (Abstr).

Quiniones, S.,S. (1980). Notes on diseases of forest trees in the Philippines. *Sylvatrop, the Philippine Forest Research Journal*, *5*, 263–271.

Quiniones, S. S., & Zamora, R. A. (1987). Forest pests and diseases in the Philippines. In E. D. De Guzman, & S. T. Nuhamara (Eds.), *Forest pests and diseases in Southeast Asia*. BIOTROP Special Publ. No. 26.

Quitco, R. T. (1989). *Investigations on the cause and effects of peanut (Arachis hypogaea L.) seed discoloration*. MS thesis. UPLB.

Ramos, E. B., Sales, A. R., & Ancheta, C. O. (1970). A guide to apple growing in Baguio and environs. *Philippine Journal of Plant Industry*, *35*, 177–202.

Reinking, O. A. (1918). Philippine economic plant diseases. *Philippine Journal of Science*, *13*, 165–216.

Reinking, O. A. (1919). Host index of diseases of economic plants in the Philippines. *Philippines Agriculturist*, *8*, 35–54.

Reinking, O. A. (1920). Diseases of sugarcane in the Philippines. *Sugar News*, *1*, 22–39.

Reinking, O. A. (1921). Citrus diseases of the Philippines, Southern China, Indo-China, and Siam. *Philippines Agriculture*, *9*, 121–179.

Ren, C. M., Gao, B. D., & He, Y. C. (2001). Advance in rice resistance to rice sheath blight. *Plant Protection*, *27*(1), 32–36.

Reyes, G. M. (1933). The black smut or bunt of rice (*Oryza sativa* L.) in the Philippines. *The Philippine Journal of Agriculture*, *4*, 241–270.

Reyes, G. M. (1939). Rice diseases and methods of control. *Philippines Journal of Agriculture*, *10*, 419–436.

References

Reyes, G. M. (1941). Notes on diseases affecting maize in the Philippines. *The Philippine Journal of Agriculture, 12,* 61–71.

Reyes, G. M. (1956). A note on the occurrence of a species of corn rust new to the Philippines. *Araneta Journal of Agriculture, 3,* 68–71.

Reyes, T. T., Escober, J. T., & Pusag, C. (1980). Sugarcane diseases in the Philippines. In *Proceeding of Symposium Philippines Phytopathology 1917–1977* (pp. 329–368).

Reyes, G. M., & Orillo, F. T. (1957). The problem of citrus diseases in the Philippines. *Plant Industry Digest, 51,* 30–33.

Reyes, T. T., & Quebral, F. C. (1967). Common diseases of sugarcane in the Philippines and their control. *UPCA Technical Bulletin, 19,* 37 p.

Rimando, R. (1988). Survey of diseases affecting some essential oil-bearing species. In *PCARRD Highlights, Los Baños* (p. 191).

Rivera, J. R., & Cano, I. B. (1966). Phytopathological note: Leaf blight, purple spot, and chlorotic streak of sugarcane in Negros Island, Philippines. *Philippines Phytopathology, 2,* 53–58.

Robinson, C. B. (1908). Sugarcane smut. *Philippine Agricultural Review, 1,* 295–297.

Rodrigo, P. A., & Seggay, L. V. (1953). The culture and propagation of roses in the Philippines. *Philippines Journal of Agriculture, 18,* 145–168.

Roldan, E. F., & Tecson, J. P. (1935). The red rot of sugarcane caused by *Colletotrichum falcatum* Went. *Philippines Agriculturist, 24,* 126–141.

Sacay, F. M. (1939). *Plant diseases in A handbook of Philippine agriculture* (pp. 233–342). UPCA.

Sales, C. P. (1966). *Rhizoctonia disease of corn. I. A study of the disease, its causal fungus and methods of control.* BS thesis. UPCA.

Samuels, G. J., Ismaiel, A., Rosmana, A., Junaid, M., Guest, D., Mcmahon, P., Keane, P., Purwantara, A., Lambert, S., Rodriguez-Carres, M., & Cubeta, M. A. (2012). Vascular streak dieback of cacao in Southeast Asia and Melanesia: In planta detection of the pathogen and a new taxonomy. *Fungal Biology, 116*(1), 11–23.

San Juan, M. O. (1980). Endemic diseases of bananas in the Philippines. In *Proceeding of Symposium Philippines Phytopathology 1917–1977, PPS, Inc., December 14–15, 1977* (pp. 133–151).

San Juan, M. O. (1983). Vascular streak dieback of cacao. *Philippines Phytopathology, 19,* 1. Abstr.

Sandoval, R. F. C., & Cumagun, C. J. R. (2019). Phenotypic and molecular analyses of Rhizoctonia spp. associated with rice and other hosts. *Microorganisms, 7*(3), 88.

Serrano, F. B., & Marquez, S. L. (1926). The red rot disease of sugarcane and its control. *Philippine Agricultural Review, 19,* 263–265.

Shahjahan, A. K. M., & Mew, T. W. (1986). Sheath spot of rice in the Philippines. *International Rice Research Newsletter (Philippines), 11,* 7.

Spatafora, J. W., Aime, M. C., Grigoriev, I. V., Martin, F., Stajich, J. E., & Blackwell, M. (2017). The fungal tree of life: From molecular systematics to genome-scale phylogenies. *The Fungal Kingdom,* 1–34.

Stevens, F. L., & Atienza, J. O. (1931). Diseases of cultivated ginger. *Philippines Agriculturist, 30,* 171–176.

Taguiam, J. D., Evallo, E., & Balendres, M. A. (2021a). Reduction of Selenicereus stem cuttings weight by fungal plant pathogens during storage. *Journal of Phytopathology, 169*(9), 577–580.

Taguiam, J. D., Evallo, E., & Balendres, M. A. (2021b). Epicoccum species: Ubiquitous plant pathogens and effective biological control agents. *European Journal of Plant Pathology, 159*(4), 713–725.

Taguiam, J. D., Evallo, E., Bengoa, J., Maghirang, R., & Angelo, M. (2020c). Detection of *Nigrospora sphaerica* in the Philippines and the susceptibility of three *Hylocereus* species to reddish-brown spot. *Journal of the Professional Association for Cactus Development, 22,* 49–61.

Taguiam, J. D., Evallo, E., Bengoa, J., Maghirang, R., & Balendres, M. A. (2020a). Pathogenicity of *Epicoccum sorghinum* towards dragon fruits (*Hylocereus* species) and in vitro evaluation of chemicals with antifungal activity. *Journal of Phytopathology, 168*(6), 303–310.

Taguiam, J. D., Evallo, E., Bengoa, J., Maghirang, R., & Balendres, M. A. (2020b). Susceptibility of the three dragon fruit species to stem canker and growth inhibition of *Neoscytalidium dimidiatum* by chemicals. *Journal of Plant Pathology, 102*(4), 1077–1084.

Tandang, A. G. (1992). *On the pathogenicity test of* Trichoderma *spp. on selected cultivated crops.* BS thesis. UPLB.

Tangonan, N. (1999). *Host index of plant diseases in the Philippines* (3rd ed.). Department of Agriculture, Philippine Rice Research Institute.

Tangonan, N. G., Kakishima, M., Sebastian, F. A., & Empanado, R. A. (1997). Rust of kalachuchi (Plumeria acuminata Ait.) caused by Coleosporium plumeriae Pat. *USM R & DJ, 5*(1), 1—4.

Taylaran, A. D., Bagsic-Posada, I., Cueva, F. D., & Balendres, M. A. (2021). First report of *Lasiodiplodia theobromae* causing jackfruit (*Artocarpus heterophyllus*) brown rot in the Philippines. *Indian Phytopathology, 74*(4), 1151—1153.

Teferi, M. H. (1988). *Taxonomy, distribution, and pathogenicity of Fusarium spp. isolated from different soils.* MS thesis. UPLB.

Teodoro, N. G. (1923). A study of Macrosporium disease of onions. *Philippine Agricultural Review, 16,* 233—275.

Teodoro, N. G. (1926). Rubber tree diseases and their control. *Philippine Agricultural Review, 19,* 63—73.

Teodoro, N. G. (1958). Plant diseases. *Agricultural Industrial Life, 20.*

Teodoro, N. G. (1959). Plant diseases. *Agricultural Industrial Life, 21.*

Teodoro, N. G. (1960). Plant diseases. *Agricultural Industrial Life, 22.*

Teodoro, N. G. (1961). Plant diseases. *Agricultural Industrial Life, 23.*

Teodoro, N. G., & Bogayong, J. R. (1926). Rice diseases and their control. *Philippine Agricultural Review, 19,* 237—241.

Teodoro, N. G., & Gomez, E. T. (1926). Coffee diseases and their control. *Philippine Agricultural Review, 19,* 249—257.

Teodoro, N. G., & Serrano, F. B. (1926). Abaca heart and bunchy-top diseases and their control. *Philippine Agricultural Review, 19,* 243—248.

The World Bank. (2021). Philippines economic Update December 2021 Edition. In *Regaining lost ground, revitalizing the Filipino workforce.* World Bank's Macroeconomics, Trade, and Investment (MTI) Global Practice (GP).

Uichanco, L. B. (1959). *Philippine agriculture* (2nd ed., p. 920). UPCA.

Van Dillewizn, C., & Lopez, M. E. (1964). Leaf scorch of sugarcane. *Sugar News, 30,* 73—74.

Viswanathan, R., & Rao, G. P. (2011). Disease Scenario and management of major sugarcane diseases in India. *Sugar Tech, 13,* 336—353.

Wharton, P. S., & Diéguez-Uribeondo, J. (2004). The biology of Colletotrichum acutatum. *Anales del jardín botánico de Madrid, 61,* 3—22.

Yago, J. I., & Chung, K. R. (2011). First report of Twig blight disease of citrus caused by *Haematonectria haematococca* in the Philippines. *Plant Disease, 95*(12), 1590-1590.

Yap, L. R., & Kulshreshtha, D. D. (1978). A preliminary survey of seedborne fungi of maize from Visayas, Philippines. *Philippine Journal of Plant Industry, 42—43,* 1—5.

CHAPTER 8

Epidemiology of fungal plant diseases in the Philippines

Ireneo B. Pangga, John Bethany M. Macasero and Joselito E. Villa

Institute of Weed Science, Entomology and Plant Pathology, College of Agriculture and Food Science, University of the Philippines Los Baños, College, Laguna, Philippines

1. Introduction

Plant disease epidemiology deals with host—pathogen interactions that are dependent on abiotic and biotic environments causing disease and crop losses under man's influence (Zadoks & Schein, 1979). These interacting factors are discussed in this chapter focusing on the epidemiology, assessment of yield loss, disease and yield loss modeling, and climate change responses of several major fungal diseases of economically important crops in the Philippines.

Rice blast, sheath blight and brown spot are major diseases of rice in the Philippines that have caused severe yield losses (Castilla et al., 2020; Teng et al., 1990). Rice blast caused yield losses in the Bicol region of 50%—95% in 1962 (Olivarez et al., 1977) and 50%—60% in several thousand hectares in Leyte in 1963 (Nuque, 1963). Yield loss due to panicle blast ranged from 70% to 85% in Laguna and Quezon provinces in 1969—70 (Nuque, 1970; Teng & Revilla, 1996). Sheath blight had caused 25%—50% yield loss in the Philippines (Olivarez et al., 1977). Sheath blight has assumed economic importance due to the semidwarf and nitrogen-responsive rice cultivars that may have created a conducive microclimate for sheath blight (Raymundo, 2006; Teng & Revilla, 1996). Brown spot caused the Great Bengal Famine in 1942—43 in India that inflicted up to 80% yield loss (Padmanbhan, 1973). Rice disease surveys in the 1990s showed that sheath blight and brown spot were significant yield reducers causing 5%—10% and 0%—10% yield losses, respectively, in tropical Asia (Savary, Ficke, et al., 2012).

Fusarium and Aspergillus ear rots, Stenocarpella disease complex, banded leaf and sheath blight (BLSB) and Southern leaf blight are economically important diseases of maize in the Philippines with epidemics and yield losses described as follows. Fusarium ear rot incidence had been increasing in past years in many maize-growing areas with *Fusarium verticillioides*

Mycology in the Tropics
https://doi.org/10.1016/B978-0-323-99489-7.00007-X

189

© 2023 Elsevier Inc. All rights reserved.

(Saccardo) Nirenberg as the predominant fumonisin-producing *Fusarium* species (Pascual et al., 2016). The total annual social cost of aflatoxin contamination in maize caused by *Aspergillus flavus* Link was about AU$ 86.13 million in 1991 (Lubulwa & Davis, 1994). Diplodia ear rot, now known as Stenocarpella ear rot caused by *Stenocarpella macrospora* (Earle) Sutton, caused considerable damage to maize with 15% commonly observed incidence in northern Mindanao (Dalmacio, 1998). Southern corn leaf blight (SCLB) caused an epidemic in Texas cytoplasmic male sterile corn (Tcms) in the US corn belt in 1970−71 inflicting 15% yield loss amounting to US$ 1 billion (Brun, 2017). However, the susceptibility of cytoplasmic male sterile corn to SCLB was discovered 10 years earlier by Mercado and Lantican (1961) in the Philippines. Maize foliar diseases cause yield losses estimated at 15%−20%. BLSB of corn has attained importance in many maize-growing areas even if considered a minor disease previously (Dalmacio, 1998; Raymundo, 2006), and no high degree of resistance has been found based on BLSB screening of advanced maize populations, inbred lines and parentals (Pascual & Salazar, 2001).

In perennial crops, mango anthracnose and banana black Sigatoka, Panama wilt and Freckle diseases are economically important in the Philippines. The mango industry in Mindanao where prolonged wet periods prevail, experienced 20% loss due to anthracnose and stem-end rot (Esguerra et al., 2006). In Guimaras, estimated losses from harvest to fruit retail could reach 19%−20% and have been attributed to anthracnose disease and mishandling at retail end (Galvan, 2020). In general, losses in banana "Cavendish" plantations due to black Sigatoka are 30%−40% due to reduced yield and premature or uneven ripening of the fruit (Jackson, 2014). In 2015, 15,000 ha of banana plantations in Mindanao were infected by Panama or Fusarium wilt from a report from the Department of Agriculture (de la Cruz & Jansen, 2018). Freckle disease of banana is becoming a serious problem in "Cavendish" plantations because blemished fruit is not accepted by importing countries (Corcolon & Raymundo, 2008a).

Empirical and mechanistic models of disease development and/or yield loss of major fungal crop diseases in the Philippines are discussed in this chapter. Empirical models are descriptive or correlative such as regression models while mechanistic models are explanatory dealing with the processes of the system. These models could be used for disease management, risk prediction or forecasting and estimation of yield loss (Campbell & Madden, 1990). The availability of yield loss data is critical in plant disease management and research policy making but reliable estimates of yield losses in important crops in the Philippines are limited (Raymundo, 2006).

Climate change may affect yield losses due to plant diseases, efficacy of disease management strategies and geographical distribution of plant diseases (Chakraborty et al., 2000). Limited studies have focused on the direct effects of climate change variables on individual diseases but indirect effects of global changes via cropping practices and socio-economic factors are more rapid than climate-induced effects. Approaches to study climate change impacts on plant diseases include simulation modeling, ecological niche modeling, and geographic information system (GIS) risk mapping (Savary et al., 2011), of which several studies in the Philippines are discussed. Bioclimatic envelope or ecological niche modeling predicts species distribution using climate variables or by understanding the physiological responses of the species to climate change (Guisan & Zimmermann, 2000).

2. Epidemiological studies

2.1 Rice blast

Basic studies on the rice blast disease cycle were done at the International Rice Research Institute (IRRI) including dispersal, colonization and sporulation. Spore trapping of rice blast fungus showed that trapped conidia have a unimodal pattern across the season with peaks before and at the middle of the cropping season. Sedimentation rather than impaction was the major factor for blast spore deposition because few spores were observed on glass rods. Spore catch results were significantly correlated with several weather variables (Pinnschmidt et al., 1993). Under upland conditions, mean infectious period was 11.5, 8, and 7.5 days on lesions on leaf blades, leaf sheaths, and panicle necks, respectively (Pinnschmidt, Bonman, & Kranz, 1995).

The effects of climatic factors on leaf blast were studied at IRRI. Conidial release occurred at an optimum temperature of 20°C under relative humidity (RH) > 80% at 5 h after inoculation. Appressorial formation was maximum at 16−24°C but it was inhibited below 80% RH. Maximum sporulation potential, the capacity of the fungus to sporulate at optimum conditions, was observed at 11 and 12 days after lesion appearance at 20 and 25/15°C (day/night), respectively. Leaf blast infection requires a minimum of 6 h of leaf wetness. The number of lesions increased linearly up to 24 h at constant leaf wetness at 19−26°C. An average dew period per night of 9.34 h was observed at 10 × 10 cm spacing, which was higher than those at 20 × 20 and 40 × 40 cm spacing with 6.64 and 6.34 h, respectively (El Refaei, 1977).

Water management affects rice blast development with variable results. Preinfection water stress at 22 days after sowing (DAS) prolonged the infectious period and increased relative infection efficiency and lesion size, but it did not affect latent period. Leaf blast lesions on rice plants with severe leaf roll produced 3.5 times more conidia than on water-unstressed plants (Gill & Bonman, 1988). However, another study showed that preinoculation drought stress increased the relative infection efficiency of both leaf blast resistant and susceptible lesions but there was no effect on lesion size and sporulation. Latent period was significantly shorter on drought-stressed plants than unstressed plants (Klein-Gebbinck, 1995). Leaf and neck blast severities of upland rice were increased by water deficit at the vegetative stage of 42−46 DAS in the dry season (Bonman et al., 1988). On the other hand, flooding effect on leaf blast depended on the genotype. Flooding was effective in reducing apparent infection rate and terminal disease severity of a partially resistant cultivar, IR36, but not in a susceptible cultivar, IR50 (Sah & Bonman, 1992).

The relationship between leaf blast and neck/panicle blast is still unclear but few studies have been done. The effect of blast isolate origin on the leaf or neck was determined on the aggressiveness of the rice blast pathogen on leaves and necks in IRRI glasshouse experiments. The results suggested that there is no specialization within the blast population on the type of organ infected. Blast isolates originating on leaves can infect necks. Infection efficiency, sporulation intensity and lesion area on leaves are positively correlated with lesion length and sporulation intensity on necks indicating similar mechanisms in the nondispersive phase of the disease cycle (Ghatak et al., 2013). Based on regional surveys of rice crop health in South and Southeast Asia from 1997 to 2011, chi-square tests showed that the probability of neck blast occurrence conditional to leaf blast occurrence is 0.45 while the probability of neck blast occurrence if leaf blast did not occur is 0.10 (Savary et al., 2022).

2.2 Rice sheath blight

Sheath blight spread has been investigated in several studies. In *R. solani*-inoculated field experiments at IRRI, the focus expansion of sheath blight was wider and faster in the rainy season than in the dry season (Castilla et al., 1994). The spatiotemporal structure of rice sheath blight was studied in a naturally-infected farmer's field in Laguna. In rainy and dry seasons, the epidemics were initiated from a random distribution of inoculum and sheath blight showed clumping of infected hills as the season continued. The higher rate of disease increase in the rainy season was associated with spatial structure at 1 m scale indicating aggregation and occurrence of polycyclic epidemic phase while the lower rate of disease increase in the dry season was barely limited to spatial structure beyond the scale of one hill showing limited aggregation (Savary et al., 2001).

There are several studies on the effect of crop establishment factors on sheath blight development. High crop density influenced the sheath blight severity on tillers, in which 15×15 cm and 15×20 cm spacing produced 20% and 19% infection, respectively, which were significantly higher than the low crop density treatments (Leaño et al., 1993). The effects of direct-seeding and plant spacing on sheath blight development were studied in Pila, Laguna. The results showed that 25×25 cm spacing showed a delay in the onset of epidemic and lower disease incidence compared to the shorter spacing treatments. Direct-seeded rice had a lower apparent infection rate based on disease incidence and terminal severity than transplanted rice regardless of plant spacing (Willocquet et al., 2000).

The effect of applied nitrogen (N) on sheath blight development was investigated in several studies. An experiment on the effect of two N rates (60 and 120 kg/ha) on sheath blight showed that increasing nitrogen rates increased the development of sheath blight in the field (Shahjahan & Mew, 1992). Focus expansion of sheath blight was faster as N increased from 0 to 180 kg/ha and in plots where source plants were inoculated with high inoculum levels at the upper layer (Castilla et al., 1994). The sheath blight foci expanded most rapidly at the highest level of 120 kg N/ha as compared to the medium level of 80 kg N/ha and no N treatment. The focal expansion of sheath blight was driven by N via increased tissue contacts in the canopy and higher leaf wetness. Inoculation at the leaf level of the canopy on the source hills resulted in faster spread of the disease than that at the sheath level (Savary et al., 1995).

2.3 Rice brown spot

There are a few epidemiological studies on rice brown spot in the Philippines. No brown spot lesion was observed at leaf wetness duration of less than 8 h. Brown spot lesions increased as leaf wetness duration increased from 8 to 20 h and decreased at 24 h. The effect of constant temperature on the number of brown spot lesions at 100% RH showed that 25°C had the highest lesion number of 39.7 as compared to 5.2 and 5.3 at 15 and 20°C, respectively, at 3 weeks after inoculation (Obusan et al., 1982). A study in Benguet and Ilocos Sur in northern Philippines determined the effect of rice straw compost amendment on copper toxicity and rice brown spot disease. Mine tailings contamination of the rice field caused the high soil copper content and high brown spot disease severity. There was a positive correlation between soil copper concentration and area under the rice brown spot severity curve. Rice straw compost application reduced soil copper concentration and rice brown spot severity in both locations by 9%—23% and 12%—37%, respectively (Malamnao, 2017).

2.4 Maize Stenocarpella disease complex

A study was conducted to determine the effect of site of inoculation (leaf, stalk, ear and combinations) and inoculum concentration (1, 2, 3, and 4×10^4 spores mL^{-1}) of *Stenocarpella macrospora* on the disease progress of the Stenocarpella disease complex in maize in the first (November–March) and second (June–October) seasons (Alovera & Raymundo, 2004a). Leaf blight severity increased as the inoculum concentration increased in both seasons. Inoculation on the ear and other plant parts caused the highest ear rot severity of 100% and 97% in the first and second seasons, respectively. Leaf inoculation did not cause stalk rot infection in both seasons but caused ear severity of 12.67% and 18.3% in the first and second seasons, respectively, at the highest inoculum concentration of 4×10^4 spores mL^{-1}.

2.5 Perennial crops

Basic epidemiological studies on mango anthracnose caused by *Colletotrichum gloeosporioides* (Penz.) Penz. and Sacc. have been conducted in the Philippines. The optimal temperature for growth of mycelia and formation of germ tubes of C. *gloeosporioides* was at 25–30°C while spore germination and fungal growth increased as RH increased from 90% to 100% but was inhibited at below 90% (Estrada & Ilag, 1991). Isolates of C. *gloeosporioides* were observed to show differences in their stimulatory temperature to produce appressoria with isolates I-2 and I-4 being stimulated at 25 and 20°C, respectively (Estrada et al., 2000). In a 7-year-old "Carabao" mango tree, infected leaves constitute the most significant source of inoculum followed by infected branch terminals indicating that sanitation by pruning could reduce initial inoculum (Opina & Eusebio, 1997). The anthracnose pathogen can infect young leaves only up to 10 days after bud break indicating a relatively small window for leaf infection. Protecting the leaves during this time by protective fungicides and cultural practices can reduce initial inoculum (Opina et al., 1997).

Weather factors critical to *Mycosphaerella fijiensis* M. Morelet sporulation and black Sigatoka disease progress were studied in a "Cavendish" banana plantation in Davao del Norte. Correlation analysis showed that number of trapped ascospores and conidia was highly correlated with maximum temperature at 34–36°C at previous 4–8 weeks and the correlation coefficient was higher than that of amount of rainfall. Black Sigatoka severity was highly correlated with cumulative number of spores. Maximum temperature of 34–36°C at previous 3–8 weeks had the highest correlation with disease severity among the weather variables. Highly significant multiple linear regression models of black Sigatoka severity as affected by weather variables at previous 4, 5 and 6 weeks were generated (Ordoyo et al., 2009).

The influence of weather factors on the expression of resistance to black Sigatoka in eight banana varieties was studied under natural infection in Los Baños, Laguna. Across varieties, the number of necrotic spots was positively correlated with adjusted rainfall, minimum temperature and RH with the highest correlation coefficient at previous 8 weeks, and negatively correlated with solar radiation and maximum temperature with highest correlation coefficients at previous 1 and 5 weeks, respectively. Rainfall and minimum temperature were positively correlated with infection index but evaporation showed the reverse effect. Multiple linear regression equations of the number of necrotic spots on six necrotic spot-forming varieties had rainfall at previous 8 weeks as the common significant weather factor (Galang, 2013).

194   8. Epidemiology of fungal plant diseases in the Philippines

Recent studies on *Fusarium oxysporum* f.sp. *cubense* Tropical Race 4 (Foc TR4) investigated its occurrence in soil and water and alternate hosts in Mindanao. Using Taqman probe-based qPCR assay, inoculum of Foc TR4 was detected as deep as 1 m below the surface of soils planted to banana cvs. Cavendish and Lacatan and in surface water of rivers and reservoirs inside and at the borders of a commercial banana plantation indicating the risk of Foc TR4 reaching nearby farms. Using LAMP TR4 diagnostics, Foc TR4 was detected in eight symptomless weed species indicating its survival in nonhost plants (Salacinas et al., 2019). Using realtime PCR, five most dominant weed species in "Cavendish" farms were colonized after artificial inoculation by Foc TR4 and *Eleusine indica* L. was found to be naturally colonized by Foc TR4 in the field. Reisolated Foc TR4 isolates from artificially inoculated and naturally colonized but asymptomatic weeds were pathogenic and more virulent when reintroduced back to Cavendish banana (Catambacan et al., 2021).

Freckle disease on different banana cultivars caused by *Phyllosticta musarum* (Cooke) Van der Aa appears 2—4 weeks after bud emergence and becomes most noticeable at harvest. Using artificial spray inoculation, different ranges of incubation period on the leaves and fruits were observed from 7 to 38 days under field condition and 4—16 days on detached fruits (Table 8.1) (Corcolon & Raymundo, 2008b).

3. Crop loss studies

3.1 Rice blast

Several studies have determined yield losses to blast in the Philippines. A yield loss of 15.2% was observed at 75% blast disease severity in rice cv. IR8 in the 1968 dry season under artificial inoculation (Exconde & Raymundo, 1970). A yield loss of 20.9% was observed in IR50 with a 45.8% severe neck blast incidence in 1989 (Bonman et al., 1991). An empirical damage function of blast disease-yield loss relationship was developed by Torres & Teng (1993). From three experimental trials, the equation was as follows:

$$Y = 5.7909 + 1.3854X2 + 0.4740X3$$

where Y = percent yield loss, X2 = leaf blast severity and X3 = panicle blast incidence with an R^2 of 0.75**. The predictive model explained 75% of the variation in yield loss. The effect of leaf blast is close to three times the effect of panicle blast on percent yield loss.

TABLE 8.1   Incubation period (number of days) of Freckle disease caused by *Phyllosticta musarum* on leaves and fruits of three banana cultivars using artificial spray inoculation (Corcolon & Raymundo, 2008b).

Cultivar	Leaf (field)	Fruit (field)	Fruit (detached)
Cavendish	16—28	17—27	4—11
Lacatan	9—44	7	9—11
Cardava	11—28	38	12—16

II. Fungi in agriculture, health and environment

3.2 Rice sheath blight

Several studies at IRRI explored the effect of crop management factors on yield losses to sheath blight. Dilla (1993) determined yield losses to sheath blight due to crop density, nitrogen and amount of inoculum. Percent yield loss increased as the amount of inoculum increased. The highest yield loss was 19.4 and 20.4% in the wet and dry seasons, respectively, at the highest inoculum level of 120 g m^{-2}. Shahjahan and Mew (1992) studied the effect of N and crop spacing on yield loss of three cultivars, with N influencing yield loss in some cultivars. Yield losses were 23.4% and 22.9% under upland and lowland conditions, respectively, as means of all cultivars. Cu et al. (1996) studied the effect of sheath blight on yield in an intensive rice production system as affected by N. Grain yield was increased by N, but reduced at the highest N which could be due to the increasing sheath blight incidence as N increased. Yield loss due to sheath blight ranged from 8% to 20% and 30% and 42% in the 1993 and 1994 dry seasons, respectively. Another field study determined the relationship between sheath blight development and yield loss at different N levels in four rice varieties. Sheath blight index, relative lesion height, and yield loss due to sheath blight increased as N level increased but the level of yield loss varied among varieties (Tang et al., 2007).

3.3 Multiple rice fungal diseases

Yield losses due to rice pests, diseases and weeds in a range of production situations (PS) were quantified in a series of lowland rice experiments in farmer's fields in Laguna and IRRI experimental farm using different crop management factors representing different PS characterized in surveys in tropical Asia The results showed that sheath blight, brown spot and leaf blast caused yield losses between 1% and 10% in tropical Asia. Correspondence analysis showed that medium yield losses were associated with low sheath blight level while high yield losses were associated with medium to high levels of sheath blight and neck blast and any level of leaf blast and brown spot. A principal component regression model showed that yield losses to sheath blight increased with increasing attainable yield (Savary, Willocquet, Elazegui, Castilla et al., 2000).

3.4 Maize banded leaf and sheath blight

Yield loss of maize to the banded leaf and sheath blight caused by *Rhizoctonia solani* Kuhn was assessed in two studies. Four maize varieties were inoculated with *R. solani* by varying the number of inoculated plants to establish five disease intensities and highest relative lesion height (HRLH) and yield loss were assessed. As disease intensity increased from 25% to 100%, yield loss increased from 5.72% and 6.94% to 19.19% and 21.39% in the dry and wet seasons, respectively. Multiple linear regression analysis showed the best predictor equations of percent yield loss based on HRLH were between midwhorl to silking growth stages with R^2 values of 0.93 and 0.83 in the dry and wet seasons, respectively (Evangelista et al., 2001). Twenty maize varieties were inoculated with sorghum grains inoculated with *R. solani* mycelia and determined yield losses. Mean lesion length differed among varieties and was higher in the wet season than in the dry season per variety. Yield losses varied among varieties ranging from 1.29% to 18.53% and 4.87% to 16.78% in the dry and wet seasons, respectively (Pangga, 1987).

3.5 Maize Stenocarpella disease complex

Yield loss of maize to Stenocarpella disease complex was assessed under the effects of inoculation site (leaf, stalk, ear and combinations) and inoculum levels (1, 2, 3 and 4×10^4 spores mL^{-1}) in the first (November to March) and second (June to October) seasons (Alovera & Raymundo, 2004b). Yield decreased as the inoculum level increased. Yield loss ranged from 8.86% to 100% and 5.51% to 100% in the first and second seasons, respectively. Ear inoculation and in combination with inoculation at other plant parts had 100% yield loss regardless of inoculum concentration. Response surface models of yield loss (%) based on disease severities at different assessment dates were obtained for the first and second seasons.

3.6 Perennial crops

Two studies have estimated yield losses to major diseases of banana. Scheerer et al. (2018) used a quantified approach to estimate production losses caused by Foc TR4 spread in different countries over a 25-year period via rate of internal spread. The factors determining the rate of internal spread were the quality of internal quarantine measures, importance of "Cavendish" and importance of banana for research investment and public policy. Assuming a 50% increase in rate of loss in production area in successive 5-year periods, the Philippines will have 8%, 19%, 34%, 52% and 71% loss in banana production area due to Foc TR4 every 5 years up to 25 years, respectively, from 2018. Corcolon and Raymundo (2008b) estimated yield loss due to freckle disease in banana in the absence of fungicide control. Fruits downgraded from Class A to Class B due to freckle infection were as high as 78% while 59% were rejected. The freckle infection severity on the leaves was highly correlated with the number of fruits which were either rejected or downgraded to a lower class.

4. Survey studies

A "Survey Portfolio" was developed to produce a standardized procedure designed for large scale assessment and analysis of rice crop health and yield losses in farmer's fields (Castilla et al., 2020; Savary & Castilla, 2009). Using this survey protocol, pest surveys were conducted in farmers' fields in six sites in tropical Asia including Central Luzon, Laguna, and Iloilo in the Philippines from 1987 to 1998 to characterize the injury profiles (IP) of insect pests, diseases and weeds in relation to PS and yield. Cluster analysis was used to categorize the different IP and PS across the six sites and correspondence analysis was used to analyze linkages among IP, cropping practices, weather and yield. Stem rot and sheath blight were in IP1 (common in Central Luzon and Iloilo) that was associated with high fertilizer input, long fallow period, low pesticide use and good water management in mostly transplanted rice in a rice—rice rotation. Leaf blast and brown spot were in IP2 (common in Central Luzon, Iloilo, and Laguna) that was associated with low fertilizer and pesticide input with poor water management or fertilizer and pesticide input were high with adequate water management in direct seeded rice in rice—rice rotation. Yield levels were not associated with any particular IP and the relationship between yield level and IP is indirect (Savary, Willocquet, Elazegui, Teng et al., 2000).

A recent study analyzed a first phase of survey data in tropical Asia from Savary, Willocquet, Elazegui, Teng, et al., (2000) and a second phase of survey data from 2009 to 2011 in the same sites but including Cotabato, Philippines (Savary et al., 2022). Multiple correspondence analysis showed that Central Luzon, Laguna and Cotabato survey data were linked with transplanted rice with rice—rice rotation and synchronized planting in PS1 (medium duration of fallow period and frequent planting of hybrids) and PS2 (one-month fallow period using only high-yielding varieties), and IP1 (moderate but highly variable sheath blight and grain discoloration, low but variable brown spot and neck blast, and moderate-low but variable leaf blast). A decrease in sheath blight was observed in comparing the two phases which can be due to its reduced spread in direct-seeded rice. Glume discoloration intensity nearly doubled between the two phases which can be due to hybrid rice being more infected by the brown spot pathogen and the significant positive correlation between brown spot and glume discoloration (Savary et al., 2022). The second phase of surveys in the 2000s indicated increased levels of leaf and neck blast in tropical Asia (Castilla et al., 2020).

A national pest survey was conducted from 2014 to 2018 in 1367 farmers' fields in all rice-growing regions in the Philippines to characterize and analyze the relationship between PS, pest injuries and yield (Macasero, 2018). Across regions, the incidence of sheath blight was 4.90%. High sheath blight incidence was observed in Regions IV (9.38%), XIII (8.21%) and X (6.33%). Results also indicate that dirty panicle is an emerging disease, particularly in Cordillera Administrative Region (CAR) (Table 8.2). Correspondence analysis showed that rice yield was positively associated with sheath blight in PS with high attainable yield. Brown spot of rice was identified to be most influenced by the type of ecosystem, crop establishment method and fertilization in this survey. There was a positive relationship between the increase in brown spot incidence and shifts from irrigated to rainfed ecosystem and transplanting to direct-seeding based on principal component analysis (Fig. 8.1). Across all regions, high incidence of brown spot was associated with rainfed ecosystem and direct-seeded rice but it decreased yield in both irrigated and rainfed ecosystems (Fig. 8.2) (Macasero, 2018), which follow the results of Savary, Willocquet, Elazegui, Teng et al., 2000; Savary, Castilla, et al., 2005 and Savary et al. 2022 indicating that high brown spot was found in areas where rice was direct-seeded and rainfed with low fertilization. Water management has a strong effect on brown spot using two-way analysis of variance of survey data in the 1990s where there was a significant increase in brown spot incidence in poorly irrigated fields compared to well-irrigated fields (Savary et al., 2005).

5. Plant disease modeling

Empirical and mechanistic models of major crop diseases have been developed (Table 8.3). They can be used to understand epidemics or predict risk to inform disease management or policy formulation (Savary, Ficke et al., 2012).

5.1 Rice blast

Empirical models to forecast leaf blast 5 days before at seedling and tillering stages were developed by El Refaei (1977) based on IRRI nursery experiments. The equations were based

198   8. Epidemiology of fungal plant diseases in the Philippines

TABLE 8.2   Average incidence of major rice fungal diseases by region in the Philippines in the national rice pest survey from 2014 to 2018[a] (Macasero, 2018).

Region	BS[b]	DP[c]	LB[b]	NB[c]	SHB[c]	SHR[c]
CAR[d]	286	16.41	92	1.97	1.46	0.51
Region I	261	8.16	23	1.58	5.49	1.23
Region II	105	2.41	115	3.26	5.23	1.10
Region III	126	5.77	52	2.65	5.88	1.62
Region IV	31	1.85	18	0.59	9.38	2.25
MIMAROPA[e]	166	8.72	73	0.45	3.84	0.16
Region V	125	0.24	5	0.07	2.17	1.11
Region VI	614	1.28	370	3.16	4.86	1.48
Region VII	982	6.08	111	3.91	4.46	4.25
Region VIII	104	4.92	72	3.16	5.13	1.47
Region IX	107	0.92	26	0.44	1.51	0.63
Region X	91	0.32	14	0.16	6.33	2.16
Region XI	68	5.09	49	2.71	5.83	1.28
Region XII	137	5.78	40	0.28	5.58	0.63
Region XIII	44	1.17	146	2.13	8.21	0.94
ARMM[f]	283	1.96	90	1.91	3.05	1.47

[a]*Meaning of acronyms of rice diseases in table: BS, Brown spot; DP, Dirty panicle; LB, Leaf blast; NB, Neck blast; SHB, Sheath blight; SHR, Sheath rot.*
[b]*Disease incidence used the unit of measure %-development stage unit (%-DSU) for Area under the disease progress curve (AUDPC).*
[c]*Disease incidence in percentage.*
[d]*CAR, Cordillera Administrative Region.*
[e]*MIMAROPA, Mindoro, Marinduque, Romblon, and Palawan.*
[f]*ARMM, Autonomous Region of Muslim Mindanao.*

on the exponential relationship between the number of leaf blast lesions per seedling with leaf wetness and leaf blast spore concentration in the air. However, negative coefficients in the equations cannot be interpreted biologically and the effects of ontogenetic changes on susceptibility were not considered (Katsantonis et al., 2017).

Empirical models for forecasting leaf and panicle blast were developed by Calvero et al. (1996b) using multiple regression models using weather variables identified by the Window Pane software as highly correlated with the blast disease parameters. Different weather factors were included in the models of IR50 and C22 rice cultivars in Cavinti, Laguna indicating that the developed models were cultivar and location-specific. Generally, validation tests showed that the models predicted leaf blast well except panicle blast severity on IR50. However, several weather variables such as precipitation frequency were negatively correlated with leaf and panicle blast.

5. Plant disease modeling 199

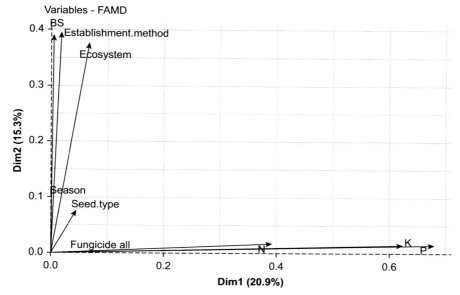

FIGURE 8.1 Relationship of production practices and brown spot incidence (BS) based on survey data from 2014 to 2018 in Philippine rice fields analyzed using Principal Component Analysis (Macasero, 2018). *Seed.type*, hybrid or inbred rice; *Fungicide all*, frequency of fungicide application; *N, P, K*, nitrogen, phosphorus and potassium application rate, respectively.

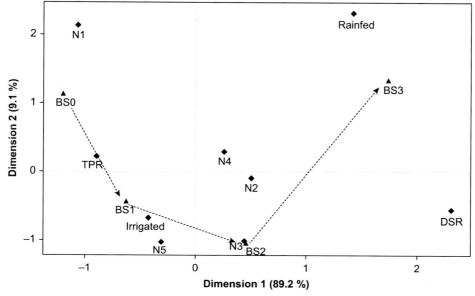

FIGURE 8.2 Plot showing the relationship between brown spot incidence (BS), nitrogen fertilization (N), crop establishment method, and rice ecosystem in Philippine rice fields based on survey data from 2014 to 2018 using correspondence analysis (Macasero, 2018). *DSR*, direct-seeded rice; *TPR*, transplanted rice.

II. Fungi in agriculture, health and environment

200   8. Epidemiology of fungal plant diseases in the Philippines

TABLE 8.3   Mechanistic simulation models and machine learning methods used in modeling major fungal plant diseases of economically important crops in the Philippines.

Host	Disease	Pathogen	Model name	Development/ Language or software	Application	Reference
Rice	Leaf blast	*Magnaporthe oryzae*	BLASTSIM.2	Disease cycle/ FORTRAN[b]	Prediction	Calvero and Teng (1991)
					Fungicide application	Calvero and Teng (1997)
Rice	Leaf blast	*Magnaporthe oryzae*	BLASTSIM.2/ CERES—Rice[a]	Coupled to crop model/ FORTRAN	Knowledge generation	Luo et al. (1997)
					Climate change	Luo et al. (1998)
Rice	Leaf blast	*Magnaporthe oryzae*	MACROS-L1D	Coupled to crop model/ FORTRAN	Knowledge generation	Bastiaans (1993)
Rice	Leaf blast Brown spot Sheath blight	*Magnaporthe oryzae* *Bipolaris oryzae* *Rhizoctonia solani*	EPIRICE	SEIR[c]/STELLA[d] & R programming language	Knowledge generation	Savary, Nelson et al. (2012)
Rice	Leaf blast	*Magnaporthe oryae*	MODEPIRICE-LB	HLIR[e]/STELLA	Climate change	Caniamo (2017)
Rice	Rice blast	*Magnaporthe oryzae*		ANN[f]/SVM[g]	Prediction	Malicdem and Fernandez (2015)
Rice	Sheath blight	*Rhizoctonia solani*	BLIGHT	Coupled to crop model/ FORTRAN	Knowledge generation	Elings et al. (1994)
Rice	Sheath blight	*Rhizoctonia solani*	BLIGHTASIRRI	Disease cycle/ FORTRAN	Prediction	Kobayashi et al. (1995)
Rice	Sheath blight	*Rhizoctonia solani*	Sheath blight simulation model	Disease cycle/ STELLA	Knowledge generation	Savary et al. (1997)
Rice	Rice blast Brown spot Sheath blight	*Magnaporthe oryae* *Bipolaris oryzae* *Rhizoctonia solani*	Pest module in CERES rice	Coupled to crop model/ FORTRAN	Knowledge generation	Pinnschmidt, Batchelor et al. (1995)
Rice	Rice blast Brown spot Sheath blight	*Magnaporthe oryzae* *Bipolaris oryzae* *Rhizoctonia solani*	RICEPEST	Coupled to crop model/Fortran Simulation Translator	Crop loss	Willocquet et al. (2002)
Maize	Southern leaf blight	*Helminthosporium maydis*		Disease cycle/ STELLA	Knowledge generation	Ciron (2004)
Maize	Southern leaf blight	*Helminthosporium maydis*	maydiSim	Disease cycle/ STELLA	Climate change	Raymundo and Bactong (2014)

II. Fungi in agriculture, health and environment

5. Plant disease modeling

TABLE 8.3 Mechanistic simulation models and machine learning methods used in modeling major fungal plant diseases of economically important crops in the Philippines.—cont'd

Host	Disease	Pathogen	Model name	Development/ Language or software	Application	Reference
Maize	Southern leaf blight	*Helminthosporium maydis*		HLIR/STELLA	Climate change	Dimasingkil (2019)
Maize	Banded leaf and sheath blight	*Rhizoctonia solani*		Disease cycle/ STELLA	Knowledge generation	Reodique (2004)
Maize	Fusarium ear rot	*Fusarium verticillioides*		Fuzzy logic	Climate change	Salvacion et al. (2015)
	Aspergillus ear rot	*Aspergillus flavus*		Bioclimatic niche model		Pangga et al. (2015)
Coffee	Coffee rust	*Hemileia vastatrix*	COFRUS	HLIR/STELLA	Climate change	Arevalo (2014)
Mango	Mango anthracnose	*Colletotrichum gloeosporioides*	MangoMan	Disease cycle/ Visual Basic	Fungicide application	Ditan (2005)
Banana	Fusarium wilt	*Fusarium oxysporum* f.sp. *cubense* tropical race 4		Maxent[h] and Fuzzy logic Maxent	Climate change Prediction	Salvacion et al. (2019a) Salvacion et al. (2019b)

[a]*CERES-Rice—Crop Environment Resource Synthesis—rice growth model.*
[b]*FORTRAN—Formula Translation programming language.*
[c]*SEIR—Suscept-Exposed-Infectious-Removed modeling approach.*
[d]*STELLA—Structural Thinking Experiential Learning Laboratory and Animation Modeling Software.*
[e]*HLIR—Healthy-Latent-Infectious-Removed modeling approach.*
[f]*ANN—Artificial Neural Network.*
[g]*SVM—Support Vector Machine.*
[h]*MaxEnt—Maximum Entropy Modeling Software.*

A multivariate procedure was developed to assess the temporal risk of tropical leaf blast (Calvero et al., 1996a). Regression models of leaf and panicle blast for IR50 and C22 developed in Cavinti and IRRI blast nursery were used to generate blast disease data using weather data from SIMMETEO weather generator. These data sets were incorporated in a matrix of diseased leaf area, panicle blast severity and sowing dates which were grouped to from blast-prone groups (BPGs) through cluster analysis. The proneness was further characterized by Principal Component Analysis (PCA) in relation to weather variables. Discriminant analysis (DA) was used to generate empirical equations for predicting risk by matching sowing dates to BPGs. IR50 was prone to leaf blast and panicle blast for July to December sowing and panicle blast for January to May sowing in Cavinti. IR50 at IRRI blast nursery was highly prone to leaf blast and panicle blast all throughout the year.

BLASTSIM.2, a leaf blast simulation model, was developed by Calvero and Teng (1991) for tropical conditions. BLASTSIM.2 has two submodels: a leaf blast disease cycle model using a state variable modeling approach and DEWFOR, a dew period simulation model (Weihong & Goudriaan, 1991). BLASTSIM.2 has been used in several simulation studies. BLASTSIM.2

II. Fungi in agriculture, health and environment

was coupled with CERES—RICE, a rice growth simulation model, by incorporating the effects of leaf blast on rice leaf photosynthesis and biomass production. Sensitivity analysis showed that temperature was a sensitive variable in the coupled model but precipitation was not a sensitive variable that may be due to no water limitation and rainfall only led to small changes in dew period (Luo et al., 1997). BLASTSIM.2 was used to analyze the effect of weather factors on simulated leaf blast development on two cultivars with different canopy structure, IR72 and an IRRI new rice plant type (NPT) line, as affected by N and plant spacing in comparison with actual field leaf blast progress. Simulated and actual leaf blast severities were higher on IR72 than on the NPT line. Simulated leaf blast severity increased as plant spacing increased. There was no significant effect of N on simulated and observed leaf blast development (Pangga et al., 1996). A modified BLASTSIM.2 model with a fungicide submodule FUNGICID was coupled to the rice growth model ORYZA1 to evaluate spray strategies for leaf blast management. Multivariate analysis was used to obtain optimum management strategies in 1991 and 1992 dry and wet season trials (Calvero & Teng, 1997).

Malicdem and Fernandez (2015) developed Artificial Neural Network (ANN) and Support Vector Machine (SVM) binary classifiers for the prediction of rice blast occurrence or nonoccurrence in Northern Philippines. The models used weather variables such as rainfall, minimum and maximum temperature, and humidity that were identified by PCA as the most important. The SVM model provided a more accurate prediction of rice blast than the ANN model with a higher R^2 of 0.7758 and lower mean square error of 0.2374.

A generic simulation model for rice diseases, EPIRICE, was developed using STELLA modeling software, converted to R programming language, and coupled with a GIS to map potential epidemics in rice worldwide. EPIRICE is a simple Suscept-Exposed-Infectious-Removed (SEIR) model (Zadoks, 1971) for five diseases parameterized using data from the literature, including leaf blast, brown spot and sheath blight. It was tested against published disease progress curves (DPCs) showing concurrence to the DPCs using visual and statistical validation methods (Savary, Nelson et al., 2012; Savary et al., 2018).

A leaf blast simulation model MODEPIRICE-LB was developed following the healthy-latent-infectious-removed (HLIR) epidemic modeling approach where diseased plants were partitioned into nonoverlapping states that is the same as the SEIR modeling approach (Madden et al., 2007). MODEPIRICE-LB model is a modification of EPIRICE for leaf blast (Savary, Nelson et al., 2012). It was parameterized using data from the literature in tropical conditions, combined with a simple rice crop growth submodel, and used leaf wetness duration output from the DEWFOR dew period model (Weihong & Goudriaan, 1991). Model validation results using rice cvs. IR50 and IR72 showed that the model-generated leaf blast disease progress curves satisfactorily simulated the observed leaf blast epidemics in four Philippine sites with different climate types: Los Baños and Calauan, Laguna (Type 3), Rosario, Batangas (Type 1), and Lucban, Quezon (Type 4) (Caniamo, 2017).

5.2 Rice sheath blight

A few simulation models of rice sheath blight disease have been developed. Kobayashi et al. (1995) developed a rice sheath blight forecasting model, BLIGHTASIRRI, by calculating disease incidence from vertical and horizontal disease development using Hashiba's formula.

Vertical development was estimated from temperature, RH and susceptibility of leaf sheath while horizontal development was estimated from temperature, RH and number of sclerotia and tillers. Savary et al. (1997) developed a rice sheath blight epidemic model based on primary and secondary infections of tillers and aggregation of infected tillers. Other factors included in the model were relative growth rate of tillers, senescence of tillers and rate of recovery from infection. The model adequately simulated actual epidemics except for the decrease in disease incidence at the latter stage of the epidemics.

5.3 Maize

A multiple linear regression model to predict fumonisin concentration in maize using insect damage to ears and weather variables was developed by de la Campa et al. (2005). The model accounted for 82% of the variability in Argentina and Cauayan and Bukidnon sites in the Philippines in 2 years of experiments. Further studies are needed for sensitivity analysis and model validation.

Mechanistic models of the Southern leaf blight of maize have been developed at UPLB. Ciron (2004) developed a Helminthosporium leaf spot of corn model based on the EPIMAY model (Waggoner et al., 1972) and SCLB model of Massie (1973). Simulated outputs compared well with the disease progress curves in Waggoner et al. (1972) and Massie (1973). A simulation model of Southern leaf blight, maydiSim, was developed by combining important components from existing models and local data (Raymundo & Bactong, 2014). A SCLB simulation model was developed following the HLIR epidemic modeling approach. The simulated SCLB disease severity was assessed in surveys in 2018 dry and wet seasons in four validation sites each in Bay, Laguna using maize cv. IPB Var six and Angadanan, Isabela using maize cv. Pioneer 3774 YHR. Model validation results indicated a satisfactory model and sensitivity analysis showed that the model was more sensitive to average daily temperature than daily RH both in one site in Bay, Laguna and Bunnay site in Angadanan, Isabela (Dimasingkil, 2019).

5.4 Perennial crops

Empirical models of mango anthracnose were developed by logistic regression of the percentage of conidia forming appressoria modified from the model by Fitzell et al. (1984) using temperature, leaf wetness and the addition of RH. Model II using 100% RH could lessen the number of fungicide spray as observed in two trials (Dodd et al., 1991). This model was tested in a field trial in 1991−92 which resulted in the application of five fewer fungicide sprays compared with a standard spray program (Estrada et al., 1996). A linear relationship was found between the number of disseminated conidia within a "Carabao" mango tree in the leaf flush stage and rainfall collected per sampler with an R^2 of 0.75 (Dodd et al., 1991). Rainfall intensity greatly influenced mango anthracnose during fruit development and rainfall >20 mm within a month before harvest caused high anthracnose infection after harvest (Estrada et al., 1996).

A forecasting system for mango anthracnose, MangoMan, was developed in Visual Basic by using daily average temperature and hours of RH $\geq$ 90 as input variables and translated into relative disease severity values (SV) which were used as basis for fungicide spray recommendations. MangoMan used the modified predictive model of Estrada et al. (1996) and modified BLITECAST (Krause et al., 1975) that converted temperature and RH into SV. MangoMan was evaluated in two seasons in Calamba, Laguna indicating that MangoMan-assisted spray program required less fungicide application to achieve the same disease and fruit yield levels with commonly used calendar-spray schedule of a fungicide (Ditan, 2005).

Environmental determinants of Foc TR4 occurrence on banana in Southcentral Mindanao were determined using Maximum Entropy (MaxEnt) using topographic, bioclimatic and edaphic variables tested against Foc TR4 occurrence data (Salvacion et al., 2019b). MaxEnt is a general purpose machine learning method that estimates a target probability distribution based on distribution of maximum entropy (i.e., closest to uniform) subject to the constraints brought about by incomplete information about the target distribution (Phillips et al., 2006). MaxEnt modeling results showed that precipitation during the driest month, during the wettest month, and of the warmest quarter; slope; and elevation were the most important variables for predicting the probability of Foc TR4 occurrence (Salvacion et al., 2019b).

6. Disease-yield loss modeling

Several mechanistic yield loss models due to major crop diseases have been developed (Table 8.3). Empirical models of yield loss are data set-specific that cannot be applied across cropping and pest conditions, thus mechanistic models are needed that simulate yield loss using a crop model coupled with damage mechanisms or "coupling points" of the disease (Pinnschmidt & Teng, 1994; Savary, Ficke, et al., 2012). The reduction in leaf photosynthetic rate, increase in maintenance respiration, assimilate uptake and premature leaf senescence were used as damage mechanisms in simulating the effect of leaf blast on rice crop growth using MACROS-L1D, a mechanistic crop growth model. Simulation results showed that 50%−70% in yield reduction are due to lesion coverage and premature senescence causing direct effects on radiation use efficiency (Bastiaans, 1993). A generic multiple pest damage module was coupled to the CERES−Rice, a rice growth simulation model, using physiological coupling points for the different types of pest or disease damage on yield including blast and sheath blight. The coupling points represent a crop rate or state variable that are influenced by the pest or disease. Multiple pest damage effects are simulated mechanistically in this approach. Validation in three field trials showed closeness of the simulated yield to observed yield with an R^2 of 0.71−0.86 indicating satisfactory validation results. Simulation results showed the dominant yield-reducing effect of panicle blast (Pinnschmidt & Teng, 1994).

RICEPEST is a production situation-driven crop growth model simulating yield losses to multiple pest injuries for tropical Asia. It simulated rice yield losses to several pests and fungal diseases including sheath blight, brown spot and sheath rot for irrigated or rainfed lowland rice using equations of the damage mechanisms. RICEPEST was validated successfully in Laguna based on grain yield and relative yield loss. It can be used to understand the mechanisms of interactions between injuries, improve pest management in a multiple pest system and for research prioritization (Willocquet et al., 2002).

7. Climate change and plant diseases

Several mechanistic plant disease models and machine learning methods have been used for simulation of the effects of climate change on plant diseases in the Philippines (Table 8.3).

7.1 Rice

A few modeling studies have explored the effects of climate change on rice blast. The BLASTSIM.2 leaf blast simulation model coupled with CERES−RICE, a rice crop growth simulation model, was used to study the effects of global temperature changes in several locations in the Philippines using daily weather data from WGEN and WMAK weather generators. An increase in temperature caused a significant decrease in rice blast severity while a decrease by −3°C caused more severe blast epidemics as indicated by high AUDPC values (Luo et al., 1998). MODEPIRICE-LB, a leaf blast HLIR model, was also used to assess the impact of climate change based on Intergovernmental Panel on Climate Change (IPCC) Representative Concentration Pathways (RCP) 4.5 (median range scenario) and 8.5 (high end scenario) using IR50. Climate change results showed that rice leaf blast will have decreasing disease severity and lesion number in RCP 4.5 and 8.5 in dry and wet seasons in Calauan, Laguna and Lucban, Quezon in 2049−2050 (Caniamo, 2017).

7.2 Maize

The SCLB simulation model by Dimasingkil (2019) was used to analyze the risk of SCLB epidemics under future IPCC climate change scenarios. Climate change simulation using this SCLB model showed that SCLB severity will increase in Bunnay site in Angadanan, Isabela under IPCC climate change scenarios of RCP 4.5 (median range scenario) and 8.5 (high end scenario) in both dry and wet seasons in 2030 and 2050 (Dimasingkil, 2019).

Bioclimatic niche modeling was used to predict the preharvest risk of Aspergillus and Fusarium ear rots in dry and wet cropping seasons in the Philippines based on the projected temperature increase in 2050 under the RCP 8.5 climate change scenario using temperature ranges from the literature. Aspergillus ear rot will increase in 2050 in the dry season than in the wet season due to the high temperature requirement of Aspergillus ear rot but medium to high risk varied in different maize growing seasons. In 2050, Fusarium ear rot will have medium to high risks in both dry and wet seasons in different maize growing areas of the country because this disease is favored by high temperature (Pangga et al., 2015).

Fuzzy logic is a soft-computing methodology that can tolerate vagueness, imprecision, and suboptimality but can provide simple, quick, and sufficient solutions (Dubey et al., 2013). Fuzzy logic methodology was used to assess the contamination risks of aflatoxin and fumonisin on maize in the Philippines under climate change using temperature and rainfall requirements. Under RCP 8.5 climate change scenario in 2050, aflatoxin contamination risk in the country will be reduced due to increased rainfall because Aspergillus ear rot requires high temperature and dry conditions. However under the same scenario, most parts of the country will be at a very high risk for fumonisin concentration due to high temperature and ample rainfall conditions that are favorable to Fusarium ear rot (Salvacion et al., 2015).

7.3 Perennial crops

A coffee rust mechanistic simulation model, COFRUS, was developed using the HLIR epidemic modeling approach. It was satisfactory validated in Silang, Cavite and Jamboree, UPLB. The effect of climate change on coffee rust using COFRUS was performed by adding 1, 2 and 3°C to the average daily temperature and simulation results showed that simulated coffee rust severity will increase and onset of coffee rust will occur earlier as temperature increases (Arevalo, 2014).

Banana suitability and Foc TR4 distribution in the Philippines were determined under baseline and future climate conditions in 2050. Banana suitability mapping was done via fuzzy logic methodology while Foc TR4 distribution was done using Maximum Entropy modeling. Climate change may have limited effect on banana suitability but projected changes in rainfall in the future will increase Foc TR4 occurrence from 20% in baseline condition to 27% in 92.5% and 28.5% of highly and moderately suitable areas for banana, respectively (Salvacion et al., 2019a).

8. Concluding remarks and future directions

Epidemiological studies generate the knowledge that constitutes the interaction of the components of the plant disease tetrahedron serving as a basis for sustainable disease management (Savary, Horgan et al., 2012; Teng, 1994). A rice blast management "tool kit" integrates epidemiological principles in knowledge, physical, communication, and policy tools that affect the amount of initial inoculum, Xo, apparent infection rate, and duration of the epidemic (Teng et al., 1991). This can also be applied to other major diseases with sufficient epidemiological knowledge but management of the interaction may be different among pathosystems. For example, rice sheath blight involves crop management practices as the main option that interacts with microclimate, crop density and partial resistance while rice blast is mainly managed by host plant resistance interacting with crop management practices such as fungicide use, physical environment and pathogen diversity (Savary, Horgan et al., 2012). Another example is an epidemiology-based integrated approach to manage black Sigatoka of banana based on Xo reduction, fungicides, cultural operations such as deleafing or removal of infected plant parts, and biological control to reduce *M. fijiensis* survival and growth (Raymundo, 2008).

Plant disease modeling can be used to understand the pathosystem and use the understanding for disease management (Savary, Ficke et al., 2012; Savary et al., 2018; Teng & Savary, 1992). A rice sheath blight SEIR model showed that rice sheath blight is influenced largely by Xo but is delayed by strong secondary spread. Thus, reduction of primary inoculum and slowing of the rate of disease increase can be used to manage the disease (Savary, 2014). A rice brown spot SEIR model showed that terminal disease severity can be greatly reduced by reducing the corrected basic infection rate, Rc, and latent period through partial resistance, microclimate, and water and nutrient management, and reducing Xo through seed cleaning (Barnwal et al., 2013). Several mechanistic simulation models (e.g., Caniamo, 2017) or machine learning methods (e.g., Malicdem & Fernandez, 2015; Salvacion et al., 2019a) could project future plant disease risk and guide disease management strategies under climate change conditions.

Limited yield loss estimates due to diseases in economically important crops in the Philippines have been emphasized (Raymundo, 2006). However in addition to crop loss reviews (Pangga, 1996; Raymundo et al., 1996; Raymundo, 2006; Teng et al., 1990), this chapter discussed available yield loss estimates from experiments, models, and surveys of several important diseases of major crops, mainly in rice and maize. Rice crop health and yield loss surveys in six sites in tropical Asia (Castilla et al., 2020; Savary, Willocquet, Elazegui, Castilla et al., 2000; Savary, Willocquet, Elazegui, Teng, et al. 2000; Savary et al., 2022) and in all major rice-growing regions of the Philippines (Macasero, 2018) found brown spot and dirty panicle as important, emerging diseases. Empirical models of yield loss may be static and data-set specific. Nevertheless, empirical models will still remain as a useful tool for applications that cannot rely on simulation models (Pinnschmidt & Teng, 1994). Advances in coupling plant disease and crop growth models to simulate yield loss have led to several mechanistic coupled disease-growth models in rice (Bastiaans, 1993; Pinnschmidt, Batchelor, & Teng, 1995; Willocquet et al., 2002).

Crop loss research in the Philippines should focus on generation of additional crop loss data on major and emerging plant diseases, estimation of crop loss to multiple pests and diseases and postharvest diseases, and development and/or use of new quantitative approaches to model yield losses but the lack of research funding may hamper progress (Raymundo, 2006). However, there are recent rice health surveys generating crop loss data such as the rice pest surveillance in farmers' fields under the Philippine Rice Information System project in 2013 and continued as the Pest Risk Identification and Management project in 2017. These projects used a modified pest survey protocol due to recent changes in rice cropping practices such as planting of new varieties and the shift from transplanting to direct seeding (Castilla et al., 2021).

References

Alovera, R. B., & Raymundo, A. D. (2004a). Stenocarpella disease of corn: I. Disease progression as affected by site of inoculation and inoculum concentration. *Journal of Tropical Plant Pathology, 40,* 1–13.

Alovera, R. B., & Raymundo, A. D. (2004b). Stenocarpella disease of corn: II. Losses as affected by site of inoculation and inoculum concentration. *Journal of Tropical Plant Pathology, 40,* 14–25.

Arevalo, L. R. (2014). *Simulation modeling of coffee rust epidemics caused by Hemileia vastatrix* Berk & Br. Unpublished BS Thesis (p. 75). University of the Philippines Los Baños, Los Baños.

Barnwal, M. K., Kotasthane, A., Magculia, N., Mukherjee, P. K., Savary, S., Sharma, A. K., Singh, H. B., Singh, U. S., Sparks, A. H., Variar, M., & Zaidi, N. (2013). A review on crop losses, epidemiology and disease management of rice brown spot to identify research priorities and knowledge gaps. *European Journal of Plant Pathology, 136,* 443–457.

Bastiaans, L. (1993). *Understanding yield reduction in rice due to leaf blast.* PhD Thesis (p. 127). Wageningen University.

Bonman, J. M., Estrada, B. A., Kim, C. K., Ra, D. S., & Lee, E. J. (1991). Assessment of blast disease and yield loss in susceptible and partially resistant rice cultivars in two irrigated lowland environments. *Plant Disease, 75,* 142–146.

Bonman, J. M., Sanchez, L. M., & Mackill, A. O. (1988). Effects of water deficit on rice blast II. Disease development in the field. *Journal of Plant Protection in the Tropics, 5*(2), 67–74.

Brun, H. A. (2017). Southern corn leaf blight: A story worth retelling. *Agronomy Journal, 109,* 1218–1224.

Calvero, S. B., Jr., Coakley, S. M., & Teng, P. S. (1996a). A procedure to assess temporal risk of tropical rice blast. *Philippines Phytopathology, 32,* 1–17.

Calvero, S. B., Jr., Coakley, S. M., & Teng, P. S. (1996b). Development of empirical forecasting models or rice blast based on weather factors. *Plant Pathology, 45,* 667–678.

Calvero, S. B., & Teng, P. S. (1991). BLASTSIM.2 a model for tropical leaf blast-rice pathosystem. *Philippine Phytopathology, 27*, 46. Abstr.

Calvero, S. B., & Teng, P. S. (1997). Use of simulation models to optimize fungicide use for managing tropical rice blast disease. In M. J. Kropff, P. S. Teng, P. K. Aggarwal, J. Bouma, B. A. M. Bouman, J. W. Jones, et al. (Eds.), *Proceedings of the second international symposium on systems approaches for agricultural development, held at IRRI, Los Baños, Philippines, 6—8 December 1995: Vol. 2. Applications of systems approaches at the field level* (pp. 305—320). Springer.

de la Campa, R., Hooker, D. C., David Miller, J., Schaafsma, A. W., & Hammond, B. G. (2005). Modeling effects of environment, insect damage, and Bt genotypes on fumonisin accumulation in maize in Argentina and the Philippines. *Mycopathologia, 159*(4), 539—552.

Campbell, C. L., & Madden, L. V. (1990). *Introduction to plant disease epidemiology* (p. 532). John Wiley & Sons.

Caniamo, J. R. C. (2017). *Simulation modeling of rice leaf blast epidemics caused by Magnaporthe oryzae B.C. Couch under present and future climate*. Unpublished MS Thesis (p. 139). University of the Philippines Los Baños, Los Baños.

Castilla, N.P., Duque, U.G., Marquez, L.V., Martin, E.C., Callejo, A.M.L., Montecillo, J.D., & Laborte, A.G. (2021). Pest surveillance system for food security: A case study in the Philippines. In Ganguly P., Siddiqui M.W., Goswami T.N., Ansar M., Sharma S.K., Anwer M.A., Prakash N., Vishwakarma R., & Ghatak A. (Eds.), *Souvenir — international web conference on ensuring food safety, security and sustainability through crop protection, August 5—6, 2020* (pp. 43-49). Bihar Agricultural University.

Castilla, N. P., Elazegui, F. A., & Savary, S. (1994). Rice sheath blight spread: A case study on focal expansion in plant disease epidemiology. *Philippine Phytopathology, 30*(1), 1—15.

Castilla, N. P., Macasero, J. B., Villa, J. E., Sparks, A. H., Willocquet, L., & Savary, S. (2020). The impact of rice diseases in tropical Asia. In P. Scott, R. N. Strange, L. Korsten, & M. L. Gullino (Eds.), *Plant diseases and food security in the 21st century* (pp. 97—126). ISPP and Springer.

Catambacan, D. G., Pangga, I. B., Ardales, E. Y., Diaz, M. G. Q., & Cumagun, C. J. R. C. (2021). Weeds as alternate hosts of *Fusarium oxysporum* f. sp. *cubense* Tropical race 4 causing Fusarium wilt in Cavendish banana. In *Paper presented at PMCP 53rd Anniversary and annual scientific conference, July 6—7, 2021, Virtual conference*.

Chakraborty, S., Tiedemann, A. V., & Teng, P. S. (2000). Climate change: Potential effects on plant diseases. *Environmental Pollution, 108*(3), 317—326.

Ciron, P. E. A. (2004). *Modeling epidemics of leaf spot caused by Helminthosporium maydis Nis. & Miy. in corn (Zea mays L.)*. Unpublished BS Thesis (p. 69). University of the Philippines Los Baños.

Corcolon, B. M., & Raymundo, A. D. (2008a). Estimating yield losses in banana due to freckle disease caused by *Phyllosticta musarum* (Cke.) Van der Aa. *Philippine Journal of Crop Science, 33*(2), 75—85.

Corcolon, B. M., & Raymundo, A. D. (2008b). Quantifying the expression of freckle disease, caused by *Phyllosticta musarum* (Cke) Van der Aa, on banana leaves and fruits. *Philippines Phytopathology, 44*(1&2), 13—34.

de la Cruz, J., & Jansen, K. (2018). Panama disease and contract farming in the Philippines: Towards a political ecology of risk. *Journal of Agrarian Change, 18*, 249—266.

Cu, R. M., Mew, T. W., Cassman, K. G., & Teng, P. S. (1996). Effect of sheath blight on yield in tropical, intensive rice production systems. *Plant Disease, 80*, 1103—1108.

Dalmacio, S. C. (1998). Importance of and growing concerns for maize diseases in the Asian region. In S. K. Vasal, F. Gonzales Ceniceros, & F. Xingming (Eds.), *Proceedings of the seventh Asian regional maize workshop. Strengthening hybrid maize technology and public-private partnership to accelerate maize production in the Asian region, February 23—27, 1998*. PCARRD.

Dilla, E. (1993). *Yield loss due to sheath blight in direct-seed rice as affected by planting density, nitrogen level and amount of inoculum*. Unpublished PhD Thesis. University of the Philippines Los Baños.

Dimasingkil, S. F. D. (2019). *Modeling epidemics of Southern corn leaf blight of maize caused by Bipolaris maydis (Nisikado & Miyake) Shoemaker under climate change*. Unpublished MS Thesis (p. 125). University of the Philippines Los Baños.

Ditan, M. R. (2005). *MangoMan: A computerized forecast system for anthracnose on 'Carabao' mango (Mangifera indica Linn.)*. Unpublished BS Thesis (p. 80). University of the Philippines Los Baños.

Dodd, J. C., Estrada, A. B., Matcham, J., Jeffries, P., & Jeger, M. J. (1991). The effect of climatic factors on *Colletotrichum gloeosporioides*, causal agent of mango anthracnose, in the Philippines. *Plant Pathology, 40*, 568—575.

Dubey, S., Pandey, R. K., & Gautam, S. S. (2013). Literature review on fuzzy expert system in agriculture. *International Journal of Soft Computing and Engineering, 2*(6), 289—291.

References

Elings, A., Rossing, W. A., & Teng, P. S. (1994). Structure and development of BLIGHT, a model to simulate the effects of bacterial leaf blight and sheath blight on rice. In A. Elings, & E. G. Rubia (Eds.), *Analysis of damage mechanisms by pests and diseases and their effects on rice yield* (pp. 9–30). SARP Research Proceedings, DLO & WAU.

El Refaei, M. I. (1977). *Epidemiology of rice blast disease in the tropics with special reference to the leaf wetness in relation to disease development*. PhD Thesis (p. 214). Indian Agricultural Research Institute.

Esguerra, E. B., Chavez, S. M., & Traya, R. V. (2006). A modified and rapid heat treatment for the control of postharvest diseases of mango (*Mangifera indica* Linn. cv. Carabao) fruits. *Philippine Agricultural Scientist, 89*, 125–133.

Estrada, A. B., Dodd, J. C., & Jeffries, P. (2000). Effect of humidity and temperature on conidial germination and appressorium development of two Philippine isolates of the mango anthracnose pathogen *Colletotrichum gloeosporioides*. *Plant Pathology, 49*, 608–618.

Estrada, A. B., & Ilag, L. L. (1991). Effects of temperature and humidity on germination and infection of *Colletotrichum gloeosporioides* (Penz.) Sacc. on 'Carabao' mango (*Mangifera indica* L.). *Philippine Phytopathology, 27*(1&2), 26–35.

Estrada, A. B., Jeffries, P., & Dodd, J. C. (1996). Field evaluation of predictive model to control anthracnose disease of mango in the Philippines. *Plant Pathology, 45*(2), 294–301.

Evangelista, R. B., Franje, N. S., Pava, H. M., Corey, F. M., Jr., & Olazo, I. M. (2001). Yield loss assessment in corn due to the banded leaf and sheath blight caused by *Rhizoctonia solani* Kuhn. *USM Research & Development Journal, 9*, 107–146.

Exconde, O. R., & Raymundo, A. D. (1970). Further study on the assessment of yield losses due to rice blast. *Philippine Phytopathology, 6*, 66–74.

Fitzell, R. D., Peak, C. M., & Darnell, R. E. (1984). A model for estimating infection levels of anthracnose disease of mango. *Annals of Applied Biology, 104*, 451–458.

Galang, R. L. (2013). *Expression of resistance and susceptibility to black Sigatoka (Mycospaerella fijiensis Morelet) in the different Musa x paradisiaca L. varieties as influenced by environmental parameters*. Unpublished PhD Thesis (p. 111). University of the Philippines Los Baños, Los Baños.

Galvan, S. (2020). Postharvest losses of mango (*Mangifera indica* L.) in Iloilo and Guimaras, Philippines. *The International Journal of Innovation, Creativity and Change, 13*, 1228–1239.

Ghatak, A., Willocquet, L., Savary, S., & Kumar, J. (2013). Variability in aggressiveness of rice blast (*Magnaporthe oryzae*) isolates originating from rice leaves and necks: A case of pathogen specialization? *PLOS One, 8*, 1–7.

Gill, M. A., & Bonman, J. M. (1988). Effects of water deficit on rice blast I. Influence of water deficit on components of resistance. *Journal of Plant Protection in the Tropics, 5*(2), 61–66.

Guisan, A., & Zimmermann, N. E. (2000). Predictive habitat distribution models in ecology. *Ecological Modelling, 135*, 147–186.

Jackson, G. (2014). *Black sigatoka of banana. Mycosphaerella fijiensis*. http://africasoilhealth.cabi.org/wpcms/wp-content/uploads/2015/02/1-banana-black-sigatoka.pdf.

Katsantonis, D., Kadoglidou, K., Dramalis, C., & Puigdollers, P. (2017). Rice blast forecasting models and their practical value: A review. *Phytopathologia Mediterranea, 56*(2), 187–216.

Klein-Gebbinck, H. W. (1995). *The effects of pre-inoculation drought stress on the components of resistance or rice (Oryza sativa L.) to leaf blast (Magnaporthe grisea [Hebert] Barr)*. Unpublished PhD Thesis (p. 174). University of Alberta.

Kobayashi, T., Ijiri, T., Mew, T. W., Maningas, G., & Hashiba, T. (1995). Computerized forecasting system (BLIGHT-ASIRRI) for rice sheath blight disease in the Philippines. *Annals of the Phytopathological Society of Japan, 61*, 562–568.

Krause, R. A., Massie, L. B., & Hyre, R. A. (1975). BLITECAST: A computerized forecast of potato late blight. *Plant Disease Reports, 59*, 95–98.

Leaño, R. M., Lapis, D. B., & Savary, S. (1993). Analyzing the monocyclic process in sheath blight of rice under semi-controlled conditions. *Philippine Phytopathology, 29*, 1–16.

Lubulwa, A. S. G., & Davis, J. S. (1994). Estimating the social costs of the impacts of fungi and aflatoxins in maize and peanuts. In E. Highley, E. J. Wright, H. J. Banks, & B. R. Champ (Eds.), *Stored product protection. Proceedings of the 6th international working conference on stored-product protection, April 17–23, 1994* (Vol. 2, pp. 1017–1042). CAB International.

Luo, Y., Teng, P. S., Fabellar, N. G., & Tebeest, D. O. (1997). A rice-leaf blast combined model for simulation of epidemics and yield loss. *Agricultural Systems, 53*, 27–39.

Luo, Y., Teng, P. S., Fabellar, N. G., & Tebeest, D. O. (1998). Risk analysis of yield losses caused by rice leaf blast associated with temperature changes above and below for five Asian countries. *Agriculture, Ecosystems & Environment, 68,* 197–205.

Macasero, J. B. M. (2018). *Analysis of the relationship between production situation, pest injuries and yield of rice in the Philippines.* Unpublished MS Thesis (p. 99). University of the Philippines Los Baños. Los Baños.

Madden, L. V., Hughes, G., & Van den Bosch, F. (2007). *The study of plant disease epidemics* (p. 421). APS Press.

Malamnao, R. G. E. (2017). *Effects of rice straw compost amendment on brown spot disease caused by Bipolaris oryzae* (Breda de Haan) Shoemaker *and soil copper in paddy fields contaminated with mine tailings in Mankayan, Benguet and Cervantes, Ilocos Sur.* Unpublished MS Thesis (p. 79). University of the Philippines Los Baños. Los Baños.

Malicdem, A. R., & Fernandez, P. L. (2015). Rice blast disease forecasting for northern Philippines. *WSEAS Transactions on Information Science and Applications, 12,* 120–129.

Massie, L. B. (1973). *Modelling and simulation of Southern corn leaf blight disease caused by* Helminthosporium maydis *Nis. and Miy.* Unpublished Ph.D Thesis. Pennsylvania State University.

Mercado, A. C., & Lantican, R. M. (1961). The susceptibility of cytoplasmic male sterile lines of corn to *Helminthosporium maydis* Nish and Miy. *Philippine Agriculturist, 45,* 235–243.

Nuque, F. L. (1963). *Panicle blast outbreak in Leyte provinces* (p. 3). Report to the IRRI Director General.

Nuque, F. L. (1970). *Panicle blast outbreaks in Laguna and Quezon provinces* (p. 2). Report to the IRRI Director General.

Obusan, A., Nuque, F. L., Vergel de Dios, T., & Crill, P. (1982). Effect of leaf wetness and constant temperatures on infection by *Helminthosporium oryzae* B. de Haan. In *Paper presented at the 19th annual meeting of the Philippine Phytopathological Society, May 5–8, 1982, Baguio city,, 18(1&2)* pp. 14–15). Philippine Phytopathology. abstr.

Olivares, F. M., Jr., Lapis, D. B., & Nuque, F. M. (1977). Diseases of rice and wheat. In *Proceedings of the symposium on Philippine phytopathology 1917–1977. December 14–15, 1977* (pp. 76–97). University of the Philippines Los Baños.

Opina, O. S., & Eusebio, A. A. (1997). Quantification of inoculum of *Colletotrichum gloeosporioides* Penz. within a 'Carabao' mango orchard. *Philippine Phytopathology, 33*(1), 9–16.

Opina, O. S., Eusebio, A. A., & Basio, N. A. M. (1997). Temporal susceptibility of 'Carabao' mango flushes to *Colletotrichum gloeosporioides* Penz. *Philippine Phytopathology, 33*(1), 45–48.

Ordoyo, R. F., Generalao, L. C., Dionio, B. T., Ugay, V. P., & Raymundo, A. D. (2009). Epidemiology of black Sigatoka of banana caused by *Mycosphaerella fijiensis* Morelet: I. Climatic factors critical to pathogen sporulation and disease progression. *Southeastern Philippines Journal of Research and Development, 18*(2), 58–94.

Padmanabhan, S. Y. (1973). The great Bengal Famine. *Annual Review of Phytopathology, 11,* 11–26.

Pangga, I. B. (1987). *Evaluation of corn varieties and assessment of yield loss due to banded leaf and sheath blight.* Unpublished BS Thesis (p. 59). University of the Philippines Los Baños.

Pangga, I. B. (1996). Methods in quantifying blast effects on rice yield. In E. G. Rubia, G. C. Santiago, K. L. Heong, H. D. Justo, V. P. Gapud, E. Benigno, P. S. Teng, R. Fernandez, & R. F. Barroga (Eds.), *Improving regional decisions in rice pest management* (pp. 68–79). Philippine Rice Research Institute.

Pangga, I. B., Salvacion, A. R., & Cumagun, C. J. R. (2015). Climate change and plant diseases caused by mycotoxigenic fungi: Implications for food security. In L. M. Botana, & M. J. Sainz (Eds.), *Climate change and mycotoxins* (pp. 1–20). De Gruyter.

Pangga, I. B., Teng, P. S., & Raymundo, A. D. (1996). Blast development on the new rice plant type in relation to canopy structure, microclimate, and crop management practices. *Philippine Phytopathology, 32*(1), 18–34.

Pascual, C. B., Barcos, A. K. S., Mandap, J. A. L., & Ocampo, E. T. M. (2016). Fumonsin-producing *Fusarium* species causing ear rot of corn in the Philippines. *Philippine Journal of Crop Science, 41*(1), 12–21.

Pascual, C. B., & Salazar, A. M. (2001). Development of multiple disease resistant corn populations. *Journal of Tropical Plant Pathology, 37*(2), 8–15.

Phillips, S. J., Anderson, R. P., & Schapire, R. E. (2006). Maximum entropy modeling of species geographic distributions. *Ecological Modelling, 190,* 231–259.

Pinnschmidt, H. O., Batchelor, W. D., & Teng, P. S. (1995). Simulation of multiple species pest damage in rice using CERES-rice. *Agricultural Systems, 48,* 193–222.

Pinnschmidt, H. O., Bonman, J. M., & Kranz, J. (1995). Lesion development and sporulation of rice blast. *Journal of Plant Diseases and Protection, 102*(3), 299–306.

Pinnschmidt, H. O., Klein-Gebbinck, H. W., Bonman, J. M., & Kranz, J. (1993). Comparison of aerial concentration, deposition, and infectiousness of conidia of *Pyricularia grisea* by spore-sampling techniques. *Phytopathology, 83,* 1182–1189.

References

Pinnschmidt, H. O., & Teng, P. S. (1994). Advances in modeling multiple insect-disease-weed effects on rice and implications for research. In P. S. Teng, K. L. Heong, & K. Moody (Eds.), Rice pest management. *Selected papers from the international rice research conference* (pp. 101–128). International Rice Research Institute.

Raymundo, A. D. (2006). *Lost harvest: Quantifying the effect of plant diseases on major crops in the Philippines* (p. 236). Multinational Printers.

Raymundo, A. D. (2008). *Black sigatoka of banana: Analyzing the main threat to an important export crop* (p. 44). David Murdock/DOLE Asia Professorial Lecture. University of the Philippines Los Baños.

Raymundo, A. D., & Bactong, M. A., Jr. (2014). Modeling corn leaf blight epidemics under climate change. In *Poster presented at the 45th pest management council of the Philippines Anniversary and annual scientific conference, Cebu city, May 6-8, 2014*. Abstr.

Raymundo, A. D., Pantua, M. R., & Teng, P. S. (1996). Quantifying crop loss due to sheath blight in rice. In E. G. Rubia, G. C. Santiago, K. L. Heong, H. D. Justo, V. P. Gapud, E. Benigno, P. S. Teng, R. Fernandez, & R. F. Barroga (Eds.), *Improving regional decisions in rice pest management* (pp. 82–85). Philippine Rice Research Institute.

Reodique, B. A. (2004). *Modeling epidemics of sheath blight caused by Rhizoctonia solani* Kuhn *in corn (Zea mays L.)*. Unpublished BS Thesis (p. 35). University of the Philippines Los Baños.

Sah, D. N., & Bonman, J. M. (1992). Effects of seedbed management on blast development in susceptible and partially resistant rice cultivars. *Journal of Phytopathology, 136*, 73–81.

Salacinas, M. A., Stoorvogel, J., Mendes, O., Schoen, C., Rebuta, A. M., Catambacan, D. G., Corcolon, B., Tuba, J., Mora, J., Bacus, L., Truggelmann, L. T., Mamora, H. S., Meijer, H. J. G., & Kema, G. H. J. (2019). Epidemiology and management of Panama disease in the Mindanao Cavendish banana belt of the Philippines. PhD Thesis. In M. Salacinas (Ed.), *Spot on: Managing Panama disease of banana in the Philippines* (pp. 61–90). Wageningen University.

Salvacion, A. R., Cumagun, C. J. R., Pangga, I. B., Magcale-Macandog, D. B., Sta Cruz, P. C., Saludes, R. B., Solpot, T. C., & Aguilar, E. A. (2019a). Banana suitability and Fusarium wilt distribution in the Philippines under climate change. *Spatial Information Research, 27*(3), 339–349.

Salvacion, A. R., Cumagun, C. J. R., Pangga, I. B., Magcale-Macandog, D. B., Sta Cruz, P. C., Saludes, R. B., Solpot, T. C., & Aguilar, E. A. (2019b). Exploring environmental determinants of Fusarium wilt occurrence on banana in South Central Mindanao, Philippines. *Hellenic Plant Protection Journal, 12*, 78–90.

Salvacion, A. R., Pangga, I. B., & Cumagun, C. J. R. (2015). Assessment of mycotoxin risk on corn in the Philippines under current and future climate change conditions. *Reviews in Environmental Health, 30*(3), 135–142.

Savary, S. (2014). The roots of crop health: Cropping practices and disease management. *Food Security, 6*, 819–831.

Savary, S., & Castilla, N. P. (2009). *A survey portfolio to characterize yield-reducing factors in rice.*. Discussion Paper series no. 18 (p. 33). International Rice Research Institute.

Savary, S., Castilla, N. P., Elazegui, F. A., McLaren, C. G., Ynalvez, M. A., & Teng, P. S. (1995). Direct and indirect effects of nitrogen supply and disease source structure on rice sheath blight spread. *Phytopathology, 85*, 959–965.

Savary, S., Castilla, N. P., Elazegui, F. A., & Teng, P. S. (2005). Multiple effects of two drivers of agricultural change, labour shortage and water scarcity, on rice pest profiles in tropical Asia. *Field Crops Research, 91*, 263–271.

Savary, S., Castilla, N. P., & Willocquet, L. (2001). Analysis of the spatiotemporal structure of rice sheath blight epidemics in a farmer's field. *Plant Pathology, 50*, 53–68.

Savary, S., Ficke, A., Aubertot, J.-N., & Hollier, C. (2012). Crop losses due to diseases and their implications for global food production losses and food security. *Food Security, 4*, 519–537.

Savary, S., Horgan, F., Willocquet, L., & Heong, K. L. (2012). A review of principles for sustainable pest management in rice. *Crop Protection, 32*, 54–63.

Savary, S., Nelson, A., Sparks, A. H., Willocquet, L., Duveiller, E., Mahuku, G., Forbes, G., Garrett, K., Hodson, D., Padgham, J., Pande, S., Sharma, S., Yuen, J., & Djurle, A. (2011). International agricultural research tackling the effects of global and climate changes on plant diseases in the developing world. *Plant Disease, 95*(10), 1204–1216.

Savary, S., Nelson, A., Willocquet, L., Pangga, I., & Aunario, J. (2012). Modeling and mapping potential epidemics of rice diseases globally. *Crop Protection, 34*, 6–17.

Savary, S., Pangga, I. B., Willocquet, L., & Teng, P. S. (2018). Section 5. Advances in research Chapter 2. The epidemiology of rice diseases. In T. W. Mew, H. Hibino, S. Savary, C. M. Vera Cruz, R. Opulencia, & G. P. Hettel (Eds.), *Rice diseases: Biology and selected management practices* (pp. 1–21). International Rice Research Institute.

Savary, S., Willocquet, L., Castilla, N. P., Nelson, A., Singh, U. S., Kumar, J., & Teng, P. S. (2022). Wither rice health in the lowlands of Asia: Shifts in production situations, injury profiles, and yields. *Plant Pathology, 71*, 55—85.

Savary, S., Willocquet, L., Elazegui, F. A., Castilla, N. P., & Teng, P. S. (2000). Rice pest constraints in tropical Asia: Quantification of yield losses due to rice pests in a range of production situations. *Plant Disease, 84*, 357—369.

Savary, S., Willocquet, L., Elazegui, F. A., Teng, P. S., Du, P. V., Zhu, D., Tang, Q., Huang, S., Lin, X., Singh, H. M., & Srivastava, R. K. (2000). Rice pest constraints in tropical Asia: Characterization of injury profiles in relation to production situations. *Plant Disease, 84*, 341—356.

Savary, S., Willocquet, L., & Teng, P. S. (1997). Modelling sheath blight epidemics on rice tillers. *Agricultural Systems, 55*(3), 359—384.

Scheerer, L., Pemsl, D., Dita, M., Perez Vicente, L., & Staver, C. (2018). A quantified approach to projecting losses caused by Fusarium wilt Tropical race 4. In I. Van den Bergh, J.-M. Risède, & V. Johnson (Eds.), *Proceedings of X International Symposium on banana: ISHS-ProMusa symposium on agroecological approaches to promote innovative banana production systems, October 10, 2016, Montpellier.* (pp. 211—218). Acta Horticulturae 1196.

Shahjahan, A. K. M., & Mew, T. W. (1992). Response of sheath blight development to rice crop management in lowland and upland environments. *Philippine Phytopathology, 28*, 34—44.

Tang, Q., Peng, S., Buresh, R. J., Zou, Y., Castilla, N. P., Mew, T. W., & Zhong, X (2007). Rice varietal difference in sheath blight development and its association with yield loss at different levels of N fertilization. *Field Crops Research, 102*, 219—227.

Teng, P. S. (1994). The epidemiological basis for blast management. In R. S. Zeigler, S. A. Leong, & P. S. Teng (Eds.), *Rice blast disease* (pp. 409—434). CAB International and International Rice Research Institute.

Teng, P. S., Klein-Gebbinck, H. W., & Pinnschmidt, H. (1991). An analysis of the blast pathosystem to guide modeling and forecasting. In *Rice blast modeling and forecasting. Selected papers from the Inernational Rice Research Conference, August 27-30, 1990, Seoul, South Korea* (pp. 1—30). International Rice Research Institute.

Teng, P. S., & Revilla, I. M. (1996). Technical issues in using crop-loss data fore research prioritization. In R. E. Evenson, R. W. Herst, & M. Hossain (Eds.), *Rice research in Asia: Progress and priorities* (pp. 261—276). CAB International and International Rice Research Institute.

Teng, P. S., & Savary, S. (1992). Implementing the systems approach in pest management. *Agricultural Systems, 40*, 237—264.

Teng, P. S., Torres, C. Q., Nuque, F. L., & Calvero, S. B., Jr. (1990). Current knowledge on crop losses in tropical rice.. In *Crop loss assessment in rice. Papers given at the International workshop on crop loss assessment to improve pest management in rice and rice-based cropping systems in South and Southeast Asia, October 11-17, 1987* (pp. 39—53). International Rice Research Institute.

Torres, C. Q., & Teng, P. S. (1993). Path coefficient and regression analysis of the effects of leaf and panicle blast on tropical rice yield. *Crop Protection, 12*, 296—302.

Waggoner, P. E., Horsfall, J. G., & Lukens, R. J. (1972). EPIMAY, a simulator of southern corn leaf blight. *Connecticut Agricultural Experiment Station Bulletin, 729.*

Weihong, L., & Goudriaan, J. (1991). Leaf wetness in the rice crops caused by dew formation: A simulation study. In W. F. T. Penning de Vries, H. H. Van Laar, & M. J. Kropff (Eds.), *Simulation and systems analysis for rice production (SARP)* (pp. 320—327). Pudoc.

Willocquet, L., Fernandez, L., & Savary, S. (2000). Effect of various crop establishment methods practiced by Asian farmers on epidemics of rice sheath blight, caused by *Rhizoctonia solani. Plant Pathology, 49*, 346—354.

Willocquet, L., Savary, S., Fernandez, L., Elazegui, F. A., Castilla, N., Zhu, D., Tang, Q., Huang, S., Lin, X., Sing, H. M., & Srivastava, R. K. (2002). Structure and validation of RICEPEST, a production situation-driven, crop growth model simulating rice yield response to multiple pest injuries for tropical Asia. *Ecological Modelling, 153*, 247—268.

Zadoks, J. C. (1971). Systems analysis and the dynamics of epidemics. *Phytopathology, 61*, 600—610.

Zadoks, J. C., & Schein, R. D. (1979). *Epidemiology and plant disease management* (p. 427). Oxford University Press.

CHAPTER 9

Mycosis in the Philippines: Epidemiology, clinical presentation, diagnostics and interventions

Kin Israel R. Notarte[1,4], Adriel M. Pastrana[1], Abbygail Therese M. Ver[1], Jacqueline Veronica L. Velasco[1], Ma. Margarita Leticia D. Gellaco[1,3] and Melissa H. Pecundo[2]

[1]Faculty of Medicine & Surgery, University of Santo Tomas, Manila, Philippines; [2]Research Center for Natural and Applied Sciences, University of Santo Tomas, Manila, Philippines; [3]University of Santo Tomas Hospital, Manila, Philippines; [4]Department of Pathology, Johns Hopkins University School of Medicine, Baltimore, Maryland, United States

1. Overview of mycosis

Fungal infections affect more than a billion people globally, with severity ranging from asymptomatic-mild mucocutaneous infections to life-threatening systemic ones (Bongomin et al., 2017). Around 1.5 million deaths are attributed to serious fungal infections, most of which could have been prevented with early recognition and treatment (Bongomin et al., 2017). The burden of fungal pathogens remains of little importance to public health authorities (Tan et al., 2020). Worldwide, knowledge on the incidence of fungal infections continues to be hampered by a lack of regular national surveillance systems, no obligatory reporting of fungal diseases, poor clinician suspicion outside specialized units, poor diagnostic test performance (especially for culture) and few well-designed published studies (Bongomin et al., 2017). The Philippines is no exception to this. Very few, if any, studies on fungal disease surveillance or estimates of fungal disease burden have been done, and epidemiological data is sparse (Batac & Denning, 2017).

Mycology in the Tropics
https://doi.org/10.1016/B978-0-323-99489-7.00005-6

© 2023 Elsevier Inc. All rights reserved.

As a low-middle income country located in Southeast Asia, the Philippines has a tropical marine climate. It is the 13th most populous country in the world, with an average population density of 337 persons per square kilometer (Philippine Statistics Authority, 2016). Due to rapid economic growth and improvements in living conditions, the country is facing the triple burden of disease (Dayrit et al., 2018). The effects of climate change and natural disasters, increasing lifestyle risk factors, and the continued prevalence of infectious diseases have placed Filipinos in a vulnerable position. The country's top causes of morbidity and mortality include malignant neoplasia, tuberculosis, chronic obstructive pulmonary disease (COPD), and diabetes mellitus (Department of Health, 2010). All of these are recognized as major drivers of fungal infections (Bongomin et al., 2017). The HIV epidemic is also another concern of the country. Despite national prevalence remaining below 0.1%, there is an alarming rise in HIV incidence with a 174% increase from 2010 to 2017 (Gangcuangco, 2019). It is estimated that among those living with HIV, 33% are unaware of their status, and 55% have a CD4 count of less than 200 at the time of their diagnosis (Gangcuangco, 2019). This puts Filipinos living with HIV at greater risk of contracting opportunistic infections, to which fungal pathogens are a major contributor (Limper et al., 2017).

Due to the combination of increasing health problems and the rising population of immunocompromised individuals, Filipinos now have a greater susceptibility to developing fungal infections. This susceptibility is further supported by environmental risk factors such as heat and humidity, and socioeconomic risk factors such as overcrowding, poverty, and poor hygiene (Husain et al., 2018).

Historically, most studies on mycoses in the Philippines have focused on superficial fungal infections. Dermatophytosis continues to rank as one of the most common reasons for consultation in dermatology clinics (Handog & Dayrit, 2005). This is associated with the warm and humid weather and increased sweating during the summer months, and increased and persistent soaking of feet during the rainy season (Handog & Dayrit, 2005). Among the dermatophyte infections, pityriasis versicolor, tinea corporis, tinea cruris, and tinea pedis were the most common (Handog & Dayrit, 2005). Diagnosis of these infections is mostly clinical, and treatment includes prescribing a topical treatment of imidazole, terbinafine or sulfur-salicylic creams (Handog & Dayrit, 2005). Clinicians cited financial constraints as a reason why superficial infections are rarely sent for laboratory investigation (Handog & Dayrit, 2005).

With the emergence of the immunocompromised host brought about by various health issues, there has been a greater focus on serious fungal infections. Still, there is barely any data on the actual prevalence and incidence of serious fungal infections. In 2016, there was an estimated total of 1,852,137 severe fungal infections in the Philippines (Batac & Denning, 2017). Eighty percent of this is due to recurrent vulvovaginal candidiasis, although this is considered to be a very high estimate according to local practicing obstetricians (Batac & Denning, 2017). Among those with HIV/AIDS, a 2012 study noted that a third of those seen in a tertiary hospital presented with opportunistic infections (Salvana et al., 2012). Of these, 32% were due to *Pneumocystis* pneumonia, 4% were due to cryptococcal meningitis, and 3% were due to oral candidiasis. Across the entire Philippine population, however, candidiasis was found to be the most common opportunistic fungal infection, followed by coccidiodomycosis, cryptomycosis, and aspergillosis (Handog & Dayrit, 2005).

The Philippines has one of the highest estimated incidences and burden of chronic pulmonary aspergillosis (CPA) globally with a prevalence rate of 78 per 100,000 individuals (Bongomin et al., 2017). This is related to the fact that the tuberculosis is endemic to the Philippines, and has a high rate of COPD, both of which are associated with CPA (Batac & Denning, 2017). Because of the high prevalence, it can be surmised that there are many undiagnosed cases of CPA (Batac & Denning, 2017). Some of which may be misclassified and improperly managed as resistant TB cases (Batac & Denning, 2017). It is therefore important to make efforts to properly detect CPA and establish an actual baseline prevalence (Batac & Denning, 2017).

Subcutaneous mycosis is very rare in the Philippines. Only a handful of cases have been reported, with most resulting in disfiguring lesions that caused great distress for years (Batac & Denning, 2017). No estimates of their prevalence have been calculated (Batac & Denning, 2017).

Because of the great burden of serious fungal infections in the Philippines, the country has to focus efforts on increasing its capability to identify and manage these. Currently, the country is hindered by several challenges. These include lack of timely diagnosis, lack of proper antifungal intervention, decreased patient compliance with long-term treatment, and high cost of antifungal treatment (Batac & Denning, 2017). There is almost no access to advanced diagnostic tests such as galactomannan, β-D-glucan, and PCR in the Philippines, which is further hindered by high capital investment (Chindamporn et al., 2018). There are also very few independent mycology laboratories, and very few weekly samples are received for fungal diagnosis (Chindamporn et al., 2018). The lack of facilities hinders the management of patients afflicted with fungal infections, as well as impairs epidemiological and outbreak investigations (Chindamporn et al., 2018). Significant efforts are needed to increase laboratory capabilities. Clinicians also face challenges apart from a lack of access to advanced diagnostics. These include a lack of formal training and local guidelines on the management of fungal infections and inadequate access to antifungal agents (Tan et al., 2020). Clinicians cited the most common reasons for not using the antifungal drug of choice were due to cost and nonavailability of the drug (Tan et al., 2020). Patients also face the same challenges. The high cost of antifungal treatment and lack of easy access serve as barriers to treatment, with some preferring to explore alternative treatment options (Eusebio-Alpapara et al., 2020). There are still quite a number who do not consult their physicians and would rather self-medicate (Batac & Denning, 2017). In order to increase the country's response to fungal infections, these issues must be addressed.

2. Common fungal infections

2.1 Aspergillosis

2.1.1 Epidemiology

Aspergillus Micheli is one of the most common causes of fungal infections, primarily in the healthcare setting (Gregg & Kauffman, 2015). *Aspergillus* spp. is ubiquitous in the environment and is likely inhaled frequently by humans. They are present in water, soil, dust and food sources, and are in abundant concentrations in decaying vegetations (Zakaria et al., 2020). Among the hundreds of *Aspergillus* spp., only a few cause mycosis in humans. *A. fumigatus* Fresenius (~90%) is by far the most common species causing all forms of invasive aspergillosis (IA). This is true for both individual sporadic cases and outbreak-related cases, and if whether the infection is community-acquired or hospital-acquired (Nicolle et al., 2011). The invasiveness of *A. fumigatus* is brought about by its virulence factors, such as its synthesis of proteases, phospholipases, dismutases and mycotoxins, favoring its dissemination in the lungs and other body systems (Patterson et al., 2000). Although in the

United States *A. fumigatus* remained to be the most common fungal infection, *A. flavus* Link is the leading cause of IA in the Middle East and is greatly associated with sinusitis and endophthalmitis (Zakaria et al., 2020). In Hong Kong, *A. avus* is a more common cause of infection associated with bone marrow transplantation as compared to *A. fumigatus* (Yuen et al., 1997). *A. terreus* Thom, on the other hand, is mostly implicated with IA in immunocompromised individuals, particularly those with hematologic malignancies. The same species has also shown resistance to amphotericin B, but remains susceptible to azoles (Steinbach et al., 2004). Meanwhile, *A. niger* van Tieghem rarely causes invasive infection, but is more often found as a colonizer of abnormal airways and is associated with chronic pulmonary aspergillosis syndromes (Panackal et al., 2006).

Globally, the prevalence of aspergillosis is increasing as a result of developing advanced medical practices with a rise in the proportion of immunocompromised populations due to cancer treatment, organ transplantation, and prolonged immunosuppressive therapy (Gregg & Kauffman, 2015). A ten-fold increase in the frequency of IA was seen over the last two decades. The prevalence of IA in the Philippines, in particular, has never been studied (Batac & Denning, 2017). However, several cases of aspergillosis were documented over the past years in individuals diagnosed with tuberculosis, COPD, diabetes mellitus, and even those who underwent renal transplantation and who gave birth by caesarean section (Batac & Denning, 2017). With the limited clinical epidemiology data available, it is encouraging to determine the prevalence of IA among Filipinos considering that TB and AIDS are widespread in the Philippines, which can be risk factors for acquiring fungal infections.

2.1.2 Diagnostic strategies

An accurate and early diagnosis of IA is crucial to improve patient survival and to withhold expensive tests with a high negative predictive value and administration of potentially toxic antifungal drugs (Hope et al., 2005). A positive culture or demonstration of invasive hyphae histologically from a normally sterile environment (*e.g.*, pleural fluid) is equated as a proven invasive fungal disease. On the other hand, "probable" IA requires a combination of host factors that predispose to invasive aspergillosis (*e.g.*, prolonged neutropenia, transplantation) and clinical (*e.g.*, evidence of pneumonia) and mycological criteria. Although originally designed for clinical research, these consensus criteria can also be applied to clinical practice with the understanding that therapy for suspected IA is commonly initiated prior to fulfilling these intentionally restrictive guidelines (De Pauw et al., 2008).

In neutropenic patients, persistent fever may be the only sign of invasive fungal disease. A chest CT scan is more sensitive than radiographs for detecting early pulmonary aspergillosis and should be considered in patients with 10–14 days of neutropenia (neutrophil count <500/uL) and with a persistent or recurrent fever of unknown etiology unresponsive to empirical antibacterial treatment (Desoubeaux et al., 2014). The earliest radiological sign of IA is a nodule. A halo sign characterized as a macronodule surrounded by a perimeter of ground-glass opacity corresponding to alveolar hemorrhage, may suggest IA in patients with compatible host factors. Initiation of treatment based on this sign has been associated with a better prognosis as compared to when therapy was initiated for more advanced fungal disease (Sherif & Segal, 2010). However, other molds and bacteria capable of angioinvasion can also present with a similar radiologic appearance. Other radiographic findings associated with IA are consolidation, wedge-shaped infarcts, and cavitation. Contrary to adults, children with invasive pulmonary aspergillosis frequently do not manifest cavitation or the air crescent or halo signs (Burgos et al., 2008).

Mycological criteria require either isolation of *Aspergillus* spp. from the sinopulmonary tract or positive antigen-based laboratory markers. Bronchoalveolar lavage fluid (BALF) cultures have at best 50% sensitivity in focal pulmonary lesions (Hope et al., 2005). Antigen-based diagnosis involves the serum detection of galactomannan or beta-*D*-glucan, which are fungal cell wall constituents. The galactomannan assay is relatively specific for IA, whereas the beta-*D*-glucan assay also detects other invasive fungi, such as *Candida*, other mold pathogens (excluding zygomycetes), and *Pneumocystis (carinii) jirovecii* J.K. Frenkel (Marty et al., 2007). In addition to serving as a diagnostic adjunct for IA, a falling or rising serum concentration of galactomannan may be useful as an early marker of therapeutic success or failure, respectively. The role of galactomannan in this aspect requires more research and validation in different patient populations with IA (Segal et al., 2008).

Although the use of PCR-based techniques can be promising, this approach remains investigational in diagnosing invasive fungal infections. Potential advantages include rapidity, low cost, the ability to establish a diagnosis at the species level and the detection of genes that confer antifungal resistance. The lack of standardized methods, difficulty in reliably distinguishing fungal colonization from disease, and the possible contamination with fungal DNA are some of the limitations of PCR-based techniques (Einsele & Loeffler, 2008).

2.1.3 Clinical presentation and management

The most common portal of entry of *Aspergillus* spores is by inhalation, thus IA mainly involves the sinopulmonary tract (Sherif & Segal, 2010). Although rare, other sites of entry may include the gastrointestinal tract or skin. Fever, cough, and dyspnea are frequent, although nonspecific, clinical findings of pulmonary aspergillosis, which is the most common site of IA (Desoubeaux et al., 2014). Due to pulmonary infarction or hemoptysis, the vascular invasion may also manifest as pleuritic chest pain. There could also be central nervous system affectation as a devastating consequence of disseminated aspergillosis, and patients may manifest with seizures or focal neurological signs from mass effect or stroke (Gregg & Kauffman, 2015). Premature neonates can also develop aspergillosis with the skin as the primary site of entry (Sherif & Segal, 2010) (Fig. 9.1).

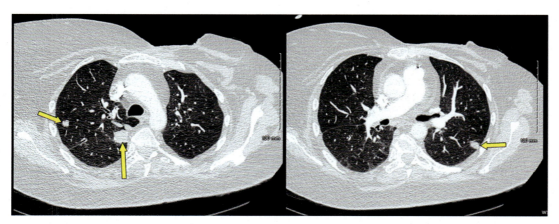

FIGURE 9.1 Chest CT scan during invasive pulmonary aspergillosis showing three nodules in a parenchymal window (yellow arrows) (Desoubeaux et al., 2014).

218   9. Mycosis in the Philippines: Epidemiology, clinical presentation, diagnostics and interventions

The gold standard primary treatment for IA is voriconazole (Sherif & Segal, 2010). In a randomized clinical trial, voriconazole was found to be more effective than amphotericin B deoxycholate (AmB-D) as initial therapy for IA and was associated with significantly improved survival (Herbrecht et al., 2002). The poorest prognosis occurred in extrapulmonary aspergillosis and in individuals who received hematopoietic stem cell transplantation. Therapeutic drug monitoring for this agent is also advised, particularly in cases of refractory fungal disease or drug toxicity (Smith, Khatcheressian et al., 2006). For individuals diagnosed with IA refractory to voriconazole or who are intolerant to this drug, a lipid formulation of amphotericin B or an echinocandin can be considered. Lipid formulations are generally preferred over AmB-D for IA to avoid nephrotoxicity (Maertens et al., 2004). If possible, immunosuppressive therapy such as corticosteroids should be reduced or discontinued. In cases of severe aspergillosis and neutropenia, adjunctive myeloid colony-stimulating factors can be administered (Smith, Safdar et al., 2006). Adjunctive recombinant interferon-γ, which may play a role in mobilizing neutrophils and monocytes can also be considered in cases of severe or refractory aspergillosis (Safdar et al., 2006).

2.2 Candidiasis

2.2.1 Epidemiology

Candida albicans (C.-P. Robin) Berkhout and non-*albicans Candida* spp., such as *C. krusei* (Castellani) Berkhout, *C. glabrata* (H.W. Anderson) S.A. Mey & Yarrow, *C. parapsilosis* Langeron & Talice and *C. tropicalis* (Castellani) Berkhout are greatly involved in the major causes of systemic infections and are considered the most common cause of opportunistic fungal infections worldwide (Dabas, 2013; Dadar et al., 2018). Among all *Candida* spp., *C. albicans* is the most studied and well-understood in terms of virulence and pathogenicity. It is part of the normal microbiota of the mouth, intestinal tract, and vagina as well as other mucous membranes. Under normal conditions, *Candida* does not cause disease, but the absence of appropriate immune recognition and response gives rise to the organism's colonization and infection, which is termed as candidiasis or candidosis (Kashem & Kaplan, 2016). It can originate from epithelia or from the biofilm formation of fungi and can present in multiple forms of clinical manifestations. *Candida* colonizes several host niches and mainly behaves as opportunistic pathogens, especially in severely immunocompromised patients. It may also be of iatrogenic origin, largely from contaminated central venous catheters and surgical operations. The organism may be disseminated by passing through the epithelial barrier and endothelium, spreading *via* the bloodstream and can infect virtually every internal organ. It may also disseminate by gaining access to the bloodstream *via* open injuries, such as severe burns and gastrointestinal damage (Dadar et al., 2018; Woods & Höfken, 2016; Sardi et al., 2013). The organism's ability to adapt to the differences in various environments in the host's system is the organism's key feature to thrive. It can adapt by using alternative carbon sources, altering its cellular proteome, inducing gene polymorphism, creating biofilm, and through its ability to adhere to and remodel its cell wall (da Silva Dantas et al., 2016).

C. albicans has the ability to transition morphologically into yeast, pseudohyphae, and hyphae, which is stimulated by various growth patterns and environmental conditions. The hyphae contain Spitzenkörper, which is made up of exocytic and endocytic vesicles and was said to promote growth at the hyphal tip by functioning as a vesicle supply center (Höfken,

II. Fungi in agriculture, health and environment

2013; Thompson et al., 2011). The organism's yeast morphology is believed to be important for colonization and dissemination to systemic organs, while the hyphal growth plays a vital role in enhancing virulence through leukocyte lysis, thigmotropism, biofilm formation, antigenic variation, and tissue invasion (Kadosh, 2019).

Hyphae are the more invasive form of the organism and are frequently isolated in infected tissues. The hyphal formation of *C. albicans* promotes virulence by its ability to invade and damage the epithelium, penetrate individual cells, lyse neutrophils and macrophages, and ability to contact sense, also known as thigmotropism. Non-*albicans candida* spp. that are only capable of developing yeast and pseudohyphae forms are said to have a reduced adherence to host cells, and reduced ability to secrete protease. In contrast to the hypha-forming *C. albicans, C. tropicalis,* and *C. dubliniensis* Sullivan *et al.* have an improved adherence and protease secretion (Thompson et al., 2011).

Several studies indicate that the virulence factors of pseudohyphae are expressed at significantly lower levels. The existence of pseudohyphal morphology may be credited to the host's specific microenvironments and niches that necessitates a decreased level of invasion and virulence factor expression, thereby allowing *C. albicans* to regulate its activity in nonoptimal environment within the host, such as low resource and nutrient levels (Mukaremera et al., 2017; Thompson et al., 2011).

In addition to the morphological transitions of *Candida* spp., its virulence can also be attributed to its ability to form biofilms, express cell surface adhesins, phenotypic switching, thigmotropism, secretion of hydrolytic enzymes and candidalysin, ability to escape from phagocytosis, and ability to counter host nutritional immunity. Studies suggest that the organism can utilize these factors in order to survive in different microenvironments and cause opportunistic infections leading to the development of antifungal resistance (Dadar et al., 2018; da Silva Dantas et al., 2016; Sumalapao et al., 2020; Tsui et al., 2016).

Biofilm formation is associated with the majority of *C. albicans* infections. They are complex structured communities that adhere to surfaces and are surrounded by an extracellular polysaccharide matrix. It involves a series of successive steps starting with the initiation step, which involves the proliferation of cells across the surface and the production of filamentations creating hyphal forms. Then proceeds the maturation of the biofilm as the extracellular polysaccharide matrix accumulates. The final step involves the release of the nonadhering yeast cells thereby spreading to the target organs (Höfken, 2013; Tsui et al., 2016).

Worldwide, *Candida* emerged as one of the most important causes of nosocomial bloodstream infection in both adults and children over the last 20 years. Almost 50% of all invasive candidiasis infections are associated with *C. albicans.* Though, there is also an ongoing increase in the infections associated with non-*albicans Candida* spp. (Alkharashi et al., 2019). The estimated annual incidence of oral/esophageal, vulvovaginal, disseminated, and intraabdominal candidiasis is 2.3 million, 134 million, 650,000, and 80,000 cases, respectively. Even though there had been medical advancements, the mortality rate of invasive candidiasis still remains at approximately 40% (Gabaldón, 2019).

In the Philippines, the estimated number of recurrent vulvovaginal candidiasis is 1.4 million. In a local setting, *C. albicans* and *C. dubliniensis* were the species isolated in vaginal discharges, particularly affecting the older age group of women >55 years old (Juayang et al., 2019). While the estimated number of oral, esophageal, candidemia and intraabdominal candidiasis are 3,467, 1,522, 1,968, and 246, respectively (Batac & Denning,

2017). Take note that only the estimates were provided, considering that the Philippines have a low number of physicians relative to the number of patients, and many Filipinos do not consult their doctors. The rising cases of HIV and AIDS in the country may also contribute to the further increase in the incidence of *Candida* infections.

2.2.2 Diagnostic strategies

Candidiasis can be diagnosed by visualization of fungal elements (*e.g.*, budding yeast cells, pseudohyphae, hyphae) on direct microscopy with 10% KOH or Gram stain. It can also be identified by their ability to metabolize and/or ferment carbohydrate substrates and nitrogen compounds. Reporting of vulvovaginal candidiasis (VVC) would require that the pH of vaginal secretion is less than 4.5, with sprout spores and pseudohyphae found under the oil microscope, and the Gram staining is positive (Zeng et al., 2018). Gynecologists routinely perform Papanicolaou smear during an annual gynecologic check-up. Candidal infection was reported to be characterized by alteration in the normal flora, showing moderate to heavy neutrophilic infiltrates and variably distributed histiocytic infiltrates (Sabu et al., 2017). *Candida* can also be isolated and identified using several cultural media (*e.g.*, Sabouraud dextrose agar, CHROMagar Candida, Biggy, CandiSelect 4). Several studies emphasize that phenotypic tests alone are not definitive for the identification of *C. albicans*, hence genotypic tests are important in order to precisely identify the isolated organisms. Laboratory diagnosis is deemed difficult because of the circulating antibodies against *Candida* spp. detected may indicate mere colonization of the mucosae. In cases of deep-seated candidiasis, 50% may be negative or it may only be positive in late infections. However, specific conditions and new diagnostic techniques paved the way for the identification of the organism. Blood culture samples collected through sterile methods are acknowledged as a gold standard in diagnosing invasive candidiasis. Modern diagnostic techniques, such as detection assays and polymerase chain reaction (PCR), can now be used as supplementary procedures to culture-based methods. In addition, T2 magnetic resonance (T2 MR), DNA-based techniques, fluorescence *in situ* hybridization (FISH) using labeled peptide nucleic acid (PNA) probes, and matrix-assisted laser desorption—ionization—time-of-flight mass spectrometry (MALDI-TOF MS) are currently being introduced (Dabas, 2013; Dadar et al., 2018; Moron et al., 2017).

2.2.3 Clinical presentation and management

2.2.3.1 Candidal paronychia

The normal flora of nails does not include *Candida* species. *Staphylococcus aureus* Rosenbach is the most common pathogen involved in acute paronychia after the protective nail barrier has been breached. However, nonbacterial pathogens such as *C. albicans* have also been identified as a possible causative agent of acute paronychia. The presence of *C. albicans* may also signify secondary colonization of the nail fold inducing more inflammation and leading to the development of chronic paronychia. The clinical signs and symptoms of candidal paronychia include inflammation of the nail fold that is more or less painful, presence of edema, and redness. It is also distinguished through the presence of progressive striations, discoloration, and gradual thickening, which eventually becomes crumbly. It is frequently observed in the hands and fingers and is commonly associated with occupations that involve frequent water contact. Management of the nail infection involves the prevention of trauma, contact irritants,

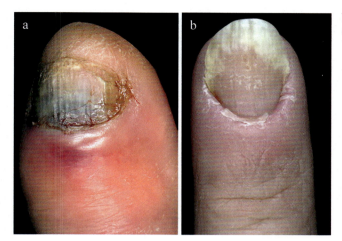

FIGURE 9.2 Acute paronychia (A) and chronic paronychia (B) (Dinulos, 2020).

exposure to water and medications. Antifungal therapy may include oral (efinaconazole 10% or tavaborole 5%), topical (amorolfine 5% or ciclopirox olamine 8% nail lacquer) or a combination of both (Martins et al., 2014; Shafritz & Coppage, 2014) (Fig. 9.2).

2.2.3.2 Mucocutaneous candidiasis

Oral candidosis (OC) typically presents as thrush, however, clinicians must also be familiar with the other presentations of this disease (Singh et al., 2014). The most commonly isolated *Candida* sp. in a healthy oral mucosa and in OC is *C. albicans*, which is found in more than 80% of the lesions of the mouth. The presentation of OC can be pseudomembranous, erythematous, and chronic hyperplastic candidosis. Pseudomembranous candidiasis or thrush presents as white, soft, slightly elevated plaques on the tongue, buccal mucosa, soft palate, hard palate, and oral pharynx (Millsop & Fazel, 2016). The plaques consist of the overgrowth of yeast mixed with desquamated epithelium, keratin, necrotic tissue, leukocyte, fibrin, bacteria, and fungal hyphae. It is important for the therapy of OC to remove or treat the underlying cause or risk factors. In mild cases of OC, topical antifungal agents (*e.g.*, gentian violet, nystatin cream, clotrimazole troche) alone may be adequate. However, for patients at risk for disseminated infection or for immunocompromised patients, systemic antifungal agents (*e.g.*, ketoconazole, itraconazole, fluconazole) are needed (R & Rafiq, 2020) (Fig. 9.3).

2.2.3.3 Vulvovaginal candidiasis

Vulvovaginal candidiasis (VVC), considered the second cause of genital infection among women, occurs by excessive multiplication of yeast, found in the vaginal microbiota of women in the reproductive phase (Pereira et al., 2021). VVC is accompanied by nonspecific symptoms, such as vulvar pruritus and burning, dyspareunia, soreness, vaginal and vulvar erythema, edema, and vaginal discharge, which may be described as white "cottage cheese-like" in character (R & Rafiq, 2020). Some physiological and tissue changes, caused by

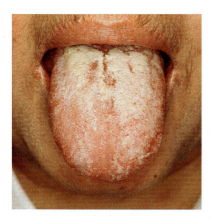

FIGURE 9.3   Clinical presentation of pseudomembranous candidiasis (Millsop & Fazel, 2016).

reproductive hormones increase the susceptibility of women in the active reproductive age group to *Candida* infection (Zeng et al., 2018). Anatomically, the anus lies in close proximity to the vagina thus providing an avenue for the migration of *Candida* from the gut to the vagina. Several factors may be related to the development of VVC: history of uncontrolled diabetes mellitus, use of tight clothing, frequent intravaginal douching and history of lower genital tract infection (Zeng et al., 2018). Among these risk factors listed, vaginal microbiota dysbiosis has been most recently described as directly associated with infection pathogenesis (Pereira et al., 2021).

Topical antimycotic agents and oral therapy are available for the treatment of VC. In uncomplicated VVC, short-term local therapy or single-dose oral treatment is effective. Short-term therapy with local azoles is approximately 3 days in duration. Topical azoles are an alternative treatment and can be used in combination with vaginal suppositories. Single-dose 150 mg oral fluconazole may also be an option for treatment. In complicated VVC, prolonged treatment is necessary. Local azoles are applied daily for the duration of at least 1 week, while oral fluconazole can be given three times with 72-hour gaps in between (Dovnik et al., 2015; Johal et al., 2016). It should be remembered that the cure of VVC is marked merely by the control of symptoms, rather than the eradication of all *Candida* organisms from the genital system, which is virtually impossible (Zeng et al., 2018) (Fig. 9.4).

2.2.3.4 Invasive candidiasis

The advances in medical technology and therapy strategies have greatly changed the epidemiological landscape of candidiasis, leading to alteration of the *Candida* spp. causing invasive infections. Invasive candidiasis may refer to candidemia (bloodstream infection caused by *Candida* spp.) and deep-seated infections or osteomyelitis - with or without candidemia. The major predisposing factors that increase the risk for developing invasive candidiasis are the long term and/or repeated use of broad-spectrum antibiotics, breach of cutaneous and gastrointestinal barriers due to inflammation, and iatrogenic immunosuppression. In addition to this, the organism's ability to form biofilms leads to the development of antifungal drug resistance. Several organs or sites including the bones, muscles, joints, eyes,

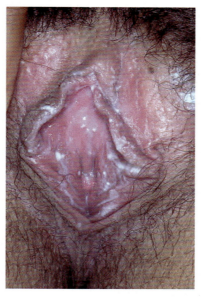

FIGURE 9.4 Common presentation of acute vaginal candidiasis is a red inflamed vulva and vagina and a white, thick discharge (Dinulos, 2020).

kidneys, liver, spleen, heart and brain may be affected (McCarty & Pappas, 2016; Pappas et al., 2018). The main options for treatment of invasive candidiasis are composed of three major drug classes — polyenes, azoles and echinocandins. In CNS candidiasis and *Candida* endocarditis, flucytosine is used as an adjunct in treatment.

2.3 Other fungal infections

2.3.1 Epidemiology

Other notable fungal infections are the dermatophytes, *Pneumocystis* pneumonia and *Cryptococcus neoformans (San Felice) Vuill*. Dermatophytes specifically tinea corporis, tinea cruris and tinea pedis are among the most frequently encountered cases in the Philippines. On the other hand, *Pneumocystis (carinii) jiroveci* and *Cryptococcus neoformans* are opportunistic microorganisms commonly encountered among immunocompromised patients.

Fungal infection ranks second as the leading cause of consultation in dermatology in the Philippines with a prevalence of 12.98% (Handog & Dayrit, 2005). Tinea corporis, tinea cruris and tinea pedis account for 22.63%, 16.7% and 16.38% of the cases, respectively (Handog & Dayrit, 2005). Moisture, warmth, the composition of sebum and perspiration, heavy exposure and genetic predisposition are factors that promote susceptibility of dermatophytes (Carroll et al., 2016). Thus, the Philippines with a tropical climate and warm and humid environment has a high prevalence of these infections, especially during the early summer and early rainy season. Tinea pedis, in particular, shows increased incidence from July to September probably due to persistent soaking of feet during floods. Retrospective studies in other Asian countries (*e.g.*, Thailand, Taiwan, Hong Kong) showed a similar prevalence of tinea pedis (Cheng and Chong, 2002; Smith et al., 2002; Ungpakorn et al., 2004). Living conditions are also associated with exposure to these infections such as in the military. Other predisposing factors include age and occupation. Previous studies on the prevalence of infection among different age groups were quite high among pediatric and elderly patients. Tinea corporis and tinea pedis were more pervasive in pubertal age groups (10—14 years of age) (Handog & Dayrit, 2005).

Among opportunistic infections, 32% were due to *Pneumocystis* pneumonia and 4% to cryptococcal meningitis (Batac & Denning, 2017). Predisposed patients are mostly on immunosuppressive agents or immunocompromised due to medical conditions, such patients are more common in the urban centers of the Philippines. This increases the severity of the infections brought about by opportunistic agents. With the advent increase in AIDS in the Philippines through the years, there is a rapid increase in the incidence of opportunistic infections. It is important to note though that the incidence of *Pneumocystis* pneumonia and cryptococcal meningitis are based mostly on HIV/AIDS cases in the recent past (Handog & Dayrit, 2005). *Cryptococcus* spp. has a wide distribution and has five serotypes. Serotype A identified in strains of *C. neoformans* is predominantly found in patients with AIDS and is mostly identified in Asia (Khayhan et al., 2013). *P. (carinii) jirovecii* has a worldwide distribution and serologic testing suggests that most individuals are infected in their early childhood. While there is a low culture positivity rate for these infections, it might be attributed to diagnostic skills, sample collection difficulties or even the increased application of self-prescribed medication.

2.3.2 Diagnostic strategies

Dermatophyte infection is spread by contact and is readily diagnosed by history taking and physical examination. However, it is important to take note that there are many clinical variants and diagnosis may be difficult if there were previous use of other medications (Leung et al., 2020). The use of KOH microscopy and culture may be helpful in identifying the specific fungal species responsible.

Specimens may be obtained from scrapings of infected skin, nails or hair. In KOH preparation, branching hyphae or chain arthroconidia (arthrospores) are observed. Depending on the specific species that caused the infection, ectothrix or endothrix may be observed. *Microsporum* spp. form dense sheaths of spores around the hair — ectothrix and impart a greenish or silvery fluorescence under Wood's light. On the other hand, *Trichophyton tonsurans* Malmsten and *T. violaceum* Sabouraud produce spores inside the hair shaft — endothrix and do not fluoresce under Wood's light (Carroll et al., 2016).

Inhibitory Mold Agar (IMA) and Sabouraud Dextrose Agar (SDA) slants are used and contain cycloheximide and chloramphenicol to suppress mold and bacterial growth. Cultures are incubated for 1—3 weeks at room temperature. Identification can be based on either colonial or microscopic morphology. Colonial morphology's basis is the growth rate, surface texture and pigmentation, while microscopic morphology examines the presence of macroconidia or microconidia (Carroll et al., 2016). Nutritional requirements also help determine the species of the cause of fungal infection. Fungal cultures remain the gold standard in diagnosing dermatophytosis. However, this technique would entail 1—2 weeks and is relatively expensive (Leung et al., 2020).

For opportunistic infections, more specific and sensitive tests could be performed. For direct microscopy and culture of *Cryptococcus neoformans*, specimens used are CSF, tissues, exudates, sputum, blood, cutaneous scrapings and urine. Examination using microscopy is aided with the use of India ink, which delineates the capsule. Cultures are also used and colonies develop within 1−2 days on most media at 37°C. Their growth at 37°C and the presence of urease are important identifying factors. On diphenolic substrate, phenol oxidase of *C. neoformans* produces melanin in cell walls thus colonies develop brown pigmentation (Carroll et al., 2016).

Diagnosis of *C. neoformans* is the first application of antigen detection in fungal infection and received wide clinical use. Through latex agglutination, the assay is able to detect glucuronoxylomannan (GXM). GXM is the major capsular polysaccharide of *C. neoformans* that is shed in large amounts into the blood and CSF during cryptococcal meningitis. GXM has four major serotypes: A-D and a hybrid AD. As mentioned earlier, serotype A is predominant in patients with AIDS and is quite prevalent in Asia (Khayhan et al., 2013). More recent advancement in diagnostic studies is the development of lateral flow immunoassay or dipstick format known as CrAg LFA. It is reactive to all GXM serotypes and is greatly applicable in resource-limited settings because it is inexpensive, requires no power, clean water, or refrigeration and can be performed even by personnel with limited training. Recent studies have also shown that CrAg could be used for prospective testing among asymptomatic, but high-risk patients for cryptococcosis (Kozel & Wickes, 2014).

Pneumocystis pneumonia can be diagnosed through samples from lung biopsy, bronchoalveolar lavage or induced sputum. They may be present as either trophozoites or cysts. Cysts can be stained with giemsa stain, toluidine blue, methenamine silver or calcofluor white. However, they are not amenable to culture (Carroll et al., 2016). Blood tests can also be performed to determine the presence of β-D-glucan (BDG), a polysaccharide component of most fungal cell walls. This is mostly used to screen for presumptive diagnosis of invasive fungal infection (IFI). It has a strong negative predictive value, which is useful in excluding IFI (Kozel & Wickes, 2014). However, there are several limitations to this test, which include the possibility of false-positive reactions, typically performed at reference laboratories not applicable for routine use. PCR also provides enhanced sensitivity over other conventional methods with a sensitivity of $\geq97\%$ and a negative predictive value of $\geq99\%$. A combination of BDG and PCR is recommended for optimal diagnosis, however, given the limited resources in the Philippines, it is not cost-efficient (Chindamporn et al., 2018; White et al., 2018). Serologic testing can be performed to establish its prevalence, but it is not clinically significant. Instead, a computerized tomography (CT) is a sensitive modality and may be used for early diagnosis, especially when a chest X-ray seems normal (Fig. 9.5).

2.3.3 Clinical presentation and management

Dermatophytes are transmitted through contact. Among the dermatophytes, tinea pedis also known as athlete's foot is the most prevalent. It is usually characterized by a vesicular, ulcerative or moccasin type of lesion with hyperkeratosis in the webs of the feet. Initially, it presents with itching and the development of small vesicles, which release thin fluid when ruptured. Skin in the affected area becomes macerated and peels. A crack may also develop, which predisposes that patient to bacterial infection. Chronic infection presents with peeling and cracking of the skin accompanied by pain and pruritus (Carroll et al., 2016). Meanwhile,

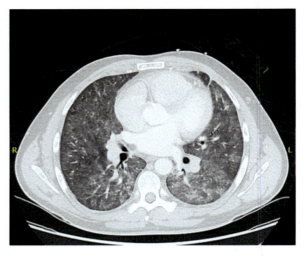

FIGURE 9.5 Computerized tomography (CT) of centrilobular ground glass opacification in HIV patients with *Pneumocystis* pneumonia (White et al., 2018).

tinea corporis has an incubation period of 1—3 weeks. This is characterized by a clearing, scaly center surrounded by a red advancing border, which may be vesicular, which gives its characteristic "ringworm" appearance (Leung et al., 2020). While these only grow within dead, keratinized tissue, fungal metabolites, enzymes and antigen diffuse through viable layers of the epidermis and cause erythema, vesicle formation and pruritus. As hyphae age, they form chains of arthroconidia and lesions expand centrifugally. Thus, hyphal growth is more active at the periphery and the preferred site to obtain samples. The persistent infection leads to penetration of the stratum corneum of the thick plantar and palmar surfaces. Lesions are dry and involve the skin other than the hands, feet, groin, scalp and face. In general, lesions caused by anthropophilic species are less inflammatory or erythematous compared to its geophilic and zoophilic counterparts (Leung et al., 2020). Tinea cruris has similar clinical manifestations but presents in the groin area and is also called jock's itch. It starts on the scrotum and spreads to the groin (Fig. 9.6).

The most effective management is the removal of infected and dead epithelial structures and the application of antifungal drugs. Topical and oral antifungals may be prescribed to the patient. Systemic antifungals are usually prescribed upon failure of topical therapy or widespread infection. Most commonly used and efficacious drugs for dermatophytes include itraconazole and terbinafine (Sahoo & Mahajan, 2016). Miconazole nitrate, tolnaftate, and clotrimazole may also be given. Medications should be applied for at least two to 4 weeks for a cure rate of 70—100%. Even after clearing, treatment should be continued for another one to 2 weeks (Carroll et al., 2016). To prevent infection, body surfaces should be kept dry and the source of infection should be avoided.

Infection of *Cryptococcus* is initiated by inhalation of yeast cells, which are easily aerosolized. It is not transmitted from person to person. During infection, polysaccharide or GXM is solubilized in the spinal fluid, serum or urine and may be detected with the use of enzyme immunoassay or agglutination latex. Usually, primary pulmonary infections are asymptomatic or may

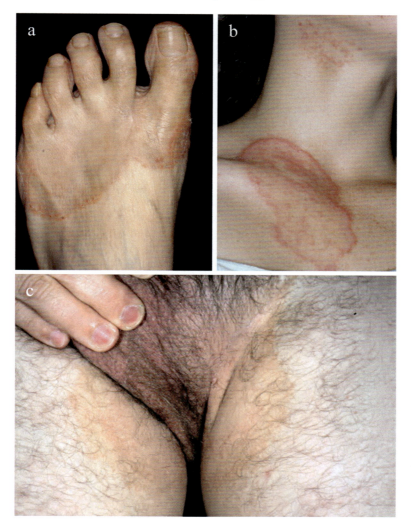

FIGURE 9.6 Clinical presentation of tinea pedis (A), tinea corporis (B), and tinea cruris (C) (Dinulos, 2020).

mimic influenza-like respiratory infections, but resolve spontaneously. However, among immunocompromised patients, yeasts multiply and spread to other parts of the body such as the CNS and may present as cryptococcal meningoencephalitis. Other organs that may be affected are the adrenals, bone, prostate gland, skin and eyes. Inflammatory reactions are usually minimal or granulomatous (Carroll et al., 2016).

Upon infection of the CNS, CSF pressure, protein and cell count increase while glucose level may remain normal or decrease. The patient may also experience headache, neck stiffness or disorientation. Lesions in the skin, lung or other organs may also be present. The entire course of the infection may fluctuate over long periods of time, but can be fatal when left untreated. Combination therapy of amphotericin B and flucytosine is the standard treatment for cryptococcal meningitis. However, the benefit of adding flucytosine is unclear. Amphotericin B alone is curative in most nonAIDS patients. Inadequate treatment in AIDS patients will always lead to relapse once medications are withdrawn and would require suppressive therapy of fluconazole because of its property to penetrate the CNS. Most of the time, highly active antiretroviral therapy (HAART) among AIDS patients lowers the incidence of cryptococcal infection. However, in a third of the patients using HAART with cryptococcal meningitis, immune reconstitution inflammatory syndrome (IRIS) may develop and exacerbate the illness. Sometimes, IRIS may also unmask undiagnosed cryptococcal infection (Carroll et al., 2016).

Compared to dermatophytes, *Pneumocystis* fungi are present in the lungs of many animals but rarely cause disease unless in immunocompromised individuals. It is an obligate member of microflora and its mode of infection is unclear, but transmission by aerosol may be possible. Until the advent of AIDS, they were confined to the interstitial plasma cell pneumonitis in malnourished infants or immunosuppressed patients as plasma cells are absent in AIDS patients. This could lead to blockage of the oxygen exchange interface and consequently cyanosis. Since then, it has become a major cause of death among AIDS patients until the use of chemoprophylactic agents. *Pneumocystis* spp. may take the form of thin- walled trophozoites or thick-walled cysts. They are usually found in tight mass among specimens, which reflects their mode of growth. As an extracellular pathogen, its growth is limited to the surfactant layer about the alveolar epithelium. In the absence of immunosuppression, *P. (carinii) jirovecii* causes no clinical manifestation. Cell-mediated immunity plays an important role in the resistance to the disease. *Pneumocystis* pneumonia is not usually seen in AIDS patients until CD4 lymphocytes drop below 400/uL (Carroll et al., 2016).

Chemoprophylaxis are now available and dramatically decreases the incidence of pneumonia, but infection of other organs such as the spleen, lymph nodes and bone marrow may increase. Prophylaxis is the mainstay for the prevention of at-risk patients and trimethoprim-sulfamethoxazole is the current drug of choice. HIV patients with *Pneumocystis* pneumonia should be given secondary prophylaxis with trimethoprim-sulfamethoxazole until immune reconstitution is achieved. In acute cases, trimethoprim-sulfamethoxazole or pentamidine isethionate may also be given (White et al., 2017).

3. Concluding remarks

The Philippines is a hotspot for TB and HIV/AIDS infections. This, coupled with advancing medical practices has led to an increased number of immunocompromised hosts. As a result, the burden of fungal infections has continued to rise. This trend is evident globally wherein pathogens such as *Aspergillus*, *Candida*, dermatophytes, and opportunistic microorganisms have significantly increased in frequency and severity of infections. Despite this, the burden of fungal pathogens is still not a priority for public health officials (Tan et al., 2020). Knowledge on the incidence of fungal infections is limited by the lack of regular national surveillance systems, no obligatory reporting of fungal diseases, poor clinician suspicion outside specialized units, poor diagnostic test performance (especially for culture) and few well-designed published studies (Bongomin et al., 2017).

In the Philippines, these limitations are evident in the very sparse and outdated epidemiological data. Very few studies have been conducted on fungal disease surveillance (Batac & Denning, 2017). Because of the absence of timely local data, we are unable to accurately assess and consequently prioritize the burden of fungal infections. Apart from surveillance, the management of fungal infections also has several limitations. These are the lack of available and accessible diagnostic tools, antifungal intervention, formal training and local guidelines, and decreased patient compliance with long-term treatment (Batac & Denning, 2017; Tan et al., 2020).

If these shortcomings are not addressed, fungal infections will continue to be an unrecognized burden until they can no longer be ignored. Evidently, it has already become more prominent because of the COVID-19 pandemic that is devastating the entire world. There has been an associated rise in fungal infections with COVID-19 infections, most commonly pulmonary aspergillosis, and invasive candidiasis (Centers for Disease Control and Prevention [CDC], 2021). These secondary infections contribute to increased illness severity and death (CDC, 2021). This trend has been reported in several countries, but it is most notable in India, wherein COVID-19 associated mucormycosis has become an emerging syndemic (Gandra et al., 2021). Major risk factors for mucormycosis include poorly controlled diabetes and immunosuppression (Gandra et al., 2021). These risk factors have become aggravated by the wave of COVID-19 infections that overwhelmed the country's healthcare system. COVID-19 is treated with corticosteroids, and this drug has been noted to suppress the immune system and increase blood glucose levels (Gandra et al., 2021). Misuse of corticosteroids and failure to control elevated blood glucose levels have contributed to the increase in COVID-19 associated mucormycosis (Gandra et al., 2021). Other factors that have contributed to this syndemic include shortages of antifungal medications, overuse of antibiotics, and shortages of hospital facilities and equipment, such as beds and oxygen (Gandra et al., 2021).

The current situation has highlighted the importance of recognizing and responding to the increasing burden of fungal disease. Fungal disease surveillance and management must be a priority of public health officials. Systemic changes must be implemented in order to tackle the emerging problem. These include initiating and encouraging epidemiological studies on fungal infections and providing training and guidelines for healthcare workers. Diagnostic facilities, antifungal medications, and other resources should also be made available and accessible. The public should be educated on the importance of consulting medical practitioners and adherence to antifungal treatment to decrease improper self-medication, decrease treatment costs, and improve patient outcomes (Weinberg, 2009). Through these measures, the burden of fungal infections can be properly addressed before our health system becomes overwhelmed.

Acknowledgments

The authors acknowledge the numerous authors whose papers are cited in this paper.

References

Alkharashi, N., Aljohani, S., Layqah, L., Masuadi, E., Baharoon, W., Al-Jahdali, H., & Baharoon, S. (2019). *Candida* bloodstream infection: Changing pattern of occurrence and antifungal susceptibility over 10 years in a tertiary care Saudi hospital. *Canadian Journal of Infectious Diseases and Medical Microbiology, 2019.* https://doi.org/10.1155/2019/2015692

Batac, M. C., & Denning, D. (2017). Serious fungal infections in the Philippines. *European Journal of Clinical Microbiology & Infectious Diseases, 36*(6), 937−941. https://doi.org/10.1007/s10096-017-2918-7

Bongomin, F., Gago, S., Oladele, R., & Denning, D. (2017). Global and multi-national prevalence of fungal diseases: Estimate precision. *Journal of Fungi, 3*(4), 57−86. https://doi.org/10.3390/jof3040057

Burgos, A., Zaoutis, T. E., Dvorak, C. C., Hoffman, J. A., Knapp, K. M., Nania, J. J., Prasad, P., & Steinbach, W. J. (2008). Pediatric invasive aspergillosis: A multicenter retrospective analysis of 139 contemporary cases. *Pediatrics, 121*, e1286−1294. https://doi.org/10.1542/peds.2007-2117

Carroll, K. C., Hobden, J. A., Miller, S., Morse, S. A., Mietzner, T. A., Detrick, B., Mitchell, T. G., McKerrow, J. H., & Sakanari, J. A. (2016). In Jawetz, Melnick, & Adelberg (Eds.), *Medical microbiology* (27th ed., pp. 657−683). New York, NY: Lange Medical Books/McGraw-Hill, Medical Pub.

Centers for Disease Control and Prevention. (2021). *Fungal diseases and COVID-19.* https://www.cdc.gov/fungal/covid-fungal.html#ref-24.

Cheng, S., & Chong, L. (2002). A prospective epidemiological study on tinea pedis and onychomycosis in Hong Kong. *Chinese Medical Journal, 115*(6), 860−865.

Chindamporn, A., Chakrabarti, A., Li, R., Sun, P. L., Tan, B. H., Chua, M., Wahyuningsih, R., Patel, A., Liu, Z., Chen, Y. C., & Chayakulkeeree, M. (2018). Survey of laboratory practices for diagnosis of fungal infection in seven Asian countries: An Asia Fungal Working Group (AFWG) initiative. *Medical Mycology, 56*(4), 416−425. https://doi.org/10.1093/mmy/myx066

Dabas, P. S. (2013). An approach to etiology, diagnosis and management of different types of candidiasis. *Journal of Yeast and Fungal Research, 4*(6), 63−74. https://doi.org/10.5897/JYFR 2013.0113

Dadar, M., Tiwari, R., Karthik, K., Chakraborty, S., Shahali, Y., & Dhama, K. (2018). *Candida albicans*: Biology, molecular characterization, pathogenicity, and advances in diagnosis and control: An update. *Microbial Pathogenesis, 128*−138. https://doi.org/10.1016/j.micpath.2018.02.028

Dayrit, M., Lagrada, L., Picazo, O., Pons, M., & Villaverde, M. (2018). *The Philippines health system review* (2nd ed.). World Health Organization, Regional Office for South-East Asia.

De Pauw, B., Walsh, T. J., Donnelly, J. P., Stevens, D. A., Edwards, J. E., Calandra, T., Pappas, P. G., Maertens, J., Lortholary, O., & Kauffman, C. A. (2008). Revised definitions of invasive fungal disease from the European organization for research and treatment of cancer/invasive fungal infections cooperative group and the national institute of allergy and infectious diseases mycoses study group (EORTC/MSG) consensus group. *Clinical Infectious Diseases, 46*, 1813−1821. https://doi.org/10.1086/588660

Department of Health. (2010). *The 2010 Philippine Health Statistics.* Retrieved from https://doh.gov.ph/sites/default/files/publications/PHS2010_March13.compressed.pdf.

Desoubeaux, G., Bailly, É., & Chandenier, J. (2014). Diagnosis of invasive pulmonary aspergillosis: Updates and recommendations. *Medecine et Maladies Infectieuses, 44*(3), 89−101. https://doi.org/10.1016/j.medmal.2013.11.006

Dinulos, J. (2020). *Habif's clinical dermatology − a color guide to diagnosis and therapy* (7th ed.). Elsevier.

Dovnik, A., Golle, A., Novak, D., Arko, D., & Takac, I. (2015). Treatment of vulvovaginal candidiasis: A review of the literature. *Acta Dermatovenerologica Alpina, Pannonica et Adriatica, 24*(1), 5−7. https://doi.org/10.15570/actaapa.2015.2

Einsele, H., & Loeffler, J. (2008). Contribution of new diagnostic approaches to antifungal treatment plans in high-risk haematology patients. *Clinical Microbiology and Infection, 14*(Suppl. 4), 37−45. https://doi.org/10.1111/j.1469-0691.2008.01980.x

Eusebio-Alpapara, K., Dofitas, B., Balita-Crisostomo, C., Tioleco-Ver, G., Jandoc, L., & Frez, M. (2020). *Senna (Cassia) alata* (Linn.) Roxb. leaf decoction as a treatment for *Tinea imbricata* in an indigenous tribe in Southern Philippines. *Mycoses, 63*(11), 1226−1234. https://doi.org/10.1111/myc.13159

Gabaldón, T. (2019). Recent trends in molecular diagnostics of yeast infections: From PCR to NGS. *FEMS Microbiology Reviews, 43*(5), 517−547. https://doi.org/10.1093/femsre/fuz015

Gandra, S., Ram, S., & Levitz, S. (2021). The "black fungus" in India: The emerging syndemic of COVID-19−associated mucormycosis. *Annals of Internal Medicine, 174*(9), 1301−1302. https://doi.org/10.7326/M21-2354

Gangcuangco, L. (2019). HIV crisis in the Philippines: Urgent actions needed. *The Lancet Public Health, 4*(2), e84. https://doi.org/10.1016/s2468-2667(18)30265-2

Gregg, K. S., & Kauffman, C. A. (2015). Invasive aspergillosis: Epidemiology, clinical aspects and treatment. *Seminars in Respiratory and Critical Care Medicine, 36*(5), 662−672. https://doi.org/10.1055/s-0035-1562893

References

Handog, E., & Dayrit, J. (2005). Mycology in the Philippines, revisited. *Nippon Ishinkin Gakkai Zasshi, 46*(2), 71−76. https://doi.org/10.3314/jjmm.46.71

Herbrecht, R., Denning, D. W., Patterson, T. F., Bennett, J. E., Greene, R. E., Oestmann, J. W., Kern, W. V., Marr, K. A., Ribaud, P., & Lortholary, O. (2002). Voriconazole versus amphotericin B for primary therapy of invasive aspergillosis. *The New England Journal of Medicine, 347*, 408−415. https://doi.org/10.1056/NEJMoa020191

Höfken, T. (2013). *Candida* and candidiasis. *Microbial Pathogenesis*, 82−114.

Hope, W. W., Walsh, T. J., & Denning, D. W. (2005). Laboratory diagnosis of invasive aspergillosis. *The Lancet Infectious Diseases, 5*(10), 609−622. https://doi.org/10.1016/S1473-30 99(05)70238-3

Husain, A., Fattah, N., Afify, A., & Mostafa, N. (2018). Epidemiology and risk factors of superficial fungal infections in Toukh primary health care centre. *The Egyptian Journal of Hospital Medicine, 72*(7), 4898−4902. https://doi.org/10.12816/EJHM.2018.10171

Johal, H. S., Garg, T., Rath, G., & Goyal, A. K. (2016). Advanced topical drug delivery system for the management of vaginal candidiasis. *Drug Delivery, 23*(2), 550−563. https://doi.org/10.3109/10717544.2014.928760

Juayang, A. C., Lim, J. T., de los Reyes, Z. B., Tuante, M. B., Batiles, Z. P., Guino-o, J. V., Villanueva, F. A., & de los Reyes, G. B. (2019). Antifungal resistance of *Candida species* in bacolod city, Philippines. *Journal of Infectious Diseases and Epidemiology, 5*, 076. https://doi.org/10.23937/2474-3658/1510076

Kadosh, D. (2019). Regulatory mechanisms controlling morphology and pathogenesis in Candida albicans. *Current Opinion in Microbiology, 52*, 27−34. https://doi.org/10.1016/j.mib.2019. 04.005

Kashem, S. W., & Kaplan, D. H. (2016). Skin immunity to Candida albicans. *Trends in Immunology, 37*(7), 440−450. https://doi.org/10.1016/j.it.2016.04.007

Khayhan, K., Hagen, F., Pan, W., Simwami, S., Fisher, M., Wahyuningsih, R., Chakrabarti, A., Chowdhary, A., Ikeda, R., Taj-Aldeen, S., Khan, Z., Ip, M., Imran, D., Sjam, R., Sriburee, P., Liao, W., Chaicumpar, K., Vuddhakul, V., Meyer, W., Trilles, L., van Iersel, L., Meis, J., Klaassen, C., & Boekhout, T. (2013). Geographically Structured Populations of Cryptococcus neoformans Variety grubii in Asia Correlate with HIV Status and Show a Clonal Population Structure. *PLoS ONE, 8*(9), e72222.

Kozel, T. R., & Wickes, B. (2014). Fungal diagnostics. *Cold Spring Harbor perspectives in medicine, 4*(4), a019299. https://doi.org/10.1101/cshperspect.a019299

Leung, A. K., Lam, J. M., Leong, K. F., & Hon, K. L. (2020). Tinea corporis: An updated review. *Drugs in Context, 9*. https://doi.org/10.7573/dic.2020-5-6

Limper, A., Adenis, A., Le, T., & Harrison, T. (2017). Fungal infections in HIV/AIDS. *The Lancet Infectious Diseases, 17*(11), e334−e343. https://doi.org/10.1016/s1473-3099(17)30303-1

Maertens, J., Raad, I., Petrikkos, G., Boogaerts, M., Selleslag, D., Petersen, F. B., Sable, C. A., Kartsonis, N. A., Ngai, A., & Taylor, A. (2004). Efficacy and safety of caspofungin for treatment of invasive aspergillosis in patients refractory to or intolerant of conventional antifungal therapy. *Clinical Infectious Diseases, 39*, 1563−1571. https://doi.org/10.1086/423381

Martins, N., Ferreira, I. C., Barros, L., Silva, S., & Henriques, M. (2014). Candidiasis: Predisposing factors, prevention, diagnosis and alternative treatment. *Mycopathologia, 177*(5−6), 223−240. https://doi.org/10.1093/femspd/ftw018

Marty, F. M., Koo, S., Bryar, J., & Baden, L. R. (2007). (1−>3)beta-D-glucan assay positivity in patients with *Pneumocystis (carinii) jiroveci* pneumonia. *Annals of Internal Medicine, 147*, 70−72. https://doi.org/10.7326/0003-4819-147-1-200707030-00018

McCarty, T. P., & Pappas, P. G. (2016). Invasive candidiasis. *Infectious Disease Clinics of North America, 30*(1), 103−124. https://10.1056/NEJMra1315399.

Millsop, J. W., & Fazel, N. (2016). Oral candidiasis. *Clinics in Dermatology, 34*(4), 487−494. https://doi.org/10.1016/j.clindermatol.2016.02.022

Moron, L. S., Oyong, G. G., Chua, J. C., & Cabrera, E. C. (2017). Polyphasic identification of clinical Candida albicans in the Philippines. *Current Research in Environmental Sustainability, 7*(4), 346−355. https://doi.org/10.5943/cream/7/4/11

Mukaremera, L., Lee, K. K., Mora-Montes, H. M., & Gow, N. A. (2017). *Candida albicans* yeast, pseudohyphal, and hyphal morphogenesis differentially affects immune recognition. *Frontiers in Immunology, 7*(8), 629. https://doi.org/10.3389/fimmu.2017.00629

Nicolle, M. C., Benet, T., & Vanhems, P. (2011). Aspergillosis: Nosocomial or community-acquired? *Medical Mycology, 49*(Suppl. 1), S24−S29. https://doi.org/10.3109/13693786. 2010.509335

Panackal, A. A., Imhof, A., Hanley, E. W., & Marr, K. A. (2006). *Aspergillus ustus* infections among transplant recipients. *Emerging Infectious Diseases, 12*(3), 403−408. https://doi.org/10.3201/eid1203.050670

Pappas, P. G., Lionakis, M. S., Arendrup, M. C., Ostrosky-Zeichner, L., & Kullberg, B. J. (2018). Invasive candidiasis. *Nature Reviews Disease Primers, 4*(1), 1–20. https://doi.org/10.1038/nrdp.2018.26

Patterson, T. F., Kirkpatrick, W. R., White, M., Hiemenz, J. W., Wingard, J. R., Dupont, B., Rinaldi, M. G., Stevens, D. A., & Graybill, J. R. (2000). Invasive aspergillosis disease spectrum, treatment practices and outcomes. *Medicine, 79*(4), 250–260.

Pereira, L. C., Correia, A. F., da Silva, Z. L., de Resende, C. N., Brandão, F., Almeida, R. M., & Nóbrega, Y. D. (2021). Vulvovaginal candidiasis and current perspectives: New risk factors and laboratory diagnosis by using MALDI TOF for identifying species in primary infection and recurrence. *European Journal of Clinical Microbiology & Infectious Diseases, 40*. https://doi.org/10.1007/s10096-021-04199-1

Philippine Statistics Authority. (2016). *Philippine population density (based on the 2015 census of population)*. https://psa.gov.ph/sites/default/files/attachments/hsd/pressrelease/Press%20Release_Philippine%20Population%20Density.pdf.

R, A. N., & Rafiq, N. B. (2020). Candidiasis. In StatPearls. *Treasure Island (FL)*. StatPearls Publishing. PMID:32809459.

Sabu, S., Nayak, D. M., Nair, S., & Shetty, R. (2017). Role of Papanicolaou smear in the diagnosis of pathologic flora in asymptomatic patients in rural health care set-up. *Journal of Clinical and Diagnostic Research, 11*(10), EC10–EC13. https://doi.org/10.7860/JCDR/2017/25044.10733

Safdar, A., Rodriguez, G. H., Lichtiger, B., Dickey, B. F., Kontoyiannis, D. P., Freireich, E. J., Shpall, E. J., Raad, I. I., Kantarjian, H. M., & Champlin, R. E. (2006). Recombinant interferon gamma 1b immune enhancement in 20 patients with hematologic malignancies and systemic opportunistic infections treated with donor granulocyte transfusions. *Cancer, 106*, 2664–2671. https://doi.org/10.1002/cncr.21929

Sahoo, A. K., & Mahajan, R. (2016). Management of tinea corporis, tinea cruris, and tinea pedis: A comprehensive review. *Indian Dermatology Online Journal, 7*(2), 77–86. https://doi.org/10.4103/2229-5178.178099

Salvana, E., Leyritana, K., Alejandria, M., Lim, J., Destura, R., & Tenorio, A. (2012). *Opportunistic infections in Filipino HIV patients. Poster session presented at: Poster Abstract Session: HIV-Associated Infections and Malignancies*. San Diego, CA: Infectious Disease week 2012 Conference; 2012 Oct 17–21.

Sardi, J. C., Scorzoni, L., Bernardi, T., Fusco-Almeida, A. M., & Giannini, M. M. (2013). *Candida* species: Current epidemiology, pathogenicity, biofilm formation, natural antifungal products and new therapeutic options. *Journal of Medical Microbiology, 62*(1), 10–24. https://doi.org/10.1099/jmm.0.0 45054-0

Segal, B. H., Herbrecht, R., Stevens, D. A., Ostrosky-Zeichner, L., Sobel, J., Viscoli, C., Walsh, T. J., Maertens, J., Patterson, T. F., & Perfect, J. R. (2008). Defining responses to therapy and study outcomes in clinical trials of invasive fungal diseases: Mycoses study group and European organization for research and treatment of cancer consensus criteria. *Clinical Infectious Diseases, 47*, 674–683. https://doi.org/10.1086/590566

Shafritz, A. B., & Coppage, J. M. (2014). Acute and chronic paronychia of the hand. *Journal of the American Academy of Orthopaedic Surgeons, 22*(3), 165–174. https://doi.org/10.5435/JAAOS-22-03-165

Sherif, R., & Segal, B. H. (2010). Pulmonary aspergillosis: Clinical presentation, diagnostic tests, management and complications. *Current Opinion in Pulmonary Medicine, 16*(3), 242–250. https://doi.org/10.1097/MCP.0b013e328337d6de

da Silva Dantas, A., Lee, K. K., Raziunaite, I., Schaefer, K., Wagener, J., Yadav, B., & Gow, N. A. (2016). Cell biology of *Candida albicans*: Host interactions. *Current Opinion in Microbiology, 34*, 111–118. https://doi.org/10.1016/j.mib.2016.08.006

Singh, A., Verma, R., Murari, A., & Agrawal, A. (2014). Oral candidiasis: An overview. *Journal of Oral and Maxillofacial Pathology, 18*(Suppl. 1), S81. https://doi.org/10.4103/0973-029X.141325

Smith, D. R., Guo, Y. L. L., Lee, Y. L., Hsieh, F. S., Chang, S. J., & Sheu, H. M. (2002). Prevalence of skin disease among nursing home staff in southern Taiwan. *Industrial health, 40*(1), 54–58.

Smith, T. J., Khatcheressian, J., Lyman, G. H., Ozer, H., Armitage, J. O., Balducci, L., Bennett, C. L., Cantor, S. B., Crawford, J., & Cross, S. J. (2006). 2006 update of recommendations for the use of white blood cell growth factors: An evidence-based clinical practice guideline. *Journal of Clinical Oncology, 24*, 3187–3205. https://doi.org/10.1200/JCO. 2006.06.4451

Smith, J., Safdar, N., Knasinski, V., Simmons, W., Bhavnani, S. M., Ambrose, P. G., & Andes, D. (2006). Voriconazole therapeutic drug monitoring. *Antimicrobial Agents and Chemotherapy, 50*, 1570–1572. https://doi.org/10.1128/AAC.50.4.1570-1572.2006

Steinbach, W. J., Benjamin, D. K., Jr., Kontoyiannis, D. P., Perfect, J. R., Lutsar, I., Marr, K. A., Lionakis, M. S., Torres, H. A., Jafri, H., & Walsh, T. J. (2004). Infections due to *Aspergillus terreus*: A multicenter retrospective analysis of 83 cases. *Clinical Infectious Diseases, 39*(2), 192–198. https://doi.org/10.1086/421950

Sumalapao, D. E., Villarante, N. R., Salazar, P. B., Alegre, F. M., Altura, M. T., Sia, I. C., Flores, M. J., Amalin, D. M., & Gloriani, N. G. (2020). Polymeric compositions of medical devices account for the variations in Candida albicans biofilm structural morphology. *Current Research in Environmental Sustainability, 10*(1), 1–9. https://doi.org/10.5943/cream/10/1/1

Tan, B., Chakrabarti, A., Patel, A., Chua, M., Sun, P., & Liu, Z. (2020). Clinicians' challenges in managing patients with invasive fungal diseases in seven Asian countries: An Asia Fungal Working Group (AFWG) Survey. *International Journal of Infectious Diseases, 95*, 471–480. https://doi.org/10.1016/j.ijid.2020.01.007

Thompson, D. S., Carlisle, P. L., & Kadosh, D. (2011). Coevolution of morphology and virulence in *Candida* species. *Eukaryotic Cells, 10*(9), 1173–1182. https://doi.org/10.1128/EC. 05085-11

Tsui, C., Kong, E. F., & Jabra-Rizk, M. A. (2016). Pathogenesis of *Candida albicans* biofilm. *Pathogens and Disease, 74*(4), ftw018. https://doi.org/10.1093/femspd/ftw018

Ungpakorn, R., Lohaprathan, S., & Reangchainam, S. (2004). Prevalence of foot diseases in outpatients attending the Institute of Dermatology, Bangkok, Thailand. *Clinical and Experimental Dermatology, 29*(1), 87–90. https://doi.org/10.1111/j.1365-2230.2004.01446.x

Weinberg, J. M. (2009). Increasing patient adherence in antifungal infection treatment: Once-daily dosing of sertaconazole. *The Journal of Clinical and Aesthetic Dermatology, 2*(2), 38–42. PMID:20967180.

White, P., Backx, M., & Barnes, R. (2017). Diagnosis and management of Pneumocystis jirovecii infection. *Expert Review of Anti-infective Therapy, 15*(5), 435–447.

White, P. L., Price, J. S., & Backx, M. (2018). Therapy and management of. *Pneumocystis Jirovecii Infection. J Fungi (Basel), 4*(4), 127. https://doi.org/10.3390/jof4040127

Woods, K., & Höfken, T. (2016). The zinc cluster proteins U pc2 and E cm22 promote filamentation in Saccharomyces cerevisiae by sterol biosynthesis-dependent and-independent pathways. *Molecular Microbiology, 99*(3), 512–527.

Yuen, K. Y., Woo, P. C., Ip, M. S., Liang, R. H., Chiu, E. K., Siau, H., Ho, P. L., Chen, F. F., & Chan, T. K. (1997). Stage-specific manifestation of mold infections in bone marrow transplant recipients: Risk factors and clinical significance of positive concentrated smears. *Clinical Infectious Diseases, 25*(1), 37–42. https://doi.org/10.1086/514492

Zakaria, A., Osman, M., Dabboussi, F., Rafei, R., Mallat, H., Papon, N., Bouchara, J. P., & Hamze, M. (2020). Recent trends in the epidemiology, diagnosis, treatment, and mechanisms of resistance in clinical *Aspergillus* species: A general review with a special focus on the middle eastern and north african region. *Journal of Infection and Public Health, 13*(1), 1–10. https://doi.org/10.1016/j.jiph.2019.08.007

Zeng, X., Zhang, Y., Zhang, T., Xue, Y., Xu, H., & An, R. (2018). Risk factors of vulvovaginal candidiasis among women of reproductive age in Xi'an: A cross-sectional study. *BioMed Research International*, 1–8. https://doi.org/10.1155/2018/9703754. Article ID 9703754.

Further reading

Ben-Ami, R. (2018). Treatment of invasive candidiasis: A narrative review. *Journal of Fungi, 4*(3), 97. https://doi.org/10.3390/jof4030097

Iorizzo, M. (2015). Tips to treat the 5 most common nail disorders: Brittle nails, onycholysis, paronychia, psoriasis, onychomycosis. *Dermatologic Clinics, 33*(2), 175–183. https://doi.org/10.1016/j.det.2014.12.001

Mayeaux, E. J. (2013). Candida vulvovaginitis. In R. P. Usatine (Ed.), *The color Atlas and Synopsis of family medicine* (2nd ed.). McGraw Hill Medical Publishing.

Nobile, C. J., & Johnson, A. D. (2015). *Candida albicans* biofilms and human disease. *Annual Review of Microbiology, 69*, 71–92. https://doi.org/10.1146/annurev-micro-091014-104330

Yapar, N. (2014). Epidemiology and risk factors for invasive candidiasis. *Therapeutics and Clinical Risk Management, 10*, 95. https://doi.org/10.2147/TCRM.S40160

CHAPTER 10

Environmental mycology in the Philippines

Jonathan Jaime G. Guerrero[1], Charmaine A. Malonzo[2], Ric Ryan H. Regalado[3] and Arnelyn D. Doloiras-Laraño[4]

[1]College of Medicine, University of the Philippines Manila, Manila, Philippines; [2]Department of Biology, College of Science, Bicol University, Legazpi City, Philippines; [3]National Institute of Molecular Biology and Biotechnology, College of Science, University of the Philippines, Diliman, Philippines; [4]Graduate School of Science and Engineering, Ehime University, Matsuyama, Ehime, Japan

1. Introduction: the growing environmental concern in Southeast Asia

Countries in Southeast Asia have experienced rapid urbanization in the last 30 years. The total population is estimated to reach 700 million in 2025 (United Nations Population Division, 2020), and that 65% of this will live in urban centers (Thuzar, 2011). This scenario is far from the historically low-density Southeast Asia before 1750 (Hirschman & Bonaparte, 2012). Although megacities, like Manila and Bangkok, differ in urban land changes, they face the potential loss of spaces (Estoque & Murayama, 2015) and the occurrence of many urban heat islands (Estoque et al., 2017). The vibrant economic dynamism in this part of the world is also met with a dilemma such as pollution, congestion, and poor urban environmental conditions (Thuzar, 2011). Expansion to the countryside (Melia, 2020; Redman & Jones, 2005), as well as meeting the needs of the growing population, requires the increase and better-quality food supply, transport, and storage (Hammond et al., 2015; McClements et al., 2020), smart transport systems (Kumar et al., 2018), and efficient manufacturing (Siemieniuch et al., 2015), among others. All of these put pressure on a finite healthy environment.

This, in turn, creates more waste that goes into soil and water systems. It is estimated that 275 million tons of plastic wastes were produced worldwide in 2010, of which 31.9 million tons were mismanaged and ended up either in land or water systems (Jambeck et al., 2015; Ritchie & Roser, 2018). At least 80% of the plastic problem originates from Asia (Marks et al., 2020). Vietnam, Indonesia, and the Philippines are among the three Southeast Asian countries in the top 5 marine plastic polluters globally (Garcia et al., 2019). This scenario is greatly complexified by our increasing understanding of microplastics (Rochman & Hoillien, 2020), and quite recently, nanoplastics (Mitrano et al., 2021), and the risks they pose to the environment and biosphere.

Similarly, 1.7–8.8 million metric tons of petroleum hydrocarbons are estimated to have been released into the ocean (Ossai et al., 2020) and find their way into shores and land areas. The entry of petroleum hydrocarbon into the environment alters the ecosystem's functionality (Truskewycz et al., 2019). They have been shown to cause alterations to soil characteristics (Osuji & Nwoye, 2007; Wang et al., 2013) and cause harm among marine animals (Khan et al., 2005), birds (Albers, 2006), microbiota (Klimek et al., 2016), and humans (Adipah, 2019). Countries like Thailand, Malaysia, and the Philippines have recognized the increasing petroleum hydrocarbon in water systems (Balce, 1997; Sakari et al., 2012; Wattayakorn, 2012).

In 2019, 56.3 million metric tons of e-waste were generated worldwide (Houessionon et al., 2021), which translates to heavy metals finding their way into landfills and sewage systems every year. In addition, Southeast Asia has recorded heavy metal contamination in soil and crops (Zarcinas, Ishak et al., 2004; Zarcinas, Pongsakul et al., 2004), groundwater (Rahman et al., 2009), food supply (Agusa et al., 2007, and eventually to the human population (Kladsomboom et al., 2020). This is besides the pressure applied in extracting raw materials from mines and deposits deep beneath the Earth's surface.

With the strong reliance on agriculture in Southeast Asian countries, fertilizers and pesticides are common. Small-scale farmers rely on pesticides (Schreinemachers et al., 2017), with over 77% considered overuse (Schreinemachers et al., 2020). Pesticide-contaminated foods have also been reported (Lam et al., 2017), such as in the Philippines (Del Prado, 2015) and Cambodia (Wang et al., 2011). Part of the sustainable trajectory of countries is to minimize consumption, both at the individual and country levels, and to provide solutions to the mounting pollution of soil and water systems. As the waste accumulates, more technologies are evaluated for their potential application to clean up contaminated soils and water environments.

2. Mycoremediation

Bioremediation has been performed worldwide to restore the functionality of heavily contaminated soils and water systems. The process usually utilizes indigenous plant species or a consortium of microorganisms to expedite the cleaning process. The choice of an organism depends largely on the pollutant to be remediated, the organism's tolerance to that specific pollutant and the environmental factors to optimize the bioremediation mechanism. It is also important to consider the potential harm on nontarget organisms and the safety of handling by humans.

Microbial bioremediation research in the Philippines is largely along bacterial isolates (Adriano et al., 2018; Dela Cruz & Halos, 1997; Lim & Halos, 1995; Su, 2016; Villegas et al., 2018). However, fungi also have characteristics that make them good candidates for remediation. Unlike bacteria, they have longer life cycles, greater biomass, and an extensive hyphal network (Singh et al., 2015). The ramification of the hyphal networks provides mechanical support and a larger surface area through which enzymes can be released to break down chemical pollutants (Singh & Gauba, 2014). The fungal cell wall is also helpful in binding pollutants (Igir et al., 2018). Further, the diversity of fungi allows one to mix and match so that the correct species are used to target a particular pollutant (Rhodes, 2014). How the fungal cells interact with different pollutants, especially recalcitrant or persistent ones, have been researched extensively, and Yadav et al. (2021) summarize these in Fig. 10.1.

Environmental mycology, particularly on remediating and rehabilitating soil and water ecosystems, is a viable field to address mounting environmental problems. Several researchers harnessed the potentials of fungi in remediating soils and water contaminated with heavy metals, pesticides, and fertilizers, plastics, and hydrocarbons. This chapter compiles these studies to present better the current picture of environmental mycology in the Philippines.

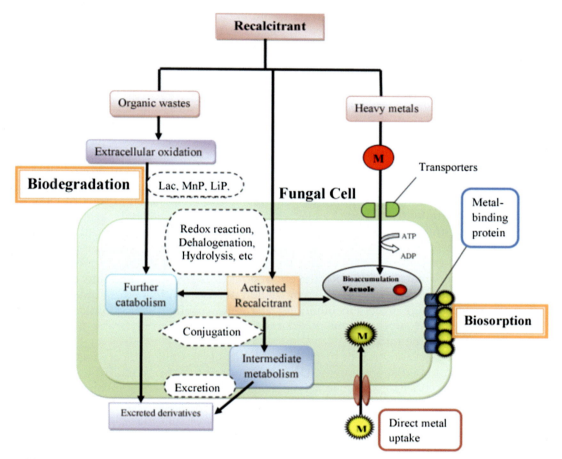

FIGURE 10.1 The fungal cell is an active player in the biodegradation and biosorption of chemical wastes. *Figure adapted from Yadav, P., Rai, S.N., Mishra, V., & Singh, M.P. (2021). Mycoremediation of environmental pollutants: A review with special emphasis on mushrooms. Environmental Sustainability, 4(4), 605-618..*

3. Mycoremediation of heavy metals

An intrinsic advantage of the Philippines is its being an archipelago. This gives rise to a diversity of soil taxa, and these have been comprehensively presented by Carating et al. (2014). In addition, this provides the country with a diversity of environmental conditions that gives rise to a corresponding microbial diversity. For example, geothermal springs (Lacap et al., 2005), underwater plateaus (Gajigan et al., 2018), and various farming systems (Monsalud et al., 2009) across the country have unique microbiota.

The Philippines is also a metal-rich country. As such, mining companies are spread across many islands targeting particular metals. The country has one of the largest copper and gold resources globally and is rich in nickel and other precious metals (Hicks et al., 2012). Continuous mining, however, contributes to environmental degradation. Eventually, mining companies leave sites after the expiration of contracts leaving the soil nutrient-poor and unviable for use due to the highly extractive nature of mining. As a result, abandoned and inactive sites are high in toxic wastes. They have altered soil characteristics such as high pH, high salinity, low water retention capacity, and high heavy metal concentrations (Samaniego et al., 2020). As of 2019, there are 27 abandoned and inactive mines in the Philippines, according to the Mines and Geosciences Bureau (Aggangan et al., 2019).

Several studies have tapped on this opportunity to isolate fungal species tolerant to high levels of heavy metals and potentially revitalize the abandoned mines through remediation, including mycoremediation. Most isolation areas are mining sites, tailings, or nearby areas where effluents flow. However, few species were isolated *ex situ*, such as from banana and citrus peels (Casamorin et al., 2014) and the mangrove *Avicennia* L. (Marcelo et al., 2018) and tested for their tolerance and mycoremediation ability (Table 10.1).

In all the researches presented in the Philippines, most isolates are from already-contaminated soil or wastewater. Autochthonous fungi, or those which are indigenous to a particular area, have shown to be important catalysts in the remediation of soils and water systems polluted with hydrocarbons (Covino et al., 2015; D-Annibale et al., 2006), leachates from municipal landfills (Zegzouti et al., 2020), and heavy metals (Muñoz et al., 2012; Prigioni et al., 2009). Aside from the innate morphological and physiological characteristics, these native fungal isolates perform well because they are well-adapted to the site's conditions. It is important to note that the soil and water environments are complex, and thus organisms first need to acclimate before starting the remediation process. On this note, autochthonous fungi have an advantage over allochthonous or foreign counterparts. Hosokawa et al. (2009) also detailed that allochthonous microorganisms tend to be only stable at the start of the process and decline midway.

The creation of fungal consortia or fungal combinations to optimize efficacy has also been seen as a good mycoremediation strategy. However, most of these were limited to the use of mycorrhizal isolates, either the commercially available MYKOVAM and similar products or indigenous isolates from contaminated areas. Moreover, because these are members of Glomeromycota, these are used with host plants to improve survivability in contaminated soil or improve their metal uptake. Some of the host plants are *Desmodium cinereum* (Adiova et al., 2013; Aggangan & Cortes, 2018; Aggangan et al., 2019), *Chrysopogon zizanioides* (L.) Roberty (Bretaña et al., 2019), and *Paraserianthes falcataria* (Rollon et al., 2017). In addition, the species

3. Mycoremediation of heavy metals

TABLE 10.1 Fungal isolates were tested for tolerance against metals and remediating ability.

Species	Place of isolation	Substrate	Activity	Reference
Glomus etunicatum W.N. Becker & Gerd *Glomus macrocarpum* Tul & C. Tul, *Gigaspora margarita* W.N. Becker & I.R. Hall	—	Mycorrhizal inoculant from the BIOTECH, University of the Philippines Los Baños, Laguna	Improved uptake of Cu by *Desmodium cinereum* (Kunth) DC at 800, 1200, and 1600 ppm	Adiova et al. (2013)
Glomus etunicatum Glomus macrocarpum Gigaspora margarita + *Glomus* spp.	—	MYKOVAM from BIOTECH, University of the Philippines Los Baños, Laguna	Increased uptake rate of mercury by *Chrysopogon zizanioides* (L.) Roberty at high Hg concentration	Bretaña et al. (2019)
Debaryomyces hansenii (Zopf) Lodder & Kreger-van Rij *Candida parapsilosis* Langeron & Talice	Mankayan, Benguet	Cu-rich soil	Alleviate heavy metal stress in plant host (*Phragmites australis* Cav.)	Hipol et al. (2015)
Salispilia tartarea S1YP1	Samal Island, Davao del Norte	Mangrove (*Avicennia* sp.)	With Fe-chelating activity with IC50 of 541.32 ± 5.43 µg/mL	Marcelo et al. (2018)
Aspergillus sp.	Coto Chromite Deposit, Masinloc, Zambales	Soil	Tolerated 1440 mg/L of Cr(VI)	De Sotto et al. (2015)
Aspergillus sp.	Motolite Battery Plant, Novaliches, Quezon City	Soil	Tolerated 1800 mg/L of Cr(VI)	De Sotto et al. (2015)
Acyra cinerea (Bull.) Pers.	Central Luzon Region	Soil	High Bioconcentration factor for Mn	Rea-Maminta et al. (2015)
Physarum album (Bull.) Chevall. *Physarum pusillum* (Berk. & M.A. Curtis) G. Lister	Central Luzon Region	Soil	High bioconcentration factor for Mn and Cr	Rea-Maminta et al. (2015)
Vanrija sp.	Philex Mining Site, Benguet Province	Soil	Multi-metal biosorption for Cu, Cr, Mn, Ni, and Zn	Coronado et al. (2016)
Aspergillus sp. (fungal isolate 3)	Brgy. Ibo, Lapu-Lapu City, Cebu	Effluents of industrial plants	Cd biosorption efficiency of 13.87% in 100 mL potato dextrose broth with 10 mL CdSO$_4$	Manguilimotan and Bitacura (2018)
Aspergillus sp. (fungal isolate 4)	Brgy. Ibo, Lapu-Lapu City, Cebu	Effluents of industrial plants	Cd biosorption efficiency of 10.71% in 100 mL potato dextrose broth with 10 mL CdSO$_4$	Manguilimotan and Bitacura (2018)

(*Continued*)

240   10. Environmental mycology in the Philippines

TABLE 10.1 Fungal isolates were tested for tolerance against metals and remediating ability.—cont'd

Species	Place of isolation	Substrate	Activity	Reference
Penicillium sp. (fungal isolate 6)	Brgy. Ibo, Lapu-Lapu City, Cebu	Effluents of industrial plants	Cd biosorption efficiency of 11.46% in 100 mL potato dextrose broth with 10 mL $CdSO_4$	Manguilimotan and Bitacura (2018)
Aspergillus flavus Link	Meycauayan, Bulacan, Guimaras, Iloilo	Heavy metal-contaminated soil and hydrocarbon-contaminated soil	Able to reduce hexavalent Cr from Cr^{6+} to Cr^{3+} through reduced-coupled biotransformation	Bennett et al. (2013)
Aspergillus niger van Tieghem	Meycauayan, Bulacan, Guimaras, Iloilo	Heavy metal-contaminated soil and hydrocarbon-contaminated soil	Able to reduce hexavalent Cr from Cr^{6+} to Cr^{3+} through reduced-coupled biotransformation	Bennett et al. (2013)
Aspergillus sp.	Meycauayan, Bulacan, Guimaras, Iloilo	Heavy metal-contaminated soil and hydrocarbon-contaminated soil	Able to reduce hexavalent Cr from Cr^{6+} to Cr^{3+} through reduced-coupled biotransformation	Bennett et al. (2013)
Penicillium spp. *Fonsecaea* sp. *Aspergillus* spp. *Fusarium* sp. *Trichoderma* sp.	Calancan Bay, Marinduque	Copper-laden sediments	With possible bioaccumulation activities	Su et al. (2014)
Aspergillus unguis Weill & L. Gaudin *Penicillium griseofulvum* Dierckx, R.P.	Guimaras, Iloilo	Oil-contaminated soils	In consortia degraded $72 \pm 1.3\%$ Nickel protoporphyrin disodium and $90 \pm 2.8\%$ Vanadium oxide octaethylporphyrin, both at 20 mg/L	Cordero et al. (2015)
Penicillium canescens Sopp, O.J. *Penicillium* sp. *Talaromyces macrosporus* (Stolk & Samson) Frisvad et al. *Talaromyces* sp.	Marilao River within the Meycauayan-Marilao-Obando river system	Pb-contaminated Soil and water	Tolerant to 500 µg/mL of Pb; removal efficiency of 35.75%—99.5% of Pb at 3000 µg/mL	Zomesh et al. (2019)
Trichoderma harzianum Rifai *Trichoderma virens* Pers. *Trichoderma saturnisporum* Hammill *Trichoderma gamsii* Samuels & Druzhinina	Mine tailing sites in Itogon, Benguet	Wastewater with Cr, Cu, and Pb exceeding allowable standards	All tolerant to 1000 ppm of Cr and Pb; *T. harzianum* and *T. virens* tolerates up to 1000 ppm Cu; *T. virens* able to remove Pb 91% —96% in liquid media	Tansengco et al. (2018)
Rhodotorula toruloides Banno *Candida tropicalis* (Castellani) Berkhout *Papiliotrema laurentii* (Kuff.) X.Z. Liu, F.Y. Bai, M. Groenew & Boekhout *Candida maltosa* Komag.,	Six mine tailing sites in Itogon, Benguet	Wastewater with Cr, Cu, and Pb exceeding allowable standards	Cu and Pb adsorption capacity at 50 mg/L; *Nodulisporium* sp. Capable of treating Ni from wastewater	Gacho et al. (2019)

TABLE 10.1 Fungal isolates were tested for tolerance against metals and remediating ability.—cont'd

Species	Place of isolation	Substrate	Activity	Reference
Nakase & Katsuya) *Nodulisporium* sp. *Candida guilliermondii* (Castell.) Langeron & Guerra *C. lusitaniae* Uden & Carmo Souza	—	Banana and citrus peels	Tolerant to Cd concentrations; Cr absorption capacity	Casamorin et al. (2014)
Pleurotus ostreatus (Jacq.) P. Kumm.	Mandaluyong and Tagaytay	Soil	Adsorptive capacity for Pb and Mn in contaminated soils	Llarena and Solidum (2012)
Glomus sp. *Gigaspora* sp. *Acaulospora* sp. *Scutellospora* sp. *Entrophospora* spp.	—	MYKOVAM from BIOTECH, University of the Philippines Los Baños, Laguna	Enhanced seedling survival and tree growth in mined-out areas (Mogpog, Marinduque, Cu)	Aggangan et al. (2019)
AMF associated with ferns	Abandoned Copper mine in Mogpon, Marinduque	Soil	Effects comparable with commercially available AMF	Aggangan and Cortes (2018)
Coprinus comatus (O.F. Mull.) Pers.	—	Culture collection of the Center for Tropical Mushroom Research and Development, Nueva Ecija	Accumulated high levels of copper in its fruiting bodies	Dulay et al., (2015)
Trichoderma harzianum	Trichoderma microbial inoculant	Soil	Increased yield of rice in Cu-rich rice paddies	Cuevas et al. (2019)
Glomus etunicatum *Glomus* sp. *Gigaspora margarita*	—	MYKOVAM BIOTECH, University of the Philippines Los Baños, Laguna	In combination with carbonized rice hull, Improved nutrient uptake of *Paraserianthes falcataria* (L.) in copper-contaminated soil	Rollon et al. (2017)
Trichoderma spp.	Mt. Talipanan, Oriental Mindoro, La Mesa Ecopark Sorsogon province Las Piñas-Parañaque Ecotourism Park	Soil and leaf litter, marine substrates (seawater, seafoam, decayed seagrass, and seaweeds)	Moderate to high tolerance to Ni at 50–1200 ppm Ni uptake by 6 species range from 66% to 68%	De Padua (2021)

Aspergillus unguis and *Penicillium griseofulvum* were also used in consortia to degrade compounds containing Ni and Va (Cordero et al., 2015).

Microbial consortia present advantages, especially on higher metal scavenging capacity and resilience against environmental fluctuations (Mishra & Malik, 2014). In addition, different species may take advantage of unique optimum ranges so that the remediation process remains active across the tolerance spectrum. Interkingdom consortia, which contain fungi, bacteria, and other microorganisms, may also improve the process. Zhang et al. (2018) said that there is a synergistic division of resources or labor, enhanced tolerance of inhibitors or toxicants, antagonistic interactions that lead to the production of beneficial metabolites, and optimized efficiency and consortia robustness through assembled biotransformation.

At least two studies were recorded using fungal allies: *Salispilia tartarea*, an oomycete (Marcelo et al., 2018), and the myxomycetes *Acyria cinerea*, *Physarum album* and *P. pusillum* (Rea-Maminta et al., 2015). The limited studies among the potentials of fungal allies for remediation may lead to an opportunity for expanding future works.

4. Fungi as bioremediation agents for pesticides

The Philippines' steady pace in modernizing its agri-fishery sector has conventionally encouraged the practice of applying a diverse range of synthetic agrochemicals. However, its massive and indiscriminate use consequently created public health and environmental repercussions, including disruption of the ecosystems (Abreo et al., 2015), high rates of bioaccumulation (Tingson et al., 2018), and large-scale soil contamination (Navarrete et al., 2017). These detrimental events render a large group of beneficial nontarget organisms (Mahmood et al., 2016) vulnerable to the adverse effects of these chemicals, which disrupts soil quality and the pivotal process of pedogenesis (Samal & Mishra, 2021). In addition, the high cost of physical and chemical methods has necessitated the development of various bioremediation strategies in the country to remove contaminants effectively and sustainably, especially in pesticide-polluted areas (Abo-Amer, 2012; Carascal et al., 2017; Mercado et al., 2012; Poncian et al., 2019).

Among the most commonly used pesticide are several chemical subgroups, namely carbamates, pyrethroids, organophosphates, and organochlorines (Cubelo & Cubelo, 2021; Lu, 2010; Lu et al., 2010; Tirado & Bedoya, 2008). While rice is the largest consumer of pesticides in terms of volume (due to a larger production area), the pesticide application in high-value crops is more aggressive (Bajet, 2015). In an analysis by Lu (2010) on the brands of pesticide used in the largest vegetable producing area in the Philippines, Tamaron was the most prevalent type of pesticide used, an organophosphate pesticide. Consequently, pesticide fate is governed by transfer and degradation processes in any agricultural ecosystem, which physicochemical or biological agents can reduce upon reaching the soil, sediment, or water ecosystems. Following a schematic diagram (Fig. 10.2) adopted from Barik (1984), decomposition of xenobiotic compounds in such environments undergo photometabolism, oxidation, reduction, and hydrolysis, which are always driven by the changes of many physicochemical forces such as pH, temperature, ion concentration, and redox potential (Díaz, 2004).

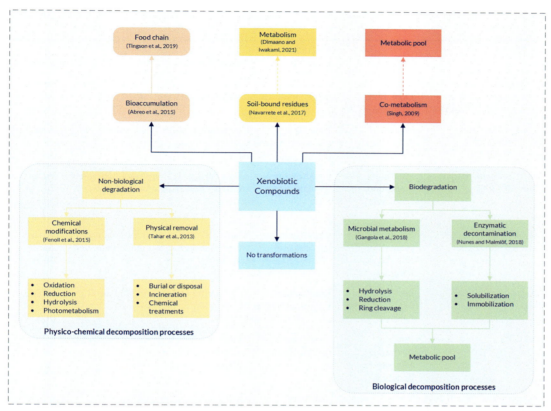

FIGURE 10.2 Fate of xenobiotic compounds in soil, aquatic, and microbial systems by Barik (1984). The diagram is grouped according to transfer and degradation processes; rounded rectangles indicate transfer while regular rectangles indicate degradation. Physico-chemical and biological degradation pathways were also highlighted. Moreover, selected case studies were included to highlight examples from each process. Studies conducted in the Philippines were used, whenever possible. *Adapted by permission from Springer Nature Customer Service Center GmbH: Springer* Insecticide Microbiology *Metabolism of Insecticides by Microorganisms, Sudhakar Barik, Copyright by Springer, Berlin, Heidelberg (1984).*

In the Philippines, mycoremediation studies of pesticide-polluted soil and water systems have not been extensively explored. However, renewed interests in recent years (Carascal et al., 2017; Mercado et al., 2012; Poncian et al., 2019) have advanced the understanding, applicability, and regard of the strategy in the country. The small fraction of fungal species documented in the Philippines with pesticide degradation potential (summarized in Table 10.2) belongs to the group of Ascomycetes found in the provinces of Batangas, Benguet, and Laguna.

Some fungal species were documented to degrade organochlorines, specifically butachlor (Carascal et al., 2017) and endosulfan (Mercado et al., 2012) pesticides. Organochloride pesticides are synthetic, making them recalcitrant, resistant to biodegradation, and characterized mainly by a slow breakdown rate (Jayaraj et al., 2016; Newton, 2018). The findings by

244

10. Environmental mycology in the Philippines

TABLE 10.2 Pesticide-degrading fungi isolated from contaminated sites with degradation potential of more than 50%.

Strain	Source	Degraded pesticide	Degradation potential	References
Acremonium crassum Petch 4PULB05 *Aspergillus fumigatus* Fresenius 8BINM03 *Fusarium* sp.1 9MATK02	Surface water, Lake Taal, Batangas	Butachlor (organochloride)	Degradation of up to 100 mg/L. No percentage data indicated	Carascal et al. (2017)
Aureobasidium sp. MATK04 *Lecythophora* sp. BINM02-1 *Phialemonium* sp.2 ALA03	Submerged wood, Lake Taal, Batangas	Butachlor (organochloride)	Degradation of up to 100 mg/L. No percentage data indicated	Carascal et al. (2017)
Aspergillus heteromorphus Batista & H. Maia strain SF-6390	Potato plantation, Mankayan, Benguet	Cypermethrin (pyrethroid)	About 20 mg/L 1-naphthol equivalent concentration indicating CES enzyme activity	Poncian et al. (2019)
Aspergillus sp. BDP3	Potato plantation, Mankayan, Benguet	Cypermethrin (pyrethroid)	About 58 mg/L 1-naphthol equivalent concentration indicating CES enzyme activity	Poncian et al. (2019)
Candida tropicalis (Castellani) Berkhout P601	Lowland rice field, Calamba, Laguna	Endosulfan (organochloride)	91.35% total endosulfan (100 mg/L) degradation in 14 days	Mercado et al. (2012)
Fusarium sp. BDP3	Potato plantation, Mankayan, Benguet	Cypermethrin (pyrethroid)	About 32 mg/L 1-naphthol equivalent concentration indicating CES enzyme activity	Poncian et al. (2019)
Neodeightonia subglobosa Booth IFM 63572	Surface water, Lake Taal, Batangas	Butachlor (organochloride)	94.68% degradation (100 mg/L) in 5 days using mycelial mat and 73.4% degradation (50 mg/L) in 5 days using mycelial balls	Carascal et al. (2017)
Penicillium sp. BDP1	Potato plantation, Mankayan, Benguet	Cypermethrin (pyrethroid)	About 35 mg/L 1-naphthol equivalent concentration indicating CES enzyme activity	Poncian et al. (2019)
Penicillium sp. BDP10	Potato plantation, Mankayan, Benguet	Cypermethrin (pyrethroid)	About 54 mg/L 1-naphthol equivalent concentration indicating CES enzyme activity	Poncian et al. (2019)
Penicillium sp. BDP12	Potato plantation, Mankayan, Benguet	Cypermethrin (pyrethroid)	About 30 mg/L 1-naphthol equivalent concentration indicating CES enzyme activity	Poncian et al. (2019)
Sclerotium hydrophilum Saccardo IFM 63573	Submerged wood, Lake Taal, Batangas	Butachlor (organochloride)	89.64% degradation (100 mg/L) in 5 days using mycelial mat and 55.6% degradation (50 mg/L) in 5 days using mycelial balls	Carascal et al. (2017)
Unidentified mold species P701	Lowland rice field, Calamba, Laguna	Endosulfan (organochloride)	About 65% total endosulfan (100 mg/L) degradation in 14 days	Mercado et al. (2012)

Carascal et al. (2017) pursued the mycodegradation of a widely used organochloride herbicide butachlor from the surface water and submerged wood collected in Taal Lake, Batangas. Among the best isolates with a significantly high growth capacity on butachlor as the sole carbon sources were *Neodeightonia subglobosa* Booth and *Sclerotium hydrophilum* Saccardo, which showed increased mycelial biomass and decreased butachlor concentration. Moreover, Mercado et al. (2012) showed the capacity of *Candida tropicalis* (Castellani) Berkhout to degrade α- and β-endosulfan, a cyclodiene organochlorine insecticide, in soil showing 91.35% total endosulfan degradation (93.85% for alpha-endosulfan and 90.29% for beta-endosulfan). Despite such efforts, the number of other popular organochloride pesticides used in the country (Cubelo & Cubelo, 2021; Lu, 2010) are still widely uncharacterized such as lindane, aldrin/dieldrin, heptachlor, DDT, heptachlor epoxide, and 4,4-DDE, which necessitates further studies on biodegradation and remediation strategies especially since the majority of these pesticides were reported to be present as residues.

Similarly, the Philippines' principal management approach for pests and diseases also depends on applying pyrethroids—a class of more than 1000 powerful, broad-spectrum insecticides. Compared to organochlorines, pyrethroids biodegrade in the environment easily (Ostrea et al., 2014). One of the main commercially available pyrethroids is cypermethrin. Poncian et al. (2019) reported that the carboxylesterase activity of several filamentous soil fungi from a potato plantation in Mankayan, Benguet is effective in cypermethrin degradation. In addition, the enzyme carboxylesterase in the study was examined as an indicator of pyrethroid degradation since this enzyme has the specific function to mediate pyrethroid cleavage (Cycoń & Piotrowska-Seget, 2016). Among the collected isolates, unidentified Ascomycete species, BDP3 and BDP10 exhibited the greatest carboxylesterase activity and were identified as the most viable candidates for mycoremediation. According to the authors, both unidentified species were likely related to species of *Aspergillus* or *Penicillium* based on morphology and close homology of their D1/D2 sequences. Parallel studies on other pesticide families like anthranilic diamides, neonicotinoids, and thiourea should likewise be pursued since they have been gradually gaining traction in the country's agricultural setting in recent years (Almarinez et al., 2020; Del Prado-Lu, 2015; Lu, 2011a, 2011b, 2012).

5. Fungi as biocontrol agents in agriculture

Although pesticide management and residue risk assessment on food safety is regulated by the Fertilizer and Pesticide Authority (FPA) in the country, challenges in pesticide residues in water (Navarrete et al., 2018; Varca, 2012) and soil (Lu, 2011a, 2011b) and other environmental concerns related to pesticide usage (Cubelo & Cubelo, 2021; Lu et al., 2010) in agriculture remains increasingly pervasive. To date, biological control involving fungal species as biological control agents (BCAs) forms part of one of the widely utilized disease management approaches in Philippine crop protection mechanisms.

Several studies have been carried out in the recent decade to identify effective biocontrol candidates for pre- and postharvest pest-disease management (Table 10.3). National scientist Romulo G. Davide III pioneered considerable progress in biocontrol, evidenced by his numerous contributions to nematology. Perhaps the earliest report in biocontrol was

246
10. Environmental mycology in the Philippines

TABLE 10.3 Documented fungal species and commercially available root inoculants, in the recent decade, with biological control potential against several zoo- and phytopathogens and their reported biological activity.

Strain/Biocontrol agent/Commercial inoculant	Source	Pest/disease and host	Target pathogen	Potential mode of action	References
Beauveria bassiana (Bals.-Criv.) Vuill.	Los Baños, Laguna	Mite infestation on papaya	*Tetranychus kanzawai* Kishida	Predation, hyperparasitism, and enzymatic activity	Sanjaya et al. (2013b, 2014, 2015, 2016)
	University of the Philippines Los Baños, Laguna	Asian corn borer infestation	*Ostrinia furnacalis* Guenée	Predation and hyperparasitism	Nicolas et al. (2013)
Ceratobasidium spp. TDC037, TDC241, TDC474	Banana farms in Mindanao	Fusarium wilt (FocTR4) on Cavendish variety of banana	*Fusarium oxysporum* f. sp. *cubense* (E.F. Smith) Snyder & Hansen	Competition, growth enhancement, antibiosis, and induced systemic resistance	Catambacan and Cumagun (2021)
Chaetomium globosum Kunze	Not indicated	Blast disease in rice	*Pyricularia oryzae* Cavara	Antibiosis (visualized using crude extracts)	Gandalera et al. (2013)
Exophiala sp. NLE 03	Needle leaves of *Casuarina equisetifolia* L. growing in Tagaytay City, Cavite	Fusarium wilt diseases on various crops	*Fusarium oxysporum* Schltdl., *F. solani* (Mart.) Sacc., & *F. moniliforme* J. Sheld	Competition, antibiosis, mycoparasitism, enzymatic activity, and induced systemic resistance	De Mesa et al. (2020)
Fusarium sp. CGP150	Mt. Apo rainforest, Davao	Fusarium wilt (FocTR4) on Cavendish variety of banana	*Fusarium oxysporum* f. sp. *cubense*	Antibiosis, enzymatic activity, mycoparasitism, and competition	Puig and Cumagun (2019)
Geotrichum sp. EF-ds104-16	Lowland rice fields of Nueva Ecija	Sheath blight in rice	*Rhizoctonia solani*	Mycoparasitism, antibiosis, and enzymatic activity	Donayre and Dalisay (2015)
Guignardia spp. NLE 08, NLE 09, NLE 11	Needle leaves of *Casuarina equisetifolia* growing in Tagaytay City, Cavite	Fusarium wilt diseases on various crops	*Fusarium oxysporum*, *F. solani*, *F. moniliforme*	Competition, antibiosis, mycoparasitism, enzymatic activity, and induced systemic resistance	De Mesa et al. (2020)

II. Fungi in agriculture, health and environment

5. Fungi as biocontrol agents in agriculture

247

TABLE 10.3 Documented fungal species and commercially available root inoculants, in the recent decade, with biological control potential against several zoo- and phytopathogens and their reported biological activity.—cont'd

Strain/Biocontrol agent/Commercial inoculant	Source	Pest/disease and host	Target pathogen	Potential mode of action	References
Lasiodiplodia theobromae (Patouillard) Griffon & Maublanc	Banana farms in Mindanao	Fusarium wilt (FocTR4) on Cavendish variety of banana	*Fusarium oxysporum* f. sp. *cubense*	Competition, growth enhancement, antibiosis, and induced systemic resistance	Catambacan and Cumagun (2021)
Metarhizium anisopliae (Metschn.) Sorokīn	Los Baños, Laguna	Mite infestation on papaya	*Tetranychus kanzawai*	Predation, hyperparasitism, and enzymatic activity	Sanjaya et al. (2013a, 2013b, 2016)
	Department of Agriculture General Santos City, South Cotabato	Tick infestation on cows	*Rhipicephalus microplus* Canestrini	Hyperparasitism	Alagos et al. (2015)
Metarhizium flavoviride Gams & Rozsypal var. *flavoviride*	Los Baños, Laguna	Earworm infestation on corn	*Helicoverpa armigera* Hübner	Hyperparasitism	Belen et al. (2011)
Metarhizium rileyi (Farlow) Kepler, S.A. Rehner & Humber	Infested onion fields from San Jose City, Nueva Ecija	Fall armyworm infestation on corn	*Spodoptera exigua* Hübner	Hyperparasitism, antibiosis, and enzymatic activity	Montecalvo and Navasero (2021)
Metarhizium sp. AB001	Abaca aphid at Visayas State University, Southern Leyte	Mealworm infestation	*Tenebrio molitor* L.	Hyperparasitism	Pajar et al. (2013)
Metarhizium sp. UP001	BIOTECH, University of the Philippines Los Baños, Laguna	Mealworm infestation	*Tenebrio molitor*	Hyperparasitism	Pajar et al. (2013)
Paecilomyces lilacinus (Thom) Samson	Los Baños, Laguna	Mite infestation on papaya	*Tetranychus kanzawai*	Predation, hyperparasitism, and enzymatic activity	Sanjaya et al. (2013b, 2016)
Pestalotiopsis sp. CGP117	Mt. Apo rainforest, Davao	Fusarium wilt (FocTR4) on Cavendish variety of banana	*Fusarium oxysporum* f. sp. *cubense*	Antibiosis, enzymatic activity, and mycoparasitism	Puig and Cumagun (2019)

(Continued)

II. Fungi in agriculture, health and environment

248   10. Environmental mycology in the Philippines

TABLE 10.3 Documented fungal species and commercially available root inoculants, in the recent decade, with biological control potential against several zoo- and phytopathogens and their reported biological activity.—cont'd

Strain/Biocontrol agent/Commercial inoculant	Source	Pest/disease and host	Target pathogen	Potential mode of action	References
Phyllosticta sp. NLE 06	Needle leaves of *Casuarina equisetifolia* growing in Tagaytay City, Cavite	Fusarium wilt diseases on various crops	*Fusarium oxysporum, F. solani & F. moniliforme*	Competition, antibiosis, mycoparasitism, enzymatic activity, and induced systemic resistance	De Mesa et al. (2020)
Plectosphaerella sp. NLE 02	Needle leaves of *Casuarina equisetifolia* growing in Tagaytay City, Cavite	Fusarium wilt diseases on various crops	*Fusarium oxysporum, F. solani, & F. moniliforme*	competition, antibiosis, mycoparasitism, enzymatic activity, and induced systemic resistance	De Mesa et al. (2020)
Ramalina farinacea (L.) Acharius	Guimaras, Iloilo	Caustic pathogens and weed infestation in several crops	Pathogenic microbes and weeds	Antibiosis (visualized using crude extracts)	Gazo et al. (2019)
Ramalina nervulosa (Müller Arg.) Abbayes	Guimaras, Iloilo	Caustic pathogens and weed infestation in several crops	Pathogenic microbes and weeds	Antibiosis (visualized using crude extracts)	Gazo et al. (2019)
Ramalina roesleri (Hochstetter ex Schaerer) Nylander	Brgy. Hoskyn, Jordan Guimaras, Iloilo	Caustic pathogens and weed infestation in several crops	Pathogenic microbes and weeds	Antibiosis (visualized using crude extracts)	Gazo et al. (2019)
Schizophyllum commune Fries	Mt. Apo rainforest, Davao	Fusarium wilt (FocTR4) on Cavendish variety of banana	*Fusarium oxysporum* f. sp. *cubense*	antibiosis, enzymatic activity, and mycoparasitism	Puig and Cumagun (2019)
Trichoderma asperellum Samuels, Lieckfeldt & Nirenberg	Banana farms in Mindanao	Fusarium wilt (FocTR4) on Cavendish variety of banana	*Fusarium oxysporum* f. sp. *cubense*	Competition, growth enhancement, and antibiosis	Catambacan and Cumagun (2021)
Trichoderma ghanense Yoshim. Doi, Y. Abe & Sugiy.	University of the Philippines Los Baños, Laguna	Root diseases in aerobic rice variety Apo	*Pythium arrhenomanes* Drechsler	Growth enhancement, antibiosis, enzymatic activity, and parasitism	Banaay et al. (2012)

II. Fungi in agriculture, health and environment

5. Fungi as biocontrol agents in agriculture

249

TABLE 10.3 Documented fungal species and commercially available root inoculants, in the recent decade, with biological control potential against several zoo- and phytopathogens and their reported biological activity.—cont'd

Strain/Biocontrol agent/Commercial inoculant	Source	Pest/disease and host	Target pathogen	Potential mode of action	References
Trichoderma strain KA	Buguias, Benguet	Clubroot disease on crucifers	*Plasmodiophora brassicae* Woronin	Induced systemic resistance, growth enhancement, mycoparasitism, and competition	Bulcio and Nagpala (2014)
Trichoderma sp. CGP106	Mt. Apo rainforest, Davao	Fusarium wilt (FocTR4) on Cavendish variety of banana	*Fusarium oxysporum* f. sp. *cubense*	Antibiosis, enzymatic activity, and mycoparasitism	Puig and Cumagun (2019)
Trichoderma sp.	Central Luzon State University-RMCARES, Muñoz, Nueva Ecija	Not indicated (conducted in vitro)	*Fusarium verticillioides*	Mycoparasitism, antibiosis, and competition	Santos et al. (2017)
Xylaria sp. NLE 04	Needle leaves of *Casuarina equisetifolia* growing in Tagaytay City, Cavite	Fusarium wilt diseases on various crops	*Fusarium oxysporum, F. solani & F. moniliforme*	Competition, antibiosis, mycoparasitism, enzymatic activity, and induced systemic resistance	De Mesa et al. (2020)
SM EFds61-73, SM EF-ds68-129, and SM EF-ds375-97	Lowland rice fields of Nueva Ecija	Sheath blight in rice	*Rhizoctonia solani*	Antibiosis and enzymatic activity	Donayre and Dalisay (2015)
Commercial microbial inoculants used for biocontrol					
Bio-Quick composting inoculant containing spores of *Trichoderma harzianum*	Developed by BIOTECH, University of the Philippines Los Baños, Laguna	Fusarium wilt (FocTR4) on "Lakatan" variety of banana	*Fusarium oxysporum* f. sp. *cubense*	Induced systemic resistance, growth enhancement, and competition	Castillo et al. (2019)
BIOSPARK *Trichoderma* microbial inoculant containing strains of *T. ghanense & T. harzianum*	Developed by the University of the Philippines Los Baños, Laguna	Scale insect infestation on lanzones	*Unaspis mabilis*	Induced systemic resistance and growth enhancement	Silva et al. (2019)
		Clubroot disease on crucifers	*Plasmodiophora brassicae*	Growth enhancement and induced systemic resistance	Cuevas et al. (2011, 2012)

(*Continued*)

II. Fungi in agriculture, health and environment

250

10. Environmental mycology in the Philippines

TABLE 10.3 Documented fungal species and commercially available root inoculants, in the recent decade, with biological control potential against several zoo- and phytopathogens and their reported biological activity.—cont'd

Strain/Biocontrol agent/Commercial inoculant	Source	Pest/disease and host	Target pathogen	Potential mode of action	References
MYKOVAM arbuscular mycorrhizal inoculant containing spores and chopped roots colonized by species of *Glomus* and *Gigaspora*	Developed by BIOTECH, University of the Philippines Los Baños, Laguna	Bacterial wilt disease in hot peppers	*Ralstonia solanacearum* (Smith) Yabuuchi	Growth enhancement, competition, antibiosis, and induced systemic resistance	Agoncillo (2018)
		Parasitic nematode infestation on tissue-cultured "Lakatan" variety of banana	*Radopholus similis* (Cobb) Thorne, & *Meloidogyne incognita* (Kofoid & White) Chitwood	Induced systemic resistance, growth enhancement, and competition	Aggangan et al. (2013)
VAMRI vesicular-arbuscular mycorrhizal root inoculant composed of chopped dried corn roots infected with *Glomus mosseae* (T.H. Nicolson & Gerd.) Gerd. & Trappe and/or *Glomus fasciculatum* (Thaxt.) Gerd. & Trappe	Developed by BIOTECH, University of the Philippines Los Baños, Laguna	Fusarium wilt (FocTR4) on "Lakatan" variety of banana	*Fusarium oxysporum* f. sp. *cubense*	Induced systemic resistance, growth enhancement, and competition	Castillo et al. (2019)
		Fungal onion root rot	*Sclerotium rolfsii* Sacc., *Fusarium oxysporum* & *Rhizoctonia solani*	Growth enhancement, induced systemic resistance, and antibiosis	Nepomuceno et al. (2019)

documented in a collaborative work by Cortado and Davide (1968) using nematode-trapping fungal species of *Dactylella*, *Arthrobotrys*, and *Harposporium* from rice straw compost and cow manure which showed how the fungi held nematodes in captivity. Subsequent pioneering works by Dr. Davide in 1979 on nematode biocontrol immediately followed upon discovering *Paecilomyces lilacinus* (Thom) Samson as a biocontrol agent against burrowing nematodes causing disease on tomato, potato, banana, and other agronomically important crops (Davide, 1988).

Renewed interest in studying nematophagous fungi have been recently reported by Aggangan et al. (2013) involving arbuscular mycorrhizal fungi (AMF) species of *Glomus* and *Gigaspora* from a soil-based mycorrhizal biofertilizer (under the product name MYKO-VAM) which showed effectiveness in controlling nematode population and infestation with decreased galled-roots in tissue-cultured banana (var Lakatan) under screenhouse conditions. Additionally, Oclarit and Cumagun (2009) demonstrated the reapplication of *P. lilacinus* obtained from the original culture used by Villanueva and Davide (1984),

II. Fungi in agriculture, health and environment

Generalao and Davide (1986), Orolfo and Davide (1986) as an effective biocontrol agent against *Meloidogyne incognita* (Kofoid & White) Chitwood attacking tomatoes. Furthermore, Nicolas et al. (2013) reported the potency of a lepidopteran-associated *Beauveria bassiana* (Balsamo-Crivelli) Vuillemin obtained from the University of the Philippines Los Baños (UPLB), Laguna in suppressing the Asian corn borer *Ostrinia furnacalis* Guenée under field conditions. In another study, Santiago et al. (2001) initially documented the virulence of *Metarhizium anisopliae* (Metschnikoff) Sorokīn against the nymphs of the Oriental migratory locust *Locusta migratoria manilensis* Meyen, which have been causing major infestations in Central Luzon and Negros Island. In recent work, Alagos et al. (2015) reported the higher acaricidal capacity of *M. anisopliae* than *Trichoderma viride* Persoon against different developmental stages of cow ticks *Rhipicephalus microplus* Canestrini that were collected in General Santos City. Interrelated experiments by Sanjaya et al. (2016; 2013a; 2013b) evaluated the same virulence of *M. anisopliae* and other fungal isolates of *B. bassiana* and *P. lilacinus* against the red spider mite *Tetranychus kanzawai* Kishida. Other *Metarhizium* species were also explored in the studies of Belen et al. (2011), Pajar et al. (2013), and more recently by Montecalvo and Navasero (2021, 2020).

For the majority of it, the potential of *Trichoderma* spp. as a potent BCA has received the most attention as a fungal BCA because of the ability of some of its species to function not just as microbial antagonists of many phytopathogenic fungi but also as avirulent plant symbionts (Vinale et al., 2008). Cuevas et al. (2011, 2012) reported the potential economic benefits of using *Trichoderma* spp. on vegetable farmers' profit for the field control of clubroot disease of crucifers caused by *Plasmodiophora brassicae* Woronin. The growth-promoting activity and potential antibiotic activity of *T. ghanense* Yoshim. Doi, Y. Abe & Sugiyama isolate CDO (TgCDO) from UPLB, Laguna was also seen against the virulent *Pythium arrhenomanes* Drechsler (Banaay et al., 2012). Bulcio and Nagpala (2014) published results on soil incorporation of *Trichoderma* strain KA against the spread of clubroot infection in highland cabbages. Their research conducted in Buguias, Benguet further discovered that lime (CaO) combined with the isolate significantly reduced clubroot infection, hence lowered disease severity. In another study by Santos et al. (2017), a particular *Trichoderma* sp. from Nueva Ecija demonstrated antibiosis, mycoparasitism, and competition for space and nutrients as suppression mechanisms against *Fusarium verticillioides* (Saccardo) Nirenberg in vitro. Induced plant systemic resistance by *Trichoderma* Persoon was also documented by Silva et al. (2019) to increase lanzones (*Lansium domesticum* Corrêa) defense and resistance against the scale-insect *Unaspis mabilis* Lit & Barbecho. Results indicated that applying the *Trichoderma* microbial inoculant (under the product name BIOSPARK) in lanzones plants demonstrated better resistance when challenged by *U. mabilis* infestation.

In other studies, Castillo et al. (2019) documented the delayed disease progression of Panama wilt (causal organism: Tropical Race four of *Fusarium oxysporum* f.sp. *cubense* (E.F. Smith) Synder & Hansen) on young 'Lakatan' banana seedlings using the combined treatment of two commercial root inoculants, Bio-Quick and VAMRI. Puig and Cumagun (2019) also documented the antibiosis of five rainforest fungal endophytes from Mt. Apo in Davao on *Foc*TR4. Among the isolates, *Schizophyllum commune* Fries has efficient eradicative ability against *Foc*TR4 in the Cavendish variety GCTCV 219 through numerous lytic enzymes. This result was corroborated in a much recent work of Catambacan and Cumagun (2021), who also documented the antagonistic activity (antibiosis and competition) of five weed-associated fungal endophytes, *Lasiodiplodia theobromae* (Patouillard) Griffon & Maublanc, *Trichoderma asperellum* Samuels, Lieckfeldt & Nirenberg and three species of *Ceratobasidium* against *Foc*TR4 infecting the same banana variety.

Aside from bananas, rice (*Oryza sativa* L.) proves to be the most cultivated and highly valued commodity in Asia, especially in Southeast Asia. Because of this, intensive crop protection and heightened pest-disease management of rice are particularly fundamental in the Philippines. In biocontrol, Donayre and Dalisay (2015) performed bio-efficacy screenings in vitro of potential endophytic fungi of barnyard grass weed for biocontrol of the rice sheath blight pathogen, *Rhizoctonia solani* J.G. Kühn. Their findings revealed that *Geotrichum* sp. demonstrated the most effective antagonistic activity against *R. solani* by penetration, coiling and disruption of its hyphae. In another study, Gandalera et al. (2013) investigated the biocontrol of *Pyricularia oryzae* Cavara, the causative fungus of the rice blast disease. Using the crude ethanol extract of its antagonistic fungi *Chaetomium globosum* Kunze, the disease was significantly inhibited at a certain concentration purportedly by the bioactive molecules present in the extract. A study on the growth-promoting activity in seedlings of aerobic rice variety Apo and subsequent disease control activity through antibiosis of *Trichoderma ghanense* was also discussed previously (Banaay et al., 2012).

Meanwhile, De Mesa et al. (2020) isolated fungal endophytes from needle-leaf trees along Cavite and Batangas and tested their potential antagonistic activities against *Fusarium* pathogens. Several endophytes exhibited antagonistic activity against *F. oxysporum* Schltdl. on contact via the preventive, eradicative, and simultaneous approaches. However, the mechanism of action by which the isolates utilize remains to be ascertained. In another study, Nepomuceno et al. (2019) reported a reduction in disease incidence and severity of onion root rot and the potential of enhancing the biocontrol activity of AMF through coinoculation with plant growth-promoting microorganisms such as *Pseudochrobactrum asaccharolyticum* Kämpfer. In a related study, Agoncillo (2018) discussed the importance of applying the AMF inoculum at the early seedling stage of hot pepper to colonize its roots and protect it from entry by its pathogenic microorganism *Ralstonia solanacearum* (Smith) Yabuuchi. Meanwhile, in a unique study by Gazo et al. (2019), the fruticose lichen *Ramalina* collected within the Guimaras Island in Iloilo was tested for its potential antimicrobial and herbicidal activities. Effective antiproliferative activity of the crude acetone extracts of various *Ramalina* species was documented, attributed to their lichen acids. Finally, Pascual et al. (2004, 2000) explored the suppressive ability of the hypovirulent, binucleate *Rhizoctonia* sp. (Rhv7) against the virulent *Rhizoctonia solani* AG1-1A, which is the causative pathogen of the banded leaf and sheath blight on corn.

The efficiency of a particular biocontrol agent against potential, known, and emerging pathogens still greatly depends upon many influencing factors such as ecological/abiotic factors, host-agent-pathogen trophic interactions, suitable time of application, frequency of treatment, nature or technique of treatment, pathogen resistance, and persistence or population maintenance of the agent (Gang et al., 2013; Haïssam, 2011; Pascual et al., 2000; Seehausen et al., 2021; Thambugala et al., 2020). The same biocontrol agent can also demonstrate variable responses in vitro, in vivo, and *in planta* conditions (Besset-Manzoni et al., 2019; Padder & Sharma, 2011), making it complex to obtain consistent results for the same biocontrol agent in the field versus the laboratory applications. Despite the gaining popularity of several locally developed bio-pesticides (Javier & Brown, 2007), which are already commercially available, the majority of the potential BCAs documented in this review, albeit promising results, remain relatively underdeveloped and warrant further studies both in vitro and *in planta* before scaling up its possible applications. Additionally, reducing pathogen levels below the damage threshold should only constitute a portion of the overall control strategies in our integrated pest-disease management program. The proper and regulated use of pesticides (perhaps in reduced doses and in conjugation with compatible biotic organisms), stringent pesticide residue monitoring, a shift to more sustainable options, and other cultural practices should also be crucially considered if we hope to achieve both food security and safety and a sustainable environment.

6. Plastic degrading fungi in the Philippines

The 42nd annual scientific meeting of the National Academy of Science and Technology called on their resolution that single-use nonbiodegradable plastics should be phased out (NAST, 2019). The resolutions formulated the recommendations to support research and development efforts in biodegradable plastics.

As commonly applied, "plastic" refers to a group of synthetic polymers. Thermoplastics and thermosets are the two categories of plastics. Thermoplastics are polyethylene (PE), polyethylene terephthalate (PET), polystyrene (PS), polypropylene (PP), polyurethane (PUR), polyester polyurethane (PU), high-density polyethylene (HDPE), and low-density polyethylene (LDPE) (Wei & Zimmermann, 2017). PE has been utilized in making grocery bags, food packaging film, and toys; PET has been used for bottles for water and other drinks; PS has been used in making disposable food trays and laboratory plastic wares; PP for the creation of straws, car seats and container cups; HDPE has been used for shampoo containers, milk bottles and ice cream containers and LDPE has been used for food packaging film. Despite the importance of plastics, plastic pollution, mainly PE and PP, has been a major ecological challenge in the Philippines.

Based on the global estimates in 2015, 79% of plastic wastes ever produced accumulated in landfills, while 12% had been incinerated and only 9% had been recycled (Geyer et al., 2017). However, landfills and plastic waste incineration have environmental and health impacts. Therefore, one of the appropriate methods is plastic biodegradation. This is the process of converting organic carbon into biogas and biomass associated with the activity of a community of microorganisms (bacteria, fungi, and actinomycetes) capable of using plastic as a carbon source (Shah et al., 2008).

Some species of fungi are known to degrade plastics by utilizing plastic polymers as carbon and energy sources. In the Philippines, few studies of these organisms have been published that potentially degrade plastics (Table 10.4).

The genera *Aspergillus* and *Trichoderma* demonstrate the capability to degrade polyethylene (Swift, 1997). Previous studies reported that fungal species are degrading plastics in the Philippines. As early as 1997, a study on these organisms was done by (Cuevas and Manaligod, 1997) in a forest environment (Mount Makiling, Laguna). The study identified several fungal species such as *Chaetomium globosum* Kunze, *Trichoderma* sp., *Penicillium funiculosum* Thom, and *Aspergillus niger* Tieghem and reported potential plastic degrading ability. Plastic

TABLE 10.4 Fungal isolates from the Philippines with plastic-degrading ability.

Species	Plastic type	Source	References
Xylaria sp.	Polyester Polyurethane	Laguna	Urzo et al. (2017)
Xylaria sp.	Cellulose and lignin	Laguna	Cuevas and Manaligod (1997); Clutario and Cuevas (2001); Cuevas et al. (2008)
Aspergillus, Penicillium, and *Paecilomyces*	Low-density Polyethylene	Tondo, Manila	Vaghaye and Dogma (1998)
Phanerochaete chrysosporium Burdsall	Low-density Polyethylene and Oxybiodegarable	Culture Collection of BIOTECH, University of the Philippines Los Banos, Laguna	Gutierrez et al. (2018)
Xylaria sp.	Polystyrene	Mount Makiling, Laguna	Abecia et al. (2019)
Aspergillus sp.	Polyhydroxybutyrate	San Mateo and Carmona, Cavite	Tansengco and Dogma (1999)
Pleurotus ostreatus (Jacq.) P. Kumm	Polyethylene	Novaliches, Quezon City	Bermundo et al. (2019) (*unpublished*)

biodegradation showed physical holes in plastic sheets buried in soil and litter for 4 months. In the study of (Vaghaye and Dogma, 1998), fungal species of the genera *Aspergillus, Penicillium,* and *Paecilomyces* were isolated from the dumpsite in Smokey Mountain in Tondo, Manila. The ability of these fungi to degrade PE was determined by growing them in a mineral salt medium and 1% glucose (MSG) in shake flasks for 15 days at room temperature. Percent dry weight loss of plastic was monitored to determine the degradation activity of fungal isolates. Species of *Aspergillus* and *Penicillium* yielded significantly high weight losses of PE after the incubation period. Different combinations of fungal isolates were tested for the possible synergistic effect on the degradation of PE. The maximum synergism was among two species of *Aspergillus* and one of the *Penicillium* in MSG (2.8% weight loss) and between one species of *Aspergillus* and *Penicillium* in MS (1.2% weight loss). After 6 months of burial in soil, samples revealed possible biodegradation in natural soil conditions, as shown in microscopic examination. Moreover, a weight loss of 20.52% was also measured.

Tansengo and Dogma (1999) tested five fungal isolates for plastic degradation collected from the landfills in San Mateo, Rizal, and Carmona, Cavite. Scanning Electron Microscope (SEM) microphotograph revealed the attachment of the microbial cells and fungal mycelium and spores on the surfaces. Physical holes and cavities were noted due to the microbial degradation processes. In 2001, Clutario and Cuevas screened the fungal isolates on solid mineral medium (MM/S) by the clearance assay. They conducted the study to show the physical evidence of colonization of PE plastics strips by *Xylaria* sp. via SEM. The first reported crystal-like structure associated with plastic degradation was done by Urzo et al. (2017, pp. 572–580). The researchers implied that this could be a component of a novel mechanism of plastic degradation. From 18 test fungi, four had the potential for degrading PU. The identified four fungal isolates were *Lasiodiplodia theobromae* (Pat.) Griffon & Maubl., *Penicillium janthinellum* Biourge, *Fusarium verticillioides* (Sacc.) Nirenberg, and *Paecilomyces puntonii* (Vuill. Nann.). They were able to utilize DNA sequencing to identify the fungal species; however, they did not conduct tests to identify enzymes that can catalyze the lysis of recalcitrant synthetics polymers.

From the plastic bag in the forest soil of Mount Mailing, Laguna, fungal species formed surface biofilms, colonized and degraded PS (Abecia et al., 2019). They assessed the ability of local *Xylaria* sp. to grow and penetrate and damage the surface and structures of PE using SEM. Unpublished research work by Bemundo et al. (2019) screened *Pleurotus ostreatus* (Jacquin) F. Kummer as a potential agent for biodegradation of LDPE and HDPE. Results showed a segment of the plastic sheets depleted via an electronic single pan balance test in 1 month.

Gutierrez et al. (2018) studied the fungal biodegradation of so-called biodegradable plastics used in several establishments in Baguio City and Metro Manila utilizing the white rot fungus *Phanerochaete chrysosporium* Burdsall. This fungus was tested by incubating strips of biodegradable plastic with pure isolates in Petri dishes and determining their weight loss through time. The biodegradation was shown in the weight loss of plastics when incubated in the laboratory condition for specific periods using pure cultures of fungal species. The results indicated that pure fungal species gave the ability to break down plastics. The fungal species with more significant percent weight loss were found in LDPE over the oxybiodegarable (OBD) incubated. SEM analyses showed signs of degradation like holes, cracks, striations, and flakes on the surface of LDPE and OBD. Enzymes like amylases, cellulase peroxidase, and other ligninolytic enzymes are produced by this organism. This may potentially support the plastic degradation of several aromatic compounds of LDPE and OBD. Recently, a study enhancing the plastic-degrading ability of *Xylaria* sp. was conducted by (Cuevas et al., 2008). The promising result of the study was the production of albino fungal mutants that have better capability to degrade and utilize synthetic polymers. These mutants produced the wild-type (black-pigmented) xylarious fungus.

Most of the findings and insights of the previous studies focused on the identification and physical biodegradation of the fungal species for plastic degradation via light microscopy and SEM analyses. Results showed cracks, holes, and crystal-like structures at the surface of plastics used in the study. Essentially, biodegradation was shown in the surfaces of plastic sheet samples. Biomass was also measured in the previous studies, inferring the loss in weight as an indication of plastic biodegrading. Most fungal degrading plastics studies in the Philippines were conducted in terrestrial environments (soil landfills and forest environments). Whereas very few studies were conducted in freshwater and marine environments. Studies have recently isolated, identified, and screened potential fungi degrading plastic from collections done in diverse mangrove ecosystems (Apurillo et al., 2019; Calabon et al., 2018; Guerrero et al., 2018; Moron et al., 2018; Ramirez et al., 2020). Far from environmental conditions such as terrestrial and marine environments, most studies were based on selecting and testing fungal isolates in laboratory conditions. Moreover, most of the studies focused on LDPE types of plastics. A focus on PE and PP should be considered since these are mainly mismanaged plastics.

DNA sequencing has been utilized for identification and fungal diversity study in the Philippines (Urzo et al., 2017, pp. 572–580). However, the mechanism of fungal biodegradation is yet to be explored. Identifying enzymes and metabolic pathways responsible for plastic biodegradation is an interesting field. This is one of the major gaps of the previous studies on the potential application of fungi in plastic degradation. Essentially, a recent study by (Cuevas et al., 2008) used DNA mutation to enhance the capability to degrade the identified fungal species. The production of albino mutants was a promising result of the study. Hence, understanding the mutant's (albino) better capability of degrading plastic than wild type (black-pigmented) is important to gain more insights into the differences between two organisms. Studies on whether the potential fungal isolates and mutants from previous can completely degrade plastic could also be conducted. The study of long-term degradation to know how long plastics degrade is one of the research challenges since it will take time. Future studies may explore the duration of research from 6 months to several years. Comparison with other organisms such as bacteria is an interesting avenue to explore. These opportunities may help in the complete understanding of the fungal species degrading plastics. These studies may contribute to the application of the research and development of biodegradable plastics and the reduction of plastic pollution.

7. Mycoremediation of hydrocarbons

Current published mycoremediation studies in the Philippines are limited, and all are just exploratory researches dwelling on the isolation of fungi and bioassay of these locally isolated strains on hydrocarbons (Table 10.5). Philippines is a maritime gateway and has a high dependence on natural gas and coal to drive its domestic consumption (Rein & Cruz, 2008). The Philippines, as a source of hydrocarbons and the natural gas reservoir, has been explored as early as the 1800s, particularly in the waters off Palawan (Tamesis, 1981). These activities make the Philippines vulnerable to terrestrial and aquatic hydrocarbon pollution. For example, a high dissolved/dispersed petroleum hydrocarbon concentration was monitored in the waters near an offshore oil production site in Palawan. At the same time, contamination with DPPH along the western coast was also inferred to have come from the shipping activities in the area (Saramun & Wattayakorn, 2000).

Evaluation of the biosurfactant production of *Saccharomyces cerevisiae* (Desm.) Meyen 2031 from Nipa (*Nypa fruticans* Wurmb) sap from Bulacan, Philippines, was an offshoot of research conducted on the ethanol production of *S. cerevisiae* 2031 (Alcantara et al., 2010). This was the first report of a locally isolated *S. cerevisiae* strain with bioremediation potential, although there have been reported strains of *S. cerevisiae* with bioemulsifier and hydrocarbon-degrading potentials in other countries (Abioye et al., 2013; Cameron et al., 1988; Ilori & Adebusoye, 2008). The partially purified biosurfactant from this isolate emulsified aromatic hydrocarbon (benzene toluene and xylene) better than aliphatic hydrocarbons (pentane, hexane, hexadecane), suggesting that its biosurfactant activity is dependent on the hydrocarbon length (Alcantara et al., 2010). While the high emulsification indices of the biosurfactant showed that it is a promising candidate for bioremediation of petroleum-based pollutants, no further attempt was made to elucidate its mechanism of hydrocarbon emulsification.

The sinking of M/T Solar I off the coast of Guimaras Island on August 11, 2006, which caused the release of around 350,000 tons of bunker oil, posed a great concern on the environmental impacts of petroleum hydrocarbons, particularly the polycyclic aromatic

7. Mycoremediation of hydrocarbons

257

TABLE 10.5 Fungal strains isolated from hydrocarbon-contaminated sources with potential mycoremediation activities.

Strain	Source	Activity	Reference
Saccharomyces cerevisiae (Desm.) Meyen 2031	Nipa sap from Bulacan	Production of biosurfactant with emulsification activities on oils and hydrocarbons	Alcantara et al. (2010)
Phialophora sp. *Penicillium* sp. *Cladosporium* sp.	Baywalk, Manila Nasugbu, Batangas Calatagan, Batangas	Decolorization of Congo red and crystal violet	Torres et al. (2011)
Aspergillus sp. 1	Coastal sediments from Ormoc City Port Area	Degradation of engine oil	Bitacura et al. (2012)
Aspergillus sp. 2			
Aspergillus sp. 3			
Penicillium sp.			
Ganoderma lucidum (Curtis) P. Karst. *Pleurotus florida* Singer	Not indicated	Utilization of diesel in growth medium	Enriquez (2015)
Aspergillus fumigatus Fresen. *Aspergillus cf. repens* (Corda) de Bary *Aspergillus niger* Tiegh. *Paecilomyces* sp. 1 *Penicillium* sp. 1	Sediments from the oil-contaminated beach and mangrove areas in Estancia and Batad, Iloilo	Degradation of TPH, PAH, and alkanes	Sadaba and Niego (2017)

hydrocarbons (PAH), on marine life (Uno et al., 2010). Two years after the incident, PAHs are still within detectable limits in the sediments and shellfishes in the affected areas. However, there was a significant decrease in the level of PAH (Pahila et al., 2010). To further determine the biological impacts of oil spills on the microorganisms, specifically the fungal community in the affected habitats, Sadaba and Sarinas (2010) conducted a 3-year (2006 and 2009) monitoring of the fungal composition in the contaminated sites. They observed an increase in fungal density in oil-contaminated sites in 2009, indicating a possible recovery and reestablishment of autochthonous fungal species. *Aspergillus* species dominated the isolates, possibly due to their ability to utilize hydrocarbons as energy sources (Asemoloye et al., 2020; Barnes et al., 2018). This assumption, however, was not further investigated.

Another oil spill incident in the Philippines occurred when the Power Barge 103 operated by the National Power Corporation (NAPOCOR) broke loose from its moor at the height of Typhoon Haiyan (locally as Typhoon Yolanda) on November 8, 2013. This caused the spillage of around 800,000L of bunker oil, which consequently contaminated the coastlines of Barangay Botongan in the town of Estancia, Iloilo, and neighboring areas (Joint UNEP/OCHA Environment Unit, 2013, pp. 1–16). Hydrocarbon-degrading fungi were isolated from the

II. Fungi in agriculture, health and environment

contaminated beach and mangrove areas of Batad and Estancia showed promising results in the degradation of total petroleum hydrocarbon (TPH), PHA, and alkanes either axenically or as consortia (Sadaba & Niego, 2017). The study showed that the efficiency of degradation of hydrocarbons either by single culture or as consortia depends on the complexity of the hydrocarbon involved. In this case, the more complex PAH was degraded efficiently by fungal consortia.

Strains of *Aspergillus* spp. isolated from coastal sediments in the port area of Ormoc City were also documented to be efficient in the degradation of hydrocarbons, this time using used engine oil as substrate (Bitacura et al., 2012). However, mean fat loss as the measure of the biodegradative abilities of the isolates was only determined by growing the isolates individually on the substrate. No attempt was made to evaluate their effects if grown in concert. Just like in the fungal strains collected by Sadaba and Niego (2017), the putative identification of the isolates was based on morphological and other phenotypic features. Enriquez (2015) also conducted a bioremediation study using diesel as the substrate. However, unlike the rest of the studies, which utilized filamentous fungi isolated from contaminated sites, he used the white rot fungi *Ganoderma lucidum* and *Pleurotus florida*. The mushrooms could tolerate and grow in the substrates with as high as 40% diesel content for *P. florida*. The use of white rot fungi as a bioremediator has been suggested by Reddy (1995), attributing their biodegradative potential to side reactions of their lignin-degrading enzyme systems. Moreover, the white rot fungus *Phanerochaete chrysosporium* Burdsall has been demonstrated to degrade bunker oil under nonlignocellulytic conditions attributed to the cytochrome P-450 enzyme system of this basidiomycete (Kanaly & Hur, 2006).

The hydrocarbon mycoremediation studies cited above focused on the degradation of PAH and oils. Torres et al. (2011) looked into the possibility of using marine-derived fungi to degrade synthetic dyes, which are hydrocarbon-derivatives (Jarman & Ballschmiter, 2012). *Phialophora* sp. was shown to decolorize Congo red. At the same time, species of *Penicillium* and *Cladosporium* were able to decolorize both Congo red and crystal violet, possibly due to the production of extracellular enzymes and biosorption activities.

8. Moving forward

The majority of the isolates presented in this review have only been tested in the laboratory. While this is necessary to ascertain their tolerance and mycoremediating ability, much needs to be done to validate their potentials in the field. Among those with established ability are the mycorrhizae due to their unique association with plants, and thus provides improved capacity for phytoremediation. It is likewise important that isolates remain viable so that follow-up researches may be done. A deposit in a mycological museum is a move forward.

Acknowledgment

Authors acknowledge the contributions of fellow researchers in the country toward a better understanding of environmental mycology and to the overall development of the science of mycology in the Philippines.

References

Abecia, J. E. D., Egloso, M. B., Tavanlar, M. A., & Santiago, A. T. A. (2019). Scanning microscopy investigation of polystyrene damage due to colonization by locally isolated *Xylaria* sp. *Philippine Journal Health Research and Development, 23*(1), 64–70.

Abioye, O. P., Akinsola, R. O., Aransiola, S. A., Damisa, D., & Auta, S. H. (2013). Biodegradation of crude oil by *Saccharomyces cerevisiae* isolated from fermented zobo (locally fermented Beverage in Nigeria). *Pakistan Journal of Biological Sciences, 16*(24), 2058–2061.

Abo-Amer, A. E. (2012). Characterization of a strain of *Pseudomonas putida* isolated from agricultural soil that degrades cadusafos (an organophosphorus pesticide). *World Journal of Microbiology and Biotechnology, 28*(3), 805–814.

Abreo, N. A. S., Macusi, E. D., Cuenca, G. C., Ranara, C. T. B., Andam, M. B., Cardona, L. T., & Arabejo, G. F. P. (2015). Nutrient enrichment, sedimentation, heavy metals and plastic pollution in the marine environment and its implications on Philippine marine biodiversity: A review. *IAMURE International Journal of Ecology and Conservation, 15*(1), 111–167.

Adiova, J. M., Pampolina, N. M., & Aggangan, N. S. (2013). Effect of Arbuscular Mycorrhizal Fungi inoculation on growth and Cu uptake and toxicity of Desmodium cinereum (Kunth) D.C. *Philippine Journal of Science, 142*(1), 87–96.

Adipah, S. (2019). Introduction of petroleum hydrocarbons contaminants and its human effects. *Journal of Environmental Science and Public Health, 3*(1), 1–9.

Adriano, I. S., Oyong, G. G., Cabrera, E. C., & Janairo, J. I. B. (2018). Screening of silver-tolerant bacteria from a major Philippine landfill as potential bioremediation agents. *Ecological Chemistry and Engineering, 25*(3), 469–485.

Aggangan, N. S., Anarna, J. A., & Cadiz, N. M. (2019). Tree legume — microbial symbiosis and other soil amendments as rehabilitation strategies in mine tailings in the Philippines. *Philippine Journal of Science, 148*(3), 481–491.

Aggangan, N. S., & Cortes, A. D. (2018). Screening mine-out indigenous mycorrhizal fungi for the rehabilitation of mine tailing areas in the Philippines. *Reforesta, 6*, 71–85.

Aggangan, N. S., Tamayao, P. J. S., Aguilar, E. A., Anarna, J. A., & Dizon, T. O. (2013). Arbuscular mycorrhizal fungi and nitrogen fixing bacteria as growth promoters and as biological control agents against nematodes in tissue-cultured banana var. Lakatan. *Philippine Journal of Science, 142*(2), 153–165.

Agoncillo, E. S. (2018). Control of bacterial wilt disease caused by *Ralstonia solanacearum* in pepper using Arbuscular Mychorrhizal Fungi (Mykovam). *Journal of Natural Sciences Research, 8*(6), 62–66.

Agusa, T., Kunito, T., Sudaryanto, A., Monirith, I., Kan-Atireklap, S., Iwata, H., Ismail, A., Sanguansin, J., Muchtar, M., Tana, T. S., & Tanabe, S. (2007). Exposure assessment for trace elements from consumption of marine fish in Southeast Asia. *Environmental Pollution, 145*(3), 766–777.

Alagos, N. J., Teofilo, R. C., Par, L. G., Requieron, E. A., Torres, M. A., Amalin, D. M., Caranding, J. S., & Flores, M. J. (2015) Effectivity test of the fungi *Trichoderma viride* and *Metarhizium anisopliae* as biocontrol agents against cow ticks *Rhipicephalus microplus*. *Animal Biology & Animal Husbandry, 7*(2), 141–150.

Albers, P. H. (2006). Birds and polycyclic aromatic hydrocarbons. *Avian and Poultry Biology Reviews, 17*(4), 125–140.

Alcantara, V. A., Pajares, I. G., Simbahan, J. F., Villarante, N. R., & Rubio, M. L. D. (2010). Characterization of biosurfactant from *Saccharomyces cerevisiae* 2031 and evaluation of emulsification activity for potential application in bioremediation. *The Philippine Agricultural Scientist, 93*(1), 22–30.

Almarinez, B. J. M., Barrion, A. T., Navasero, M. V., Navasero, M. M., Cayabyab, B. F., Carandang, J. S. R., Legaspi, J. C., Watanabe, K., & Amalin, D. M. (2020). Biological control: A major component of the pest management program for the invasive coconut scale insect, *Aspidiotus rigidus* Reyne, in the Philippines. *Insects, 11*(11), 745.

Apurillo, C. C. S., Cai, L., & dela Cruz, T. E. E. (2019). Diversity and bioactivities of mangrove fungal endophytes from Leyte and Samar, Philippines. *Philippine Science Letters, 12*, 33–48.

Asemoloye, M. D., Tosi, S., Daccò, C., Wang, X., Xu, S., Marchisio, M. A., Gao, W., Jonathan, G. S., & Pecoraro, L. (2020). Hydrocarbon degradation and enzyme activities of *Aspergillus oryzae* and *Mucor irregularis* isolated from Nigerian crude oil-polluted sites. *Microorganisms, 8*(1912), 1–19.

Bajet, C. M. (2015). Pesticide residues in food and the environment in the Philippines: Risk assessment and management. *FFTC Agricultural Policy Platform*, 1–8.

Balce, G. R. (1997). Issues and challenges of oil spill response structure in the Philippines: DOE perspective. *Ocean L. & Pol'y Series, 1*, 50.

Banaay, C. G. B., Cuevas, V. C., & Cruz, C. V. (2012). *Trichoderma ghanense* promotes plant growth and controls disease caused by *Pythium arrhenomanes* in seedlings of aerobic rice variety apo. *The Philippine Agricultural Scientist, 95*(2).

Barik, S. (1984). Metabolism of insecticides by microorganisms. In R. Lal (Ed.), *Insecticide microbiology* (pp. 87–130). Berlin, Heidelberg: Springer.

Barnes, N. M., Khodse, V. B., Lotlikar, N. P., Meena, R. M., & Damare, S. R. (2018). Bioremediation potential of hydrocarbon-utilizing fungi from select marine niches of India. *3 Biotech, 8*(21), 1–10.

Belen, J. M., Ocampo, V. R., & Caoili, B. L. (2011). Pathogenicity and biological characterization of entomopathogenic fungi isolated from corn earworm, *Helicoverpa armigera* (Hübner)(Lepidoptera: Noctuidae). *The Philippine Entomologist, 25*(1), 48–63.

Bennett, R. M., Cordero, P. R. F., Bautista, G. S., & Dedeles, G. R. (2013). Reduction of hexavalent chromium using fungi and bacteria isolated from contaminated soil and water samples. *Chemistry and Ecology, 29*(4), 320–328.

Bermundo, M. A., Medina, R. L., & Calaramo, F. E. (2019). *Potential degrading capability of white Oyster mushroom* (Pleurotus ostreatus) *to conventionally used polyethylene plastics.* Unpublished.

Besset-Manzoni, Y., Joly, P., Brutel, A., Gerin, F., Soudière, O., Langin, T., & Prigent-Combaret, C. (2019). Does in vitro selection of biocontrol agents guarantee success *in planta*? A study case of wheat protection against *Fusarium* seedling blight by soil bacteria. *PloS One, 14*(12), e0225655.

Bitacura, J. G., Balala, A. C., & Abit, P. P. (2012). Fungi from coastal sediments as potential agents in biodegrading used engine oil. *Annals of Tropical Research, 34*(2), 112–125.

Bretaña, B. L. P., Salcedo, S. G., Casim, L. F., & Manceras, R. S. (2019). Growth performance and inorganic mercury uptake of vetiver (*Chrysopogon zizanoides* Nash) inoculated with arbuscular mycorrhiza fungi (AMF): Its implication to phytoremediation. *Journal of Agricultural Research, Development, Extension and Technology, 25*(1), 39–47.

Bulcio, J. M., & Nagpala, A. L. (2014). Management of clubroot (*Plasmodiophora brassicae* wor.) on cabbage using *Trichoderma* KA and lime in Natubleng, Buguias, Benguet. *Mountain Journal of Science and Interdisciplinary Research (Formerly Benguet State University Research Journal), 71*, 23–31.

Calabon, M. S., Sadaba, R. B., & Campos, W. L. (2018). Fungal diversity of mangrove-associated sponges from New Washington, Aklan, Philippines. *Mycology, 10*(1), 6–21.

Cameron, D. R., Cooper, D. G., & Neufeld, R. J. (1988). The mannoprotein of *Saccharomyces cerevisiae* is an effective bioemulsifier. *Applied and Environmental Microbiology, 54*(6), 1420–1425.

Carascal, M. B., del Rosario, M. J. G., Notarte, K. I. R., Huyop, F., Yaguchi, T., & dela Cruz, T. E. E. (2017). Butachlor biodegradation potential of fungi isolated from submerged wood and surface water collected in Taal Lake, Philippines. *Philippine Science Letters, 10*(2), 81–88.

Carating, R. B., Galanta, R. G., & Bacatio, C. D. (2014). *The soils of the Philippines.* Springer Science & Business.

Casamorin, J. A., Bennett, R. B., & Dedeles, G. (2014). Biosorption of Cd(II) by yeasts from ripe fruit peels in the Philippines. *Journal of Health Pollution, 7*, 14–24.

Castillo, A. G., Puig, C. G., & Cumagun, C. J. R. (2019). Non-synergistic effect of *Trichoderma harzianum* and *Glomus* spp. in reducing infection of *Fusarium* wilt in banana. *Pathogens, 8*(2), 43.

Catambacan, D. G., & Cumagun, C. J. R. (2021). Weed-associated fungal endophytes as biocontrol agents of *Fusarium oxysporum* f. sp. *cubense* TR4 in Cavendish Banana. *Journal of Fungi, 7*(3), 224.

Clutario, M. T. P., & Cuevas, V. C. (2001). Colonization of plastic by *Xylaria* sp. *Philippine Journal of Science, 130*(2), 89–96.

Cordero, P. R. F., Bennett, R. M., Bautista, G. S., Aguilar, J. P. P., & Dedeles, G. R. (2015). Degradation of nickel protoporphyrin disodium and vanadium oxide octaethylporphyrin by Philippine microbial consortia. *Bioremediation Journal, 19*(2), 92–103.

Coronado, F. F., Unciano, N. M., Cabacang, R. M., & Hernandez, J. T. (2016). Removal of heavy metal compounds from industrial wastes using a novel locally-isolated *Vanjira* sp. HMAT2. *Philippine Journal of Science, 145*(4), 327–338.

Cortado, R. V., & Davide, R. G. (1968). Nematode-trapping fungi in the Philippines. *Philippine Phytopathology, 4*(1&2), 4.

Covino, S., D'Annibale, A., Stazi, S. R., Cajthaml, T., Čvančarová, M., Stella, T., & Petruccioli, M. (2015). Assessment of degradation potential of aliphatic hydrocarbons by autochthonous filamentous fungi from a historically polluted clay soil. *Science of the Total Environment, 505*, 545–554.

Cubelo, J. E. C., & Cubelo, T. A. (2021). A survey on the pesticide application practices and presence of pesticide residues on mangoes in Negros Oriental, Philippines. *Journal of Nature Studies, 20*(1), 25–43.

Cuevas, V. C., & Manaligod, R. (1997). Isolation of decomposer fungi with plastic degrading ability. *Philippine Journal Science, 126*(2), 117–130.

Cuevas, V. C., Lagman, C. A., Jr., Anupo, X., Orajay, J. I., & Malamnao, F. G. (2019). Yield improvement with compost amendment and *Trichoderma* microbial inoculant (TMI) in rice paddies inundated by copper-rich mine tailings. *Philippine Science Letters, 12*(1), 31–38.

Cuevas, V. C., Lagman, C. A., Jr., Cammagay, G. E., & Cuevas, A. C. (2012). *Trichoderma* inoculant as disease biocontrol agent for high value crops: Potential financial impact. *Philippine Journal of Crop Science, 37*(3), 64–75.

Cuevas, V. C., Lagman, C. A., Jr., & Cuevas, A. C. (2011). Potential impacts of the use of *Trichoderma* spp. on farmers' profit in the field control of club root disease of crucifers caused by *Plasmodiophora brassicae* Wor. *Philippine Agricultural Scientist, 94*(2), 171–178.

Cuevas, V. C., Tavanlar, M. A. T., Lat, E. C., & Clemencia, H. C. (2008). *Protoplast fusion of Xylaria sp. for enhanced plastic-degrading ability and use of mixed culture of Trichoderma activator for rapid composting of biodegradable municipal water.* Retrieved https://agris.fao.org/agris-search/search.do?recordID=PH2009001476.

Cycoń, M., & Piotrowska-Seget, Z. (2016). Pyrethroid-degrading microorganisms and their potential for the bioremediation of contaminated soils: A review. *Frontiers in Microbiology, 7*, 1463.

D'Annibale, A., Rosetto, F., Leonardi, V., Federici, F., & Petruccioli, M. (2006). Role of autochthonous filamentous fungi in bioremediation of a soil historically contaminated with aromatic hydrocarbons. *Applied and Environmental Microbiology, 72*(1), 28–36.

Davide, R. G. (1988). Nematode problems affecting agriculture in the Philippines. *Journal of Nematology, 20*(2), 214–218.

De Mesa, R. B. C., Espinosa, I. R., Agcaoili, M. C. R. R., Calderon, M. A. T., Pangilinan, M. V. B., De Padua, J. C., & dela Cruz, T. E. E. (2020). Antagonistic activities of needle-leaf fungal endophytes against *Fusarium* spp. *MycoAsia, 6*, 1–11.

De Padua, J. C. (2021). Isolation and characterization of nickel-tolerant Trichoderma strains from marine and terrestrial environments. *Journal of Fungi, 7*(8), 591.

De Sotto, R., Monsanto, R. Z., Edora, J. L., Bautista, R. H., Bennett, R. M., & Dedeles, G. R. (2015). Reduction of Cr (VI) using indigenous *Aspergillus* spp. isolated from heavy metal contaminated sites. *Mycosphere, 6*(1), 53–59.

Del Prado-Lu, J. L. (2015). Insecticide residues in soil, water, and eggplant fruits and farmers' health effects due to exposure to pesticides. *Environmental Health and Preventive Medicine, 20*(1), 53–62.

Dela Cruz, J., & Halos, P. M. (1997). Isolation, identification and bioremediation potential of oil-degrading bacteria from Manila Bay and Pasig River [Philippines]. In *Seminar-workshop on microbial-based technologies for pollution abatement of Laguna de Bay, Manila (Philippines), 25–27 Feb 1997.*

Díaz, E. (2004). Bacterial degradation of aromatic pollutants: A paradigm of metabolic versatility. *International Microbiology, 7*(3), 173–180.

Donayre, D. K. M., & Dalisay, T. U. (2015). Potential of endophytic fungi of barnyard grass weed for biological control of the rice sheath blight pathogen, *Rhizoctonia solani* Kühn. *International Journal of Philippine Science and Technology, 8*(2), 48–51.

Dulay, R. M. R., Pascual, A. H. L., Constante, R. D., Tiniola, R. C., Areglo, J. L., Arenas, M. C., Kalaw, S. P., & Reyes, R. G. (2015). Growth response and mycoremediation activity of *Coprinus comatus* on heavy metal contaminated media. *Mycosphere, 6*(1), 1–7.

Enriquez, V. A. (2015). Growth, fructification of reishi (*Ganoderma lucidum*) and oyster (*Pleurotus florida*) mushrooms, and physic-chemical changes of medium treated with different concentrations of diesel. *BIMP-EAGA Journal for Sustainable Tourism Development, 4*(2), 82–93.

Estoque, R. C., & Murayama, Y. (2015). Intensity and spatial pattern of urban land changes in the megacities of Southeast Asia. *Land Use Policy, 48*, 213–222.

Estoque, R. C., Murayama, Y., & Myint, S. W. (2017). Effects of landscape composition and pattern on land surface temperature: An urban heat island study in the megacities of Southeast Asia. *Science of the Total Environment, 577*, 349–359.

Gacho, C. G., Coronado, F. T., Tansengco, M. L., Barcelo, J. R., Borromeo, C. C., & Guiterrez, B. J. M. (2019). Isolation, identification and heavy metal biosorption assessment of yeast isolates indigenous to abandoned mine sites of Itogon Benguet, Philippines. *Journal of Environmental Science and Management, 22*(1), 109–121.

Gajigan, A. P., Yñiguez, A. T., Villanoy, C. L., Jacinto, G. S., & Conaco, C. (2018). Diversity and community structure of marine microbes around the Benham Rise underwater plateau, northeastern Philippines. *PeerJ, 6*, e4781.

Gandalera, E. E., Divina, C. C., & Dar, J. D. (2013). Inhibitory activity of *Chaetomium globosum* Kunze extract against Philippine strain of *Pyricularia oryzae* Cavara. *Journal of Agricultural Technology, 9*(2), 333–348.

Gang, G., Bizun, W., Weihong, M., Xiaofen, L., Xiaolin, Y., Chaohua, Z., ... Huicai, Z. (2013). Biocontrol of *Fusarium* wilt of banana: Key influence factors and strategies. *African Journal of Microbiology Research, 7*(41), 4835–4843.

Garcia, B., Fang, M. M., & Lin, J. (2019). Marine plastic pollution in Asia: All hands on deck. *Chinese Journal of Environmental Law, 3*(1), 11–46.

Gazo, S. M. T., Santiago, K. A. A., Tjitrosoedirjo, S. S., & Dela Cruz, T. (2019). Antimicrobial and herbicidal activities of the fruticose lichen *Ramalina* from Guimaras Island, Philippines. *Biotropia, 26*(1), 23–32.

Generalao, L. C., & Davide, R. G. (1986). Biological control of *Radopholus similis* on banana with three nematophagous fungi. *Philippine Phytopathology, 22*, 36–41.

Geyer, R., Jambeck, J. R., & Law, K. L. (2017). Production, use, and fate of all plastics ever made. *Science Advances, 3*(7), e1700782.

Guerrero, J. J. G., General, M. A., & Serrano, J. E. (2018). Culturable foliar fungal endophytes of mangrove species in Bicol Region, Philippines. *Philippine Journal of Science, 147*(4), 563–574.

Gutierrez, R. M., Daupan, S. M. M. A. V., Fabian, A. V., & Miclat, C. C. (2018). Microbiological investigation on some biodegradable plastics used as packaging materials. *American Journal of Applied Sciences, 9*(1).

Haïssam, J. M. (2011). *Pichia anomala* in biocontrol for apples: 20 years of fundamental research and practical applications. *Antonie Van Leeuwenhoek, 99*(1), 93–105.

Hammond, S. T., Brown, J. H., Burger, J. R., Flanagan, T. P., Fristoe, T. S., Mercado-Silva, N., Nekola, J., & Okie, J. G. (2015). Food spoilage, storage, and transport: Implications for a sustainable future. *BioScience, 65*(8), 758–768.

Hicks, R. M., Acosta, N. O., & Candelaria, S. M. (2012). Crafting a sustainable mining policy in the Philippines. *Natural Resources & Environment, 27*, 43.

Hipol, R. M., Dalisay, T. U., Ardales, E. Y., Cedo, M. L. O., & Cuevas, V. C. (2015). Endophytic yeasts possibly alleviate heavy metal stress in their host *Phragmites australis* Cav. (Trin) ex Stued. Through the production of plant growth promoting hormones. *Bulletin of Environment, Pharmacology, and Life Sciences, 4*(3), 82–86.

Hirschman, C., & Bonaparte, S. (2012). Population and society in Southeast Asia: A historical perspective. In *Demographic change in Southeast Asia: Recent histories and future directions.* Cornell Southeast Asia Program Publications.

Hosokawa, R., Nagai, M., Morikawa, M., & Okuyama, H. (2009). Autochthonous bioaugmentation and its possible application to oil spills. *World Journal of Microbiology and Biotechnology, 25*(9), 1519–1528.

Houessionon, M. G., Ouendo, E. M. D., Bouland, C., Takyi, S. A., Kedote, N. M., Fayomi, B., Fobil, J. N., & Basu, N. (2021). Environmental heavy metal contamination from electronic waste (E-waste) recycling activities worldwide: A systematic review from 2005 to 2017. *International Journal of Environmental Research and Public Health, 18*(7), 3517.

Igiri, B. E., Okoduwa, S. I., Idoko, G. O., Akabuogu, E. P., Adeyi, A. O., & Ejiogu, I. K. (2018). Toxicity and bioremediation of heavy metals contaminated ecosystem from tannery wastewater: A review. *Journal of Toxicology, 2018*, 1–16.

Ilori, M. O., & Adebusoye, S. A. (2008). Isolation and characterization of hydrocarbon-degrading and biosurfactant-producing yeast strains obtained from a polluted water. *World Journal of Microbiology and Biotechnology, 24*, 2539–2545.

Jambeck, J. R., Geyer, R., Wilcox, C., Siegler, T. R., Perryman, M., Andrady, A., Narayan, R., & Law, K. L. (2015). Plastic waste inputs from land into the ocean. *Science, 347*(6223), 768–771.

Jarman, W. M., & Ballschmiter, K. (2012). From coal to DDT: The history of the development of the pesticide DDT from synthetic dyes till silent spring. *Endeavour, 36*(4), 131–142.

Javier, P. A., & Brown, M. B. (2007). Bio-fertilizers and bio-pesticides research and development at UPLB. *Food & Fertilizer Technology Center, 602*, 1–22.

Jayaraj, R., Megha, P., & Sreedev, P. (2016). Organochlorine pesticides, their toxic effects on living organisms and their fate in the environment. *Interdisciplinary Toxicology, 9*(3–4), 90.

Joint UNEP/OCHA Environment Unit. (2013). *Oil spill in Estancia Iloilo province, western Visayas, Philippines resulting from Typhoon Haiyan (Yolanda) joint assessment report.*

Kanaly, R. A., & Hur, H. (2006). Growth of *Phanerochaete chrysosporium* on diesel fuel hydrocarbons at neutral pH. *Chemosphere, 63*, 202–211.

Khan, M. Q., Al-Ghais, S. M., Catalin, B., & Khan, Y. H. (2005). Effects of petroleum hydrocarbons on aquatic animals. *Developments in Earth and Environmental Sciences, 3*, 159–185.

References

Kladsomboon, S., Jaiyen, C., Choprathumma, C., Tusai, T., & Apilux, A. (2020). Heavy metals contamination in soil, surface water, crops, and resident blood in Uthai District, Phra Nakhon Si Ayutthaya, Thailand. *Environmental Geochemistry and Health, 42*(2), 545–561.

Klimek, B., Sitarz, A., Choczyński, M., & Niklińska, M. (2016). The effects of heavy metals and total petroleum hydrocarbons on soil bacterial activity and functional diversity in the Upper Silesia industrial region (Poland). *Water, Air, & Soil Pollution, 227*(8), 1–9.

Kumar, H., Singh, M. K., & Gupta, M. P. (April 2018). Smart mobility: Crowdsourcing solutions for smart transport system in smart cities context. In *Proceedings of the 11th international conference on theory and practice of electronic governance* (pp. 482–488).

Lacap, D. C., Smith, G. J., Warren-Rhodes, K., & Pointing, S. B. (2005). Community structure of free-floating filamentous cyanobacterial mats from the Wonder Lake geothermal springs in the Philippines. *Canadian Journal of Microbiology, 51*(7), 583–589.

Lam, S., Pham, G., & Nguyen-Viet, H. (2017). Emerging health risks from agricultural intensification in Southeast Asia: A systematic review. *International Journal of Occupational and Environmental Health, 23*(3), 250–260.

Lim, S. T., & Halos, P. S. M. (1995). Isolation, characterization and determination of bioremediation potential of oil-degrading bacteria from the Manila Bay [Philippines]. *Philippine Journal of Biotechnology, 6*(1), 1–12.

Llarena, Z. M., & Solidum, J. N. (2012). Mycoremediation of toxicants from chosen sites in the Philippine setting. *International Journal of Chemical and Environmental Engineering, 3*(5), 339–343.

Lu, J. L. (2010). Analysis of trends of the types of pesticide used, residues and related factors among farmers in the largest vegetable producing area in the Philippines. *Journal of Rural Medicine, 5*(2), 184–189.

Lu, J. L. (2011a). Farmers' exposure to pesticides and pesticide residues in soil and crops grown in Benguet Philippines. *Philippine Journal of Crop Science, 36*(3), 19–27.

Lu, J. L. (2011b). Insecticide residues in eggplant fruits, soil, and water in the largest eggplant-producing area in the Philippines. *Water, Air, & Soil Pollution, 220*(1), 413–422.

Lu, J. L. (2012). Pesticide residues in eggplant during dry and wet seasons in Sta. Maria, Pangasinan. *Philippine Journal of Crop Science, 37*(3), 93–98.

Lu, J. L., Cosca, K. Z., & Del Mundo, J. (2010). Trends of pesticide exposure and related cases in the Philippines. *Journal of Rural Medicine, 5*(2), 153–164.

Mahmood, I., Imadi, S. R., Shazadi, K., Gul, A., & Hakeem, K. R. (2016). Effects of pesticides on environment. In K. Hakeem, M. Akhtar, & S. Abdullah (Eds.), *Plant, soil and microbes* (pp. 253–269). Cham: Springer.

Manguilimotan, L. C., & Bitacura, J. G. (2018). Biosorption of Cadmium by filamentous fungi isolated from coastal water and sediments. *Journal of Toxicology, 2018*, 1–6.

Marcelo, A., Geronimo, R. M., Vicente, C. J. B., Callanta, R. B. P., Bennett, R. M., Ysrael, M. C., & Dedeles, G. R. (2018). TLC screening profile of secondary metabolites and biological activities of *Salisapilia tartarea* S1YP1 isolated from Philippine mangroves. *Journal of Oleo Science, 67*(12), 1585–1595.

Marks, D., Miller, M. A., & Vassanadumrongdee, S. (2020). The geopolitical economy of Thailand's marine plastic pollution crisis. *Asia Pacific Viewpoint, 61*(2), 266–282.

McClements, D. J., Barrangou, R., Hill, C., Kokini, J. L., Ann Lila, M., Meyer, A. S., & Yu, L. (2020). Building a resilient, sustainable, and healthier food supply through innovation and technology. *Annual Review of Food Science and Technology, 12*, 1–28.

Melia, S. (2020). Urban expansion, road building and loss of countryside—a non-linear Relationship. *World Transport Policy & Practice, 26*(2).

Mercado, J. V. L., Borja, J., & Gallardo, S. (2012). Isolation and screening of endosulfan degrading fungi from soil. *ASEAN Engineering Journal, 1*(1), 5–13.

Mishra, A., & Malik, A. (2014). Novel fungal consortium for bioremediation of metals and dyes from mixed waste stream. *Bioresource Technology, 171*, 217–226.

Mitrano, D. M., Wick, P., & Nowack, B. (2021). Placing nanoplastics in the context of global plastic pollution. *Nature Nanotechnology, 16*(5), 491–500.

Monsalud, F. C., Monsalud, R. G., Brown, M. B., & Badayos, R. B. (2009). Distribution of soil microorganisms in selected farming systems in Laguna, Philippines. *Philippine Journal of Crop Science, 34*(3), 90–101.

Montecalvo, M. P., & Navasero, M. M. (2020). Effect of entomopathogenic fungus *Metarhizium (Nomuraea) rileyi* (Farl.) Samson on the third instar larvae of the onion armyworm, *Spodoptera exigua* Hübner (Lepidoptera: Noctuidae), under laboratory conditions. *Philippine Agricultural Scientist, 103*(2), 159–164.

Montecalvo, M. P., & Navasero, M. M. (2021). *Metarhizium (= Nomuraea) rileyi* (Farlow) Samson from *Spodoptera exigua* (Hübner) Cross infects Fall armyworm, *Spodoptera frugiperda* (J.E. Smith) (Lepidoptera: Noctuidae) larvae. *Philippine Journal of Science, 150*(1), 193–199.

Moron, L., Lim, Y., & dela Cruz, T. (2018). Antimicrobial activities of crude culture extracts from mangrove fungal endophytes collected in Luzon island, Philippines. *Philippine Science Letters, 11*, 28–36.

Muñoz, A. J., Ruiz, E., Abriouel, H., Gálvez, A., Ezzouhri, L., Lairini, K., & Espínola, F. (2012). Heavy metal tolerance of microorganisms isolated from wastewaters: Identification and evaluation of its potential for biosorption. *Chemical Engineering Journal, 210*, 325–332.

National Academy of Science and Technology (NAST). (2019). *NAST calls for action on key sustainable development areas.* Retrieved July 15, 2021.

Navarrete, I. A., Gabiana, C. C., Dumo, J. R. E., Salmo, S. G., Guzman, M. A. L. G., Valera, N. S., & Espiritu, E. Q. (2017). Heavy metal concentrations in soils and vegetation in urban areas of Quezon City, Philippines. *Environmental Monitoring and Assessment, 189*(4), 145.

Navarrete, I. A., Tee, K. A. M., Unson, J. R. S., & Hallare, A. V. (2018). Organochlorine pesticide residues in surface water and groundwater along Pampanga River, Philippines. *Environmental Monitoring and Assessment, 190*(5), 1–8.

Nepomuceno, R. A., Brown, C. M. B., Mojica, P. N., & Brown, M. B. (2019). Biological control potential of Vesicular Arbuscular Mycorrhizal Root Inoculant (VAMRI) and associated phosphate solubilizing bacteria, *Pseudochrobactrum asaccharolyticum* against soilborne phytopathogens of Onion (*Allium cepa* L. var. Red Creole). *Archives of Phytopathology and Plant Protection, 52*(7–8), 714–732.

Newton, I. (2018). Organochlorine pesticides, Rachel Carson, and the environmental movement. In D. Dellasala, & M. Goldstein (Eds.), *Encyclopedia of the anthropocene* (pp. 97–104). Elsevier.

Nicolas, J. A., Tamayo, N. V., & Caoili, B. L. (2013). Improving the yield of glutinous white corn by distance of planting and use of biocontrol agents for management of Asian corn borer, *Ostrinia furnacalis* Guenee. In Recent advances in biofertilizers and biofungicides (PGPR) for sustainable agriculture. In *Proceedings of 3rd Asian conference on plant growth-promoting Rhizobacteria (PGPR) and other microbials, Manila, Philippines, 21–24 April, 2013* (pp. 50–74). Asian PGPR Society for Sustainable Agriculture.

Oclarit, E., & Cumagun, C. (2009). Evaluation of efficacy of *Paecilomyces lilacinus* as biological control agent of *Meloidogyne incognita* attacking tomato. *Journal of Plant Protection Research, 49*(4), 337–340.

Orolfo, E. B., & Davide, R. G. (1986). Biological control of root-knot nematodes attacking tomato plants through the use of mycorrhiza and nematophagous fungi. *Philippine Agriculturist, 69*(3), 307–315.

Ossai, I. C., Ahmed, A., Hassan, A., & Hamid, F. S. (2020). Remediation of soil and water contaminated with petroleum hydrocarbon: A review. *Environmental Technology & Innovation, 17*, 100526.

Ostrea, E. M., Jr., Villanueva-Uy, E., Bielawski, D., Birn, S., & Janisse, J. J. (2014). Trends in long term exposure to propoxur and pyrethroids in young children in the Philippines. *Environmental Research, 131*, 13–16.

Osuji, L. C., & Nwoye, I. (2007). An appraisal of the impact of petroleum hydrocarbons on soil fertility: the Owaza experience. *African Journal of Agricultural Research, 2*(7), 318–324.

Padder, B. A., & Sharma, P. N. (2011). *In vitro* and *in vivo* antagonism of biocontrol agents against *Colletotrichum lindemuthianum* causing bean anthracnose. *Archives of Phytopathology and Plant Protection, 44*(10), 961–969.

Pahila, I. G., Taberna, H., Jr., Sadaba, R., Jr., Gamarcha, L., Koyama, J., & Uno, S. (2010). *Assessment of residual petroleum hydrocarbon two years after the M/T Solar I oil spill in Southern Guimaras.* Central.

Pajar, J. A. L., Cabahug, D. V., Sumaya, N. H. N., Genevieve, J., Ma, T. M., Suzette, R., & Rivero, H. I. (2013). Virulence of local *Metarhizium* spp. isolates against *Tenebrio molitor* (Linn): An initial comparison with non-native and commercially available strains. *International Journal of the Computer, the Internet and Management, 21*(1), 48–52.

Pascual, C. B., Raymundo, A. D., & Hyakumachi, M. (2000). Efficacy of hypovirulent binucleate *Rhizoctonia* sp. to control banded leaf and sheath blight in corn. *Journal of General Plant Pathology, 66*(1), 95–102.

Pascual, C. B., Raymundo, A. D., & Hyakumachi, M. (2004). Suppression of *Rhizoctonia solani* in corn by hypovirulent binucleate *Rhizoctonia* and the nature of protection. *Philippine Agricultural Scientist, 87*(1), 36–40.

Poncian, M., Beray, B. J. W., Dadulla, H. C. P., & Hipol, R. M. (2019). Carboxylesterase activity of filamentous soil fungi from a potato plantation in Mankayan, Benguet. *Studies in Fungi, 4*(1), 292–303.

Prigione, V., Zerlottin, M., Refosco, D., Tigini, V., Anastasi, A., & Varese, G. C. (2009). Chromium removal from a real tanning effluent by autochthonous and allochthonous fungi. *Bioresource Technology, 100*(11), 2770–2776.

References

Puig, C. G., & Cumagun, C. J. R. (2019). Rainforest fungal endophytes for the bio-enhancement of banana toward *Fusarium oxysporum* f. sp. *cubense* Tropical Race 4. *Archives of Phytopathology and Plant Protection, 52*(7–8), 776–794.

Rahman, M. M., Naidu, R., & Bhattacharya, P. (2009). Arsenic contamination in groundwater in the Southeast Asia region. *Environmental Geochemistry and Health, 31*(1), 9–21.

Ramirez, C. S. P., Notarte, K. I. R., & dela Cruz, T. E. E. (2020). Antibacterial activities of mangrove leaf endophytic fungi from Luzon Island, Philippines. *Studies in Fungi, 5*(1), 320–331.

Rea-Maminta, M. A. D., Dagamac, N. H. A., Huyop, F. Z., Wahab, R. A., & dela Cruz, T. E. E. (2015). Comparative diversity and heavy metal biosorption of myxomycetes from forest patches on ultramafic and volcanic soils. *Chemistry and Ecolohu, 31*(8), 741–753.

Reddy, C. A. (1995). The potential for white-rot fungi in the treatment of pollutants. *Current Opinion in Biotechnology, 6*, 320–328.

Redman, C. L., & Jones, N. S. (2005). The environmental, social, and health dimensions of urban expansion. *Population and Environment, 26*(6), 505–520.

Rein, A., & Cruz, K. (2008). Philippines energy policy and development. *The Journal of Energy and Development, 34*(1/2), 129–140.

Rhodes, C. J. (2014). Mycoremediation (bioremediation with fungi)—growing mushrooms to clean the earth. *Chemical Speciation & Bioavailability, 26*(3), 196–198.

Ritchie, H., & Roser, M. (2018). *Plastic pollution*. Our World in Data. https://ourworldindata.org/plastic-pollution.

Rochman, C. M., & Hoellein, T. (2020). The global odyssey of plastic pollution. *Science, 368*(6496), 1184–1185.

Rollon, R. J. C., Galleros, J. E. V., Galos, G. R., Villasica, L. J. D., & Garcia, C. M. (2017). Growth and nutrient uptake of *Paraserianthes falcataria* (L.) as affected by carbonized rice hull and arbuscular mycorrhizal fungi grown in an artificially copper contaminated soil. *Advances in Agriculture & Botanics, 9*(2), 57–67.

Sadaba, R. B., & Niego, A. G. T. (2017). Biodegradation of hydrocarbon by marine-derived fungi isolated from oil-contaminated beach and mangrove soil in Western Visayas, Philippines. *International Oil Spill Conference Proceedings, 1*, 2017357.

Sadaba, R. B., & Sarinas, B. G. S. (2010). Fungal communities in bunker C oil-impacted sites off southern Guimaras, Philippines: A post-spill assessment of solar 1 oil spill. *Botanica Marina, 53*, 565–575.

Sakari, M., Zakaria, M. P., Lajis, N. H., Mohamed, C. A. R., & Abdullah, M. H. (2012). Three centuries of polycyclic aromatic hydrocarbons and teriterpane records in Tebrau Strait, Malaysia; recent pollution concern in a pristine marine environment. *Polycyclic Aromatic Compounds, 32*(3), 364–389.

Samal, S., & Mishra, C. S. K. (2021). Agrochemical contamination of soil recent technology innovations for bioremediation. In A. Rakshit, M. Parihar, B. Sarkar, H. Singh, & L. F. Fraceto (Eds.), *Bioremediation science from theory to practice* (pp. 170–179). CRC Press.

Samaniego, J., Gibaga, C. R., Tanciongco, A., Rastrullo, R., Mendoza, N., & Racadio, C. D. (2020). Comprehensive assessment on the environmental conditions of abandoned and inactive mines in the Philippines. *ASEAN Journal on Science and Technology for Development, 37*(2), 81–86.

Sanjaya, Y., Ocampo, V. R., & Caoili, B. L. (2013a). Infection process of entomopathogenic fungi *Metarhizium anisopliae* in the *Tetranychus kanzawai* (Kishida)(Tetranychidae: Acarina). *AGRIVITA, Journal of Agricultural Science, 35*(1), 64–72.

Sanjaya, Y., Ocampo, V. R., & Caoili, B. L. (2013b). Selection of entomopathogenic fungi against the red spider mite *Tetranychus kanzawai* (Kishida)(Tetranychidae: Acarina). *Arthropods, 2*(4), 208–215.

Sanjaya, Y., Ocampo, V. R., & Caoili, B. L. (2014). Entomopathogenic characterization of *Beauveria bassiana* fungi against *Tetranychus kanzawai* (Kishida)(Tetranychidae: Acarina) spider mite by its region. *Thai Journal of Agricultural Science, 47*(1), 13–21.

Sanjaya, Y., Ocampo, V. R., & Caoili, B. L. (2015). Infection process of entomopathogenic fungi *Beauveria bassiana* in the *Tetrancyhus kanzawai* (Kishida)(Tetranychidae: Acarina). *Arthropods, 4*(3), 90–97.

Sanjaya, Y., Ocampo, V. R., & Caoili, B. L. (2016). Pathogenicity of three entomopathogenic fungi, *Metarhizium anisopliae, Beauveria bassiana,* and *Paecilomyces lilacinus,* to *Tetranychus kanzawai* infesting papaya seedlings. *Arthropods, 5*(3), 109–113.

Santiago, D. R., Castillo, A. G., Arapan, R. S., Navasero, M. V., & Eusebio, J. E. (2001). Efficacy of *Metarhizium anisopliae* (Metsch.) Sor. against the oriental migratory locust, *Locusta migratoria manilensis* Meyen. *Philippine Agricultural Scientist, 84*(1), 26–34.

Santos, A. J. C., Divina, C. C., Pineda, F. G., & Lopez, L. L. M. A. (2017). *In vitro* evaluation of the antagonistic activity of *Trichoderma* sp. against *Fusarium verticillioides*. *Journal of Agricultural Technology, 13*(7.3), 2539–2548.

Saramun, S., & Wattayakorn, G. (2000). Petroleum hydrocarbon contamination in seawater along the western coast of the Philippines. In *Proceedings of the third technical seminar on marine fishery resources survey in the south China sea, area III: Western Philippines, 13–15 July 1999* (pp. 316–320). Secretariat, Southeast Asian Fisheries Development Center.

Schreinemachers, P., Chen, H. P., Nguyen, T. T. L., Buntong, B., Bouapao, L., Gautam, S., Le, N. T., Pinn, T., Vilaysone, P., & Srinivasan, R. (2017). Too much to handle? Pesticide dependence of smallholder vegetable farmers in Southeast Asia. *Science of the Total Environment, 593*, 470–477.

Schreinemachers, P., Grovermann, C., Praneetvatakul, S., Heng, P., Nguyen, T. T. L., Buntong, B., Le, N. T., & Pinn, T. (2020). How much is too much? Quantifying pesticide overuse in vegetable production in Southeast Asia. *Journal of Cleaner Production, 244*, 118738.

Seehausen, M. L., Afonso, C., Jactel, H., & Kenis, M. (2021). Classical biological control against insect pests in Europe, North Africa, and the middle East: What influences its success? *NeoBiota, 65*, 169–191.

Shah, A. A., Hasan, F., Hameed, A., & Ahmed, S. (2008). Biological degradation of plastics: A comprehensive review. *Biotechnology Advances, 26*, 246–265.

Siemieniuch, C. E., Sinclair, M. A., & Henshaw, M. D. (2015). Global drivers, sustainable manufacturing and systems ergonomics. *Applied Ergonomics, 51*, 104–119.

Silva, B. B., Banaay, C. G., & Salamanez, K. (2019). *Trichoderma*-induced systemic resistance against the scale insect (*Unaspis mabilis* Lit & Barbecho) in lanzones (*Lansium domesticum* Corr.). *Agriculture & Forestry, 65*(2), 59–78.

Singh, A., & Gauba, P. (2014). Mycoremediation: A treatment for heavy metal pollution of soil. *Journal of Civil Engineering and Environmental Technology, 1*, 59–61.

Singh, M., Srivastava, P. K., Verma, P. C., Kharwar, R. N., Singh, N., & Tripathi, R. D. (2015). Soil fungi for mycoremediation of arsenic pollution in agriculture soils. *Journal of Applied Microbiology, 119*(5), 1278–1290.

Su, L. S. (2016). Isolation and identification of heavy metal-tolerant bacteria from an industrial site as a possible source for bioremediation of cadmium, lead, and nickel. *Advances in Environmental Biology, 10*, 10–15.

Su, G. S., Fernandez, E. L., Masigan, M. M., Sison, M. A., Su, M. L. S., Ragragio, E., & Bungay, A. A. (2014). Isolation and colonial morphological characterization of fungal species in copper laden sediments of Calancan Bay, Marinduque. *Annual Research & Review in Biology, 4*(11), 1777–1783.

Swift, G. (1997). Non-medical biodegradable polymers: Environmentally degradable polymers. In A. J. Domb, J. Kost, & D. Wiseman (Eds.), *Handbook of biodegradable polymers* (pp. 473–511). CRC Press.

Tamesis, E. V. (1981). Hydrocarbon potentials of Philippine basins. *Energy, 6*(11), 1181–1206.

Tansengco, M. L., & Dogma, I., Jr. (1999). Microbial degradation of poly(β-hydroxybutyrate) or PHB in local landfill soils. *Acta Biotechnology, 19*(3), 191–203.

Tansengco, M., Tejano, J., Coronado, F., Gacho, C., & Barcelo, J. (2018). Heavy metal tolerance and removal capacity of *Trichoderma* species isolated from mine tailings in Itogon, Benguet. *Environment and Natural Resources Journal, 16*(1), 39–57.

Thambugala, K. M., Daranagama, D. A., Phillips, A. J., Kannangara, S. D., & Promputtha, I. (2020). Fungi vs. fungi in biocontrol: An overview of fungal antagonists applied against fungal plant pathogens. *Frontiers in Cellular and Infection Microbiology, 10*.

Thuzar, M. (2011). Urbanization in Southeast Asia: Developing smart cities for the future? *Regional Outlook, 96*.

Tingson, K., Zafaralla, M., Macandog, D., & Rañola, R., Jr. (2018). Lead biomagnification in a food web of the open waters along Sta. Rosa Subwatershed, Philippines. *Journal of Environmental Science and Management, 21*(1).

Tirado, R., & Bedoya, D. (2008). Agrochemical use in the Philippines and its consequences to the environment. *Greenpeace Southeast Asia, 9*, 1–12.

Torres, J. M. O., Cardenas, C. V., Moron, L. S., Guzman, A. P. A., & de la Cruz, T. E. E. (2011). Dye decolorization activities of marine-derived fungi isolated from Manila Bay and Calatagan Bay, Philippines. *Philippine Journal of Science, 140*(2), 133–143.

Truskewycz, A., Gundry, T. D., Khudur, L. S., Kolobaric, A., Taha, M., Aburto-Medina, A., Ball, A. S., & Shahsavari, E. (2019). Petroleum hydrocarbon contamination in terrestrial ecosystems—fate and microbial responses. *Molecules, 24*(18), 3400.

United Nations Population Division. (2020). *World population prospects*. United Nations.

Uno, S., Koyama, J., Kokushi, E., Monteclaro, H., Santander, S., Cheikyula, J. O., Miki, S., Añasco, N., Pahila, I. G., Taberna, H. S., Jr., & Matsuoka, T. (2010). Monitoring of PAHs and alkylated PAHs in aquatic organisms after 1 month from the Solar I oil spill off the coast of Guimaras Island, Philippines. *Environmental Monitoring and Assessment, 165*, 501–515.

Urzo, M. L. R., Cuevas, V. C., & Opulencia, R. B. (2017). Screening and Identification of Polyester Polyurethane-Degrading Fungi. Philippine Agricultural Scientist. *Philippine Agricultural Scientist, 100*, S72–S80.

Vaghaye, C. A., & Dogma, I. J., Jr. (1998). Fungal degradation of low-density polyethylene plastic. In *27th Annual Convention of the Philippine Society for Microbiology, Inc., Manila (Philippines), 7-8 May 1998*.

Varca, L. M. (2012). Pesticide residues in surface waters of Pagsanjan-Lumban catchment of Laguna de Bay, Philippines. *Agricultural Water Management, 106*, 35–41.

Villanueva, L. M., & Davide, R. G. (1984). Evaluation of several isolates of soil fungi for biological control of root-knot nematodes. *Philippine Agriculturist, 67*, 361–371.

Villegas, L. C., Llamado, A. L., Catsao, K. V., & Raymundo, A. K. (2018). Removal of heavy metals from aqueous solution by biofilm-forming bacteria isolated from mined-out soil in Mogpog, Marinduque, Philippines. *Philippine Science Letters, 11*, 19–27.

Vinale, F., Sivasithamparam, K., Ghisalberti, E. L., Marra, R., Woo, S. L., & Lorito, M. (2008). *Trichoderma*–plant–pathogen interactions. *Soil Biology and Biochemistry, 40*(1), 1–10.

Wang, Y., Feng, J., Lin, Q., Lyu, X., Wang, X., & Wang, G. (2013). Effects of crude oil contamination on soil physical and chemical properties in Momoge wetland of China. *Chinese Geographical Science, 23*(6), 708–715.

Wang, H. S., Sthiannopkao, S., Du, J., Chen, Z. J., Kim, K. W., Yasin, M. S. M., Hashim, J. H., Wong, C. K. C., & Wong, M. H. (2011). Daily intake and human risk assessment of organochlorine pesticides (OCPs) based on Cambodian market basket data. *Journal of Hazardous Materials, 192*(3), 1441–1449.

Wattayakorn, G. (2012). Petroleum pollution in the Gulf of Thailand: A historical review. *Coastal Marine Science, 35*(1), 234–245.

Wei, R., & Zimmermann, W. (2017). Microbial enzymes for the recycling of recalcitrant petroleum-based plastics: How far are we? *Microbial Biotechnology, 10*(6), 1308–1322.

Yadav, P., Rai, S. N., Mishra, V., & Singh, M. P. (2021). Mycoremediation of environmental pollutants: A review with special emphasis on mushrooms. *Environmental Sustainability, 4*(4), 605–618.

Zarcinas, B. A., Ishak, C. F., McLaughlin, M. J., & Cozens, G. (2004). Heavy metals in soils and crops in Southeast Asia. *Environmental Geochemistry and Health, 26*(3), 343–357.

Zarcinas, B. A., Pongsakul, P., McLaughlin, M. J., & Cozens, G. (2004). Heavy metals in soils and crops in Southeast Asia 2. Thailand. *Environmental Geochemistry and Health, 26*(4), 359–371.

Zegzouti, Y., Boutafda, A., El Fels, L., El Hadek, M., Ndoye, F., Mbaye, N., … Hafidi, M. (2020). Screening and selection of autochthonous fungi from leachate contaminated-soil for bioremediation of different types of leachate. *Environmental Engineering Research, 25*(5), 722–734.

Zhang, S., Merino, N., Okamoto, A., & Gedalanga, P. (2018). Interkingdom microbial consortia mechanisms to guide biotechnological applications. *Microbial Biotechnology, 11*(5), 833–847.

Zomesh, A. N., Aribal, K. M. J., Narag, R. M. A., Jeorgina, K. L. T., Frejas, J. A. D., Arriola, L. A. M., Gulpeo, P. C. G., Navarete, I. A., & Lopez, C. M. (2019). Lead (II) tolerance and uptake capacities of fungi from polluted tributaries in the Philippines. *Applied Environmental Biotechnology, 4*(1), 18–29.

Further reading

Alvindia, D. G., & Natsuaki, K. T. (2008). Evaluation of fungal epiphytes isolated from banana fruit surfaces for biocontrol of banana crown rot disease. *Crop Protection, 72*(8), 1200–1207.

Arunakumara, K. K. I. U., Walpola, B. C., & Yoon, M. H. (2013). Current status of heavy metal contamination in Asia's rice lands. *Reviews in Environmental Science and Bio/Technology, 12*(4), 355–377.

Bauri, S., Sen, M. K., Das, R., & Mondal, S. K. (2021). In-silico investigation of the efficiency of microbial dioxygenases in degradation of sulfonylurea group herbicides. *Bioremediation Journal*, 1–13.

Bhatt, P., Gangola, S., Chaudhary, P., Khati, P., Kumar, G., Sharma, A., & Srivastava, A. (2019). Pesticide induced up-regulation of esterase and aldehyde dehydrogenase in indigenous *Bacillus* spp. *Bioremediation Journal, 23*(1), 42–52.

Boland, G. J. (2004). Fungal viruses, hypovirulence, and biological control of *Sclerotinia* species. *Canadian Journal of Plant Pathology, 26*(1), 6–18.

Bolo, N. R., Diamos, M. J. C., Sia SU, G. L., Ocampo, M. A. B., & Suyom, L. M. (2015). Isolation, identification, and evaluation of polyethylene glycol and low-density polyethylene-degrading bacteria from Payatas Dumpsite, Quezon City, Philippines. *Philippine Journal of Health Research and Development, 19*(1), 50—59.

Conrath, U., Beckers, G. J., Langenbach, C. J., & Jaskiewicz, M. R. (2015). Priming for enhanced defense. *Annual Review of Phytopathology, 53*, 97—119.

Dimaano, N. G., & Iwakami, S. (2021). Cytochrome P450-mediated herbicide metabolism in plants: Current understanding and prospects. *Pest Management Science, 77*(1), 22—32.

Fenoll, J., Garrido, I., Cava, J., Hellín, P., Flores, P., & Navarro, S. (2015). Photometabolic pathways of chlorantraniliprole in aqueous slurries containing binary and ternary oxides of Zn and Ti. *Chemical Engineering Journal, 264*, 720—727.

Gangola, S., Bhatt, P., Chaudhary, P., Khati, P., Kumar, N., & Sharma, A. (2018). Bioremediation of industrial waste using microbial metabolic diversity. In P. Bhatt, & A. Sharma (Eds.), *Microbial biotechnology in environmental monitoring and cleanup* (pp. 1—27). Hershey, PA: IGI Global.

Gedalanga, P. B., Pornwongthong, P., Mora, R., Chiang, S. Y. D., Baldwin, B., Ogles, D., & Mahendra, S. (2014). Identification of biomarker genes to predict biodegradation of 1, 4-dioxane. *Applied and Environmental Microbiology, 80*(10), 3209—3218.

Jauregui, J., Valderrama, B., Albores, A., & Vazquez-Duhalt, R. (2003). Microsomal transformation of organophosphorus pesticides by white rot fungi. *Biodegradation, 14*(6), 397—406.

Jayashree, R., & Vasudevan, N. (2007). Effect of tween 80 added to the soil on the degradation of endosulfan by *Pseudomonas aeruginosa*. *International Journal of Environmental Science & Technology, 4*(2), 203—210.

Li, W., Dai, Y., Xue, B., Li, Y., Peng, X., Zhang, J., & Yan, Y. (2009). Biodegradation and detoxification of endosulfan in aqueous medium and soil by *Achromobacter xylosoxidans* strain CS5. *Journal of Hazardous Materials, 167*(1—3), 209—216.

Martin, K. D. G., Astrero, M. F. T., Mallari, L. A. N., & Hipol, R. M. (2021). Activity of laccase enzyme present in the phenol-contaminated sediments of the Marilao-Meycauayan-Obando River System, Philippines. *Oriental Journal of Chemistry, 37*(1), 162—168.

Mondal, S., Baksi, S., Koris, A., & Vatai, G. (2016). Journey of enzymes in entomopathogenic fungi. *Pacific Science Review A: Natural Science and Engineering, 18*(2), 85—99.

Nunes, C. S., & Malmlöf, K. (2018). Enzymatic decontamination of antimicrobials, phenols, heavy metals, pesticides, polycyclic aromatic hydrocarbons, dyes, and animal waste. In C. S. Nunes, & V. Kumar (Eds.), *Enzymes in human and animal nutrition* (pp. 331—359). Academic Press.

Pieterse, C. M., Zamioudis, C., Berendsen, R. L., Weller, D. M., Van Wees, S. C., & Bakker, P. A. (2014). Induced systemic resistance by beneficial microbes. *Annual Review of Phytopathology, 52*, 347—375.

Pizzul, L., del Pilar Castillo, M., & Stenström, J. (2009). Degradation of glyphosate and other pesticides by ligninolytic enzymes. *Biodegradation, 20*(6), 751—759.

Raaijmakers, J. M., & Mazzola, M. (2012). Diversity and natural functions of antibiotics produced by beneficial and plant pathogenic bacteria. *Annual Review of Phytopathology, 50*, 403—424.

Roca, E., D'Errico, E., Izzo, A., Strumia, S., Esposito, A., & Fiorentino, A. (2009). In vitro saprotrophic basidiomycetes tolerance to pendimethalin. *International Biodeterioration & Biodegradation, 63*(2), 182—186.

Sakata, S., Mikami, N., & Yamada, H. (1992). Degradation of pyrethroid optical isomers by soil microorganisms. *Journal of Pesticide Science, 17*(3), 181—189.

Singh, B. K. (2009). Organophosphorus-degrading bacteria: Ecology and industrial applications. *Nature Reviews Microbiology, 7*(2), 156—164.

Spadaro, D., & Droby, S. (2016). Development of biocontrol products for postharvest diseases of fruit: The importance of elucidating the mechanisms of action of yeast antagonists. *Trends in Food Science & Technology, 47*, 39—49.

Tahar, A., Choubert, J. M., & Coquery, M. (2013). Xenobiotics removal by adsorption in the context of tertiary treatment: A mini review. *Environmental Science and Pollution Research, 20*(8), 5085—5095.

Wang, B. Z., Guo, P., Hang, B. J., Li, L., He, J., & Li, S. P. (2009). Cloning of a novel pyrethroid-hydrolyzing carboxylesterase gene from *Sphingobium* sp. strain JZ-1 and characterization of the gene product. *Applied and Environmental Microbiology, 75*(17), 5496—5500.

World Bank Group. (2021). *Market study for the Philippines: Plastics circularity opportunities and barriers.* East Asia and Pacific region marine plastics series. World Bank. https://openknowledge.worldbank.org/handle/10986/35295.

PART III

Fungal conservation and management

CHAPTER

11

Edible mushrooms of the Philippines: traditional knowledge, bioactivities, mycochemicals, and in vitro cultivation

Thomas Edison E. dela Cruz[1,2] and Angeles M. De Leon[3]

[1]Department of Biological Sciences, College of Science, University of Santo Tomas, Manila, Philippines; [2]Fungal Biodiversity, Ecogenomics and Systematics (FBeS) Group, Research Center for the Natural and Applied Sciences, University of Santo Tomas, Manila, Philippines; [3]Department of Biological Sciences, College of Science, Central Luzon State University, Science City of Muñoz, Philippines

1. Introduction

The Convention on Biological Diversity defined traditional knowledge as the knowledge, innovations and practices of indigenous and local communities based on their experiences that were gained over the centuries, adapted in their local culture and environment, and transmitted orally from generation to generation (https://www.cbd.int/traditional/intro. shtml). This traditional knowledge can be in a form of stories, songs, folklores, proverbs, cultural values, beliefs, rituals, community laws, and local languages including agricultural practices. An example of an agriculture practice by Filipino indigenous communities is the muyong system by the Ifugaos. Butic and Ngidlo (2003) stated that the Ifugao muyong system can be viewed as either a forest conservation strategy, a watershed rehabilitation technique, a farming system, or an assisted natural regeneration strategy. This strategy of the Ifugaos facilitates the use and conversion of natural forests or woodlots into multiple-use community areas without damaging its pristine condition and is an excellent example of sustainable use. Traditional knowledge is therefore an important source of information that can contribute significantly to the attainment of sustainable development. Traditional knowledge has also been the sources for many of our health and beauty products, particularly plant-

Mycology in the Tropics
https://doi.org/10.1016/B978-0-323-99489-7.00003-2

© 2023 Elsevier Inc. All rights reserved.

based medicines (Garcia, 2020; Ratsogi et al., 2016), and for other valuable agricultural and nonwood products. Due to these economic benefits, there are a growing appreciation of the value of and a continuous support for the conservation of traditional knowledge.

There are 110 ethno-linguistic groups with an estimated 14—17 million Indigenous Peoples in the Philippines (UNDP, 2010). They are mainly concentrated in Northern Luzon, particularly in the Cordillera Administrative Region, and in Mindanao. Some communities could also be found in the Visayas. Studies on the medicinal plants used by our indigenous communities are well documented in several publications, (e.g., Abe & Ohtani, 2013; Balangcod & Balangcod, 2011; Olowa et al., 2012; Ragrario et al., 2013; see also the review paper of Dapar & Alejandro, 2020). But, in addition to medicinal plants, fungi, particularly mushrooms, are traditionally use by indigenous communities globally, and for medicinal uses. For example, the Wixaritari and mestizos of Villa Guerrero in Jalisco, Mexico uses *Ganoderma oerstedii* (Fr.) Murrill and *Pycnoporus sanguineus* (L.) Murrill for medicine, specifically to treat stomach pain, intestinal diseases, kidney problems, skin conditions, and fever (Haro-Luna et al., 2019). However, most traditional knowledge on fungi centers on wild edible mushrooms. In the paper of Devkota (2017), wild edible mushrooms were consumed by communities in Nepal Himalayas. Santiago et al. (2016) also reported wild edible mushrooms by Mixtecs or Ñuu savi, the people of the rain, from Southeastern Mexico. In Africa, six ethnic communities in the Democratic Republic of Congo reported utilization of 73 species of mushrooms, which were utilized as food (68 species), medicine (9 species), in a recreational context (2 species), and were related to myths and beliefs (7 species) (Milenge Kamalebo et al., 2018). These are just few of the many studies worldwide that documented the traditional uses of mushrooms. Boa (2004) estimated about 2327 recorded wild useful species of mushrooms, of which 2166 are edible. He also listed *Termitomyces eurhizus* (Berk.) R.Heim, *Termitomyces microcarpus* (Berk. & Broome) R.Heim, *Termitomyces striatus* (Beeli) R.Heim and species of *Agaricus, Ganoderma, Pleurotus,* and *Polyporus* are among the few species reported as edible mushrooms in the Philippines. With more than 110 ethno-linguistic indigenous groups living in the country, surely mushrooms will find its ways to their socio-cultural lives. In this paper we review studies on the traditional knowledge and uses of wild edible mushrooms by the indigenous peoples in the Philippines. We provide a list of species of mushrooms so far documented and reported studies on their cultivation and any attempts to test for their medicinal properties.

2. Ethnomycology in the Philippines

Comandini and Rinaldi (2020) described ethnobiology as a highly interdisciplinary field, with inputs from cultural anthropology, linguistics, and archeology which somehow converge into biology. Within this field, ethnomycology studies the traditional knowledge, practices, and uses of fungi, with particular emphasis on wild edible mushrooms. Detailed ethnomycological studies in the Philippines started with the pioneering research of De Leon, Reyes, and dela Cruz in 2012. They conducted the first extensive ethnomycological survey among six Aeta communities in three provinces of Central Luzon and documented their knowledge of mushrooms and their traditional beliefs during collection and use (De Leon et al., 2012). Some of the documented traditional beliefs included the performance of dancing ritual or seeking the permission of the spirits before collection. De Leon et al. (2012) also documented the different local names used by the Aetas on their mushrooms and showed that the different tribes use different local names for the same species of mushroom. The naming of mushrooms by the Aetas is also based on the substrates where they observed the mushrooms. Fourteen species were recorded in that study, 12 of which were used as food while two were used as a house decoration and medicine. De Leon et al. (2013a) further supplemented this study by documenting other macrofungi present in the ancestral domains and resettlement areas of the Aetas and recorded 76 species, 53 of which were identified up to the species level. Interestingly these were not utilized by the Aetas, albeit three species were reported previously by other studies as edible or had medicinal use. These studies on the Aetas were followed by Lazo et al. (2015) who documented the utilization of mushrooms by the Gaddang communities in Nueva Vizcaya. The Gaddang indigenous communities reported utilization of 10 species of macrofungi, mainly as food. Just like the Aetas, the naming of mushrooms is also based on the substrates where the mushrooms are found. However, the Gaddangs use the local term Tarulok for mushrooms as opposed to the local term Kuwat by the Aetas. Furthermore, the indigenous beliefs and utilization of macrofungi by the Kalanguya indigenous tribes in Carranglan, Nueva Ecija was reported by De Leon et al. (2016). This indigenous community claimed utilization of 36 species, although only 10 specimens were collected during the survey. Interestingly, the Kalanguya tribe reported one species of mushroom as insect repellant, as most report identified the use of mushrooms either as food or medicine for human. De Leon et al. (2018) also reported the traditional utilization of mushrooms by the Ifugaos. Their survey of nine barangays in three municipalities within the Ifugao province yielded 13 species, many of which were reported as edible, and documented their traditional beliefs in relation to mushrooms. Interestingly, similar to the beliefs of the Gaddang communities in Nueva Vizcaya (Lazo et al., 2015) and the Aetas (de Leon et al., 2012), the Ifugaos also believe that mushrooms are surrounded by spirits. De Leon et al. (2019) again supplemented the study with a survey of all available mushrooms in the area and reported 109 species of macrofungi, 16 of which are reported as edible. Tantengco and Ragragio (2018) also documented the macrofungi utilized by the Aytas in Bataan Province and reported 15 species as food and medicine. Recently, Torres et al. (2020) reported the utilization of mushrooms by the Bugkalot indigenous community in Nueva Vizcaya, with 21 species used as food and eight species as medicine. Some of the medicinal uses mentioned by the Bugkalots are for the treatment of headache, skin diseases, gastric ulcer, hepatitis, and stomachache. However, plotting the provinces where these ethnomycological studies were conducted showed that most were done in Luzon Island as opposed to the numerous studies on ethnobotany conducted in the Philippines (Fig. 11.1). This certainly opens many opportunities to document the traditional knowledge of mushrooms by the other indigenous communities in the country, particularly those from Mindanao and the Visayas.

3. Species listing of wild edible mushrooms in the Philippines

There are several studies documenting macrofungi in the Philippines, for example, in Paracelis, Mountain Province (37 species, De Leon et al., 2021), in Bazal-Baubo Watershed, Aurora Province (107 species, Tadiosa et al., 2011), in La Union Province (51 species, Tadiosa

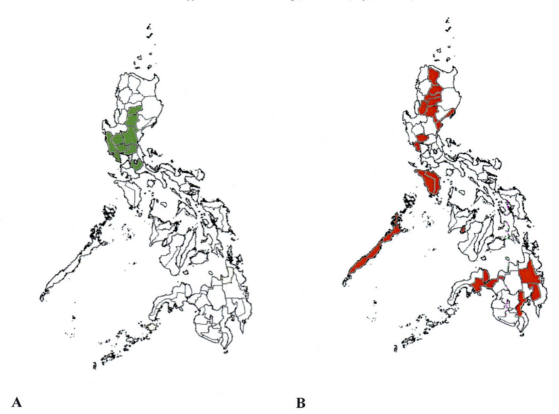

FIGURE 11.1 Provinces in the Philippines where (A) ethnomycological and (B) ethnobotanical studies were conducted. *Data for the ethnobotanical studies were derived from the review paper of Dapar, M. L. G., & Alejandro, G. J. D. (2020). Ethnobotanical studies on indigenous communities in the Philippines: Current status, challenges, recommendations and future perspectives.* Journal of Complementary Medicine Research, 11(1): 432–446. https://doi.org/10.5455/jcmr.2020.11.01.51.

& Arsenio, 2014), and in Angat Watershed Reservation, Bulacan Province (21 species, Liwanag et al., 2017). Several mountain sites were also explored for their macrofungal diversity, i.e., Mt. Bangcay in Cuyapo and Mt. Mingan in Gabaldon, both in Nueva Ecija (30 species, Dulay & Maglasang, 2017; 4 species, Guzman et al., 2018), Mt. Makiling Forest Reserve in Los Banos, Laguna (20 species, De Castro & Dulay, 2015), Mt. Maculot in Cuenca, Batangas (92 species, Arenas et al., 2018), Mt Palay-Palay — Mataas na Gulod Protected Landscape and Taal Volcano Protected Landscape in Cavite (95 species, Arenas et al., 2015; 75 species, Tadiosa & Briones, 2013), and in Mt. Malinao, Albay (9 species, Daep & Cajuday, 2003). Macrofungi were also documented from forest sites within two state universities — Central Luzon State University in Nueva Ecija (35 species, Culala & Dulay, 2018) and Bicol University in Albay (39 species, Guerrero et al., 2020). However, most studies on edible macrofungi in the country were limited to their reported traditional utilization based on the different ethnomycological surveys described earlier. These included the traditional use as reported by the

Aeta/Ayta communities in Pampanga, Zambales, Tarlac, and Bataan (De Leon et al., 2013; Tantengco & Ragragio, 2018), the Kalanguya communities in Nueva Ecija (De Leon et al., 2016), the Gaddang and Bugkalot communities in Nueva Vizcaya (Lazo et al., 2015; Torres et al., 2020), and the Ifugao communities in Ifugao Province (De Leon et al., 2018, 2019). The species of macrofungi reported by the Aetas as food are *Auricularia auricula-judae* (Bull.) Quel., *Auricularia polytricha* (Mont.) Sacc., *Ganoderma lucidum* (Curtis) P. Karst, *Lentinus tigrinus* (Bull.) Fr., *Lentinus sajor-caju* (Fr.) Fr., *Schizophyllum commune* (Fries), *Termitomyces clypeatus* R. Heim, *Trametes robustus* (P. Karst), and *Volvariella volvacea* (Bull. ex. Fr.) (de Leon et al., 2012; Tantengco & Ragragio, 2018). *Ganoderma lucidum* is also used as a house decoration (de Leon et al., 2012). On the other hand, the Gaddang indigenous community utilized as food *Auricularia auricula*, *Auriculria fuscosuccinea* (Mont.) Henn., *Schizophyllum commune*, and *Volvariella volvacea* (Lazo et al., 2015) while the Kalanguya tribe considered these species as edible: *Meripilus giganteus* (Pers.) P. Karst., *Scleroderma citrinum* Pers., *Leucoagaricus cepaestipes* (Fries), *Podocypha brasiliensis* D.A. Reid, and *Russula virescens* (Schaeffer) Fries. Among the Ifugaos, they utilized these mushrooms as food—*Auricularia auricula*, *Oudemansiella canarii* (Jungh) Hohn, *Lentinus sajor-caju*, *Coprinus disseminates* (Pers.) J.E., *Schizophyllum commune*, *Vasellum pretense* (Pers.) Kreisel, *Pleurotus ostreatus* (Jacq. Fr.) Kumm., *Pleurotus djamor* (Rumph. ex. Fr.) Boedijn, *Coprinus comatus* (O.F. Mull.) Pers., and *Trametes elegans* (Spreng) Fr. The Bugkalot indigenous community identified as food the following: *Auricularia auricula-judae*, *Auricularia polytricha*, *Coprinus atramentaria* (Bull.:Fr.) Redhead, *Coprinopsis lagopus* (Fr.) Fr., *Coprinus cinereus* Schaeff., *Lentinus tigrinus*, *Pleurotus dryinus* (Pers.) P. Kumm., *Schizophyllum commune*, and *Stereum lobatum* (Kunze ex Fr.) Fr. while they used *Ganoderma applanatum* (Pers.) Pat., *Ganoderma lucidum*, and *Polyporus picipes* Fr. as medicine. Licyayo (2018) also identified 30 species of wild edible mushrooms found in coniferous forests in Benguet and Mountain Province. Table 11.1 lists all wild edible species of mushrooms documented in the Philippines based on these ethnomycological surveys. We also included in the list those edible mushrooms that were recorded from other macrofungal studies. Fig. 11.2 shows some of the wild edible mushrooms utilized by our indigenous communities. It is important to note that a number of species of edible macrofungi in these cited ethnomycological studies were not identified up to the species level, and hence were not included in the list. Thus, the number of edible mushrooms in the country could increase when these macrofungi are fully identified.

4. In vitro culture of Philippine wild edible mushrooms

One of the early studies on the cultivation of wild edible mushrooms utilized by the indigenous communities was that of Tayamen et al. (2004). They reported the culture of *Ganoderma* sp. and *Auricularia polytricha*. Recently, many wild species of edible macrofungi were also cultivated, for example, *Schizophyllum commune* (Bulseco et al., 2005; Dulay et al., 2016; Kalaw et al., 2016), *Coprinus comatus* (Reyes, Lopez et al., 2009), *Coprinopsis cinerea* (Kalaw et al., 2016), *Collybia reinakeana* P. Henn (Reyes et al., 1997), *Lentinus sajor-caju* (Cuevas et al., 2009; De Leon et al., 2017; Kalaw et al., 2016), *Lentinus tigrinus* (De Leon et al., 2013a; Dulay et al., 2012; Kalaw et al., 2016), *Lentinus squarrosulus* (De Leon et al., 2013b, 2017), *Polyporus grammocephalus* (De Leon et al., 2013b; Dulay & Rivera, 2017), *Ganoderma lucidum* (Dulay,

TABLE 11.1 List of wild edible mushrooms documented in Luzon Island, Philippines.

Family	Species name	Collected from										References
		Bataan	Bulacan	Nueva Ecija	Nueva Vizcaya	Mt. Province	Zambales	Tarlac	Pampanga	Laguna	Ifugao	
Agaricaceae	*Leucoagaricus cepaestipes* (Fries)			✓								De Leon et al. (2016)
Amanitaceae	*Amanita javanica* (Corner & Bas) T. Oda, C. Tanaka & Tsuda					✓						Licyayo (2018)
Auriculariaceae	*Auricularia auricula* (Bull.) Quel	✓		✓	✓	✓	✓				✓	De Leon et al. (2012, 2018, 2021), Lazo et al. (2015), Dulay and Maglasang (2017) Tantengco and Ragragio (2018) Torres et al. (2020)
	Auricularia polytricha (Mont.) Sacc.	✓		✓	✓		✓	✓				De Leon et al. (2012), Dulay and Maglasang (2017) Tantengco and Ragragio (2018) Torres et al. (2020)
	Auricularia tenuis (Lev.) Farl.			✓					✓			De Leon et al. (2013)
Bolbitiaceae	*Panaeolus semiovatus* (Sowerby) S. Lundell & Nannf.			✓								Dulay and Maglasang (2017)
Boletaceae	*Boletus speciosus* Frost.					✓						Licyayo (2018)
Ganodermataceae	*Ganoderma applanatum* (Pers.) Pat.		✓									Liwanag et al. (2017)

Family	Species										References
	Ganoderma austral (Fr.) Pat.			✔							Torres et al. (2020)
	Ganoderma lucidum (Curtis) P. Karst.	✔	✔	✔					✔		Nacua et al. (2018), Dulay and Maglasang (2017) Tantengco and Ragragio (2018) Torres et al. (2020)
	Ganoderma neo-japonicum Imazeki		✔								Dulay and Maglasang (2017)
	Ganoderma sinense J.D. Zhao, L.W. Hsu & X.Q. Zhang		✔								Dulay and Maglasang (2017)
	Ganoderma tsugae Murrill			✔							Torres et al. (2020)
Gomphaceae	*Rammaria botrytis* (Pers.) Ricken				✔						Licyayo (2018)
Hygrophoraceae	*Hygrophorus russula* (Schaeff. ex Fr.) Kauffman				✔						Licyayo (2018)
Hymenochaetaceae	*Phellinus linteus* (Berk. and Curt.) Teng		✔								Dulay and Maglasang (2017)
Lycoperdaceae	*Calvatia gigantea* (Batsch) Lloyd								✔		De Leon et al. (2013)
	Vascellum pretense (Pers.) Kreisel									✔	De Leon et al. (2018)
Lyophyllaceae	*Termitomyces clypeatus* R. Heim		✔			✔	✔	✔			De Leon et al. (2012), Dulay and Maglasang (2017)

(Continued)

TABLE 11.1 List of wild edible mushrooms documented in Luzon Island, Philippines.—cont'd

Family	Species name	Bataan	Bulacan	Nueva Ecija	Nueva Vizcaya	Mt. Province	Zambales	Tarlac	Pampanga	Laguna	Ifugao	References
Lyophyllaceae	*Termitomyces robustus* (Beeli) R. Heim						✓		✓			De Leon et al. (2012)
Meripilaceae	*Meripilus giganteus* (Pers.) Karst.			✓								De Leon et al. (2016)
Phallaceae	*Dictyophora indusiate* (Vent.) Fisch			✓								Dulay and Maglasang (2017)
Physalacriaceae	*Oudemansiella canarii* (Jungh) Hohn.										✓	De Leon et al. (2018)
Pleurotaceae	*Pleurotus cystidiosus* O.K. Mill.			✓								Dulay and Maglasang (2017)
Pleurotaceae	*Pleurotus djamor* (Rumph. ex Fr.) Boedijn			✓								Dulay and Maglasang (2017)
	Pleurotus dryinus (Pers.) P. Kumm.				✓							Torres et al. (2020)
	Pleurotus ostreatus (Jacq. Fr.) Kumm										✓	De Leon et al. (2018)
	Pleurotus porrigens (Pers.) Singer							✓				De Leon et al. (2013)
	Pleurotus pulmunarius (Fr.) Quel.						✓					De Leon et al. (2013)

Family	Species										References
Pluteaceae	*Pleurotus sajor-caju* (Fr.) Singer	✓						✓			De Leon et al. (2013)
	Voltariella voltvacea (Bull. ex Fr.) Singer		✓	✓				✓		✓	De Leon et al. (2012, 2018), Lazo et al. (2015) Dulay and Maglasang (2017) Tantengco and Ragragio (2018)
Podoscyphaceae	*Podoscypha brasiliensis* D.A. Reid		✓								De Leon et al. (2016)
Polyporaceae	*Coriolus versicolor* (Yun Zhi)		✓	✓							Dulay and Maglasang (2017)
	Lentinus cladopus Lev.						✓				De Leon et al. (2013)
	Lentinus sajor-caju (Fr.) Fr.					✓				✓	De Leon et al. (2012, 2018)
	Lentinus squarrosulus (Mont.) Singer					✓					De Leon et al. (2013b)
	Lentinus strigosus Fr.				✓						De Leon et al. (2021)
	Lentinus tigrinus (Bull.) Fr.		✓	✓		✓	✓	✓	✓		De Leon et al. (2013a), De Leon et al. (2012), Nacua et al. (2018) Dulay and Maglasang (2017) Torres et al. (2020)
	Lenzites elegans (Spreng.) Pat.									✓	De Leon et al. (2018)
Polyporaceae	*Polyporus grammocephalus* Berk.		✓	✓		✓					De Leon et al. (2013b) Dulay and Maglasang (2017)

(Continued)

TABLE 11.1 List of wild edible mushrooms documented in Luzon Island, Philippines.—cont'd

Family	Species name	Collected from										References
		Bataan	Bulacan	Nueva Ecija	Nueva Vizcaya	Mt. Province	Zambales	Tarlac	Pampanga	Laguna	Ifugao	
	Polyporus picipes Fr.				✓							Torres et al. (2020)
	Trametes elegans (Spreng.) Fr.										✓	De Leon et al. (2018)
	Trametes versicolor (L.) C.G. Llyod									✓		Nacua et al. (2018)
Psathyrellaceae	*Coprinellus disseminatus* (Pers.) J.E. Lange					✓					✓	De Leon et al. (2018, 2021)
	Coprinopsis atramentaria (Bull.:Fr.) Redhead				✓							Torres et al. (2020)
	Coprinopsis lagopus (Fr.) Redhead, Vilgalys & Moncalvo				✓							Torres et al. (2020)
	Coprinus cinereus (Schaeffer: Fries) Gray				✓							Torres et al. (2020)
	Coprinus comatus (O.F. Mull.) Pers.			✓								Dulay and Maglasang (2017)
	Paneolus papilionaceus (Bull.) Quel.									✓		Nacua et al. (2018)

Family	Species										References
Russulaceae	*Lactarius volemus* (Fr.: Fr) Fr.					✓					Licyayo (2018)
	Russula virescens (Schaeff.) Fr.		✓		✓	✓					De Leon et al. (2016), Nacua et al. (2018), Licyayo (2018)
Sarcoscyphaceae	*Cookeina speciosa* (Fr.) Dennis		✓								Nacua et al. (2018)
Schizophyllaceae	*Schizophyllum commune* Fr.	✓	✓	✓	✓	✓	✓	✓	✓	✓	De Leon et al. (2012, 2018, 2021), Lazo et al. (2015) Dulay and Maglasang (2017) Tantengco and Ragragio (2018), Torres et al. (2020), Liwanag et al. (2017)
Sclerodermataceae	*Scleroderma citrinum* Pers.							✓			De Leon et al. (2016)
Sparassidaceae	*Sparassis crispa* (Wulfen) Fr.				✓						Licyayo (2018)
Stereaceae	*Stereum lobatum* (Kunze ex Fr.) Fr.						✓				Torres et al. (2020)
Tremellaceae	*Tremella foliacea* Pers.							✓			Dulay and Maglasang (2017)
	Tremella mesenterica Fries							✓			Dulay and Maglasang (2017)
Tricholomataceae	*Lyophillum fumosum* (Pers.) P.D. Orton					✓					Licyayo (2018)
	Tricholoma robostum (Alb. ex. Schw. ex. Fr.) Quel.					✓					Licyayo (2018)

FIGURE 11.2 Some edible mushrooms utilized by indigenous communities in the Philippines. (A) *Schizophyllum commune* (Aeta), (B) *Volvariella volvacea* (Gaddang), (C) *Termitomyces clypeatus* (Aeta), (D) *Oudemansiella canarii* (Ifugao), (E) *Auricularia polytricha* (Bugkalot), (F) *Lentinus tigrinus* (Ayta).

Cabalar et al., 2015; Kalaw et al., 2016; Magday et al., 2014), *Pleurotus cystidiosus* O.K. Mill (Dulay, Ray et al., 2015; Kalaw et al., 2016), *Pleurotus citrinopileatus* Sing (Jacob et al., 2015), *Pleurotus salmoneostramineus* L. Vass (Jacob et al., 2015), and *Pleurotus pumonarius* (Tolentino et al., 2016). Other species that were recently cultivated in vitro are *Trametes elegans* (Dulay, Alcazar et al., 2021), *Pleurotus djamor*, *Pleurotus sajor-caju*, and *Pleurotus florida* (Mont.) Singer (Alvarez & Bautista, 2021), *Lentinus strigosus* Fr (Dulay, Cabrera, Kalaw, & Reyes, 2020; Dulay, Cabrera, Kalaw, Reyes, & Hou, 2020), *Lentinus swartzii* Berk (Dulay, Cabrera, Kalaw,

& Reyes, 2020; Dulay, Cabrera, Kalaw, Reyes, & Hou, 2020, Dulay, Cabrera, Kalaw, & Reyes, 2021b; Tiniola et al., 2021), *Pycnoporus sanguineus* (L.) Murrill (Dulay & Damaso, 2020), *Calocybe cylindracea* (Landingin et al., 2020), *Volvariella volvacea* (Abon et al., 2020), and *Fomitopsis feei* (Fr.) Kreisel (De Leon et al., 2020a). We described below how in vitro culture of mushrooms are conducted.

Choice of mushroom species. A healthy fruiting body of wild edible mushrooms should be chosen for tissue culture. This will ensure the vigor of the mycelial culture of the desired mushroom. Any signs of damage, for example, physical aberration, spoilage, or mold contamination, should be checked prior to surface sterilization.

Isolation to pure mycelial culture. Tissue culture is an important step in spawn production and eventually, mushroom production. Initially, the fruiting bodies of the desired mushroom is brushed to remove any adhering wood particles or soil and soaked in 10% sodium hypochlorite (NaOCl, a commercial bleach) for 1 minute following the protocol of De Leon et al. (2013b). Then, the fruiting body is rinsed with sterile distilled water. Aseptically under a clean, sterile laminar hood, the fruiting body of the surfaced-sterilized mushroom is sliced with a flame-sterilized scalpel to expose the inner tissue. A $5\,mm^2$ inner tissue is cut and laid with a flame-sterilized forcep on the surface of petri plates prefilled with 15 mL Potato Dextrose Agar (PDA; 250 g potatoes, 10 g dextrose/glucose, 15 g agar in a liter of distilled water). Alternatively, PDA can also be poured on ketchup or "lapad" bottle and solidified in a slanted position. The mushroom tissue culture is then incubated at room temperature for 7–10 days (Reyes, Eguchi et al., 2009). Pure culture is obtained by subculturing the growing mycelia onto a newly prepared PDA plate. Pure culture can be distinguished by the presence of one type of mycelial growth that spreads on the medium (De Leon et al., 2013b).

Mycelial growth on different indigenous substrata. To ensure and check for the best mycelial growth performance of the cultivated mushroom, the mycelial culture is grown on different indigenous culture media: potato sucrose gelatin, rice bran decoction gelatin, crack corn decoction gelatin, and coconut water gelatin. To prepare these indigenous culture media, 1 liter decoction of either potato, rice bran, or yellow corn grits were added with 20 g shredded white gelatin bars, to which 10 g of sucrose is also added. For the coconut water media, 1 liter of fresh coconut water is filtered and boiled and to which, 20 g of shredded gelatin is added. Note that preparation of these media requires materials that can easily be acquired from local markets in the Philippines. All indigenous culture media are sterilized at 121°C or 15 psi for 15–20 min. After sterilization, the decoctions are pour-plated onto sterile petri plates. To test for the growth of the mycelial culture, a 10-mm diameter mycelial agar disc from previously cultured mushroom is inoculated centrally on the prepared indigenous culture media and incubated at room temperature for 3–5 days (De Leon et al., 2017). Mycelial growth can be measured as colony diameter over time.

Mycelial growth at different physicochemical parameters. In addition to nutritional requirements, it is also important to check the best culture conditions for the growth of the mycelial culture. The pH of the culture medium is an important factor that can influence the growth and even the morphology of fungi either directly by its action on the cell exterior or indirectly by its effect on the availability of nutrients (Abubakar et al., 2013; Gibbs et al., 2000). To determine the most ideal pH for the mycelial growth, the mycelial cultures are grown on the most suitable media (see indigenous substrata above) at different pH levels, usually ranging from 5.0 to 8.0 at 0.5 interval. These culture media with different pH levels are aseptically inoculated with 10-mm diameter fungal agar disc to allow for the ramification of the mycelia (Kalaw et al., 2021).

Oxygen and carbon dioxide are important atmospheric gases that affect the growth of mushrooms (Lee et al., 2012). According to Chang and Miles (2004, p. 477), mushrooms as aerobic organisms require sufficient supply of oxygen to promote mycelial growth. Oxygen is also important to promote normal metabolism (Carrasco et al., 2018). For the in vitro culture of mushrooms, two aeration conditions are often employed, i.e., with sealed and unsealed culture plates. Under the sealed condition, petri plates are sealed with parafilm to block or minimize the passage of air. The inoculated cultures plates are incubated at room temperature to allow for the full ramification of the mycelia (De Leon et al., 2017).

Light is another factor that can influence the growth and development of mushroom. It is a significant factor necessary for the reproduction and metabolism of fungi (Cheng et al., 2012). Light also acts as a signal that triggers various biophysical and biochemical processes leading to morphological and phototrophic reactions (Trukhonovets, 1991). The effect of light on mycelial growth is determined usually by incubating the culture mycelia under three illumination conditions: (1) with 24-hour exposure to artificial light (fluorescent light bulb), (2) in the dark by covering the culture plates with carbon paper, and (3) with exposure to normal light (day) and dark (night) condition. All treatments are incubated at room temperature until full ramification of mycelia is observed (Kalaw et al., 2021).

Finally, among the different physicochemical parameters, temperature is considered as one of the most fundamental factors. It is important for the rapid growth and for the sporulation of mushrooms. To assess the most favorable temperature for the mycelial growth of mushrooms, culture plates are incubated at three temperature regiments: room temperature (30−32°C), cool temperature (23−25°C), and low-refrigerated temperature (6−13°C) to allow ramification of mycelia.

Both the nutritional requirements and the physicochemical parameters are needed to be tested for every new culture of wild edible mushrooms.

Spawn inoculum preparation. Following the successful mycelial culture of the wild edible mushroom, it is now necessary to prepare the spawn that will serve as the primary inoculum for the mass production of fruiting bodies. The grain spawn is the starter culture to initiate the mass production of the cultivated mushroom as it facilitates the rapid colonization of the mushroom substrata (Reyes, Eguchi et al., 2009). Locally available seeds or grains can be evaluated to determine the mycelial growth performance of the cultivated mushroom. Examples of these grains are unmilled rice seeds, corn grits seeds, and sorghum seeds. These grain spawns are prepared by boiling the grains until tender, drained, and then dried until 60% humidity is attained. Forty grams of these grains are then dispensed in clear glass bottles, plugged with cotton, wrapped with aluminum foil, and then, sterilized at 121°C for 30 min. After sterilization, the bottled grain spawns are allowed to cool down and then, aseptically inoculated with 10-mm mycelial agar disc of a 7-day old pure mycelial culture. The inoculated grain spawns are again incubated at room temperature to allow ramification of the mycelia. The grain spawn with fully ramified mycelia within the shortest period is considered as the best spawning material (De Leon et al., 2013a).

Mass production. The fruiting spawn in polypropylene (PP) bags is used for the propagation of the fruiting bodies of the mushrooms. It is very important to consider all nutritional and physico-chemical factors needed for the growth of the mushroom. A fruiting spawn should also mimic the natural substrates of the mushrooms (Reyes, Eguchi et al., 2009). These could be wooden chips, sawdust or even rice straw. For initiating fruiting bodies on rice straw, initially the rice straw is soaked in water for 3 days. After this 3-day fermentation period, the water is drained, and the rice straw pile is covered with vinyl sheets to stimulate growth of natural decomposers. Composting of the rice straw could last for 7–10 days at 40–50°C with the compost pile being turned every 2 days to allow aeration. The rice straw compost is then chopped into 2.5–5.0 cm long pieces and mixed with varied parts of sawdust. The mixed substrates are then bagged in 15.2 × 30.5 cm PP bags. The openings of the bags are plugged with polyvinyl chloride (PVC) plastic rings which serve as the neck of the bags and then sealed with cotton. The bags are pasteurized for 5 h at 60–80°C and then allowed to cool down (De Leon et al., 2013b). The fruiting spawn bags are inoculated with 40 g of 7-day old grain spawn. Finally, the inoculated bags are incubated at room temperature to allow for mycelial colonization of the bagged substrates which will later developed into fruiting bodies (De Leon et al., 2013b).

Fruiting body harvesting. After attaining a luxuriant mycelial growth on the fruiting spawn, the matured spawn bags are opened after 2–3 days. This allows a transition between mycelia ramification and the initiation of fruiting initials or primordial. Matured spawn bags are now transferred to the growing house. Here, one end of the spawn bags is opened and with a mister, the opened ends are sprayed with clean water 2–3 times a day to maintain moisture and prevent the drying of fruiting bags. Fruiting bodies are ready for harvest 3–5 days after opening the bags. It is important to note that the temperature inside the growing house should be maintained between 25 and 28°C, and with minimal ventilation and diffused light (Reyes, Eguchi et al., 2009).

5. Mycochemicals and bioactivities of cultivated wild edible mushrooms

Mycochemicals are bioactive compounds that exhibit different functionalities and are naturally occurring constituents of mushrooms. In the Philippines, several cultivated edible mushrooms were assessed for their mycochemical constituents. These include *Lentinus swartzii* (Austria et al., 2021), *Calocybe cylindracea* (Landingin et al., 2021), *Pleurotus cornucopiae* (Paulet) Rolland (Landingin et al., 2021), *Xylaria papulis* Llyod (De Leon et al., 2020b), *Pycnoporus sanguineus* L. Murrill (Mendoza et al., 2020), *Pleurotus cystodiosus* (Garcia et al., 2020; Kalaw & Albinto, 2014), *Tyromyces chioneus* (Fr.) P. Karst (Capistrano et al., 2018), *Lentinus sajor-caju* (De Leon et al., 2017; Reneses et al., 2016), *Lentinus squarrosulus* (Pascua et al., 2016), *Ganoderma lucidum* (Romorosa et al., 2017), *Auricularia fuscosuccinea* (Mont.) Henn (Romorosa et al., 2017), *Trametes hirsuta* (Wulfen) Llyod (Romorosa et al., 2017), and *Panaeolus antillarium* (Fr.) Dennis (Dulay, Cabalar et al., 2015). Some of these mycochemicals were also tested for biological activities. We listed under Table 11.2 the major types of mycochemicals detected from cultivated edible mushrooms and their reported bioactivities.

286   11. Edible mushrooms of the Philippines: traditional knowledge, bioactivities, mycochemicals, and in vitro cultivation

TABLE 11.2   Mycochemicals and bioactivities detected from cultivated wild edible mushrooms from the Philippines.

Species	Mycochemicals								
	Alkaloids	Phenols	Flavonoids	Tannins	Fatty acids	Triterpenes	Essential oils	Steroids	Anthraquinones
Auricularia fuscosuccinea (Mont.) Henn.	✔			✔					
Calocybe cylindracea (DC.) Vizzini & Angelini	✔	✔	✔	✔	✔	✔	✔	✔	✔
Coprinus comatus (O.F. Mull.) Pers.	✔		✔					✔	
Ganoderma lucidum (Curtis) P. Karst.	✔			✔					
Lentinus sajor-caju (Fr.) Fr.	✔								
Lentinus swartzii Berk.		✔	✔	✔	✔	✔	✔		
Panaeolus antillarium (Fr.) Dennis	✔								
Pleurotus cornucopiae (Paulet) Quél.	✔	✔		✔	✔	✔	✔		✔
Pleurotus cystidiosus O.K. Mill.	✔	✔	✔	✔	✔	✔	✔	✔	✔
Pleurotus cystidiosus O.K. Mill.	✔		✔						
Pycnoporus sanguineus L. Murrill	✔	✔	✔	✔	✔	✔		✔	✔
Trametes hirsute (Wulfen) Llyod	✔		✔						
Tyromyces chioneus (Fr.) P. Karst.	✔	✔	✔	✔	✔	✔		✔	✔
Xylaria papulis Llyod	✔	✔	✔	✔	✔	✔	✔	✔	✔

III. Fungal conservation and management

Mycochemicals						Bioactivities				
Coumarins	Anthrones	Sugars	Terpenoids	Glycosides	Saponins	Antibacterial	Antifungal	Antidiabetic	Antioxidants	References
				✔				✔		Romorosa et al. (2017)
✔	✔	✔							✔	Landingin et al., 2021
			✔		✔					Kalaw and Albinto (2014)
				✔	✔			✔		Romorosa et al. (2017)
			✔	✔	✔	✔				De Leon et al. (2017)
		✔				✔	✔			Austria et al. (2021)
				✔	✔	✔				Dulay, Cabalar et al. (2015)
✔	✔								✔	Landingin et al., 2021
✔		✔		✔		✔				Garcia et al. (2020)
			✔		✔					Kalaw and Albinto (2014)
	✔							✔		Mendoza et al. (2020)
				✔	✔			✔		Romorosa et al. (2017)
✔	✔	✔						✔		Capistrano et al. (2018)
✔	✔					✔		✔		De Leon et al. (2020b)

6. Concluding remarks and future direction

Mushrooms have been utilized by the human society for ages. They are known as functional foods as they contain substances that have health benefits. Mushrooms are rich in proteins, carbohydrates, minerals, vitamins, unsaturated fatty acids, phenolic compounds, antioxidants, tocopherols, ascorbic acid, and carotenoids that fit the definition of food supplements. Many of our indigenous communities also record wild mushrooms as food or medicine, particularly those growing within or near their ancestral domains. The cultivation of these mushrooms offers opportunities for these local communities to augment their resources and tap into their health benefits. Therefore, it is very important to document the traditional knowledge and utilization of mushrooms by our indigenous people. Perhaps the next breakthrough drugs could be found in one of these wild edible mushrooms.

Acknowledgments

The authors acknowledge the numerous authors whose papers are cited in this paper. We also thank Krystle Angelique Santiago for her assistance in preparing the maps used in our figure.

References

Abe, R., & Ohtani, K. (2013). An ethnobotanical study of medicinal plants and traditional therapies on Batan island, the Philippines. *Journal of Ethnopharmacology, 145*, 554—565.

Abon, M. D., Dulay, R. M. R., Kalaw, S. P., Romero-Roman, M. E., Arana-Vera, L. P., Reyes-Borja, W. O., & Reyes, R. G. (2020). Effects of culture media and physical factors on the mycelial growth of the three wild strains of *Volvariella volvacea* from Ecuador. *Journal of Applied Biology and Biotechnology, 8*(6), 60—63.

Abubakar, A., Suberu, H. A., Bello, I. M., Abdulkadir, R., Daudu, O. A., & Lateef, A. A. (2013). Effect of the pH on mycelial growth and sporulation as Aspergillus parasiticus. *Journal of Plant Sciences, 1*(4), 64—67.

Alvarez, L. V., & Bautista, A. B. (2021). Growth and yield performance of *Pleurotus* on selected lignocellulosic wastes in the vicinity of PUP main campus, Philippines. *Indian Journal of Science and Technology, 14*(3), 259—269.

Arenas, M. C., Tadiosa, E. R., Alejandro, G. J. D., & Reyes, R. G. (2015). Macroscopic fungal flora of Mts. Palaypalay — Mataas na Gulod protected landscape, Southern Luzon, Philippines. *Asian Journal of Biodiversity, 6*, 1—26.

Arenas, M. C., Tadiosa, E. R., & Reyes, R. G. (2018). Taxonomic inventory based on physical distribution of macrofungi in Mt. Maculot, Cuenca, Batangas, Philippines. *International Journal of Biology, Pharmacy, 7*(5), 672—687.

Austria, A. B., Dulay, R. M. R., & Pambid, R. C. (2021). Mycochemicals, antioxidant and anti-diabetic properties of Philippine sawgill mushroom *Lentinus swartzii* (Higher Basidiomycetes). *Asian Journal of Agriculture and Biology, 2021*(2), 1—8.

Balangcod, T. D., & Balangcod, K. D. (2011). Ethnomedical knowledge of plants and healthcare practices among the Kalanguya tribe in Tinoc, Ifugao, Luzon, Philippines. *Indian Journal of Traditional Knowledge, 10*, 227—238.

Boa, E. (2004). *Wild edible fungi: A global overview of their use and importance to people.* UN Food and Agriculture Organization (FAO). https://www.fao.org/3/y5489e/y5489e00.htm.

Bulseco, M. G., Abella, E., & Reyes, R. (2005). Morphogenesis of *Schizophyllum commune*, a wild edible mushroom of Mt. Nagpale, Abucay, Bataan, Philippines. *JNS, 4*(1), 20—28.

Butic, M., & Ngidlo, R. (2003). Muyong forest of Ifugao: Assisted natural regeneration in traditional forest management. In: Advancing assisted natural regeneration (ANR) in Asia and the Pacific. In P. C. Dugan, P. B. Durst, D. J. Ganz, & P. J. McKenzie (Eds.), *Food and agriculture organization of the united nations regional office for Asia and the Pacific.* Bangkok.

Capistrano, J. G., Perez, M. V., Soriano, C. M. H., Sunga, M. K. C., Villajuan, S. G. S., Waing, K. G. D., & De Leon, A. M. (2018). Mycochemical, antioxidant, antibacterial and teratogenic activity of *Tyromyces chioneus* collected from Bambang, Nueva Vizcaya, Philippines. *International Journal of Biology, Pharmacy, 7*(7), 1263—1274.

References

Carrasco, J., Zied, D. C., Pardo, J. E., Preston, G. M., & Pardo-Gimenez, A. (2018). Supplementation in mushroom crops and its impact on yield quality. *AMB Express, 8*(146), 1—9.

Chang, S. T., & Miles, P. G. (2004). *Mushroom cultivation, nutritive value, medicinal effect, and environmental impact* (2nd ed.). CRC Press.

Cheng, C. W., Chen, C. K., Chang, C. J., & Chen, L. Y. (2012). Effect of colour LEDs on mycelial growth of *Aspergillus ficuum* and phytase production in photo-fermentation. *Journal of Photochemistry and Photobiology, 106*, 81—85.

Comandini, O., & Rinaldi, A. C. (2020). Ethnomycology in Europe: The past, the present, and the future. In J. Pérez-Moreno, A. Guerin-Laguette, R. Flores Arzú, & F. Q. Yu (Eds.), *Mushrooms humans and nature in a changing world.* Springer. https://doi.org/10.1007/978-3-030-37378-8_13

Cuevas, M. J., Reyes, R. G., & Kalaw, S. P. (2009). Biophysiology of Lentinus Sajor. *Journal of Tropical Biology, 7*, 48.

Culala, J. M., & Dulay, R. M. R. (2018). Species listing of naturally occurring mushrooms in central Luzon state university, Science city of Munoz, Nueva Ecija, Philippines. *International Journal of Biology, Pharmacy, 7*(10), 1890—1899.

Daep, N. A., & Cajuday, L. A. (2003). Mushroom diversity at Mt. Malinao, Albay. *PSSN Nature News, 2*, 57.

Dapar, M. L. G., & Alejandro, G. J. D. (2020). Ethnobotanical studies on indigenous communities in the Philippines: Current status, challenges, recommendations and future perspectives. *Journal of Complementary Medicine Research, 11*(1), 432—446. https://doi.org/10.5455/jcmr.2020.11.01.51

De Castro, M. E., & Dulay, R. M. R. (2015). Macrofungi in multistorey agroforestry system in Mt Makiling forest Reserve, Los Banos Laguna, Philippines. *Journal of Chemical, Biological and Physical Sciences, 2*, 1646—1655.

De Leon, A. M., Cruz, A. S., Evangelista, A. B. B., Miguel, C. M., Pagoso, E. J. A., dela Cruz, T. E. E., Nelsen, D. J., & Stephenson, S. L. (2019). Species listing of macrofungi found in the Ifugao indigenous community in Ifugao Province, Philippines. *Philippine Agricultural Scientist, 102*(2), 118—131.

De Leon, A. M., Dulay, A. R., Villanueva, A. L., & Kalaw, S. P. (2020a). Optimal culture conditions and toxicity assessment of *Fomitopsis feei* (Fr.): A newly documented macrofungus from the Philippines. *Studies in Fungi, 5*(1), 491—507.

De Leon, A. M., Diego, E. O., Domingo, L. K. F., & Kalaw, S. P. (2020b). Mycochemical screening, antioxidant evaluation and assessment of bioactivities of *Xylaria papulis*: A newly reported macrofungi from Paracelis mountain province, Philippines. *Current Research in Environmental & Applied Mycology, 10*(1), 300—318.

De Leon, A. M., Fermin, S. M. C., Rigor, R. P. T., Kalaw, S. P., dela Cruz, T. E. E., & Stephenson, S. L. (2018). Ethnomycological report on the macrofungi utilized by the indigenous community in the Ifugao Province, Philippines. *Philippine Agricultural Scientist, 101*(2), 194—205.

De Leon, A. M., Kalaw, S. P., Dulay, R. M. R., Undan, J. R., Alfonso, D. O., Undan, J. Q., & Reyes, R. G. (2016). Ethnomycological survey of the Kalanguya indigenous community in Caranglan, Nueva Ecija, Philippines. *Current Research in Environmental & Applied Mycology, 6*, 61—66.

De Leon, A. M., Luangsa-ard, J. J. D., Karunarathna, S. C., Hyde, K. D., Reyes, R. G., & dela Cruz, T. E. E. (2013). Species listing, distribution, and molecular identification of macrofungi in six Aeta tribal communities in Central Luzon, Philippines. *Mycosphere, 4*(3), 478—494.

De Leon, A. M., Orpilla, J. O. V., Cruz, K. V., Dulay, R. M. R., Kalaw, S. P., & dela Cruz, T. E. E. (2017). Optimization of mycelial growth and mycochemical screening of *Lentinus sajor-caju* (Fr.) from Banaue, Ifugao Province, Philippines. *International Journal of Agricultural Technology, 13*(7.3), 2549—2567.

De Leon, A. M., Pagaduan, M. A. Y., Panto, B. E., & Kalaw, S. P. (2021). Species listing of macrofungi found in Paracelis, mountain province, Philippines. *CLSU International Journal of Science & Technology, 5*(2), 22—40.

De Leon, A. M., Reyes, R. G., & dela Cruz, T. E. E. (2012). An ethnomycological survey of the macrofungi utilized by the Aeta communities in Central Luzon, Philippines. *Mycosphere, 3*(2), 251—259.

De Leon, A. M., Reyes, R. G., & dela Cruz, T. E. E. (2013a). Enriched cultivation of three wild strains of *Lentinus tigrinus* (Bull.) Fr. using agricultural wastes. *International Journal of Agricultural Technology, 9*(5), 1199—1214.

De Leon, A. M., Reyes, R. G., & dela Cruz, T. E. E. (2013b). *Lentinus squarrosulus* and *Polyporus grammocephalus*: Newly domesticated, wild edible macrofungi form the Philippines. *Philippine Agricultural Scientist, 96*(4), 411—418.

Devkota, S. (2017). Edibility of wild mushrooms in the context of Nepal: An appraisal of indigenous and local knowledge. In Karki Madhav, Hill Rosemary, Xue Dayuan, Alangui William, Ichikawa Kaoru, & Bridgewater Peter (Eds.), *Knowing our lands and resources: Indigenous and local knowledge and practices related to biodiversity and ecosystem services in Asia* (p. 200). UNESCO. Knowledges of Nature 10.

Dulay, R. M. R., Alcazar, A. A., Kalaw, S. P., Reyes, R. G., & Cabrera, E. C. (2021). Nutritional and physical requirements for mycelial growth and basidiocarp production of *Trametes elegans* from the Philippines. *Asian Journal of Agriculture and Biology, 9*(1), 1—9.

Dulay, R. M. R., Cabalar, A. C., De Roxas, M. J. B., Concepcion, J. M. P., Cruz, N. E., Esmeralda, M., Jimenez, N., Aguilar, J. C., De Guzman, E. J., Santiago, J. Q., Samoy, J. R., Bustillos, R. G., Kalaw, S. P., & Reyes, R. G. (2015). Proximate composition and antioxidant activity of *Paneolus antillarium*, a wild coprophilus mushroom. *Current Research in Environmental & Applied Mycology, 5*(1), 52—59.

Dulay, R. M. R., Cabrera, E. C., Kalaw, S. P., & Reyes, R. G. (2020). Nutritional requirement for mycelial growth of three Lentinus species from the Philippines. *Biocatalysis and Agricultural Biotechnology, 23*, 1—7.

Dulay, R. M. R., Cabrera, E. C., Kalaw, S. P., & Reyes, R. G. (2021). Optimization of culture conditions for mycelial growth and fruiting body production of naturally occurring Philippine mushroom *Lentinus swartzii* Berk. *Journal of Applied Biology and Biotechnology, 9*(3), 17—25.

Dulay, R. M. R., Cabrera, E. C., Kalaw, S. P., Reyes, R. G., & Hou, C. T. (2020). Cultural conditions for basidiospore germination of *Lentinus swartzii* and *Lentinus strigosus* and their morphogenesis. *Asian Journal of Agriculture and Biology, 8*(4), 377—385.

Dulay, R. M. R., & Damaso, E. J. (2020). The successful cultivation of Philippine wild mushroom *Pycnoporus sanguineus* (BIL7137) using rice straw and sawdust-based substrate. *Journal of Applied Biology and Biotechnology, 8*(05), 72—77.

Dulay, R. M. R., Kalaw, S. P., Reyes, R. G., Cabrera, E., & Alfonso, N. (2012). Optimization of culture conditions for mycelial growth and basidiocarp production of *Lentinus tigrinus* (Bull.) Fr., a new record of domesticated wild edible mushroom in the Philippines. *Philippine Agricultural Scientist, 95*(3), 209—214.

Dulay, R. M. R., & Maglasang, C. C. (2017). Species listing of naturally occurring mushrooms in agroecosystem of barangay Bambanaba, Cuyapo, Nueva Ecija, Philippines. *International Journal of Biology, Pharmacy, 6*(8), 1459—1472.

Dulay, R. M. R., Ray, K., & Hou, C. T. (2015). Optimization of liquid culture conditions of Philippine wild edible mushrooms as potential source of bioactive lipids. *Biocatalysis and Agricultural Biotechnology, 4*, 409—415.

Dulay, R. M. R., & Rivera, A. G. C. (2017). Mycelial growth and fruiting body production of Philippine (CLSU) strain of *Polyporus grammocephalus* (BIL7749). *Biocatalysis and Agricultural Biotechnology, 11*, 161—165.

Dulay, R. M. R., Vicente, J. J., Dela Cruz, A. G., Gagarin, J. M., Fernando, W., Kalaw, S. P., & Reyes, R. G. (2016). Antioxidant activity and total phenolic content of *Volvariella volvacea* and *Schizophyllum commune* mycelia cultured in indigenous culture media. *Mycosphere, 7*(2), 131—138.

Garcia, S. (2020). Pandemics and traditional plant-based remedies. A historical-botanical review in the era of COVID19. *Frontiers in Plant Science, 11*, 571042. https://doi.org/10.3389/fpls.2020.571042

Garcia, K., Garcia, C. J., Bustillos, R., & Dulay, R. M. R. (2020). Mycelial biomass, antioxidant, and myco-actives of mycelia of abalone mushroom *Pleurotus cystidiosus* in liquid culture. *Journal of Applied Biology and Biotechnology, 8*(02), 94—97.

Gibbs, P. A., Seviour, R. J., & Schmidt, F. (2000). Growth of filamentous fungi in submerged culture: Problems and possible solutions. *Critical Reviews in Biotechnology, 20*(1), 17—48.

Guerrero, J. J. G., Banares, E. N., General, M. A., & Imperial, J. T. (2020). Rapid survey of macrofungi within and urban forest fragment in Bicol, Eastern Philippines. Austrian Journal of Mycology (Österr. Z. Pilzk), *28*, 37—43.

Guzman, C. D., Baltazar, M. M., Sanchez, A. J., Linsangan, M. G., & Dulay, R. M. (2018). Molecular identification of four wild higher basidiomycetes collected in Mt. Mingan, Gabaldon, Nueva Ecija, Philippines. *Journal of Biodiversity and Environmental Sciences, 13*, 46—51.

Haro-Luna, M. X., Ruan-Soto, F., & Guzmán-Dávalos, L. (2019). Traditional knowledge, uses, and perceptions of mushrooms among the wixaritari and mestizos of Villa Guerrero, Jalisco, Mexico. *IMA Fungus, 10*, 16. https://doi.org/10.1186/s43008-019-0014-6

Jacob, J. K. S., Kalaw, S. P., & Reyes, R. G. (2015). Mycelial growth performance of three species of *Pleurotus* on coconut water gelatin. *Current Research in Environmental & Applied Mycology, 5*(3), 263—268.

Kalaw, S. P., & Albinto, R. F. (2014). Functional activities of Philippine wild strains of *Coprinus comatus* (O. F. Mull.: Fr.) Pers. And *Pleurotus cystidiosus* O.K. Miller grown on rice straw-based substrate formulation. *Mycosphere, 5*(5), 646—655.

References

Kalaw, S. P., Alfonso, D. O., Dulay, R. M. R., De Leon, A. M., Undan, J. Q., Undan, J. R., & Reyes, R. G. (2016). Optimization of culture conditions for secondary mycelial growth of wild macrofungi from selected areas in Central Luzon, Philippines. *Current Research in Environmental & Applied Mycology, 6*(4), 277–287.

Kalaw, S. P., De Leon, A. M., Damaso, E. J., Ramos, J. C., Del Rosario, M. A. G., Abon, M. D., Undan, J. R., Dulay, R. M. R., & Reyes, R. G. (2021). Cultivation of different strains of *Lentinus tigrinus* from selected areas of Luzon Island, Philippines. *Studies in Fungi, 6*(1), 299–306.

Landingin, H. R. R., Francisco, B. E., Dulay, R. M. R., Kalaw, S. P., & Reyes, R. G. (2020). Optimization of culture conditions for mycelial growth and basidiocarp production of *Calocybe cylindracea* (Maire). *CLSU International Journal of Science & Technology, IV*(1), 1–17.

Landingin, H. R. R., Francisco, B. E., Dulay, R. M. R., Kalaw, S. P., & Reyes, R. G. (2021). Mycochemical screening, proximate nutritive composition and radical scavenging activity of *Cyclocybe cylindracea* and Pleurotus cornucopiae. *Current Research in Environmental & Applied Mycology (Journal of Fungal Biology), 11*(1), 37–50. https://doi.org/10.5943/cream/11/1/3

Lazo, C. R. M., Kalaw, S. P., & De Leon, A. M. (2015). Ethnomycological survey of macrofungi utilized by Gaddang communities in Nueva Vizcaya, Philippines. *Current Research in Environmental & Applied Mycology, 5*(3), 256–262.

Lee, H. Y., Ham, E. J., Yoo, Y. J., Kim, E. S., Shim, K. K., Kim, M. Y., & Koo, C. D. (2012). Effects of aeration on sawdust cultivation bags on hyphal growth of *Lentinula edodes. Mycobiology, 4*(3), 164–167.

Licayayo, D. C. M. (2018). Gathering practices and actual use of wild edible mushrooms among ethnic groups in the Cordilleras, Philippines. In A. Niehof, et al. (Eds.), *Diversity and change in food wellbeing — cases in Southern Asia and Nepal* (pp. 71–86). Wegeningen Academic Publisher.

Liwanag, J. M. G., Santos, E. E., Flores, F. R., Clemente, R. F., & Dulay, R. M. R. (2017). Species listing of macrofungi in Angat watershed reservation, Bulacan province, Luzon island, Philippines. *International Journal of Biology, Pharmacy, 6*(5), 1060–1068.

Magday, J. C., Jr., Bungihan, M. E., & Dulay, R. M. R. (2014). Optimization of mycelial growth and cultivation of fruiting body of Philippine wild strain of *Ganoderma lucidum. Current Research in Environmental & Applied Mycology, 4*(2), 162–172. https://doi.org/10.5943/cream/4/2/4

Mendoza, W. C., Dulay, R. M. R., Valentino, M. J. G., & Reyes, R. G. (2020). Mycelial biomass and biological activities of Philippine mushroom *Pycnoporus sanguineus* in time-course submerge culture. *Journal of Applied Biology and Biotechnology, 8*(05), 88–93.

Milenge Kamalebo, H., Nshimba Seya Wa Malale, H., Masumbuko Ndabaga, C., Degreef, J., & De Kesel, A. (2018). Uses and importance of wild fungi: Traditional knowledge from the Tshopo province in the Democratic republic of the Congo. *Journal of Ethnobiology and Ethnomedicine, 14*, 13. https://doi.org/10.1186/s13002-017-0203-6

Nacua, A. E., Pacis, H. J. M., Manalo, J. R., Soriano, C. J. M., Tosoc, N. R. N., Padirogao, R., Clemente, K. J. E., & Deocaris, C. C. (2018). Macrofungal diversity in Mt. Makiling forest Reserve, Laguna, Philippines: With floristic update on roadside sample in Makiling Botanic Garden (MBG). *Biodiversitas, 19*(4), 1579–1585.

Olowa, L. F., Torres, M. A. J., Aranico, E. C., & Demayo, C. G. (2012). Medicinal plants used by the Higaonon tribe of rogongon, Iligan city, Mindanao, Philippines. *Advances in Environmental Biology, 6*(4), 1442–1449.

Pascua, M. S., Kalaw, S. P., & De Leon, A. M. (2016). Proximate composition, mycochemical analysis and antibacterial activity of *Lentinus squarrosulus* (Mont.) Singer. *Advances in Environmental Biology, 10*(3), 58–68.

Ragragio, E. M., Zayas, C. N., & Obico, J. J. A. (2013). Useful plants of selected Ayta communities from Porac, Pampanga, twenty years after the eruption of Mt. Pinatubo. *Philippine Journal of Science, 142*, 169–182.

Rastogi, S., Pandey, M. M., & Rawat, A. K. (2016). Traditional herbs: A remedy for cardiovascular disorders. *Phytomedicine, 23*(11), 1082–1090.

Reneses, M. A. M., Dulay, R. M. R., & De Leon, A. M. (2016). Proximate nutritive composition and teratogenic effect of *Lentinus sajor-caju* collected from Banaue, Ifugao Province, Philippines. *International Journal of Biology, Pharmacy, 5*(7), 1771–1786.

Reyes, R., Eguchi, F., Iijima, T., & Higaki, M. (1997). *Collybia reinakeana*, a wild edible mushroom from the forest of Puncan, Nueva Ecija, Philippines. *Mushroom Science and Biotechnology, 15*(2), 99–102.

Reyes, R. G., Eguchi, F., Kalaw, S. P., & Kikukawa, T. (2009). *Mushroom growing in the tropics: A practical guide.* Central Luzon State University Press. ISBN: 978-971-705-252-6.

Reyes, R., Lopez, L. L. M., Kalaw, S., Kumakura, K., Kikukawa, T., & Eguchi, F. (2009). *Coprinus comatus*, a newly domesticated wild nutraceutical mushroom in the Philippines. *International Journal of Agricultural Technology, 5*(2), 299–316.

Romorosa, E. S., De Guzman, C. T., Martin, J. R. G., & Jacob, J. K. S. (2017). Preliminary investigation on the pharmacological properties of wood-rotting mushrooms collected from Isabela State University, Echague, Isabela, Philippines. *International Journal of Agricultural Technology, 13*(7.3), 2591–2596.

Santiago, F. H., Moreno, J. P., Cázares, B. X., Suárez, J. J., Trejo, E. O., de Oca, G. M., & Aguilar, I. D. (2016). Traditional knowledge and use of wild mushrooms by Mixtecs or Ñuu savi, the people of the rain, from Southeastern Mexico. *Journal of Ethnobiology and Ethnomedicine, 12*(1), 35. https://doi.org/10.1186/s13002-016-0108-9

Tadiosa, E. R., Agbayani, E. S., & Agustin, N. T. (2011). Preliminary study on the macrofungi of Bazal-Baubo watershed, Aurora province, central Luzon, Philippines. *Asian Journal of Biodiversity, 96*(2), 149–171.

Tadiosa, E. R., & Arsenio, J. S. (2014). A taxonomic study of wood-rotting basidiomycetes at the molave forest, san Fernando city, La union province, Philippines. *Asian Journal of Biodiversity, 5*(1), 92–108.

Tadiosa, E. R., & Briones, R. U. (2013). Fungi of Taal Volcano protected landscape, Southern Luzon, Philippines. *Asian Journal of Biodiversity, 4*, 46–64.

Tantengco, O. A. G., & Ragragio, E. M. (2018). Ethnomycological survey of macrofungi utilized by Ayta communities in Bataan, Philippines. *Current Research in Environmental & Applied Mycology, 8*(1), 104–108.

Tayamen, M. J., Reyes, R. G., Floresca, E. J., & Abella, E. A. (2004). Domestication of wild edible mushrooms as nontimber forest products resources among the Aetas of Mt. Nagpale, Abucay, Bataan: *Ganoderma* sp. and *Auricularia polytricha. Journal of Tropical Biology, 3*, 49–51.

Tiniola, R. C., Pambid, R. C., Bautista, A. S., & Dulay, R. M. R. (2021). Light emitting diode enhances the biomass yield and antioxidant activity of Philippine wild mushroom *Lentinus swartzii. Asian Journal of Agriculture and Biology, 2*, 1–8.

Tolentino, J. J. V., Kalaw, S. P., Reyes, R. G., & Undan, J. R. (2016). Mycelial growth performance of three *Pleurotus* species on corn varieties in the Philippines. *Advances in Environmental Biology, 10*(7), 155–160.

Torres, M. L. S., Ongtengco, D. C., Tadiosa, E. R., & Reyes, R. G. (2020). Ethnomycological studies on the Bugkalot indigenous community in Alfonso Castaneda, Nueva Ecija, Philippines. *International Journal of Pharmaceutical Research, 9*(4), 53–54.

Trukhonovets, V. V. (1991). Effect of illumination intensity on the formation of fruiting bodies in *Pleurotus ostreatus* (Jacq. Fr.) Kumm. *Ukrainian Botanical Journal, 48*(2), 67–72.

UNDP. (2010). *Indigenous peoples in the Philippines.* https://www.ph.undp.org/content/philippines/en/home/library/democratic_governance/FastFacts-IPs.html.

CHAPTER

12

Culture collections and herbaria: Diverse roles in mycological research in the Philippines

Marian P. De Leon[1] and Maria Auxilia T. Siringan[2]

[1]Microbial Culture Collection, Museum of Natural History, University of the Philippines Los Baños, College, Laguna, Philippines; [2]University of the Philippines Culture Collection, Microbiological Research and Service Laboratory, Natural Sciences Research Institute, University of the Philippines Diliman, Quezon City, Philippines

1. Introduction

Biological resource centers (BRCs) such as microbial culture collections (MCC), biobanks, herbaria, and natural museums serve vital roles in the preservation, maintenance, and exploration of potential biological resources for a wider array of applications. The Organization for Economic Co-operation and Development (OECD) (2001) defines BRCs *"as infrastructures that serve as service providers and repositories of the living cells, genomes of organisms, and information relating to heredity and the functions of biological systems. BRCs contain collections of culturable organisms (e.g., microorganisms, and plant, animal, and human cells), replicable parts of these (e.g., genomes, plasmids, viruses, cDNAs), viable but not yet culturable organisms, cells and tissues, as well as databases containing molecular, physiological and structural information relevant to these collections and related bioinformatics"*.

BRCs provide access to valuable biological materials (VBMs) with biotechnological applications. These, in hope, aimed on the improvement of the quality of life, ensure sustainable and high-quality crops, and healthy livestocks; food safety and security; early detection, prevention and effective treatment of diseases; reduced dependence on fossils for fuel and energy sources; well-balanced and interpretation of plethora of phenomena and natural occurrences; understanding the association and interaction of organisms and evolution in the natural habitats. BRCs are research infrastructures with manpower and facilities that enable storage, maintenance, and preservation of VBMs and related information and their

Mycology in the Tropics
https://doi.org/10.1016/B978-0-323-99489-7.00001-9

© 2023 Elsevier Inc. All rights reserved.

294   12. Culture collections and herbaria: Diverse roles in mycological research in the Philippines

responsible and secured distribution in aid of research, instruction, information, industrial and biotechnological applications, and bases for policy formulations. Aside from the maintenance of the VBMs, additional functions of the BRCs include the curation and provision of the information including but not limited to phenotypic, genomic, and biochemical characteristics and sequences of VBMs including the profile of the depositors and affiliated institutions, storage, and maintenance conditions. BRCs are also service providers related to the VBMs (Mahilum-Tapay, 2009; OECD, 2001; Wang & Libum, 2009).

In the 16th century, the role of the BRCs was first described when Luca Ghini established a botanical garden and dried specimen collections in a herbarium at the University of Pisa, Italy (Pavord, 2005). This served as a repository and a library of rare or difficult-to-obtain plant specimens that were shared to other scientists including pertinent information (Wang & Libum, 2009). The establishment of the botanical herbarium by Ghini led to the science of taxonomy and encouraged and enabled more herbaria to preserve and maintain specimens of plants, mushrooms, and algae and established museums for insects, wildlife and other taxa including the microorganisms in the succeeding centuries up to present (Pavord, 2005).

2. Microbial culture collections

Microorganisms are ubiquitous and comprise the greatest number of organisms on Earth (Çaktü & Türkoğlu, 2011). These microscopic biological resources that include bacteria, filamentous fungi, yeasts, algae, viruses, and protozoans have been recognized as vital components of the world's biodiversity (Komagata, 1999a, 2000; Mahilum-Tapay, 2009). They can be found even in rare environments where almost none or few organisms can survive such as hyperthermophilic solfatara (Atomi et al., 2004), hot springs (Amo et al., 2002; Itoh et al., 1999), volcanic soils (Oliveros et al., 2021), caves (De Leon, Montecillo et al., 2018; De Leon, Park et al., 2018), marine environments (Paderog et al., 2020), among others.

Microorganisms are harnessed for potential sources of valuable products which have been proven to improve the quality of human lives. They have been explored for new and wide spectrum antimicrobial drugs, enzymes, fermented foods, biofuels, colorants or pigments, vitamins, and mineral sources, and as biocontrol agents and agents for bioremediation and wastes biodegradation (Diaz-Rodriguez et al., 2021; Hawksworth, 1985; Jozala et al., 2016; Malik et al., 1987). The increasing list of the potentials of the microorganisms has also fueled the growing interest on research to explore and attempt to answer some if not all of humans' curiosity and offer substitutes or primary sources for humans' health and food requirements and furtherance on understanding the biological system and its interaction on other life and nonlife forms.

Functions and Types of Microbial Culture Collection. MCC serves as (1) repository or "living library," (2) continuing reference source for research (Malik & Claus, 1987), and (3) preservation and maintenance of microbial strains (Çaktü & Türkoğlu, 2011) with potential patentable products, (4) conservation center of microbial diversity and genetic resources (Daniel & Prasad, 2010; Sharma & Shouche, 2014), (5) distribution center of microbial cultures for researchers, scientists, educators, and industrial partners (Smith, 2012), (6) research facility on microbial systematics, and (7) service providers in generating information for the

III. Fungal conservation and management

microbial resources. Boundy-Mills (2012) stated that the MCCs are the source of "bio" in biotechnology and the "life" in life science research.

Mahilum-Tapay (2009) enumerated the types of MCC. MCC can be a private collection maintained by individual researchers and institutions. The microbial collections are accumulated through conduct of research, teaching, or through diagnostics laboratories, hospitals, and industry. These are usually not open to the public and may be accessed through collaborative research or agreement between individuals or institutions. Specialized collections are those focused on special groups of microorganisms for specialized fields (e.g., fungi for plant pathology, actinomycetes for antibiotics production, lactic acid bacteria for probiotics, and yeasts for brewing). An example of this is the Westerdijk Fungal Biodiversity Institute (https://wi.knaw.nl), formerly known as Centraal Bureau voor Schimmelcultures (CBS), which was established in 1904, with a collection of living fungi and related metadata and leads in the performance of highly innovative mycological research that contributes to the discovery and understanding of fungi and its biodiversity and potential solutions to societal challenges (de Hoog, 1979; Hawksworth, 1985; von Arx & Schipper, 1978). Service or public collections maintain a large number of strains for public distribution and sharing of information and metadata. ATCC (American Type Culture Collection – https://www.atcc.org), a nonprofit organization established in 1925, is an example of a biological materials management and standards organization that provides authenticated, high-quality biological products, advanced model systems, and custom solutions to the global scientific community. It maintains the largest and most diverse collection of biological materials and research solutions in the world (Fig. 12.1).

The global recognition of the roles played by MCCs is also due to its valuable contribution in biotechnology. Biotechnology products of microbial origins are regulated by patent laws of most countries and thus, are required to be deposited in MCC or International Depositary Authority (IDA) recognized by member states for patent purposes (Sharma & Shouche, 2014). This is covered and mandated by an International Treaty designated as the Budapest Treaty, which was duly ratified in Budapest, Hungary in 1977 and is currently administered by World Intellectual Property Organization (WIPO) in Geneva, Switzerland (https://www.wipo.int/). Provisions of this treaty include, "(1) *deposition and designation to an MCC in its*

FIGURE 12.1 ATCC (American Type Culture Collections). Left image—One of ATCC's early headquarters located at 2112 M St, Washington, DC, NW (1954); right image—ATCC in Rockville, MD (1964). *Photos are courtesy of ATCC.*

territory that has requisite staff and facilities to perform scientific and administrative functions as an IDA, (2) requirement for an assurance of the MCCs continued existence, (3) impartiality and availability to any depositor and same conditions, (4) maintenance of live and pure form of microbes following IDA regulations and charges suitable and applicable fees, (5) confidentiality of the VBMs and release following prescribed IDA regulations; and (6) operations of IDAs follow uniform rules and regulations including forms recommended under the Budapest Treaty" (Sharma et al., 2017). The treaty clearly defines the role of the depositary (IDA or MCC) and the depositor. The former maintains the culture in pure and viable form while the latter ensure the specific and unique properties of the patented cultures. As of September 30, 2021, there are 85 countries that signed the Budapest Treaty, including the Philippines. From 2001 to 20, WIPO Statistics Database (2021) recorded a total of 6756 deposits and 950 samples from 28 countries and 48 IDAs (July 2020) as shown in Table 12.1 (https://www.wipo.int/).

History of MCC. The microorganisms are the last group of organisms gathered in a public collection (Wang & Libum, 2009). It was in the late 19th century when Prof. Frantiśek Král of the German University of Prague, Czech Republic (Kocur, 1990; Uruburu, 2003; Wang & Libum, 2009) established the first collection of microorganisms. His expertise on the isolation, cultivation, and maintenance of bacteria appointed him as an Associate Professor of Bacteriology in the same university. Professor Král published the first catalog of microorganisms from a culture collection in 1900 (Uruburu, 2003). Following Prof. Král's death in 1911, the collection was acquired by Professor Ernst Pribham of the University of Vienna who later on issued several catalogs listing the holdings of the collection (Uruburu, 2003). Many of this collection's cultures were subsequently transferred to the ATCC after Pribham's death, but others remained in the collection at Loyola University, Chicago, Illinois. The Vienna portion of the Pribham Collection was largely lost during World War II. After Král's collection, many culture collections were established. Currently, the oldest working collections are the Mycothèque de l'Universitée Catholique de Louvain established in 1894, in Louvain-la-Neuve, Belgium, and the CBS, in Utrecht, the Netherlands, now known as Westerdijk Fungal Biodiversity Institute (CBS) (Samson et al., 2004; Uruburu, 2003), both focusing on preserving and maintaining fungal cultures.

In 1970, the World Federation for Culture Collections (WFCC) emerged from the International Association of Microbiological Societies (IAMS) Section on Culture Collections which was established in 1962 during the First Conference of Culture Collections (Malik & Claus, 1987; Martin, 1963). WFCC was formerly the Committees, Commissions and Federations (COMCOF) and with dual affiliations with the International Union of Microbiological Societies (IUMS) and a scientific member of the International Union of Biological Sciences (IUBS) (Wu et al., 2017). The WFCC *"is concerned with the collection, authentication, maintenance and distribution of cultures of microorganisms and cultured cells. It aims (1) to promote and support the establishment of culture collections and related services, (2) to provide liaison and set up an information network between the collections and their users, (3) to organize workshops and conferences, publications and newsletters, and (4) to ensure the long-term perpetuation of important collections"* (WFCC, 1999). The World Data Center for Microorganisms (WDCM) was established in 1966 as the data center of the World Federation for Culture Collections (WFCC)—Microbial Resource Center (MIRCEN) (Sugawara, 2000; Takishima et al., 1989, p. 253; Wu et al., 2017). The WDCM publishes directories of culture collections and their holdings, maintains databases of member culture collections' holdings and provides integrated information services

TABLE 12.1 International depositary authorities under Budapest Treaty.

Country	Institution	Date status acquired
Australia	National Measurement Institute (NMI)	September 30, 1988
	Lady Mary Fairfax CellBank Australia (CBA)	February 22, 2010
Belgium	Belgian Coordinated Collections of Microorganisms (BCCM^{TM})	March 1, 1992
Bulgaria	National Bank for Industrial Microorganisms and Cell Cultures (NBIMCC)	October 31, 1987
Canada	International Depositary Authority of Canada (IDAC)	November 30, 1998
China	China Center for Type Culture Collection (CCTCC)	July 1, 1995
	China General Microbiological Culture Collection Center (CGMCC)	July 1, 1995
	Guangdong Microbial Culture Collection Center (GDMCC)	January 1, 2016
Chile	Colección Chilena de Recursos Genéticos Microbianos (CChRGM)	March 26, 2012
Czech Republic	Czech Collection of Microorganisms (CCM)	August 31, 1992
France	Collection nationale de cultures de microorganismes (CNCM)	August 31, 1984
Finland	VTT Culture Collection (VTTCC)	August 25, 2010
Germany	Leibniz-Institut DSMZ—Deutsche Sammlung von Mikroorganismen und Zellkulturen GmbH (DSMZ)	October 1, 1981
Hungary	National Collection of Agricultural and Industrial Microorganisms (NCAIM)	June 1, 1986
India	Microbial Type Culture Collection and Gene Bank (MTCC)	October 4, 2002
	Microbial Culture Collection (MCC)	April 9, 2011
	National Agriculturally Important Microbial Culture Collection (NAIMCC)	July 28, 2020
Italy	Ospedale Policlinico San Martino IRCCS	February 29, 1996
	Collection of Industrial Yeasts DBVPG	January 31, 1997
	Istituto Zooprofilattico Sperimentale della Lombardia e dell'Emilia Romagna "Bruno Ubertini" (IZSLER)	February 9, 2015
Japan	International Patent Organism Depositary (IPOD), National Institute of Technology and Evaluation (NITE)	May 1, 1981
	National Institute of Technology and Evaluation, Patent Microorganisms Depositary (NPMD)	April 1, 2004
Latvia	Microbial Strain Collection of Latvia (MSCL)	May 31, 1997

(Continued)

298   12. Culture collections and herbaria: Diverse roles in mycological research in the Philippines

TABLE 12.1   International depositary authorities under Budapest Treaty.—cont'd

Country	Institution	Date status acquired
Mexico	Colección de Microorganismos del Centro Nacional de Recursos Genéticos (CM-CNRG)	August 25, 2015
Morocco	Moroccan Coordinated Collections of Microorganisms (CCMM)	February 20, 2018
Netherlands	Westerdijk Fungal Biodiversity Institute (CBS)	October 1, 1981
Poland	IAFB Collection of Industrial Microorganisms	December 31, 2000
	Polish Collection of Microorganisms (PCM)	December 31, 2000
Republic of Korea	Korean Culture Center of Microorganisms (KCCM)	June 30, 1990
	Korean Collection for type Cultures (KCTC)	June 30, 1990
	Korean Cell Line Research Foundation (KCLRF)	August 31, 1993
	Korean Agricultural Culture Collection (KACC)	May 1, 2015
Russian Federation	All-Russian Collection of Industrial Microorganisms (VKPM)	August 31, 1987
	Russian Collection of Microorganisms (VKM)	August 31, 1987
Slovakia	Culture Collection of Yeasts (CCY)	August 31, 1992
Spain	Colección Española de Cultivos Tipo (CECT)	May 31, 1992
	Banco Español de Algas (BEA)	October 28, 2005
Switzerland	Culture Collection of Switzerland AG (CCOS)	January 16, 2017
United Kingdom	National Collection of Yeast Cultures (NCYC)	January 31, 1982
	National Collections of Industrial, Food, and Marine bacteria (NCIMB)	March 31, 1982
	National Collection of type Cultures (NCTC)	August 31, 1932
	Culture Collection of Algae and Protozoa (CCAP)	September 30, 1982
	CABI Bioscience, UK Centre (IMI)	March 31, 1983
	European Collection of Cell Cultures (ECACC)	September 30, 1984
	National Institute for Biological Standards and Control (NIBSC)	December 16, 2004
United States of America	Agricultural Research Service Culture Collection (NRRL)	January 31, 1931
	American Type Culture Collection (ATCC)	January 31, 1931
	Provasoli-Guillard National Center for Marine Algae and Microbiota (NCMA)	April 26, 2013

Source: WIPO (1999).

III. Fungal conservation and management

for microbial resource centers, microbiologists, researchers, and scientists all over the world (Sugawara, 2000; Takishima et al., 1989, p. 253). To date, there are about 812 culture collections from 78 countries registered in the WFCC WDCM including eight from the Philippines.

In the Philippines, the recognition of the importance and application of microbial resources has been reported as early as the inception of the Bureau of Government Laboratories in 1901. Velasco and Baens-Arcega (1984) stated that the disease-causing bacteria (pathogens), fungi and other microbes were the initial focus of the Bureau of Government Laboratories but the interest shifted toward agriculture microbiology in 1920. The Division of Botany included sections of Plant Pathology and Mycology. Studies were focused on plant pathogens, including those of sugarcane, abaca, citrus, pineapple, and tobacco. Almost 2 decades later, the Division of Industrial Mycology shifted its focus on industrial fermentation focusing on wine fermentation, industrial alcohol, vinegar, *bagoong* or fermented fish, and other fermented products of commercial value. The valuable contribution of microbial products led to the creation of the Biological Research Center in the National Institute of Science and Technology (NIST) (Science Act of 1958), now known as the Department of Science and Technology (DOST) (Velasco & Baens-Arcega, 1984).

Despite the interest in microbial resources, it was only in 1965 when the first MCC was established. This started with the collection of microorganisms of the late Dr. Flordeliz R. Uyengco, Professor of Botany at the University of the Philippines-Diliman (Fig. 12.2). It was formally established in 1971 through a funded project by the National Research Council of the Philippines (NRCP), the oldest council of DOST. The University of the Philippines Culture Collection (UPCC) with WDCM registration number of 310 is based at the

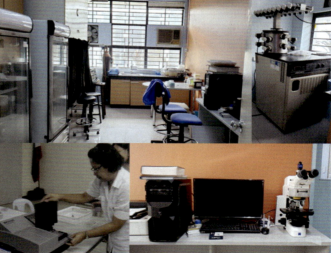

FIGURE 12.2 The University of the Philippines Culture Collection, Natural Sciences Research Institute with Dr. Flordeliz R. Uyengco as the first curator from 1971 to 86. *Photos are courtesy of Dr. Maria Auxilia T. Siringan, UPCC.*

Microbiological Research and Services Laboratory of the Natural Sciences Research Institute (NSRI) of the University of the Philippines Diliman (UPD).

Five years after and through the 877th UP Board of Regents Meeting, the Museum of Natural History (MNH) at the University of the Philippines Los Baños (UPLB), was established with seven biological sections including the MCC (MCC MNH), with Professor William L. Fernandez, Professor of Microbiology at the Institute of Biological Sciences, UPLB, as the first curator (UP Gazette 7(9):191) (Fig. 12.3). In 1984, the MCC of the UPLB MNH became a WFCC affiliate with WDCM registration number of 39, the first MCC in the Philippines to be registered in the WFCC WDCM.

In 1981, the Microbial Culture Collection and Services Laboratory (MCCSL) of the National Institute of Molecular Biology and Biotechnology (BIOTECH) in UPLB was established with Dr. Lourdes Mahilum-Tapay as the first curator (Fig. 12.4). It was elevated in 1996 as the Philippine National Collection of Microorganisms (PNCM) with WDCM registration number 620 (Monsalud, 2010). Since then, several MCCs have been established that provided and distributed microbial cultures for research and instruction as well as services and training on the isolation, identification, characterization and preservation of microbial cultures and other microbiological analyses. These MCCs in the country included the Department of Plant Pathology, UPLB, Culture Collection of Edible Mushrooms (FCUP) (WDCM registration number 103), The United Laboratories (UNILAB) Clinical Culture Collection (WDCM registration number 250), Algal Culture Collection, UPLB MNH (DBUP) (WDCM registration number 444), DOST-Industrial Technology Development Institute Microbial Culture Collection (ITDI-MCC) (WDCM registration number 503) formerly, Biological Research Center (BRC) of the NIST, and the University of Santo Tomas Collection of Microbial Strains (UST CMS) (WDCM registration number 1009) (Takishima et al., 1989, p. 253).

The growing number and increasing requirements for the management and promotion of local MCCs necessitated for a network that will link the MCCs in the country for sharing and

FIGURE 12.3 Microbial culture collection, museum of natural history, University of the Philippines Los Baños, with Professor William L. Fernandez as the first curator. *Photos are courtesy of the UPLB Museum of Natural History and Philippine Society for Microbiology, Inc.*

FIGURE 12.4 The Philippine National Collection of Microorganisms, National Institute for Molecular Biology and Biotechnology, University of the Philippines Los Baños, Laguna, Philippines with Dr. Lourdes Mahilum Nadala as the first curator. *Photo courtesy of Mr. Nik Shawn C. Tabao, PNCM and Dr. Lourdes M. Nadala, Diagnostics for the Real World Ltd., USA.*

harmonizing practices and ensuring quality, safe and secured curation, preservation and distribution of microbial cultures. And thus, the Philippine Network of Microbial Culture Collection (PNMCC), Inc. was established.

Philippine Network of Microbial Culture Collections (PNMCC, Inc.). Established in 1996, the PNMCC, Inc. is a network of culture collections in the Philippines with the PNCM, being the national repository of microorganisms in the country as its headquarters (http://pnmcc.org/). The Network was organized through the DOST-Philippine Council for Advanced Science and Technology Research and Development (PCASTRD) (currently known as Philippine Council for Industry, Energy, and Emerging Technology Research and Development (PCIEERD)). Initially, the PNMCC was composed of the following founding culture collections: Philippine National Collection of Microorganisms (PNCM) based in UPLB, College, Laguna, the UP Culture Collection (UPCC) at the Natural Sciences Research Institute, UP Diliman, Quezon City, the Bacterial Germplasm Collection (BGC) at the

International Rice Research Institute, College, Laguna, and the Microbial Culture Collection, Museum of Natural History (MCC-MNH), also in UPLB, College, Laguna.

The general objectives of the Network are to: "*(1) provide a permanent secretariat for member Philippine microbial culture collections and serve as a central contact point for Philippine scientists and any institution seeking advice and information on microbiological materials and on culture collection-related matters, (2) establish an effective liaison between individuals and organizations concerned with culture collections and among the users of the cultures, (3) collect information on the strains and services offered by the various affiliated microbial culture collections, (4) publicize the microbial resources and technical expertise within the affiliated microbial culture collections through printed and visual materials for distribution, (5) lead in the standardization and upgrading of procedures for the isolation, characterization, conservation, distribution of microorganisms and biosafety through trainings and seminars;* and *(6) maintain and update PNMCC Directory of Strains that include location of, and information about, microorganisms maintained in affiliated microbial culture collections.*" (http://pnmcc.org/). To date, there are nine (9) member affiliate culture collections, including the DOST-Industrial Technology Development Institute Microbial Culture Collection (ITDI-MCC) (WDCM 503), the Department of Environment and Natural Resources (DENR)-Ecosystems Research and Development Bureau Endomycorrhizal (ERDB) Germplasm Collection, the University of Santo Tomas Collection of Microbial Strains (UST CMS) (WDCM 1009), the United Laboratories (UNILAB) Clinical Culture Collection (WDCM 250), the Trinity University of Asia Culture Collection (TUA CC), and the Microbial Culture Collection-Research Institute for Science and Technology, Polytechnic University of the Philippines (MCC-RIST) (Table 12.2). PNMCC is a registered Network of the WFCC.

Member affiliate MCCs are academe-based (UPCC, MCC-MNH, PNCM, UST CMS, TUA MCC and MCC-RIST), industry-based (UNILAB), and government agency-based (DOST-ITDI and DENR-ERDB) (Fig. 12.5). Academe-based collections are mostly resources (microorganisms and their phenotypic and genetic information) generated from completed or on-going research and/or theses of students. In 2005, Monsalud reported that there are about 44 MCCs in the country, consisting of 12 general and 32 specialized collections. Most of the collections are project-based and may not be sustained after the termination or completion of the projects.

These MCCs espouse the cause of the Network and serve as BRCs in the country, providing and facilitating the safe and secured exchange, utilization, and disposal of the VBMs and use of genetic and related information for the advancement of science in the country and improvement of the lives of the Filipino. Membership to the Network can be categorized as follows: (1) ordinary for any individual with a declared interest in culture collections, (2) affiliate for culture collections regardless of size and geographical location represented by the MCC curator, and (3) sustaining for individuals or organizations who espouse the cause of the Network (http://pnmcc.org/). There is no restriction on the number of members from one region or institution.

The PNMCC has also released and disseminated printed and online directory of strains of affiliate member MCCs and newsletters for reference of members and researchers (Monsalud, 2010; http://pnmcc.org/) (Fig. 12.6).

2. Microbial culture collections

TABLE 12.2 PNMCC member affiliate culture collections.

Culture collections	Institution and address	Collections	Year of establishment/ Membership in PNMCC	Reference/s
UP Culture Collection (UPCC)	Natural Sciences Research Institution University of the Philippines-Diliman	Bacteria, yeasts and filamentous fungi Service provider	1971/1996	Monsalud (2010), Monsalud et al. (2012)
Microbial Culture Collection (MCC-MNH)	Museum of Natural History University of the Philippines-Los Baños	Bacteria, yeasts, molds, archaea Service provider	1976/1996	Monsalud (2010), Monsalud et al. (2012)
Philippine National Collections of Microorganisms (PNCM)	National Institute of Molecular Biology and Biotechnology University of the Philippines-Los Baños	Bacteria, yeasts, filamentous fungi, algae Service provider	1991/1996	Monsalud (2010), Monsalud et al. (2012)
Bacterial Germplasm Collection (BGC)	International Rice Research Institute, University of the Philippines Los Baños	Bacteria	___/1996	Monsalud (2010)
Industrial Technology Development Institute Microbial Culture Collection (ITDI-MCC)	Department of Science and Technology	Bacteria, yeasts, filamentous fungi Service provider	1984/1996	Monsalud (2010), Monsalud et al. (2012)
Ecosystems Research and Development Bureau Endomycorrhizal Germplasm Collection (ERDB)	Ecosystems Research and Development Bureau College of Forestry Campus, University of the Philippines Los Baños	Philippine endomycorrhizal isolates	2000/2002	Monsalud (2010), Monsalud et al. (2012), Mojica et al. (2017)
United Laboratories Clinical Culture Collection (UNILAB)	Biological Sciences Dept., Medical Affairs Division United Laboratories, Inc.	Bacterial control strains, human clinical isolates (primarily used safety and efficacy of pharmaceutical and nontraditional products)	1983/2004	Monsalud (2010), Monsalud et al. (2012), Calanasan (2021)
University of Santo Tomas Collection of Microbial Strains (UST CMS)	Research Center for the Natural Sciences Thomas Aquinas Research Complex, University of Santo Tomas	Marine luminous bacteria, clinical bacterial isolates, filamentous fungi and yeasts, microalgae and *Streptomyces* Service provider	2003/2004	Monsalud (2010), Monsalud et al. (2012)

(Continued)

III. Fungal conservation and management

TABLE 12.2 PNMCC member affiliate culture collections.—cont'd

Culture collections	Institution and address	Collections	Year of establishment/ Membership in PNMCC	Reference/s
Trinity University of Asia Culture Collection (TUA CC)	Trinity University of Asia	Bacterial strains (from environmental and clinical samples)	2013/2017	Ho and Cada (2022)
Polytechnic University of the Philippines (MCC-RIST)	Research Institute for Science and Technology Polytechnic University of the Philippines	Bacteria, filamentous fungi, algae	2016/2017	Lirio (2021)

FIGURE 12.5 Charter and active member affiliate MCCs of the Philippine Network of Microbial Culture Collections, Inc. *Logos are courtesy of ITDI-MCC (DOST), MCC-MNH (UPLB), PNCM (UPLB), MCC-RIST (PUP); UNILAB; UPCC (UPD); UST CMS (UST); TUA MCC (TUA).*

III. Fungal conservation and management

FIGURE 12.6   PNMCC newsletter (a) and printed (b) and online (c) directory of strains. *Photos are courtesy of Prof. Carlo Chris S. Apurillo (PNMCC, Inc.) and Dr. Maria Auxilia T. Siringan (UPCC).*

3. Fungal resources of Philippine MCCs

Fungal cultures have been recognized as major microbiological resources for biotechnological processes since the discovery of penicillin (Hawksworth, 1985; Kirsop, 1983). To date, fungi have been widely used for the production of the following, but not limited to: (1) industrially important enzymes, (2) fermented foods, (3) pigments or colorants, (4) antibiotics, (5) vitamins, (6) organic acids, (7) pesticides, (8) mycoprotein, (9) biocontrol agents, and (10) food (Butt & Copping, 2000; Diaz-Rodriguez et al., 2021; El-Gendi et al., 2022). The utilization and contribution of fungal resources in science and technology is limitless. The potential of fungal cultures has been explored through genetic engineering and in addressing environmental concerns, in particular, through bioremediation and symbiotic association with plant roots through mycorrhizae.

Among the MCCs in the country, only the ERDB Germplasm Collection maintains specialized fungal culture collections in the country while most MCCs maintain fungal collections along with bacteria and/or algal collections. The PNCM, the national repository of microorganisms has a total of 1200 fungal cultures with 544 filamentous fungi (molds) and 656 yeasts. Table 12.3 presents the numbers of fungal cultures maintained in PNCM and other affiliated MCCs. Fungal cultures like bacterial and algal cultures are accessed by researchers, instructors, and students both from tertiary and secondary levels and industries.

Preservation of Fungal Cultures. Most MCCs in the Philippines employ short-term and long-term preservation of fungal cultures in their collections, namely: (1) sub-culturing, (2)

TABLE 12.3 Fungal resources of Philippine MCCs.

Culture collection	Fungal cultures	Reference
ERDB	687 endomycorrhizal isolates	Monsalud et al. (2012)
ITDI-MCC	<100 molds and yeasts	Bigol (2021)
MCC-MNH	298 (163 molds and 135 yeasts)	De Leon (2021)
PNCM	1200 (544 molds and 656 yeasts)	Tabao et al. (2021)
UNILAB	<50 yeasts	Calanasan (2021)
UPCC	282 (207 molds and 75 yeasts)	Siringan (2021)
UST CMS	30 (17 molds and 13 yeasts)	https://ustcms.webs.com/

oil overlay, (3) water stock, (4) soil stock, (5) lyophilization, (6) glycerol stock, and (7) cryopreservation.

Short-term preservation involves maintenance of cultures for up to a year while long-term preservation allows more than 10 years of viability of cultures. Most fungal cultures can be maintained for that period by subculturing or periodic transfer. Subculturing is the regular transfer of microorganisms from stock cultures into fresh, sterile medium allowing it to grow continuously. It may include the transfer of culture (hyphal tip or spores) from slant to slant, slant to plate, plate to plate, plate to slant, solid medium to broth, and broth to solid media. It is done to maintain culture in its active form (prolonging life and/or increase the number of cells) for varied applications (Jain et al., 2020). This simple preservation method requires (1) storage of cultures in 5–10°C, (2) routine transfer in appropriate medium, (3) is inexpensive but time consuming and labor intensive, and (4) has no specialized equipment needed and a good option for small collections (Nakasone et al., 2004). Utmost care and selection of medium and incubation and storage conditions must be considered if this method is preferred, in order to prevent these problems: (1) loss, reduction and intensification of sporulation, activity or virulence during the course of successive subculturing on artificial media; (2) changes in the morphological and physiological characteristics of fungus over time; (3) contamination by mites or other microorganisms in between transfer especially when handling large number of samples; (4) drying of culture medium; (5) mislabeling; and (6) loss of cultures due to temperature changes (Ansari & Butt, 2011; Butt et al., 2006; Kirsop & Doyle, 1991, p. 308; Kirsop & Kurtzman, 1988; Kumar, 2020; Shah & Butt, 2005) (Fig. 12.7).

Oil overlay either using sterile paraffin or mineral oil is a low-cost and low-maintenance method for preserving cultures growing on agar slants. High-quality paraffin or mineral oil overlay is a very simple and cost-effective method of preservation of fungal cultures for a longer time even at room temperature. With this method, mycelial, or nonsporulating fungal cultures can be kept for several years or, in exceptional cases, even up to 32 years at room temperature or 15–20°C (Ajello et al., 1951; Booth, 1971). The oil also reduces mite infestations (Nakasone et al., 2004). This long-term preservation technique was first described by Lumilere and Chevrotier in 1914 when they studied gonococci (Hatsell, 1956). It has been modified through time and found to be as effective if not superior to lyophilization of most

III. Fungal conservation and management

FIGURE 12.7 Subculturing (a), overlaying fungal cultures in agar slants with mineral oil (b) and spore and mycelial slurry in cryovial (c). *Photos are courtesy of Dr. Lourdes V. Alvarez, PUP.*

strains of yeasts, bacteria, Fungi Imperfecti and certain strains of algae and basidiomycetes. Sterile and oven-dried paraffin or mineral oil is poured over the slant culture of fungal culture and stored upright at room temperature. The paraffin oil prevents dehydration of the medium, creates anaerobic condition for the growth on the fungal cultures and reduces oxygen tension, thus, slowing down its metabolic activity (Kumar, 2020) and ensures longevity and survival of bacteria, yeasts, basidiomycetes, and other fungi up to 14 years. Hartsell (1956) observed that characteristics of selected fungal strains were maintained when preserved in medium with paraffin oil overlay. This method is recommended for small collections that do not require expensive equipment and requirement for readily available cultures for use (Annear, 1956; Hartsell, 1953, 1956) (Fig. 12.7).

Another simple, inexpensive, low-maintenance and reliable method for storing fungal cultures is to immerse them in sterile distilled water (Castellani, 1939, 1967; Nakasone et al., 2004). Referred to as Castellani's procedure (McGinnis et al., 1974), the method relies on the ability of the water to suppress morphological changes in most fungi. This method has been reported to be effective in preserving ascomycetes, including their mitosporic forms (Johnson et al., 2012), basidiomycetes (Burdsall & Dorworth, 1994; Richter & Bruhn, 1989), ectomycorrhizal fungi (Marx & Daniel, 1976), human pathogenic filamentous fungi and yeasts (McGinnis et al., 1974), oomycetes (Smith & Onions, 1994), and plant pathogenic fungi (Boesewinkel, 1976). Actively sporulating fungal cultures are selected and adequate amounts of spores and pieces of hyphae are suspended in sterile distilled water for storage in vials or any appropriate container. Alternatively, a disk or a portion of the growing colony edge may be cut using sterile scalpel, sterilized straws, or Pasteur pipettes. These are then transferred to sterile cotton-plugged, screw-capped test tubes, or microvials for storage. Sealing vials or tubes with parafilm or stretched cling wrap or film is recommended to prevent water from evaporation. Sterile distilled water can also be poured onto agar slants with actively growing fungus and the spores and mycelia are gently scraped until a slurry mixture is produced. The slurry is carefully transferred into sterile, 2-dram glass vial (or cryovial) or microfuge tubes, caps tightened and/or sealed with parafilm or plastic film/cling wrap and stored at 25°C. An aliquot or a disk can be retrieved and transferred onto fresh medium prior to use (Nakasone et al., 2004).

Sporulating filamentous fungi can be preserved as soil culture. This low maintenance and cost-effective method that uses dry, sterile garden soil or sand can serve as carriers of fungal spores and is found to be appropriate in the preservation of selected groups of fungi. However, there are some disadvantages with the soil culture as a preservation method, such as, longer revival period due to the dormancy of the culture caused by dry condition and changes in the morphology in some fungi (Nakasone et al., 2004). Briefly, sand or loam soil in glass bottles or vials is sterilized by autoclaving for 20 min at 120°C twice. Spore or mycelial suspension is then added to each bottle of soil or sand. After 2—14 days of incubation at room temperature, the soil stock is then kept and stored at 4°C. To retrieve the fungus, a few grains of soil are sprinkled onto fresh agar medium and incubated at appropriate incubation conditions.

Lyophilization, or freeze-drying, is a preferred long-term preservation for many spore-forming fungi. There are several advantages of maintaining fungi as lyophilized cultures, namely: (1) low to medium cost of maintenance, (2) less storage space needed, and (3) ease of handling and transport of the lyophilized cultures (Kumar, 2020). However, it is successful in the preservation of fungi that produce a large number of spores of about 10-µm or less in diameter (Adams, 2007; Nakasone et al., 2004). Larger fungal spores tend to collapse during the lyophilization process, causing irreversible structural damage to the formation of ice crystals during freeze-drying. Rehydration will not be able to restore the morphological traits of the fungi. Lyophilization requires high-quality mechanical vacuum pump set up complete with vacuum gauge and manifold with support stand; cold trap; insulated bath; oxygen-gas torch and oxygen supply; lyophilization ampoules to contain the cell suspension; protective media for suspending the cells which can be skim milk or monosodium glutamate (MSG) (Nakasone et al., 2004). In addition, the lyophilizer must have annual or regular preventive maintenance and calibration to ensure that the machine is in good working conditions.

Freezing (−20° to −80°C) in mechanical freezers is the preferred long-term preservation method next to lyophilization (Smith, 2007). This method retains viability of most fungal cultures for more than 5 years including mitosporic ascomycetes, zygomycetes and yeasts at −20°C; medically important fungi, aerobic actinomycetes, and algae for 6 months to 13 years at −70°C (Pasarell & McGinnis, 1992). Cultures grown on agar slants in bottles or test tubes with screw caps can be placed directly in the freezer (Carmichael, 1956). Cryoprotectants with cell-penetrating capacity, such as, 10%−20% glycerol and 5% dimethyl sulfoxide (DMSO), are frequently added to protect the microbial cells by: (1) lowering the freezing point of water, (2) promoting hydrogen bond formation, (3) vitrification of solvents, and (4) preventing ice crystal formation inside the cells (Kumar, 2020; Smith, 1983). When using this as one of the long-term preservation methods, MCCs should consider the high cost of the biofreezer or ultralow freezer, the high maintenance cost and reliable and continuous power supply.

Cryopreservation methods involve the process of stabilizing biological materials at cryogenic temperatures (Smith, 1983). Ice formation occurs through the cooling process at different rates. Slow cooling allows external freezing to occur before the inside of the cell. This results in osmotic imbalance when water decreases as ice forms. This leads to water migrating out of the cell, which can be detrimental to the cell. On the other hand, when water remains inside the cell, intracellular ice formation and recrystallization after thawing can also be lethal (Mazur et al., 1972). Cryoprotective additives or chemicals are utilized to protect the cells from the detrimental effects of ice formation. Glycerol and dimethyl sulfoxide (DMSO) are the most used cryoprotective agents by minimizing the detrimental effects of increased solute concentration and ice crystal formation. Microbial cells, specifically bacterial cells and yeasts, cultivated in aerated conditions exhibit greater tolerance to the detrimental effects of cooling and freezing in contrast with nonaerated cells. In general, it is observed that a greater number of cells will give higher cell recovery. These cells are obtained during their log phase on fresh agar slants or plates, or for larger quantities, cells are inoculated in broth cultures and are obtained through centrifugation. For spore-forming fungi, spores and spore suspension in fresh medium with cryoprotectant are required for harvesting. Delay in freezing fungal spores should be avoided to prevent spore germination. Special procedure is done prior to freezing for harvesting nonspore forming fungi while agar plugs with mycelia are obtained for fungi with rough mycelia and are placed in fresh growth medium with cryoprotectant (Nakasone et al., 2004). The preservation methods are summarized in Table 12.4.

Other preservation methods that include storage in liquid nitrogen (Smith, 1982), use of organic substrates like wood, cereal grains, agar strips and insect and plant tissues and silica gel (Nakasone et al., 2004), and sawdust-freezing (Kitamota et al., 2002) will not be discussed in this chapter.

To ensure the viability and authenticity of the fungal resources, it is essential that MCCs should implement quality control of the cultures. Quality testing may include some if not all of the following procedures: (1) examination of cultural traits, (2) microscopic examination of cellular morphological traits, (3) phenotypic characterization using conventional and/or automated biochemical and physiological profiling methods, (4) ITS sequencing-based verification of taxonomic ranking, (5) generation of chemical profile using MALDI-TOF mass spectrophotometry, and (6) whole genome sequencing. These should be periodically done and properly documented to monitor the changes in the viability and characteristics of the microbial cultures.

310   12. Culture collections and herbaria: Diverse roles in mycological research in the Philippines

TABLE 12.4 Preservation methods for fungal cultures.

Preservation method	Advantages	Disadvantages
Subculture	Simple, inexpensive, no specialized equipment needed	Routinary, time-consuming, prone to contamination, and loss of activity or viability
Oil Overlay	Simple, cost-effective, low-maintenance, no specialized equipment needed; reduced mite infestation	Recommended for small collections
Water Stock	Simple, inexpensive, low-maintenance, reliable	Evaporation of water
Soil Stock	Cost-effective, low-maintenance	Longer revival period, changes in morphology
Lyophilization	Low to medium cost of maintenance, small storage space, easy handling and transport, efficient for long term preservation	Requires equipment subject to regular preventive maintenance and calibration
Glycerol stock	Efficient for long-term preservation	High cost of the biofreezer or ultralow freezer, high maintenance cost and uninterrupted power supply
Cryopreservation	Efficient for long-term preservation, decreased genetic variability, time saving, reduced labor requirements, reduced pathogenicity test, prevents contamination, high assurance for long-term preservation	High maintenance cost

4. Opportunities and challenges in microbial culture collections

The application of molecular biology in the identification, characterization and maintenance of microbial cultures facilitates the management of MCCs in the country. Faced with scarcity of resources and limited funding, MCCs explored ways to continue its operation and perform its functions. Aside from income generated from services provided and VMBs distribution, sustainability of the MCCs is also dependent on external support coming from local and international funding agencies and collaboration with local and foreign BRCs. There is a need for the standardization and harmonization of policies and guidelines as strict implementation of local and international laws on sample collections and material transfer, access, and use of resources as well as benefit sharing. Aside from biodiversity conservation and phylogenetic and taxonomic studies, biotechnological applications of fungal resources could be harnessed through collaboration between MCCs and the industrial sector. As the resource pool or biobank, MCCs will play a vital role in the advancement of biotechnology in the country. Consequently, a microbial-based bioeconomy may be developed, thus providing revenues for the sustainability of the MCCs in the country.

Authentication of microbial cultures is undertaken and facilitated through the following advanced and emerging technologies such as, (1) rapid or miniaturized phenotype-based microbial identification systems [e.g., Analytical Profile Index or API (bioMérieux, France),

III. Fungal conservation and management

VITEK system, and Biolog identification system (USA)], (2) MALDI-TOF mass spectrophotometer, (3) microbial fingerprinting using fatty acid methyl esterase (FAME), (4) sequencing of specific molecular marker, like ITS region for fungi, (5) DNA-DNA hybridization, and (6) Whole Genome Sequencing.

MCCs are also tasked to establish, maintain, and update databases of their microbial holdings. With the advances in bioinformatics, a simple, secured and reliable database system is essential, useful and helpful to ensure the storage and accessibility of the information and metadata sets of microbial cultures for reference and further analyses. This is aside from the classical and customary use of log-books that have been the practice of most if not all MCCs in the country. However, this classic data recording limits the information stored (e.g., name of microbial culture, accession code, name and affiliation of depositor/s, location, and type of specimen for the isolation and storage conditions) and do not include the sequences, biochemical and physiological characteristics and other pertinent information useful for the user.

MCC databases should include separate fields for (1) strain number or accession code, (2) genus, species and/or variety or subspecies, (3) taxonomic authority, (4) required growth conditions (cultivation medium, incubation and storage temperature, pH), (5) date and locale of isolation, (6) isolation substrate or host, (7) person who isolated the culture, (8) provenance of the strain, (9) accession number/s in other collections that reference the same strain, (10) person who identified the strain, (11) person, if anyone, who has reidentified the strain, (12) subjective synonyms, (13) databases where the sequence were deposited, (14) status of distribution, (15) manner of preservation, and (16) comments that do not fit well into other field mentioned that may include published references pertaining to the microbial culture, metabolites produced, transformations among others (Nakasone et al., 2004).

For database management, MCCs can utilize several available software such as the iCollect (https://www.biotec.or.th/). The iCollect is a software developed by National Center for Genetic Engineering and Biotechnology (BIOTEC) of Thailand (https://www.biotec.or.th/). This software has been designed to support multiple-collections and multiuser environments and can be used on Windows platform via Microsoft VB.Net and MySQL. The software provides an inventory management of biological collections, including microorganisms (http://www.anmicro.org/ & https://www.biotec.or.th/). The software can generate a barcode that contains and facilitate extraction of information regarding the VBM including information, (1) type of sample containers and storage devices, (2) type of samples and collections, (3) sample location in the storage, (4) inflow and outflow of samples, and (5) remaining aliquot or sample. Through the barcode generated, the user can monitor the total or remaining number of VBMs, graphically view the content of all levels of storages, search any information in the collection and track the VBMs for auditing or inventory. It is currently being used at BIOTEC; University of North Texas Health Science Center (Forth Worth, USA) and members of the Asian Network for Microbial Utilization (AnMicro) including Philippines MCCs like the PNCM and UST CMS.

The WDCM of the WFCC spearheaded the use of data technology (Sugawara, 2000; Wu et al., 2017). WDCM provides integrated information services including the Culture Collections Information Worldwide (CCINFO), Global Catalog of Microorganism (GCM), Analyzer of Bioresource Citation (ABC) and Reference Strain Catalog (RSC). CCINFO provides information on 708 culture collections from 72 countries and regions. GCM gathers strain catalog information and provides a data retrieval, analysis, and visualization system of >368,000 strains from 103 culture collections in 43 countries and regions. ABC is a data mining tool extracting strain related publications, patents, nucleotide sequences and genome information from public data sources to form a knowledge base. RSC maintains a database of strains listed in International Standards Organization (ISO) and other international or regional standards. RSC allocates a unique identifier to strains recommended for use in diagnosis and quality control, and hence serves as a valuable cross-platform reference. WDCM provides free access to all these services at www.wdcm.org.

The actual total number of MCCs in the world is still unknown and more institutional or personal MCCs are still undocumented. The initiatives of the WFCC and WDCM in bringing together the collections from the different BRCs are testament to the increasing number of BRCs in the world and the important role of the MCCs in biotechnology and life science research. Though some operate through government or institutional funds or supported by private agencies or foundations, many small collections are still associated and maintained by scientists and researchers in their institutions; utilized in their work and not accessible by the public.

Aside from limited funds for sustainable operation, MCCs in the Philippines faced various challenges such as: (1) engineering controls including facility design, structural integrity and state-of-the-art equipment, (2) sequencing for the authentication and updating of the nomenclature of microbial culture holdings, (3) compliance to biosafety and biosecurity requirements, (4) scarcity of technically competent staff caused by the retirement of curators and other staff, lack of security of tenure, poor staff morale, and movement of staff to other organizations, (5) lack of standardized database for all MCCs that facilitates database integration, (6) power interruption or outages, (7) limited public distribution due to pathogenicity of strains, (8) scarcity of funding and high cost of maintaining laboratory resources (i.e., culture media, laboratory supplies, equipment, etc.), (9) restrictive regulations related to bioprospecting and transporting or shipping of microbial cultures domestic and overseas, and (10) calamities and natural disasters.

Presently, none of the MCCs in the country has been elevated as an IDA for patent deposit of microbial cultures by the local researchers. To be recognized as an IDA, MCCs need to update their policies, procedures, and operations. To address the needs and gaps in the different aspects of culture collection management, the PNMCC organized and conducted several workshops on MCC management; microbial taxonomy and systematics; data management; biosafety and biosecurity and writing proposal and publications in partnership with its MCC affiliate members, WFCC, DOST and other local and international organizations and agencies. The workshops and webinars aimed to encourage institutions in the country to formalize MCCs and be engaged in the harmonization of standards and practices relevant to the maintenance and sustainability of MCCs in the country and making a united stand to protect and conserve the Philippines' diverse and abundant natural bioresources, particularly the microbial resources. Philippine MCCs will remain steadfast in its commitment to provide quality VBMs for research, instruction, extension and industrial applications on the quest for improving and providing better lives for the Filipino and the rest of the world.

5. Mycological herbarium

Mycological herbarium (MH) serves as a repository of preserved fungal specimens representing biodiversity accrued through time. MH supports research and instructions by providing data related to fungal biodiversity, ecology, conservation, biogeography, and their food, environmental, medical and agricultural applications (Väre et al., 2020). The main functions of a herbarium are as follows: (1) documentation of fungal diversity, (2) tracing of genetic information of extinct and extant fungal and oomycete plant pathogens and understand its life history, (3) resolution of disputes about taxonomy or nomenclature of fungi using authenticated herbarium specimens, and (4) determination of causal agents of historical epidemics, landscape ecology and distribution of pathogens in native plants (Ristiano, 2020). Moreover, using the existing collections and information in the herbarium, researchers, scientists and students identify and describe new and unknown specimens, determine solutions and mitigation measures on preventing proliferation of crop pests, chart impact of climate change on spread of invasive species and effects on native population, and examine historical collections for novel metabolic pathways leading to discovery of new antibiotics, traits and factors affecting interactions with other organisms (https://www.kew.org/).

History. One of the largest, oldest, and most scientifically important fungal collections in the world is the Fungarium at the Royal Botanic Garden Kew. It houses an estimated 1.25 million dried specimens reference collection of fungi from all seven continents, *"spanning the entire fungal tree of life and representing well over half of known global diversity"* (https://www.kew.org/). The oldest specimens in the Fungarium date back to the 18th century, including a rich collection of historically significant material, for example, the fungal specimens collected by renowned scientists, John Ray, Charles Darwin, and Alexander von Humboldt. It started in 1879 with the donation of Rev. Miles J. Berkeley's personal collection of around 30,000 specimens (including 6000 type specimens) and since then numerous bequests and donations including collections by its researchers contributed to its increasing collections.

Teodoro presented the historical accounts of the Philippine mycology up to 1935 in his "An Enumeration of Philippine Fungi" (1937) (Santos, 1984). Tadiosa (2012), in his paper on "The Growth and Development of Mycology in the Philippines" published in the Fungal Conservation Journal presented the timeline and interesting mycological discoveries made by American, European, and Filipino mycologists. The interest on the Philippine mycobiota began in the 16th century through the Augustinian friars and botanists who were part of the Spanish expedition. Most of these fungi belonged to the group of the Agaricales and Polyporales, and a few of the Ascomycetes (Quimio, 1986). However, it was through the discovery of Adelbert von Chamisso, a botanist of the Romanzoff Expedition to the Philippines, that the first record of a Philippine fungus, *Sphaeria eschscholzii* Ehrenb. was documented and published by Ehrenberg in Nees von Esenback's Horae Physicae Berolinensis (1820). Later in 1914, this fungus was renamed and reported as *Daldinia eschscholzii* by Rehm in Leaflets of Philippine Botany in 1914 (Teodoro, 1937). Francisco Manuel Blanco, in his Flora de Filipinas, mentioned that there are numerous species in the Philippine islands, and listed one fungus, *Sclerotium subterraneum* Tode, which was named by Graff (Mycologia 8: 254, 1916) as *Xylaria nigripes* (Klotz.) Sacc. As early as the 18th century, several Philippines fungi were discovered, studied, and published by European and American mycologists, aside from Spanish botanists from outside the Philippines. This happened before the American colonization of the Philippine islands, where a more scientific era of fungal collections began (Tadiosa, 2012).

Among those who pioneered fungal studies in the Philippines are residents and nonresident European and American mycologists and botanists who worked in the documentation of Philippine fungi. These included Abbé J. Bresadola, William H. Brown, Cumming, Paul W. Graff, Paul C. Hennings, Clarence J. Humphrey, A.H. Lee, Leveille, Curtis G. Lloyd, George E. Massee, Meyer, William A. Murrill, Narcisse T. Patouillard, Thomas Petch, Franz Petrak, Heinrich Rehm, Charles B. Robinson, Pier'Andrea Saccardo, F.L. Stevens, Hans Sydow, Paul Sydow, Roland Thaxter, Ferdinand Theissen, Warburg, Collen G. Welles, Harry S. Yates, whose works were based on the collections of Charles F. Baker, Mary Strong Clemens, Edwin B. Copeland, Elmer D. Merrill, Adolph Daniel, E. Elmer, and Otto A. Reinking (Quimio, 1986; Santos, 1984; Tadiosa, 2012).

Contemporary Filipino mycologists and plant pathologists have made considerable and important contributions to early mycobiota research in the country. Majority of these scientists were based in colleges and universities offering agriculture programs and from the Department of Agriculture, particularly from the Bureau of Plant Industry (BPI), the Bureau of Science (BS), and the former Department of Agriculture and Commerce (DAC) (Tadiosa, 2012). The Filipino pioneers in mycology include Profs. Nicanor G. Teodoro, Gerardo O. Ocfemia, Martin S. Celino, Victoria M. Ela, E.E. Roldan, and Faustino T. Orillo whose works are mainly in the field of plant pathology, while J. Mendoza and S. Palo of the Bureau of Science in Manila worked on mushrooms. Notable collections of the Filipino mycologists were then deposited in the Bureau of Soils in Manila and the Department of Plant Pathology, University of the Philippines at Los Baños (Quimio, 1986; Santos, 1984).

As early as 1903, Elmer Merrill, an American botanist, established the Bureau of Science Herbarium, the first herbarium in the country under the then Department of Agriculture and Commerce in Manila with a good collection of fungi (https://sciweb.nybg.org/). He left behind a legacy of botanical and mycological collections when he went back to the United States. These collections, unfortunately, were destroyed during World War II. It was through the efforts of Eduardo Quisumbing, Merrill's successor, that the Herbarium was reestablished and the bureau later renamed Philippine National Herbarium, now under the Botany Division of the National Museum of the Philippines (Tadiosa, 2012).

Tadiosa (2012) further mentioned that the mycology and mycological research formally began in 1901, with the establishment of the College of Agriculture (CA) at the University of the Philippines-Los Baños (UPLB), Los Baños, Laguna. Edwin B. Copeland, the first dean of the College of Agriculture and the founder of the University of the Philippines Los Baños (Santos, 1984), together with other early American botany professors taught botany where the study of fungi was incorporated. The creation of the Department of Plant Pathology in the College of Agriculture, under the leadership of its first Chairman, Otto Reinking, enabled Filipino students studying plant pathology to collect and identify fungal specimens from all over the country. The collection of fungi was made possible under the supervision of American professors who were behind the humble beginnings of the College of Agriculture, augmenting the collection during the Spanish period. However, before the first course in mycology was offered by the Department of Plant Pathology, there was already a huge collection of fungi. This collection was deposited in the Baker Herbarium, another herbarium in the country. It was named in honor of the second dean of the College of Agriculture, another botanist, Charles F. Baker. This collection, comprising 25,000 packets, was considered at that time as one of the best herbaria in the entire Indo-Malayan region (Quimio, 1986; Tadiosa, 2012). Some duplicate specimens, especially the types, were deposited in major herbaria of the world, including the Royal Botanic Garden Kew Herbarium (Tadiosa, 2012), the USDA National Fungus Collections, the Singapore Botanical Garden, and the Rijksherbarium, Netherlands (Quimio, 1986). This highlights the golden era of Philippine mycology during the period 1917—41 as documented in Teodoro's Enumeration of Philippine Fungi (1937) and together with the Baker Herbarium, the book bears witness to the grandeur of the Philippine Mycology during the pre-war era. Unfortunately, the Second World War in 1942, completely turned the fruits of all these labor to ashes. It was providential that there are duplicates distributed in herbaria abroad, and thus, bearing witness to these pioneering efforts and placing the Philippines in the annals of mycology. It took more than 20 years later before a new herbarium was rebuilt and the interest in mycology was revived (Quimio, 1986; Quimio & Quimio, 1967).

In 1963, the arrival of an American Peace Corps volunteer mycologist, Don R. Reynolds, greatly sparked the interest in study of fungi (Tadiosa, 2012). He and his students, including one of the distinguished mycologists in the country, the late Dr. Tricita H. Quimio, made extensive fungal collections in the entire Philippine archipelago from the islands of Batanes to the islands of Sulu (Dela Cruz et al., 2009). The number of fungal collections swelled from the Gerardo O. Ocfemia collection of 2200 to almost 6000. A committee was made to undertake the huge task of filing and indexing the growing collection which then was named G.O. Ocfemia Memorial Herbarium. This is a fitting tribute to the first Filipino chairman of the Department of Plant Pathology and the one instrumental to the rebuilding of a new assemblage of fungi in the Philippines (Quimio, 2001a). In 1965, the Herbarium was formally recognized when its establishment was published (Quimio & Quimio, 1967) and was subsequently listed at the Index Herbariorum in 1968, as CALP (College of Agriculture, Laguna, Philippines). The increase of the collections to 6000 added momentum with the arrival in 1967 of Richard Korf of Cornell University and his student Kent Dumont, who generously worked with financial support from the UP-Cornell Education program. They donated some mycology books, provided funds for the acquisition of cabinets and construction of driers, and allotted some funds for the acquisition of herbarium equipment (Tadiosa, 2012).

Quimio (1986) enumerated the first major collections and subsequent studies done by the Filipino mycologists on Clavariaceae (Dogma, 1966), Gasteromycetes (Reynolds, 1967), Myxomycetes (Reynolds, 1981; Uyenco, 1973), lower fungi chytrids (Dogma, 1975), and Discomycetes (Cuevas, 1978; Quimio, 1978). From these initial collections, the herbarium has evolved as the University of the Philippines Los Baños—Museum of Natural History (UPLB-MNH) MH, one of the country's premier depository institutions of fungal collections (Dela Cruz et al., 2009). The establishment of the UPLB MNH in 1976 (UP Gazette 7(9):191) provided, until now, an institution that will safeguard the priceless mycological collections that date back to the 1800s.

Mycological Herbarium Collections. Many collections of the UPLB MNH MH, presently numbering to more than 10,000 packets, were mostly derived from the majestic Mt. Makiling where UPLB is situated (Tadiosa, 2012). Among the country's fungal species, only about 4698 species belonging to 1031 genera are currently named or described (Quimio, 2002). The recent studies on Philippine Basidiomycetes by Tadiosa and Militante (2006), Tadiosa et al. (2007,

2011) substantially added more specimens to these collections. Majority of the related taxonomic studies on macrofungi in the Philippines focused on the general species listings of Basidiomycetes (Musngi et al., 2005). For example, Tadiosa and coauthors (2007, 2011) reported and described the macrofungi collected from different decaying woods from Mt. Cuenca in Batangas and Bazal–Bauso Watershed in Aurora province, updating the estimated list of 3956 fungal species and 818 genera in the Philippines. The studies highlighted the rich fungal diversity of the country, which remains to be explored.

Dela Cruz et al. (2009) discussed the significant Myxomycetes collections of the UPLB MNH MH, consisting of about 446 specimens of myxomycetes, collected and deposited mainly by D.R. Reynolds, I.L. Dogma, and T.H. Quimio. More than 50% of the said myxomycete specimens (282/446) were identified only up to the genus level. Their work enumerated the myxomycetes found as belonging to the following genera: *Arcyria, Ceratiomyxa, Cienkowskiella, Clastoderma, Comatricha, Craterium, Cribaria, Diachea, Dictydium, Diderma, Didymium, Fuligo, Hemitrichia, Lamproderma, Lycogala, Perichaena, Physarella, Physarum, Stemonitis, Trichia,* and *Tubifera*. Specimens identified to the species level included seven species of *Arcyria* and *Trichia*, five species of *Physarum*, four species of *Stemonitis*, three species of *Hemitrichia* and two species for each of *Comatricha, Diachea, Didymium, Lamproderma* and *Lycogala*. These were collected mostly from the Philippines while some from the USA, Ecuador, and Costa Rica. Some date back as old and the earlier than 1900s, i.e., 1839 to 1888.

More recently, Niem and Baldovino (2015) reported an initial checklist of macrofungi in the Cavinti Underground River and Cave Complex (CURCC), Cavinti, Laguna. A total of 507 macrofungi were collected consisting of 41 species that represent 34 genera from 20 families. Most of the species collected belong to the family Polyporaceae which accounted for 29.07% of the total relative fungal density in the collection while the most common genus was *Microporus* with three species. The highest density was recorded for *Daldinia concentrica* (Bolton) Ces. & De Not. (20.43%) followed by *Rigidoporus microporus* (Sw.), *Overeem* (14.93%), *Auricularia auricula* (L.) Underw. (11.00%), *Lentinus tigrinus* (Bull) Fr. (7.86%), *Schizophyllum commune* Fr. (5.89%), *Microporus xanthopus* (Fr.) Kuntze (5.30%), and the rest ranged from 3.93% to 0.20%. This is the first report on the macrofungal composition in CURCC. As of October 31, 2021, UPLB MNH MH has a total of 12,807 specimens in its collection with 97.98% cataloged and 9441 digitized (UPLB Museum of Natural History Annual Report, 2021).

***Preservation Methods for Fungal Herbarium Specimens*.** MH, like the botanical herbarium for plants, serve as a repository of preserved fungi in the form of pressed and dried fungal specimens mounted on a sheet of paper or placed in paper packets or in jars with alcohol or glycerin (Ristaino, 2020). Tadiosa and Soriano (2004) described the procedures from the isolation up to the preservation of fungal specimens as presented in the workbook on Tropical Fungi: Collection, Isolation, and Identification, written by T.H. Quimio in 2001. From this workbook, the following guidelines in the isolation and preservation of fungal specimens can be employed.

Important data of each fungal species should be noted, such as the host tree, substratum, form, texture, size, color, odor, and other noteworthy features at the time of collection (Tadiosa & Soriano, 2004). Specimens are separately wrapped in a newspaper together with the pertinent data and other relevant information. For fragile and fleshy specimens, these should be separated from the woody ones and placed in the same collection basket. The woody specimens should be removed with a *bolo* together with the wood tissues, while

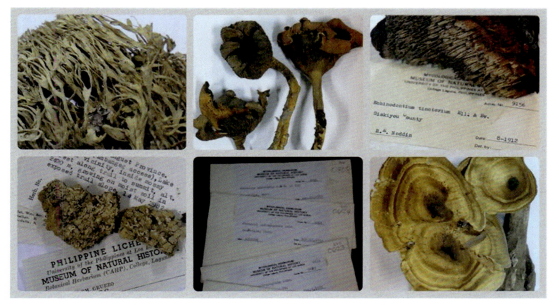

FIGURE 12.8 Fungal collections of the UPLB Museum of Natural History Mycological Herbarium. *Photos are courtesy of Mr. Florante A. Cruz, UPLB MNH.*

a knife must be used in collecting fleshy ones. Woody fungi should be dried immediately and fumigated with paradichloro-benzene crystals and ethanol in an airtight improvised fumigation chamber to kill insects, particularly the destructive larvae.

The fleshy ones and soft specimens are immersed in a 10% formalin solution to make them rigid and firm. After one or 2 days, they will be removed and washed with pure water and then transferred into tightly covered clean jars containing 70% ethanol, where they could be kept and preserved indefinitely. Specimens of small size are placed in paper packets made up of ordinary typewriting paper or in bristol board depending upon the weight and amount of the specimens.

For heavier and bigger fungi like *Ganoderma*, *Fomes* and *Daedalea*, thicker and bigger envelopes should be used as packets. The packets are then pasted on standard mounting sheets. The collected samples should be properly labeled (Tadiosa & Soriano, 2004) and placed with a card catalog containing important information (Fig. 12.8). With recent advances in genomics-based technology, phylogenetic studies of these fungi can be undertaken to establish their taxonomic rankings.

6. Opportunities and challenges in mycological herbaria

The Philippines is an archipelagic country of more than 7000 islands with a combined landmass of 300,780 square kilometers in Southeast Asia (https://www.cepf.net/). The geographic location of the many islands resulted in a number of unique flora and fauna

(Dela Cruz et al., 2009). The country's abundant and rich flora, fauna and fungal resources make it as one of the world's mega hotspots of biodiversity and the second most diverse country in the world in terms of the number of endemic species per land area, second to Madagascar despite being the second-smallest of the 17 megadiversity countries in the world (https://www.cepf.net/). Endowed with diverse habitat types from lowland rainforests and montane forests to long coastlines including the six (6) world's most colorful coral reefs, the Philippines is probably the most biologically diverse country in the world, in terms of unique terrestrial and marine plant and animal species per unit area (https://www.cepf.net/). The country has more endemic species than some larger mega diversity centers, particularly, 45%—60% plant endemism with 193 species listed in the IUCN Red List of Threatened Species (2000) and which were categorized as Critically Endangered, Endangered or Vulnerable).

With about 75,000 species known worldwide and an estimated 1.5 million species present (Hawksworth, 2001), the number of fungal species found in the Philippines is indeed low for a tropical country (Dela Cruz et al., 2009). From the earlier works of Teodoro (1937) until the list of Quimio (2002), the number of fungal species recorded for the Philippines has increased only by 36% in the past 65 years, or with an average discovery rate of 26 species per year. This number may increase with the publications of new or misidentified species and with extensive surveys of published literature or symposium proceedings (Tadiosa, 2012). Considering the reported number of species known worldwide, there are still a lot to be discovered in the vast tracts of tropical habitats, a lot of work to find the missing fungi (Hyde, 2003) before these vanish even before their discovery. The country's fungal diversity remains to be discovered and documented. Exploring these fungi in different habitats and ecological niches will engage researchers and mycologists to assess the present fungal species; to discover fungal diversity in rare environments; to study the application of the fungi and its by-products; to investigate the effect of human activities, climate change and extensive land conversion of natural areas for infrastructure, industrial, agricultural, and urban development on fungal diversity; to establish environmental protection policies and implement conservation efforts to address grave threats and loss of fungal diversity (Niem & Baldovino, 2015; Quimio, 1996; Tadiosa et al., 2011; Tadiosa & Briones, 2013).

Despite the high endemicity of the Philippine flora and fauna, there is a slow and limited progress in fungal biodiversity research in the Philippines. Dogma (1975) and later, reechoed by Dela Cruz et al. (2009, 2013), attributed the limited studies to following factors: (1) lack of personnel equipped and trained to conduct research on mycology, (2) lack of appreciation for mycology in the science curricula of most local universities and colleges and limited only to graduate level or to those subjects with practical applications (e.g., plant pathology, biotechnology, etc.), (3) continuous loss of natural resources, and (4) few MH in the country to preserve and maintain fungal specimens. These observations could also be true for the current state of mycology research in the country. In addition to this, despite the increasing number of grants and funding support from both local and international organizations, support for basic research including fungal systematics or basic mycology, especially in developing countries like the Philippines have been minimal. Funding agencies generally prioritize support to biodiversity research with economic applications or with environmental relevance while basic research, particularly those focusing on conservation and taxonomy, takes a back seat.

Appreciation and commitment to work on mycology may be addressed by offering courses on Basic and Applied Mycology in college and universities. The MH in UPLB started by engaging students enrolled in the primary course in basic mycology. After a decade, two graduate courses in mycology were instituted at the Department of Plant Pathology, resulting in the rehabilitation of the mycological collections from the ashes of the world war. Wide-ranging mycological education is developing in the Philippines. Many state colleges and universities are now offering mycology courses. Several faculty members and students have undertaken research endeavors on fungi, particularly those recognized as critical components of human life in the field of agriculture, forestry, medicine, public health, and industry (Tadiosa, 2012). These studies include fungal products of various kinds, such as enzymes, acids, and alkaloids; fungal plant and animal pathogens; mycorrhizas; biodegrading; edible fungi; and medicinal fungi. Despite these diverse areas and opportunities in fungal research, this still remains unpopular especially among the young scientists as there are few experts in the field of mycology who will serve as mentors. Higher educational institutions (i.e., universities and colleges) can bridge these gaps in mycology education and research as most if not all of the experts are university-based (Guerrero, 2020). This, in part, will enable students to gain access to institutional or personal fungal collections and herbaria for future studies. On one hand, publication and dissemination of new and recent findings on fungal diversity, systematics and biotechnology are available and accessible online, a good venue for disseminating information on mycological research in the country.

Strengthening the capability of researchers and training the next generation of mycologists are strategic and essential in the revitalizing and promoting fungal diversity research and conservation in the country. In the frontline is the Mycological Society of the Philippines (MSP), a professional organization organized on October 19, 1998 at the University of the Philippines Visayas (UPV) in Iloilo City. MSP was formally established on April 8, 1999, at the University of the Philippines at Los Baños, where Philippine mycology as a science and an academic course commenced at the start of the 20th century. MSP continues to be unwavering in its commitment to promote and advance the mycological research and development in the country.

The MH continues to serve its purpose to physically contain the fungal collections and to act as centers for research and instruction. As a repository, fungal collections continue to function as vouchers for identification and as sources of material for systematic work. Herbaria may also house numerous geographic and taxonomic references, that may aid in fungal identification. In addition to housing fungal collections, many herbaria today have initiated computerized data information systems to record and access the collection information of the fungal specimens, as well as to access information from other collections.

7. Concluding remarks and future direction

Southeast Asian countries, including the Philippines, are hotspots of biodiversity; however, micro- and macrofungi remain to be least explored and untapped. Current local research endeavors on fungal diversity in food, rare environments, clinical samples, animal and plants have been generating a considerable number of characterized potential resources with biotechnological potentials as well as references for future discoveries on novel species.

320   12. Culture collections and herbaria: Diverse roles in mycological research in the Philippines

Increasing research on biotechnology and genomics generates a tremendous amount of potential VBMs and metadata for proper curation and preservation in BRCs. There is a need for sustainable and reliable BRCs which will serve as repositories of these resources for future use and applications. Likewise, this is seen as an opportunity for the MCCs and MH to conduct and to participate in research on exploring, documenting and utilization of these biological resources. Expectedly, the involvement of these BRCs in such undertaking will result in the improvement of the operation, increase visibility and recognition of their relevance in the field of science and technology. Ultimately, MCCs and MH will be appreciated as major infrastructures for the advancement of microbiology and biotechnology in the Philippines and in Southeast Asia.

Acknowledgments

The authors would like to acknowledge the numerous authors whose papers are cited in this paper. Likewise, acknowledgment is extended to the support and pertinent materials provided by the following in the completion of this chapter: Dr. Rosario G. Monsalud, Mr. Nik Shawn C. Tabao and Mr. Eldrin DLR. Arguelles of PNCM, UPLB; Dr. Jennifer M. Niem, Mr. Florante A. Cruz, Mr. Alvin N. Fajardo, Ms. Erolyn Anne T. Benitez, Ms. Lyra Joy F. Amit and Ms. Johanna O. Sta. Ana of UPLB MNH; Mr. Juan Paolo A. Aquino of the UPLB Office of the Public Relations; Dr. Lourdes V. Alvarez and Prof. Gary Antonio C. Lirio of the PUP MCC-RIST; Dr. Ursela G. Bigol of the DOST-ITDI MCC; Dr. Gina R. Dedeles and Dr. Bernard John V. Tongol of the UST RCNAS; Dr. Howell T. Ho and Prof. Mary Christine A. Cada of the TUA CC; Mr. Mark Noe C. Ritumalta of the UNILAB; Prof. Carlo Chris S. Apurillo of the PNMCC; Dr. Donna May DC. Papa of the PSM; and Ms. Aina Downing of the ATCC.

References

Adams, G. (2007). The principles of freeze-drying. In J. D. Day, & G. Stacey (Eds.), *Methods in Molecular Biology* (2nd ed., *368* pp. 15–38). Humana Press, 1- 58829-377-7.

Ajello, L., Grant, V. Q., & Andgutzke, M. A. (1951). Use of mineral oil in the maintenance of cultures of fungi pathogenic for humans. *Archives of Dermatology and Syphilology, 63*, 747–749.

Amo, T., Paje, M. L., Inagaki, A., Ezaki, S., Atomi, H., & Imanaka, T. (2002). *Pyrobaculum calidifontis* sp. nov., a novel hyperthermophilic archaeon that grows in atmospheric air. *Archaea, 1*(2), 113–121.

Annear, A. I. (1956). Freeze-drying. III. The preservation of microorganisms. *Laboratory Practices, 5*, 102–105.

Ansari, M., & Butt, T. (2011). Effects of successive subculturing on stability, virulence, conidial yield, germination and shelf-life of entomopathogenic fungi. *Journal of Applied Microbiology, 110*, 1460–1469.

Arx, J. A. Von, & Schipper, M. A. A. (1978). The CBS fungus collection. *Advances in Applied Microbiology, 24*, 215–236.

Atomi, H., Fukui, T., Kanai, T., Morikawa, M., & Imanaka, T. (2004). Description of *Thermococcus kodakaraensis* sp. nov., a well-studied hyperthermophilic archaeon previously reported as *Pyrococcus* sp. *KOD1*. Archaea, *1*(4), 263–267.

Bigol, U. G. (2021). ITDI Microbial Culture Collection. In *Presentation during the meeting of the Philippine Network of Microbial Culture Collections, Inc. and Affiliate Culture Collections*. Philippine Network of Microbial Culture Collections, Inc.

Boesewinkel, H. J. (1976). Storage of fungal cultures in water. *Transactions of the British Mycological Society, 66*, 183–185.

Booth, C. (1971). *Methods in Microbiology* (Vol. 4). Academic Press.

Boundy-Mills, K. (2012). Yeast culture collections of the world: Meeting the needs of industrial researchers. *Journal of Industrial Microbiology and Biotechnology, 39*, 673–680.

Burdsall, H. H., Jr., & Dorworth, E. B. (1994). Preserving cultures of wood-decaying Basidiomycotina using sterile distilled water in cryovials. *Mycologia, 86*(2), 275–280.

Butt, T. M., & Copping, L. (2000). Fungal biological control agents. *Pesticide Outlook, 11*(5), 186–191. https://doi.org/10.1039/b008009h

III. Fungal conservation and management

References

Butt, T. M., Wang, C.-S., Shah, F. A., & Hall, R. (2006). Degeneration of entomogenous fungi. In J. Eilenberg, & H. M. T. Hokkanen (Eds.), *An ecological and societal approach to biological control* (p. 213). Kluwer Academic Press.

Calanasan, C. (2021). The UNILAB Clinical Culture Collection. In *Presentation during the meeting of the Philippine Network of Microbial Culture Collections, Inc. and Affiliate Culture Collections*. Philippine Network of Microbial Culture Collections, Inc.

Carmichael, J. W. (1956). Frozen storage of stock cultures of fungi. *Mycologia, 48*, 378–381.

Castellan, A. (1939). Viability of some pathogenic fungi in distilled water. *Journal of Tropical Medicine and Hygiene, 42*, 225–226.

Castellan, A. (1967). Maintenance and cultivation of common pathogenic fungi of man in sterile distilled water. Further researches. *Journal of Tropical Medicine and Hygiene, 70*, 181–184.

dela Cruz, T. E. E., Dagamac, N. H. A., Torres, J. M. O., Santiago, K. A. A., & Yulo, P. R. J. (2013). Mycology. In V. L. Barraquio (Ed.), *Proceedings of the 22nd Annual Scientific Meeting of the Philippine Academy of Microbiology* (pp. 61–68). Philippine Society for Microbiology, Inc. ISSN: 2345-8453.

dela Cruz, T. E. E., Kuhn, R. V., Javier, A. O. M., Parra, C. M., & Quimio, T. H. (2009). Status of the myxomycete collection at the UPLB-Museum of Natural History (UPLB-MNH) Mycological Herbarium. *Philippine Journal of Systematic Biology, 3*, 97–111.

Cuevas, C. M. (1978). *The taxonomy and morphology of the non-stromatic Discomycetes of the UP Land Grant (Quezon)*. MS thesis University of the Philippines.

Çaktü, K., & Türkoğlu, E. A. (2011). Microbial culture collections: The essential resources for life. *Gazi University Journal of Science, 24*(2), 175–180.

Daniel, H. M., & Prasad, G. S. (2010). The role of culture collections as an interface between providers and users: The example of yeasts. *Research in Microbiology, 161*(6), 488–496.

De Hoog, G. S. (1979). *Centraalbureau voorSchimmekuhures. 75 Years culture collection*. Centraalbureau voor Schimmelcultures.

De Leon, M. P. (2021). The UPLB Museum of Natural History Microbial Culture Collection. In *Presentation during the meeting of the Philippine Network of Microbial Culture Collections, Inc. and Affiliate Culture Collections*. Philippine Network of Microbial Culture Collections, Inc.

De Leon, M. P., Montecillo, A. D., Pinili, D. S., Siringan, M. A. T., & Park, D. S. (2018). Bacterial diversity of bat guano from Cabalyorisa cave, Mabini, Pangasinan, Philippines: A first report on the metagenome of Philippine Bat Guano. *PLOS ONE, 13*(7), e0200095. https://doi.org/10.1371/journal.pone.0200095

De Leon, M. P., Park, A. Y., Montecillo, A. D., Siringan, M. A. T., Rosana, A. R., & Kim, S. G. (2018). Near-complete genome sequences of *Streptomyces* sp. strains AC1-42T and AC1-42W isolated from bat guano of Cabalyorisa Cave, Mabini, Pangasinan, Philippines. *Microbiology Resource Announcements, 7*, e00904–e00918.

Díaz-Rodríguez, A. M., Salcedo Gastelum, L. A., Félix Pablos, C. M., Parra-Cota, F. I., Santoyo, G., Puente, M. L., Bhattacharya, D., Mukherjee, J., & de los Santos-Villalobos, S. (2021). The current and future role of microbial culture collections in food security worldwide. *Frontiers in Sustainable Food Systems, 4*, 614–739.

Dogma, I. J. (1966). Philippine Clavariaceae. The Pteruloid series. *Philippine Agricultural, 49*, 844–861.

Dogma, I. J. (1975). Of Philippine mycology and lower fungi. *Philippine Journal of Biology, 4*, 69–105.

El-Gendi, H., Saleh, A. K., Badierah, R., Redwan, E. M., El-Maradny, Y. A., & El-Fakharany, E. M. (2022). A comprehensive insight into fungal enzymes: Structure, classification, and their role in mankind's challenges. *Journal of Fungi, 8*, 23.

Guerrero, J. J. G. (2020). Insights and prospects toward the undergraduate mycological researches of Bicol University. *Philippine Journal of Science, 149*(2), 405–413.

Hartsell, S. E. (1953). The preservation of bacterial cultures under paraffin oil. *Journal of Applied Microbiology, 1*, 36–41.

Hartsell, S. E. (1956). Microbiological process report. Maintenance of cultures under paraffin oil. *Journal of Applied Microbiology, 4*(6), 350–355.

Hawksworth, D. L. (1985). Fungus culture collections as a biotechnological resource. *Biotechnology and Genetic Engineering Reviews, 3*(1), 417–453.

Hawksworth, D. L. (2001). The magnitude of fungal diversity: the 1.5 million species estimate revisited. *Mycological Research, 105*(12), 1422–1432. https://doi.org/10.1017/S0953756201004725

Ho, H. T., & Cada, M. C. A. (2022). TUA Culture Collection. In *Presentation during the meeting of the Philippine Network of Microbial Culture Collections, Inc. and Affiliate Culture Collections*. Philippine Network of Microbial Culture Collections, Inc. Personal communication.

III. Fungal conservation and management

Hyde, K. D. (2003). Mycology in Asia, the past, present, and future needs. In *Global taxonomy initiative in Asia* (pp. 201–204).

Itoh, T., Suzuki, K., Sanchez, P. C., & Nakase, T. (1999). *Caldivirga maquilingensis* gen. nov., sp. nov., a new genus of rod-shaped crenarchaeote isolated from a hot spring in the Philippines. *International Journal of Systematic and Evolutionary Microbiology, 3*, 1157–1163.

Jain, A., Jain, R., & Jain, S. (2020). Sub-culturing of bacteria, fungi and actinomycetes. In *Basic techniques in biochemistry, microbiology and molecular biology*. Humana, New York, NY: Springer Protocols Handbooks.

Johnson, D., Martin, F., Cairney, J. W. G., & Anderson, I. C. (2012). The importance of individuals: Intraspecific diversity of mycorrhizal plants and fungi in ecosystems. *New Phytologist, 194*, 614–628.

Jozala, A. F., Geraldes, D. C., Tundisi, L. L., Feitosa, V. A., Breyer, C. A., Cardoso, S. L., Mazzola, P. G., Oliveira-Nascimento, L., Rangel-Yagui, C. O., Magalhães, P. O., Marcos de Oliveira, M. A., & Pessoa, A. (2016). Biopharmaceuticals from microorganisms: From production to purification. *Brazilian Journal of Microbiology, 47*(1), 51–63.

Kirsop, B. (1983). Culture collections — their services to biotechnology. *Trends in Biotechnology, 1*(1), 4–8. https://doi.org/10.1016/0167-7799(83)90018-5

Kirsop, B. E., & Doyle, A. (1991). *Maintenance of microorganisms and cultured cells: A manual of laboratory methods*. Academic Press.

Kirsop, B. E., & Kurtzman, C. P. (1988). *Living resources for biotechnology: Yeasts* (p. 234). Cambridge University Press.

Kitamoto, Y., Suzuki, A., Shimada, S., & Yamanaka, K. (2002). A new method for the preservation of fungus stock cultures by deep-freezing. *Mycoscience, 43*(2), 143–149.

Kocur, M. (1990). History of the Kral collection. In L. Sly, & B. Kirsop (Eds.), *100 years of culture collections. Proceedings of the Kral symposium to celebrate the centenary of the establishment of the first recorded service culture collection* (pp. 4–12). Institute for Fermentation.

Komagata, K. (1999a). Microbial diversity and the role of microbial culture collections. In *Invited lecture presented at the International conference on biodiversity and bioresources: Conservation and utilization, 23–27 November 1997, Phuket, Thailand*.

Komagata, K. (1999b). Microbial research centers in Asia. In *Proceedings of the 28th annual convention of the Philippine Society for Microbiology. Palawan, Philippines*.

Komagata, K. (2000). The present and future of the asian network on microbial research. In *Proceedings of the International Workshop on Asian network on microbial research, Bangkok, Thailand. August 28–29, 2000*.

Kumar, M. (2020). *Techniques in microbiology and biotechnology: Maintenance and preservation of bacteria and fungi*. Bihar Veterinary College. Post-graduate Lecture.

Lirio, G. A. C. (2021). Polytechnic University of the Philippines Microbial Culture Collection. In *Presentation during the meeting of the Philippine Network of Microbial Culture Collections, Inc. and Affiliate Culture Collections*. Philippine Network of Microbial Culture Collections, Inc.

Mahilum-Tapay, L. M. (2009). The importance of microbial culture collections and gene banks in biotechnology. In *Biotechnology — Vol. I. Encyclopedia of life support systems (EOLSS)*. http://www.eolss.net/Eolss-sample AllChapter.aspx.

Malik, K. A., & Claus, D. (1987). Bacterial culture collections: Their importance to biotechnology and microbiology. *Biotechnology and Genetic Engineering Reviews, 5*, 137–197.

Malik, K. A., Knobloch, O., & Siefert, E. (1987). A survey of hydrogen production, nitrogen fixation and hydrogen metabolism in *Rhodospirillaceae*. In O. M. Neijssel, R. R. van der Meer, & K. Ch A. M. Luyben (Eds.), *Proceedings of the 4th European Congress on Biotechnology* (Vol. 3, pp. 558–561). Elsevier Science Publishers.

Martin, S. M. (1963). Culture collections: Perspectives and problems. *Proceeding of the First International Specialists Conference on Culture Collections*. Toronto: University of Toronto Press.

Marx, D. H., & Daniel, W. J. (1976). Maintaining cultures of ectomycorrhizal and plant pathogenic fungi in sterile water cold storage. *Canadian Journal of Microbiology, 22*, 338–341.

Mazur, P., Leibo, S. P., & Chu, E. H. Y. (1972). A two factor hypothesis of freezing injury. *Experimental Cell Research, 71*, 345–355.

McGinnis, M. R., Padhye, A. A., & Ajello, L. (1974). Storage of stock cultures of filamentous fungi, yeasts, and some aerobic actinomycetes in sterile distilled water. *Journal of Applied Microbiology, 28*(2), 218–222.

Mojica, C. M., Bonsol, F. J., Dotimas, C. N. M., & Bayot, R. (2017). Adoption of ERDB mycorrhizal technology Hi-Q VAM 1 in the national greening program. *Canopy International, 43*(2), 29–32.

References

Monsalud, R. G. (2010). *The Philippine Network of Microbial Culture Collections*. World Federation for Culture Collections. http://www.wfcc.info/index.php/news/show/387.

Monsalud, R. G., Castillo, E. P., Dedeles, G. R., De Leon, M. P., Siringan, M. A. T., Florento, L. M., Brillante, E. L., Creencia, A. R., Gana, N. H. T., Brown, C. B., Tabao, N. S. C., Ritumalta, M., Argayosa, V. B., & Bulaong, S. S. (2012). *Directory of Strains* (p. 50). Philippine Network of Microbial Culture Collections, Inc.

Musngi, R. B., Abella, E. A., Lalap, A. L., & Reyes, R. G. (2005). Four species of wild Auricularia in Central Luzon, Philippines as sources of cell lines for researchers and mushroom growers. *Journal of Agricultural Technology, 1*(2), 279–299.

Nakasone, K., Peterson, K., Stephen, W., & Jong, S. C. (2004). Preservation and distribution of fungal cultures. In *Biodiversity of fungi: Inventory and monitoring methods* (pp. 37–47). Elsevier Academic Press.

Niem, J. M., & Baldovino, M. M. (2015). Initial checklist of macrofungi in the karst area of Cavinti, Laguna. *Museum Publication Natural History, 4*.

OECD. (2001). The need for biological resource centres. In *Biological resource centres. Underpinning the future of life sciences and biotechnology*. https://www.oecd.org/.

Oliveros, K. M. P., Rosana, A. R. R., Montecillo, A. D., Opulencia, R. B., Jacildo, A. J., Zulaybar, T. O., & Raymundo, A. K. (2021). Genomic Insights into the antimicrobial and anticancer potential of *Streptomyces* sp. A1-08 isolated from volcanic soils of Mount Mayon, Philippines. *Philippine Journal of Science, 150*(6A), 1351–1377.

Paderog, M. J. V., Suarez, A. F. L., Sabido, E. M., Low, Z. J., Saludes, J. P., & Dalisay, D. S. (2020). Anthracycline shunt metabolites from Philippine marine sediment-derived *Streptomyces* destroy cell membrane integrity of multidrug-resistant *Staphylococcus aureus*. *Frontiers in Microbiology, 11*, 743.

Pasarell, L., & McGinnis, M. R. (1992). Viability of fungal cultures maintained at −70°C. *Journal of Clinical Microbiology, 30*(4).

Pavord, A. (2005). *The naming of names: The search for order in the world of plants*. Bloomsbury.

Quimio, T. H. (1978). Species of *Colletotrichum* in the Philippines. *Nova Hedwigia, 28*, 543–553.

Quimio, T. H. (1986). Records of Philippine fungi. *Proceedings of the Indian National Science, 96*, 359–362.

Quimio, T. H. (1996). Agaricales of Mt. Makiling, Laguna, Philippines. In *Proceedings of the Asian international mycological congress, Chiba, Japan* (p. 47).

Quimio, T. H. (2001). *Workbook on tropical fungi: Collection, isolation and identification*. Mycological Society of the Philippines, Inc.

Quimio, T. H. (2002). *Checklist and database of Philippine fungi (1806–2001)*. ASEAN Regional Center for Biodiversity Conservation.

Quimio, T. H., & Quimio, A. J. (1967). The Pathological and mycological herbarium of the UPCA. *Philippines Phytopathology, 3*, 22–26.

Reynolds, D. R. (1967). A key to known Philippine gasteromycetes. *Philippine Agriculturist, 50*, 268–278.

Reynolds, D. R. (1981). Southeast asian myxomycetes II. Kalikasan. *Philippines Journal of Biology, 10*, 1–25.

Richter, D., & Bruhn, J. (1989). Revival of saprotrophic and mycorrhizal basidiomycete cultures from cold storage in sterile water. *Canadian Journal of Microbiology, 35*, 1025–1060.

Ristaino, J. B. (2020). The importance of mycological and plant herbaria in tracking plant killers. *Frontiers in Ecology and Evolution, 7*, 521.

Samson, R. A., van der Aa, H. A., & de Hoog, G. S. (2004). Centraalbureau voor Schimmelcultures: Hundred years microbial resource centre. *Studies in Mycology, 50*, 1–8.

Santos, J. V. (1984). Systematic botany in the Philippines: A perspective. *ASBP Community, 1*(1), 2–21.

Shah, F. A., & Butt, T. M. (2005). Influence of nutrition on the production and physiology of sectors produced by the insect pathogenic fungus *Metarhizium anisopliae*. *FEMS Microbiology Letters, 250*, 201–207.

Sharma, S. K., Saini, S., Verma, A., Sharma, P. K., Lal, R., Roy, M., Singh, U. B., Saxena, A. K., & Sharma, A. K. (2017). National agriculturally important microbial culture collection in the global context of microbial culture collection centres. *Proceedings of the National Academy of Sciences, India Section B Biological Sciences, 89*. https://doi.org/10.1007/s40011-017-0882-8

Sharma, A., & Shouche, Y. (2014). Microbial culture collection (MCC) and international depositary authority (IDA) at national centre for cell science, Pune. *Indian Journal of Microbiology, 54*, 129–133.

Siringan, M. A. T. (2021). The UP Culture Collection. In *Presentation during the meeting of the Philippine Network of Microbial Culture Collections, Inc. and Affiliate Culture Collections*. Philippine Network of Microbial Culture Collections, Inc.

Smith, D. (1982). Liquid nitrogen storage of fungi. *Transactions of the British Mycological Society, 79*, 415–421.

Smith, D. (1983). Cryoprotectants and the cryopreservation of fungi. *Transactions of the British Mycological Society, 80,* 360—363.

Smith, D. (2007). Safe use of low temperatures in the laboratory. *Croner, UK Laboratory Manager Issue, 124,* 6—8.

Smith, D. (2012). Culture collections. *Advances in Applied Microbiology, 79,* 73—118.

Smith, D., & Onions, A. H. S. (1994). *The preservation and maintenance of living fungi* (2nd ed.). IMI Technical Handbooks Vol. 2CAB International.

Sugawara, H. (2000). *World Data Center on Microorganisms.* RIKEN.

Tabao, N. S. C., Arguelles, E. D. L. R., Capanzana, R. M. A. D., Brown, C. M. B., Creencia, A. R., Gana, N. H. T., & Monsalud, R. G. (2021). Philippine National Collection of Microorganisms. In *Presentation during the meeting of the Philippine Network of Microbial Culture Collections, Inc. and Affiliate Culture Collections.* Philippine Network of Microbial Culture Collections, Inc.

Tadiosa, E. R. (2012). The growth and development of mycology in the Philippines. *Fungal Conservation, 2,* 18—22.

Tadiosa, E. R., Agbayani, E. S., & Agustin, N. T. (2011). Preliminary study on the macrofungi of Bazal-Baubo Watersheds, Aurora Province, Central Luzon, Philippines. *AJOB, 2,* 149—171.

Tadiosa, E. R., Arsenio, J. J., & Marasigan, M. C. (2007). Macroscopic fungal diversity of Mt. Makulot, Cuenca, Batangas, Philippines. *Journal of Nature Studies, 6*(1& 2), 111—124.

Tadiosa, E. R., & Briones, R. U. (2013). Fungi of Taal volcano protected landscape, Southern Luzon, Philippines. *Asian Journal of Biodiversity, 27*(4(1)), 46—64.

Tadiosa, E. R., & Militante, E. P. (2006). Identification of important wood-decaying fungi associated with some Philippine dipterocarps at the Makiling Forest. *Sylvatrop, 16*(1—2), 17—37.

Tadiosa, E. R., & Soriano, A. P. (2004). Wood decaying fungi at the Eastern slope of Sierra Madre mountain range, Cagayan Province. In *Proceedings on the Mycological Society of the Philippines (MSP) 6th Annual Scientific Meeting and Symposium.* Laguna: Ecosystem Research and Development Bureau.

Takishima, Y., Shiura, J., Ugawa, Y., & Sugawara, H. (1989). *Guide to World Data Center on Microorganisms with list of culture collections in the world.* World Federation of Culture Collections.

Teodoro, N. G. (1937). *An enumeration of Philippine fungi.* Commonwealth of the Philippines. Department of Agriculture and Commerce. Technical Bulletin 4. Bureau of Printing.

UPLB Museum of Natural History. (2021). *The UPLB Museum of Natural History Annual Report 2021* (p. 38). Laguna, Philippines: University of the Philippines Los Baños, College.

Uruburu, F. (2003). History and services of culture collections. *International Microbiology, 6,* 101—103. World Federation of Culture Collections.

Uyenco, F. R. (1973). Myxomycetes of the Philippines. *Natural Sciences Research Center Technical Report, 12,* 1—23.

Väre, H., Myllys, L., Väinölä, R., Sihvonen, P., Kuusijärvi, A., Ståhls-Mäkelä, G., Kröger, B., Oinonen, M., Juslén, A., Schulman, L., & Hyvärinen, M.-T. (2020). Herbarium collections policy of the Finnish Museum of Natural History. *Research Ideas and Outcomes, 6,* e60470.

Velasco, J. R., & Baens-Arcega, L. (1984). National Institute of Science and Technology 1901—1982: A facet of science development in the Philippines. In *Project of the Philippine Association for the Advancement of Science.* National Science and Technology Authority.

Wang, Y., & Lilburn, T. G. (2009). Biological resource centers and systems biology. *Bioscience, 59*(2), 113—125.

WFCC Executive Board. (1999). *Guidelines for the establishment and operation of collections of cultures of microorganisms* (p. 24). Michael Grunenberg GmbH. http://www.wdcm.nig.ac.jp/wfcc/GuideFinal.html.

WIPO. (1999). *Guide to the deposit of microorganisms under the Budapest treaty.* World Intellectual Property Organization. http://www.wipo.org/treaties/registration/budapest/guide/guide.pdf.

Wu, L., Sun, Q., Desmeth, P., Sugawara, H., Xu, Z., McCluskey, K., Smith, D., Alexander, V., Lima, N., Ohkuma, M., Robert, V., Zhou, Y., Li, J., Fan, G., Ingsriswang, S., Ozerskaya, S., & Ma, J. (2017). World Data Centre for Microorganisms: An information infrastructure to explore and utilize preserved microbial strains worldwide. *Nucleic Acids Research, 45*(D1), D611—D618.

Further reading

Smith, D. (2011). Biological resource research infrastructures to drive innovation in microbiology. *Microbe, 6,* 482.

Watanabe, I., Roger, P. A., Ladha, J. K., & Van Hove, C. (1992). *Biofertilizer Germplasm collections at IRRI* (p. 74). International Rice Research Institute.

CHAPTER

13

Fungal plant pathogens of quarantine importance in the Philippines

Lilia A. Portales[1], Jonar I. Yago[2] and Amor C. Dimayacyac[3]

[1]Bureau of Plant Industry, National Plant Quarantine Services Division, Manila, Philippines; [2]Bureau of Plant Industry, Office of the Assistant Director for Operations, Technical and Production Services, Manila, Philippines; [3]Bureau of Plant Industry, Central Post-Entry Quarantine Station, Los Banos, Laguna, Philippines

1. History, plant quarantine protocols and current policy

By virtue of Presidential Decree 1433 or the Plant Quarantine Law of 1978, as amended, the Bureau of Plant Industry National Plant Quarantine Services Division (NPQSD) is mandated to protect the Philippine Agriculture by preventing the introduction of foreign pests and further spread of pests present in the country. From 1922 to 1930 to 1978, the Philippine Plant Quarantine operated under Republic Acts 3027 known as Plant Quarantine Act and 3767 or the Agricultural Pests Quarantine Act.

Currently, NPQSD has twenty-three (23) center stations and twelve (12) substations nationwide. The NPQSD-Central Office, being the center of Plant Quarantine (PQ), is composed of four different sections namely Import, Export, Domestic, and Administration. NPQSD is the National Plant Protection Organization (NPPO) of the Philippines and member of the International Plant Protection Convention (IPPC). It is committed to work with international and local organizations in establishing standards and effective Sanitary and Phytosanitary (SPS) Measures on imported and exported agricultural commodities.

The organization works with national and provincial government and local authorities for effective enforcement of PQ laws and regulations and building of technical expertise in managing pest risks. NPQSD also works with the private sector (farmers/producers, traders, exporters, and importers) to be globally competitive and provide effective SPS practices based on International Standard for Phytosanitary Measures (ISPM), and proper linkage to other NPPOs to facilitate market access. In addition, NPQSD links with the public by encouraging them to do their responsibility in managing and protecting plant resources for a better future of Philippine Agriculture.

Mycology in the Tropics
https://doi.org/10.1016/B978-0-323-99489-7.00013-5

© 2023 Elsevier Inc. All rights reserved.

The following are the major activities of NPQS implemented within the mandate of the PQ Law, International Conventions, and Agreements:

1. Plant Pest Surveillance Network
   a. Coordinates with Regional PQ Stations for nationwide detection and monitoring of pests on agricultural commodities for export and import.
   b. Conducts port inspection of all carriers (air/sea vessels), agricultural cargos, and crew/passenger luggage, postal mails for pest and illegal movement or entry of smuggled agricultural commodities.
   c. Monitors on spread/dispersal of regulated pest from infested to endangered pest-free production areas.
   d. Maintains Regional Diagnostic Laboratories to support services in the identification of intercepted potential or quarantine pest at ports of entry or exit.
2. Agricultural Commodity Treatments
   a. Supervises treatments on plants, plant products, and other materials of plant origin for export or imports (e.g., vapor heat treatment, hot-water treatment, gas fumigation, pesticide application).
   b. Develops commodity treatment schedules/protocol on exported plants or plant products to mitigate the risk of introduction of regulated pest to trading countries.
3. PQ Clearances
   a. Conducts risk assessment, for plant pest on agricultural commodities (plants, plant products, and materials of plant origin) intended for importation into the country.
   b. Regulates the transport or movement of particular agricultural commodity that may harbor quarantine pest and may introduce or spread to other areas in the country.
   c. Provides phytosanitary certification of agricultural commodities based on international standard for export.
   d. Provides establishments, declarations, and maintenance of "Pest-Free Production Areas or Sites" in the country based on rules set by the ISPM.
4. Licensing, Accreditation, and Registration Services
   a. Provides official authorizations on plant production sites in the country and overseas (e.g., backyards, farms, orchards, nurseries) which will ensure export or import requirements meet quarantine standards.
   b. Provides official registrations on plant processing, treatment, and storage facilities in the country and abroad (e.g., vapor-heat, hot-water, gas fumigation chambers, cold treatment, and storages) that meet the international standards set for agricultural export or import.
   c. Provides official recognition of importers and exporters in the country or abroad that ensures compliance with the Philippine quarantine and overseas countries import/export requirements.
5. Technical Support and Information Services
   a. Coordinates with other government agencies, departments, Local Government Units, private sector for emergency and long-term management of particular quarantine pest for eradication, suppression or containment.
   b. Provides variety of pest data, records, documents, pest lists, and newsletters for public utilization.

1.1 Quarantine protocols

International and domestic trade of plants, planting materials, and plant products require phytosanitary documents to ensure that the consignment has undergone inspection and complied with the phytosanitary requirements. Presidential Decree 1433 is further strengthened with the issuance of department circulars, quarantine, administrative, and special orders to provide regulatory guidelines to the PQ personnel in the execution of functions to prevent introduction, establishment, and further spread of pests.

1.1.1 Importation

The Department of Agriculture (DA) Department Circular 4 Series of 2016 also known as the "Guidelines on the Importation of Plants, Planting Materials, and Plant Products for Commercial Purposes" was issued as standard for the importation of agricultural commodities and products of plant origin. To ensure that the products to be imported meet the standards to protect human, animal or plant health, a Sanitary and Phytosanitary Import Clearance (SPSIC) must be secured from the Bureau of Plant Industry (BPI) prior to importation. However, not all commodities are immediately allowed to enter the country since pre- and post-shipment conditions have yet to be established through the conduct of Pest Risk Analysis. The duration of this process is not determined since this would depend on the number of pests associated with the commodity that need to be evaluated. Importers also need to apply for a license to operate especially if the importation would be on a regular basis.

Inspection of the commodity will be conducted at the port of entry. Some commodities may need Phytosanitary Certificate from the country of origin and/or treatment depending on the category of the commodity as classified in DA Quarantine Administrative Circular 1, Series of 2014, "Guidelines for Categorization of Commodities of Plant Origin."

1.1.2 Domestic transport

Domestic movement of agricultural plants and plant products are also regulated to prevent spread of pests from one place to another. Special Quarantine Orders are being issued to declare areas under quarantine to contain a pest in a certain locality.

Clearance for Domestic Transport (CDT) must be secured every shipment. Additionally, all commodities applied for transport must be presented to the PQ Officer/Inspector for inspection to verify if the commodities are free from pests and are not prohibited to be transported in the applied place of destination.

In 2018, three manuals were drafted and approved for implementation both for import and domestic quarantine. These include the Pest Risk Analysis Manual which is designed to provide practical guide to risk assessors of the National Plant Quarantine Services Division in conducting import risk assessments; Procedural Manual in the Conduct of Phytosanitary Import Inspection as guide for the PQ personnel in conducting inspection at the ports of entry and post-entry facilities of consignments of plants, plant products, and other regulated articles; Procedural Manual on Domestic Quarantine as practical reference to PQ Officers and Inspectors in the physical inspection, interceptions/confiscations, issuance of clearances, and other activities relative to domestic movement of agricultural commodities.

2. Fungal pathogens with high impact in Philippine agriculture

2.1 *Phytophthora* on coconut

In the Philippine Islands, coconut bud rot was first studied in 1908 when a serious outbreak of this disease occurred in the highland towns of Laguna province. It was caused by the oomycete, *Phytophthora palmivora* (E.J. Butler) E.J. Butler characterized by wilting and yellowing of the youngest leaves. The disease had spread very rapidly, and the coconut planters were very much alarmed, because their plantations were threatened with destruction.

Reinking (1919) showed that the cause of infected bud rot in Laguna Province is the oomycete called *Phytophthora faberi* Maubl. The fungus-like organism that causes coconut bud rot causes infection by means of the spores and mycelia. It is not known how the oomycete infects the bud of the coconut. Perhaps the spores are carried by wind or insects causing injuries. The oomycete also causes damping-off and seedling blight of cacao, canker of Para rubber, root rot and fruit rot papaya, seedling, and shoot blight of various species of citrus and seedling and shoot blight of *Cinchona*, the quinine plant.

Whenever coconut bud rot occurs, extermination or eradication is recommended as the most practical measure. This measure calls for the immediate cutting down of infected trees and complete burning of the crowns. In addition to the measure of exterminating the disease, every coconut grove should be inspected frequently. The aim of this inspection is to prevent the spread of bud rot from an occasional infection.

In 1992, the Philippine Coconut Authority (PCA) conducted nationwide surveys to determine the extent and damage caused by the pathogen, *Phytophthora palmivora* (formerly *P. faberi*). These were done in the main island groups of Luzon, Visayas and Mindanao involving two (2) methods: consolidation of incidence reports from PCA Regional offices and direct farm visits. The highest incidence was observed in Mindanao and mostly in areas planted to Malayan Yellow Dwarf X West Africa Tall (MAWA) hybrid. One of the recommendations stated was to adopt proper cultural management and proper disposal of infected palms and plant parts to eliminate possible sources of pathogen and control the spread of disease.

2.2 Panama disease of banana

Panama disease of banana or *Fusarium* wilt is a typical vascular wilt disease. The fungus invades the vascular tissue through the roots causing discoloration and wilting, eventually killing the plant. The progress of the internal symptoms can influence first appearance of the external symptoms which is the nearly yellowing/drying up of the leaves, eventually resulting to a skirt-line pattern drop. Early attack of disease on the plant results to failure to produce fruits. Infected plants failed to bear fruits or if fruits are borne, they are relatively small in size or in limited number per bunch.

The characteristic internal symptom of *Fusarium* wilt is vascular discoloration, which varies from pale yellow in the early stages to dark red or almost black in later stages. Internal symptoms first develop in the feeder roots, which are the initial infection sites. The fungus spreads to the rhizome and then the pseudostem. Externally, the first signs of disease are usually wilting and yellowing of the older leaves around the margins. The yellow leaves may remain erect or collapse at the petiole. Sometimes, the leaves remain green, except for spots on the petiole, but still snap. The collapsed leaves hang down the pseudostem like a skirt. Eventually, all the leaves fall and dry up (https://www.promusa.org/Fusarium+wilt).

Fusarium wilt of bananas caused by *Fusarium oxysporum* f. sp. *cubense* (Foc) is considered one of the most destructive banana diseases in history (Stover & Simmonds, 1987). In the Philippines, Foc Tropical Race 4 (TR4) had been reported in Mindanao since 2009 and affected a range of banana production systems located in the area. The production systems were connected as adjacent farms or via the movement of people, banana fruits or leaves, vehicles, and farm tools and machinery (Ploetz, 2006). In 2015, Foc TR4 affected an area of around 15,500 ha of banana production in Davao Region, which accounted for 32% of the total Cavendish production area (Table 13.1).

The alarming effect of Fusarium wilt devastation on "Cavendish" banana in Mindanao, Philippines in 2011 led to the implementation of programs with the concerted effort of the private and public sector. The BPI and the NPPO of the Philippines issued the Administrative Order to restrict the movement of infected plant materials from Mindanao. From the national Administrative Order, different quarantine control strategies were implemented in localities and banana plantations with infected farms. Awareness campaigns were also conducted in major banana growing areas in the country. Different information, education and communication materials were developed and distributed in strategic areas.

Due to the limited control options for TR4, the most effective form of management is prevention. Therefore, it is important to limit as much as possible the translocation of the pathogen to uninfected areas via infected hosts, soil, machinery and tools. In countries at risk, the NPPOs should make sure that TR4 is on their national list of quarantine pests, and that there is a monitoring system for identifying incursions (https://www.cabi.org/isc/datasheet/59074053).

2.3 Downy mildew on corn

Peronosclerospora philippinensis (Weston) Shaw causes a systemic infection wherein intense green and yellow stripes are observed along the entire leaf. Thick, white, woolly growth of

TABLE 13.1 Areas infected with *Fusarium oxysporum* f. sp. *cubense*, TR4.

Location	Area (ha) planted with Cavendish	Area (ha) affected with TR4	Percentage area affected with TR4
Davao Region	48,050	15,507.53	32.27
Davao del Norte	28,972	13,743.00	47.44
Davao del Sur	3642	436.00	11.97
Davao Oriental	156	36.00	23.08
Compostela Valley	11,934	1082.78	9.07
Davao City	3346	209.75	6.27
TOTAL	**96,100**	**31,015.06**	

Source: Department of Agriculture, Philippine Statistics Authority.

conidia and conidiophores can be observed underneath these areas (Gupta & Paul, 2002; Magill et al., 2006). These symptoms may appear as early as 3 days after infection. As the disease progresses, leaves become narrow and abnormally erect, and in some cases, the affected leaves can appear somewhat dried out. Tassels may also exhibit malformation hence, interrupting ear formation and sterility of seed (Magill et al., 2006). Diseased stems do not show external symptoms but are usually stunted. Movement of infected planting material is the primary reason for the introduction of a disease into a disease-free environment. Quarantine measures should be strictly enforced to prevent the entry of a disease in an area (Sugar Research Australia, undated).

Until about 1980, the annual Philippine loss in yield due to Philippine corn downy mildew caused by *P. philippinensis* was estimated at 205,470 metric tons. The low yields are mainly due to several interrelated causes of which damage by diseases particularly downy mildew was one.

The breakthrough in the chemical control of Philippine corn downy mildew with the use of Apron 35 SD has answered the long-sought solution to this most destructive disease of corn in the country. As a seed-dressing fungicide, the use of Apron 35 SD apparently agrees with the common belief that the more practical way of controlling diseases of cereals, such as corn, by chemical means is by seed treatment. Its effectiveness in controlling corn downy mildew, the ease by which this chemical is applied, and the small amount needed to sustain 100% control will surely make its application acceptable and adaptable by farmers.

2.4 South American leaf blight (SALB) on rubber

South American leaf blight (SALB) is the most serious disease of the rubber tree due to its devastating effects. Historically, SALB had destroyed several rubber plantations established in the 1930s in Central and South America. The Asian growing countries that produce more than 90% of the world's natural rubber are very concerned of the threat of SALB. This is because the climatic conditions in the major rubber producing countries are conducive to serious SALB infection (FAO, 2012).

SALB is caused by *Pseudocercospora ulei* (formerly named *Dothidella ulei* Henn. and *Microcyclus ulei* (Henn.) Arx). It infects young leaflets, those up to 12 days old being most susceptible, but thereafter becoming increasingly resistant. Lesions appear about 4—10 days after infection, on which conidia are produced and young diseased leaves are often shed. Severe conidial infections, which cause fall of young leaflets and increase the quantity of conidial inoculum, are known as the 'exploding phase' of the disease. This is the most important phase of SALB and leads to the physiological debilitation of the trees. Older leaflets that become infected do not suffer premature fall and remain on the tree. 30—60 days after infection, they develop black stromata ('sandpaper' symptoms) on which the spermogonial phase of the pathogen develops followed by the sexual, ascospore stage. After they have reached four or 5 years of age, rubber trees change their leaves. Ripe diseased leaves with 'sandpaper' symptoms, whether on the ground or still on the plants, provide ascospores as the primary source of inoculum. These are spread by wind and reach the young leaflets, where they germinate, penetrate, and colonize the tissue. Under favorable conditions for the disease, within 5 or 6 days, the infected leaflets reveal lesions covered with conidia that may also be spread by wind or rain (CABI).

The Food and Agriculture Organization (FAO) realized the importance of SALB particularly in Southeast Asia. The Philippines has been participatory of the regional trainings and workshops on SALB since 2011. In November 2011, two (2) participants from the BPI participated in a workshop that developed the training program and reference materials for protection against SALB of rubber. In July 2012, FAO conducted a workshop on Training of Trainers (TOT) on Protection against SALB of Rubber in the Asia–Pacific Region. In 2015, FAO again conducted a Workshop on Biosecurity Implementation of SALB of Rubber in the Asia–Pacific Region and in December 2018, another workshop was held on TOT for SALB. These series of training workshops paved the way to revise and update the Administrative Order No. 13, series of 1949- Regulating the Importation of Rubber Plants and Parts Thereof such as Seeds, Rubber Stumps, Budsticks, etc. from Central America in Order to Prevent the Introduction into the Philippine Islands of a Disease Known as *Dothidella ulei*, Except for Certain Purposes and Under Certain Conditions. The result was the DA Administrative Order No 03, series of 2013- Regulating the Importation of Rubber (*Hevea* sp.) Plants and Parts Thereof Such as Seeds, Stumps, Budsticks, Etc., from South American Leaf Blight (SALB) Endemic Countries in Order to Prevent the Introduction into the Philippines, the South American Leaf Blight of Rubber, a Disease Caused by *Microcyclus ulei* syn. *Dothidella ulei*, Except for Certain Purposes and Under Certain Conditions.

3. Practices in controlling entry, exit and epidemics

The BPI is the NPPO of the Philippines and is a signatory to the World Trade Organization—Agreement on the Application of Sanitary and Phytosanitary Measures. As such, it is committed to exercise more stringent quarantine measures to safeguard the Philippine agriculture from foreign pests and prevent its further spread in the country.

The National Plant Quarantine Services Division of the BPI is the regulatory arm of the Philippines' Department of Agriculture. It seeks to provide policies on matters related to import, export, domestic movement and market access of plants and plant products to strengthen its function to prevent the entry of foreign pests into the country, prevent spread of pests already existing in the country and comply with the phytosanitary requirements of the importing countries.

3.1 Preventing pest incursions

The BPI, by virtue of Presidential Decree No. 1433, also known as the PQ Law of 1978, as amended, considers the risk of introduction of foreign pests into the country through traded commodities. Formulation of phytosanitary requirements to prevent entry of pests through the importation of these commodities is based on SPS Article 5.1 which states that "Members shall ensure that phytosanitary measures are based on an assessment of risks taking into account risk assessment techniques developed by the relevant international organizations." To comply with this requirement, the BPI has crafted the BPI Quarantine Administrative Order No. 1 Series of 2018, Pest Risk Analysis Manual following the principles and standards outlined in ISPM No. 2, Framework for pest risk analysis and ISPM No. 11, Pest risk analysis for quarantine pest including analysis of environmental risk and living modified organisms. Commodities with no import records are subjected to Pest Risk Analysis to generate technically justifiable phytosanitary requirements.

Sanitary and Phytosanitary Import Clearance is required for commodities classified under Category 2, 3 and 4 (DA Department Circular No. 4 s. 2016) and shall be secured prior to importation. Likewise, a Phytosanitary (Plant Health) Certificate issued by the NPPO of the country of origin stating that the materials are free from plant pests is a common phytosanitary requirement to accompany every shipment of materials capable of harboring plant pests.

Conduct of phytosanitary import inspection is done to verify exporting country's compliance with the import requirements of the Philippines and prevent the entry, establishment and spread of pests and regulated articles. The BPI, as the IPPC-recognized NPPO in the Philippines is responsible for the inspection of consignments of plants and plant products moving in international traffic and, where appropriate, the inspection of other regulated articles, particularly the object of preventing the introduction and spread of pests (Article IV.2 (c)). Inspection procedures for imported consignments are defined in several PQ policies conceived by BPI, such as BPI Quarantine Administrative Order No. 1 s. (1981), Rules and Regulations to Implement Presidential Decree No. 1433; Department Circular No. 4, s. (2016), Guidelines on the Importation of Plants, Planting Materials and Plant Products for Commercial Purposes and BPI Quarantine Order No.02, s (2018), Procedural Manual in the Conduct of Phytosanitary Import Inspection. Along with the inspection of imported materials, document verification, consignment identity and integrity checking are also done.

As described in BPI Quarantine Administrative Order No. 1 s. (1981), "imported plants, plant products and other materials found infected/infested with plant pests shall be subjected to a prescribed commodity treatment or destroyed or returned to the country of origin at the expense of the importer."

Post-entry quarantine is undertaken to screen imported materials for pests which are difficult to detect by inspection at the points of entry. This mitigation measure ensures safe movement of plants and planting materials and prevents the introduction of injurious plant pests into new environments.

3.2 Preventing movement of pests with commodities for export

Phytosanitary certification is done to prevent the introduction of pests across international boundaries. Most importing countries require that Phytosanitary Certificate (PC) issued by the NPPO of the exporting country must accompany each consignment.

Export commodities for phytosanitary certification are subjected to official inspection and laboratory tests to determine their phytosanitary condition to meet the importing countries' plant health standards. Commodities found to be substantially free from plant pests shall be officially certified and issued the PC, while commodities infected/infested by plant pests shall either be subjected to a prescribed commodity treatment or shall be denied of certification. Plant Quarantine Officers (PQOs) appointed and/or designated by the Director of BPI are responsible for the issuance of the PC.

Plant pest surveillance is done where generated data are primarily important to ensure the absence of a particular pest in the country and further strengthen market access of Philippines agricultural produce to global markets.

3.3 Preventing pest epidemics

Domestic Quarantine aims to restrict the movement of infected and/or infested plants and plant parts or plant products to prevent further spread of indigenous quarantine pests/regulated nonquarantine pests and introduced pests to other areas in the Philippines where the pests are not yet known to occur.

To ensure that injurious plant pests will not be introduced with the movement of commodities to other parts of the country, CDT is required for domestic movement of plants, planting materials and plant products. PQOs and PQ Inspectors are responsible for the conduct of commodity inspection and application of appropriate phytosanitary measures, if necessary and issuance of CDT. The CDT is presented to PQO assigned at the points of entry for verification of the shipment against the document submitted.

The BPI has crafted the BPI Quarantine Administrative Order No. 3, s. (2018), Procedural Manual on Domestic Quarantine which aimed to rationalize and harmonize the requirements and procedures in the domestic movement, transfer and carrying of plants, planting materials, plant products, and other materials capable of harboring plant pests. It also provides surveillance procedures and guidelines to respond to emergency situations and pest outbreaks.

In situations of pest outbreaks, the Director of BPI, upon confirmation of the existence or suspected presence of dangerous "plant pest" in any place in the Philippines shall declare such localities, provinces, or island as a "quarantine area" through the promulgation of a Special Quarantine Order, containing among others, the geographical boundaries of the quarantined area, the plant pest under quarantine consideration, the host materials, and other pertinent information.

3.4 Monitoring, surveillance and forecasting of plant pest and disease epidemics

The Crop Pest Management Division (CPMD) of BPI is mandated to provide plant pest surveillance and early warning information that would ensure effective crop pest management for the protection of the Philippine crop industry and has a mission to prevent and control the spread of crop pests in the country by ensuring the availability of effective biological control agents, functional surveillance and early warning system, and regular plant health status. As of today, the Division has three pest surveillance funded projects, namely: (1) BPI-Corn Surveillance Project; (2) BPI-Cassava Surveillance Project; and (3) Pest Risk Identification and Management (PRIME) Project.

Development of Early Warning System and Database Management of Corn Arthropod Pest and Diseases in the Philippines funded by the Department of Agriculture — Bureau of Agricultural Research (DA-BAR) aims to develop an early warning system in the management of insect pest and diseases of corn, determine the major pest and diseases in genetically modified (GM) and Non-GM Corn at three levels of fertilization and establish a national pest profile and database for corn insect pests and diseases in the Philippines. Thus, relevant outputs of the project were developed. The second project entitled "National Assessment on the Incidence of Arthropod Pests and Diseases of Cassava as Influenced by Fertilization and Elevation as Basis for Forecasting Model" aims to establish an accessible online national database of major cassava arthropod pests and diseases, to determine the influence of fertilizer rate and elevation on the incidence of major cassava arthropod pests, and to develop a forecasting model on the occurrence of major cassava arthropod pests and diseases. This is relevant to the "National Survey and Early warning on Cassava Arthropod Pests and Diseases" which aims to characterize cassava phytosanitary threats and the potential drivers of their spread. The third project is the "The Pest Risk Identification and Management (PRIME)" project which is a collaborative project between DA-BPI, DA-Philippine Rice Research Institute, International Rice Research Institute, and the DA-Regional Field Offices (DA-RFOs I to XIII), Cordillera Autonomous Region (CAR) FO and Ministry of Agriculture, Fisheries and Agrarian Reform-Bangsamoro Autonomous Region in Muslim Mindanao (MAFAR-BARMM). PRIME identifies the risk factors of pest outbreaks, maps potential outbreak risks, and formulates integrated pest management strategies using field-based pest surveillance, information derived from satellite and drone images, and risk modeling. It focuses on five pest and diseases (blast, bacterial blight, rice tungro, and green leafhopper, brown plant hopper, and rodents) which cause major crop losses in the Philippines. The project standardized all field surveillance protocols for about 19 pests of rice, and developed and introduced the use of digital technology to collect, send, store, validate and analyze field data. Through this, real-time field information can be immediately and accurately translated into a comprehensive report that can be used as basis for pest management decisions and policy making. Among these ICT are the applications called "PRIME Collect," an Android based application that can be freely downloaded from Google Play Store, cloud-based storage facility is in place to store field data, and the web-based analytics and information access is already up and running called "PRIME WebApp." These three major monitoring, surveillance and forecasting applications were cascaded by CPMD of the BPI to the RFOs through Regional Crop Pest Management Centers (RCPMCs) and eventually used by our partners and stakeholders for better pests and disease management.

4. Related administrative orders

Republic of the Philippines
Department of Agriculture and Natural Resources
Bureau of Plant Industry
Manila

ADMINISTRATIVE ORDER NO. 39
Series of 1923

SUBJECT: Regulating the Eradication of Coconut Bud-rot

WHEREAS, there exists in certain parts of the Philippine Islands the disease commonly known as coconut bud-rot;

WHEREAS, this disease is menace to the agricultural interests of these Islands; and

WHEREAS, adequate measures should be adopted to prevent the spread and to effect the control of the said disease.

THEREFORE, by authority of the provisions of Legislative Act No. 3027 the disease commonly known as coconut bud-rot is hereby declared to be dangerous plant disease and shall be dealt with as herein after prescribed;

SECTION 1. Whenever an outbreak of coconut bud-rot is known to exist in any locality of the Philippine Islands it shall be the duty of the Director of Agriculture or his authorized agent to inspect all coconut trees in that locality, to mark in a suitable manner all trees ascertained to be affected by the disease, and to issue notification in writing to the owners, lessees, or persons in charge of coconut plantations, groves, or trees advising them that the disease exists among their trees and indicating to them what trees are diseased.

SEC. 2. Whenever the Director of Agriculture or his authorized agent shall have issued notification in accordance with the provisions of Section 1 hereof, it shall be the duty of the owner, lessee, or person in charge of the coconut plantation, grove, or trees where the disease exists to destroy every affected tree by cutting down and completely burning crown and other infected parts thereof.

SEC. 3. Failure to destroy affected trees within a period of 14 days from the date of receipt of written notification shall be considered prima facie evidence of an endeavor to avoid the duties imposed by virtue of this order and shall render the owner, lessee or person in charge of the coconut plantations, groves, or trees liable to the full penalties provided by Section 13 of legislative Act No. 3027, which is a fine not exceeding one thousand pesos, or imprisonment, in the discretion of the court.

SEC. 4. In order to carry out the provisions of this order the Director of Agriculture or any person acting in his behalf shall have access at all times into and upon any land occupied by any coconut tree or trees for the purpose of inspection.

336 13. Fungal plant pathogens of quarantine importance in the Philippines

BPI SPECIAL QUARANTINE ADMINISTRATIVE ORDER
No.1
Series of 2012

SUBJECT: Declaring Panama Disease of Bananas (Fusarium Wilt) caused by Fusarium oxysporum f. sp. cubense), a Dangerous and Injurious Banana and Abaca Disease; Providing its Control and Placing under Quarantine all the provinces where the Disease Already Exists to Prevent Further Spread from Infected Areas to Non-Infected Areas

WHEREAS, Panama disease of banana (Fusarium wilt) caused by a fungus Fusarium oxysporum f. sp. cubense that inhabits the soil;

WHEREAS, based on initial survey conducted by the Provincial Agricultural Office, Plant Quarantine Service, and private stakeholders, it was found out that about 620 ha of banana plantations in Mindanao were sporadically infected with Fusarium wilt;

WHEREAS, it has been reported and verified that Fusarium wilt already exists in the provinces of Davao del Norte, Davao Oriental, Davao del Sur, Davao City, Compostela Valley, Bukidnon, North and South Cotabato;

WHEREAS, banana is the country's leading fruit crop and second largest agricultural export commodity next to coconut oil, and the industry generates earnings of around $300 million annually and contributes around 11% (Php 20.9 billion) per year to the agricultural sector;

WHEREAS, if the spread of this disease is not controlled, it may eventually wipe out the entire banana and abaca industries in the country;

WHEREAS, in order to safeguard the banana and abaca industries from this disease, it is imperative that appropriate border control and quarantine measures be properly established to arrest and prevent the spread of the disease from infected to non-infected areas;

WHEREAS, the Mindanao Banana Disease Taskforce formulated the Mindanao Fusarium Action Plan to address the need to prevent further disease spread by conducting continuous monitoring/disease mapping, research and development enforcing border control and quarantine measures, instituting disease control strategies for the small farmers, and providing stakeholder education;

WHEREAS, Section 17 of PD 1433 otherwise known as the Plant Quarantine Law of 1978 empowers and authorizes the Director of the Bureau of Plant Industry to promulgate such quarantine orders, rules, and regulations to implement the provisions of the said Decree;

NOW THEREFORE, pursuant to the authority vested in me as the Director of the Bureau of Plant Industry under Section 17 of PD 1433 empowers this Special Quarantine Administrative Order is hereby promulgated as follows:

SECTION 1: The Panama Disease of Banana "Fusarium Wilt" is hereby declared an injurious and dangerous disease of banana and abaca (Musa spp.).

SECTION 2: For the purpose of this Administrative Order, the provinces of Davao del Norte, Davao Oriental, Davao del Sur, Davao City, Compostela Valley, Bukidnon, North and South Cotabato and such other areas where the disease is found to exist are hereby declared infected with Fusarium wilt and placed under quarantine in accordance with Section 8 of PD 1433 as implemented by Section 1, Rule VI of BPI Quarantine Administrative Order No. 1 series of 1981.

SECTION 3: The movement/transfer/carrying of the following materials from the infected provinces of Davao del Norte, Davao Oriental, Davao del Sur, Davao City, Compostela Valley, Bukidnon, North and South Cotabato as well as from other provinces which may hereafter be declared infected with Fusarium wilt, to non-infected provinces or sub-provinces within in Mindanao as well as to Visayas and Luzon shall be prohibited:

(a) Plants and/or its parts thereof of Musa spp. of all varieties of banana (Bungulan, Lakatan, Latundan, Saba, Cavendish, etc.);

(b) Plants and/or its parts thereof of Musa spp. of all varieties of abaca;

(c) Soil;

(d) Banana and abaca plant parts (except tissue culture-derived plants) used as packing materials for agricultural products

Exceptions from this prohibition are:

(a) Tissue-cultured banana and abaca plants and/or their parts thereof;

(b) Limited quantities of banana and abaca plants and/or their parts thereof intended for experimental purposes only, provided a prior permit for such purpose is first secured from the Director of the Bureau of Plant Industry or his/her authorized representative.

SECTION 4: The following quarantine and border control measures shall be implemented:

(a) Quarantine checkpoints shall be established in municipalities and between provinces where the disease has been identified and verified as existing;

(b) Inspection of all vehicles, farm equipment, and people passing through the quarantine checkpoints, and disinfection of the same using appropriate disinfectants;

(c) Confiscation of materials capable of spreading the disease;

(d) Installation and maintenance of tire/wheel and foot baths at every port/wharf, airport, and entry and exit points in banana and abaca farms

SECTION 5: In order to carry out effective plant quarantine measures, the Director of the BPI, in accordance with Section 16 of PD 1433, may request the national, provincial, and barangay officials and members of the Armed Forces of the Philippines as well as private stakeholders to assist the Bureau of Plant Industry-Plant Quarantine Service (BPI-PQS) in the implementation/enforcement of the prohibitions and measures stipulated in section 3 and 4 of this order to prevent the further spread of the disease.

SECTION 6: Any person who violates the provisions of this order or forges, counterfeits, alters, defaces, and destroys any document issued by virtue hereof shall be prosecuted pursuant to the penalties imposed under Section 23 of PD 1433 consisting, inter alia, of a fine of not more than PhP20,000.00 or imprisonment from prison correctional to prison mayor, or both at the discretion of the Court.

SECTION 7: This order supersedes and revokes prior orders or other issuances contrary to or inconsistent herewith and shall take effect fifteen (15) days after its publication as required by law.

(SGD.) CLARITO M. BARRON, PhD, CESO IV
Director, Bureau of Plant Industry

Approved:
(SGD.) PROCESO J. ALCALA
Secretary
Department of Agriculture
February 15, 2012

MEMORANDUM ORDER
NO. 30
Series of 2013

SUBJECT: AMENDING SECTION 3(C) OF BPI SPECIAL QUARANTINE ADMINIS-TRATIVE ORDER NO. 1, SERIES OF 2012 ENTITLED "DECLARING PANAMA DIS-EASE OF BANANAS (FUSARIUM WILT) CAUSED BY FUSARIUM OXYSPORUM F. SP. CUBENSE, A DANGEROUS AND INJURIOUS BANANA AND ABACA DISEASE; PROVIDING ITS CONTROL AND PLACING UNDER QUARANTINE ALL THE PROV-INCES WHERE THE DISEASE ALREADY EXISTS TO PREVENT FURTHER SPREAD FROM INFECTED AREAS TO NON-INFECTED AREAS".

WHEREAS, Section 3(c) of BPI Special Quarantine Administrative Order No.1, Series of 2012 need to be amended in order to express more clearly the intent of the said Order;

WHEREAS, there have been recent development and information in plant quarantine administration that necessitates adjustments in the implementation of Section 3(c) which pro-hibits the movement, transfer and carrying of soil from the infected provinces of Davao del Norte, Davao Oriental, Davao del Sur, Davao City, Compostela Valley, Bukidnon, North and South Cotabato as well as from other provinces within Mindanao as well as to Visayas and Luzon;

WHEREFORE, pursuant to Section 17 of PD 1433, the Director Plant Industry hereby pro-mulgates the amendment to Section 3(c) of BPI Special Quarantine Administrative Order No.1, Series of 2012.

Section 1. The provision of Section 3(c) should be amended as follows:

(c) Soil associated with Musa spp. Soil associated with other crops shall be allowed subject to the following conditions:

(1) The plants must be dipped in and the soil drenched with systemic fungicides, for example, Fosetyl-Al or any other fungicides that shall be prescribed by the Plant Quarantine Service

(2) Said soil and plant treatments have been performed under the supervision of a Plant Quarantine Officer.

Section 2. This Order shall take effect immediately and supersedes other Order inconsis-tent herewith.

(SGD.) CLARITO M. BARRON, PhD, CESO IV
Director, Bureau of Plant Industry

April 19, 2013

4. Related administrative orders

339

Republic of the Philippines
Department of Agriculture and Natural Resources
BUREAU OF PLANT INDUSTRY
Mannila

ADMINISTRATIVE ORDER NO. 13
Series of 1949)

August 13, 1949

SUBJECT: Regulating the importation of rubber plants and parts thereof such as seeds, rubber stumps, budsticks, etc. from Central America in order to prevent the introduction into the Philippine Islands of a disease known as *Dothidella ulei*, except for certain purposes and under certain conditions.

WHEREAS, there exists in Central America a certain disease of the rubber plant known scientifically as *Dothidella ulei*, which attacks Hevea plants;

WHEREAS, the said disease is not yet present or known to be present in the Philippines; and

WHEREAS, if this disease were introduced into this country, it would be a menace to our rubber industry;

NOW, THEREFORE, under authority, conferred upon me by Section 1 of Act No. 3027, entitled "An Act to Protect the Agricultural Industries of the Philippine Islands from Injurious Plant Pests and Diseases, etc.," the following regulations are hereby promulgated to govern and regulate the importation, bringing or introduction of rubber plants and parts thereof such as seeds, budsticks, rubber stumps, and other parts of the said plant capable of propagation or of carrying the disease into the Philippine Islands from the above mentioned places.

SECTION 1. The importation of rubber plants or parts thereof such as seeds, rubber stumps, budsticks, and other parts of the rubber plant capable of propagation and of carrying the said disease from Central America, is hereby prohibited: Provided, that the importation through the Port of Manila of small quantities of rubber seeds, seedlings or budsticks may be permitted in order to secure better varieties, new propagating stock or specimens for experimental purposes, is in accordance with Section 2 of Administrative Order No. 2 of this Bureau. Such importation must be made through the Director of Plant Industry, subject to the provisions of said Administrative Order No. 2, and to the conditions that this imported stock must be held in quarantine in an isolation station until it is evident that no plant diseases or pests are present on such plant materials.

SECTION 2. Rubber plants, seedlings, cuttings or budsticks, rubber stumps, or seeds of said plant imported in contravention of the provisions of this Order shall be seized by the Director of Plant Industry or by his duly authorized representatives and shall be either immediately returned to the country or place of origin, or completely destroyed or treated, according to the decision which the Director of Plant Industry may make in the premises; Provided: That the cost of the return, or the destruction or treatment of the said plant materials shall be borne by the importer.

SECTION 3. Any person, firm, association, or corporation, who violates or contravenes any of the provisions of this Administrative Order shall be liable to prosecution and upon

III. Fungal conservation and management

conviction shall suffer the penalty provided in Section 13 of Act No. 3027, which is a fine not exceeding one thousand pesos (P1,000.00), or imprisonment not exceeding 6 months, or both, in the discretion of the court.

SECTION 4. The provisions of this Administrative Order shall take effect on approval. January 31, 2013

ADMINISTRATIVE ORDER
NO. 03
SERIES OF 2013

SUBJECT: Regulating the importation of rubber (Hevea sp.) plants and parts thereof such as seed, stumps, budsticks, etc. from South American Leaf Blight (SALB) endemic countries in order to prevent the introduction into the Philippines, the South American Leaf Blight of Rubber, a disease caused by *Microcyclus* ulei syn. *Dothidella ulei*, except for certain purposes and under certain conditions

WHEREAS, BPI Administrative Order No. 13, Series of 1949 was promulgated and issued by the Director of the Bureau of Plant Industry (BPI) to "regulate the importation of rubber plants and plants thereof such as seed, rubber stumps, budsticks, etc., from Central America in order to prevent the introduction into the Philippine Islands of a disease caused by *Microcyclus ulei* syn. *Dothidella ulei*, except for certain purposes and under certain conditions; "

WHEREAS, the South American leaf blight (SALB) caused by the fungus *Microcyclus ulei* (P. Henn) v. Arx is the most destructive disease of the rubber plant and a major constraint in the production of rubber in South America;

WHEREAS, the rubber industry is an important component of the agricultural sector and economies of rubber-producing countries in Asia and the Pacific, including the Philippines, and the introduction of the disease into these countries will cause great economic damage;

WHEREAS, this malady was recognized by countries of the Asia and the Pacific region, prompting the Asia and Pacific Plant Protection Commission (APPPC), to advise members to take strict phytosanitary measures to prevent the entry, establishment and spread of the South American leaf blight (SALB) of Hevea sp. in the region;

WHEREAS, the International Plant Protection Convention (IPPC) of the United Nations Food and Agriculture Organization (UN FAO) has identified the following countries as SALB endemic countries:

1. Belize	8. El Salvador	15. Nicaragua
2. Bolivia	9. French Guiana	16. Panama
3. Brazil	10. Guatemala	17. Paraguay
4. Colombia	11. Guyana	18. Peru
5. Costa Rica	12. Haiti	19. Surinam
6. Dominican Republic	13. Honduras	20. Trinidad & Tobago
7. Ecuador	14. Mexico	21. Venezuela

WHEREAS, the SALB disease is not yet present in the Philippines and other Asian rubber-producing countries;

WHEREAS, in order to protect the local rubber industry, there is a need to modify Administrative Order No. 13, Series of 1949 and strengthen existing measures against this important disease of natural rubber;

NOW THEREFORE, pursuant to the authority vested in the Director of the Bureau of Plant Industry (BPI) by Presidential Decree 1433, otherwise known as the Plant Quarantine Decree of 1978 BPI Administrative Order No. 13, Series of 1949 is hereby amended as follows:

SECTION 1. The importation of rubber plants or parts thereof such as seeds, rubber stumps, budsticks, and other parts of the rubber plant from the countries identified by the IPPC as SALB endemic countries is hereby prohibited, except in the following specific cases and conditions:

(a) Importations of limited quantities of planting materials such as rubber seeds, seedlings or budsticks for the purpose of securing better varieties, new propagating stock, or specimen for experimental purposes, in accordance with BPI Administrative Order No. 1 Series of 1981;

(b) Importation shall be made only with the approval of the Director of BPI, subject to the provisions of this Administrative Order, and only through BPI-designated and authorized ports of Manila, and

(c) The imported stock must be held in quarantine in an isolation station until it is evident that no plant diseases or pests are present on such materials.

The requirements for the importation of these planting materials shall be reflected in the Plant Quarantine Clearance (PQC).

SECTION 2. The BPI Plant Quarantine Service (BPI-PQS) shall prescribe and implement the regulations and measures to:

(a) The prevention of the introduction of SALB into the Philippines;

(b) Determination of quantity of planting materials to be imported for the specific purposes stated in Section 1a of this Administrative Order

(c) Design and implementation of appropriate training programs

(d) Minimum requirements for personnel and facilities

(e) All necessary and appropriate coordination and monitoring to effectively implement this Administrative Order

The BPI-PQS shall ensure that the aforementioned activities are in line with the provisions of International Standards for Phytosanitary Measures (ISPM) of the IPPC. The activities shall also be guided by pertinent provisions and guidelines of the Regional Standards for Phytosanitary Measures (RSPM) of the APPPC.

SECTION 3. Rubber plants, seedlings, cuttings or budsticks, rubber stumps, or seeds of said plant imported in contravention of the provisions of this Order shall be seized by the BPI Director or his authorized representatives and shall be subjected to any of the following actions:

(a) Immediately returned to the country of origin, or

(b) Completely destroyed, or

(c) Treated in the premises where initially unloaded in a manner deemed appropriate for the purpose by the BPI Director

The cost of implementing these actions shall be borne by the importer.

SECTION 4. Any person, firm, association, or corporation who violates or contravenes any of the provisions of this Administrative Order shall be liable to prosecution and upon conviction shall suffer the penalty provided in P.D. 1433.

SECTION 5. This Administrative Order supersedes and revokes all Orders inconsistent herewith and shall take effect fifteen (15) days after publication in two (2) newspapers of general circulation as required by law.

RECOMMENDING APPROVAL:
(SGD.) CLARITO M. BARRON, PhD, CESO IV
Director, Bureau of Plant Industry

APPROVED:

(SGD.) PROCESO J. ALCALA
Secretary

References

BPI Quarantine Administrative Order No. 1 s. (1981). *Rules and regulations to implement presidential Decree No. 1433.*
BPI Quarantine Administrative Order No. 3, s. (2018). *Procedural manual on domestic quarantine.*
BPI Quarantine Order No.02, s. (2018). *Procedural manual in the conduct of phytosanitary import inspection.*
Bureau of Plant Industry website. https://www.buplant.da.gov.ph.
DA Department Circular No. 4 s. (2016). *Guidelines on the importation of plants, planting materials and plant products for commercial purposes.*
Food and Agriculture Organization. (2012). *Protection against South American leaf blight of rubber in Asia and the Pacific region* (Vol. II, p. 187).
Gupta, V. K., & Paul, Y. S. (2002). In V. K. Gupta, & Y. S. Paul (Eds.), *Diseases of field crops* (p. 464). Indus Publishing Company.
Magill, C., Frederiksen, R., Malvick, D., White, D., Gruden, E., & Huber, D. (2006). *Philippine downy mildew and Brown stripe downy mildew of corn.* The American Phytopathological Society (full reference citation pls).
Ploetz, R. C. (2006). Fusarium wilt of banana is caused by several pathogens referred to as *Fusarium oxysporum* f. sp. *cubense. Phytopathology, 96,* 653–656. https://doi.org/10.1094/PHYTO-96-0653
Presidential Decree No. 1433. (1978). *Promulgating the plant quarantine law of 1978, thereby revising and consolidating existing plant quarantine laws to further improve and strengthen the plant quarantine service of the Bureau of plant industry".* Malacañang.
Reinking, O. A. (1919). *Phytophthora faberi* Maubl.: The cause of the coconut bud rot in the Philippines. *The Philippine Journal of Science, 14,* 131–151.
Stover, R. H., & Simmonds, N. W. (1987). *Bananas* (3rd ed.). Longman Group.

Further reading

CAB International. (2021). *Plant wise knowledge bank.* https://www.plantwise.org/KnowledgeBank/datasheet/44646.
Montiflor, M. O., Vellema, S., & Digal, L. N. (2019). Coordination as management response to the spread of a global plant disease: A case study in a major Philippine banana production area. *Frontiers in Plant Science, 10,* 1048. https://doi.org/10.3389/fpls.2019.01048
Vezina, A. (2017). *Musapedia, the banana knowledge compendium.* http://www.promusa.org/Tropical+race+4+-+TR4.

CHAPTER
14

Innovative learning activities for teaching mycology in the Philippines

Thomas Edison E. dela Cruz[1], Reuel M. Bennett[1], Marilen P. Balolong[2], Bryna Thezza D. Leaño[3], Angeles M. De Leon[4], James Kennard S. Jacob[5], Joel C. Magday, Jr.[6], Almira Deanna Lynn C. Valencia[7], Maria Feliciana Benita M. Eloreta[8], Jocelyn E. Serrano[9], Jayzon G. Bitacura[10], Carlo Chris S. Apurillo[11], Judee N. Nogodula[12], Melissa H. Pecundo[13], Krystle Angelique A. Santiago[14] and Jeane V. Aril-dela Cruz[15]

[1]Department of Biological Sciences, College of Science, University of Santo Tomas, Manila, Philippines; [2]Department of Biology, College of Arts and Science, University of the Philippines Manila, Manila, Philippines; [3]Department of Biology, College of Science, De la Salle University, Manila, Philippines; [4]Department of Biological Sciences, College of Science, Central Luzon State University, Science City of Muñoz, Philippines; [5]Department of Biological Sciences, College of Arts and Sciences, Isabela State University, Echague, Isabela, Philippines; [6]Philippine Science High School - Cagayan Valley Campus, Bayombong, Nueva Vizcaya, Philippines; [7]Department of Biology, College of Arts and Sciences, Partido State University, Goa, Camarines Sur, Philippines; [8]Philippine Science High School - Bicol Region Campus, Goa, Camarines Sur, Philippines; [9]Department of Biology, College of Science, Bicol University, Legazpi, Albay, Philippines; [10]Department of Biological Sciences, Visayas State University, Baybay City, Leyte, Philippines; [11]Center for Research in Science and Technology (CReST), Philippine Science High School - Eastern Visayas Campus, Palo, Leyte, Philippines;

Mycology in the Tropics
https://doi.org/10.1016/B978-0-323-99489-7.00012-3

© 2023 Elsevier Inc. All rights reserved.

[12]Department of Natural Sciences, College of Arts and Sciences, University of Southeastern Philippines, Davao City, Philippines; [13]Research Center for Natural and Applied Sciences, University of Santo Tomas, Manila, Philippines; [14]School of Science, Monash University Malaysia, Bandar Sunway, Selangor Darul Ehsan, Malaysia; [15]Institute of Biology, Freie Universität Berlin, Berlin, Germany

1. Introduction

Mycology as a field of study in the Philippines is beset with many challenges. As earlier pointed out by Dogma (1975) and later, affirmed by dela Cruz, Dagamac et al. (2013), Philippine mycology, just like many basic branches or fields in biology, suffers from lack of appreciation within the science curricula among our local colleges and universities. The study of fungi is often treated as an introductory chapter within a general microbiology course, albeit programs in agriculture such as plant pathology offers a course in mycology as an integral component of its curriculum. In others, mycology is offered as a key field of specialization, for example, for the Bachelor of Science in Agriculture major in Plant Pathology at the University of the Philippines Los Baños. Dogma (1975) suggested the need to develop a departmental program committed to the development of the discipline of mycology in the Philippines. Universities that offer bachelor program in microbiology and/or biology with major or track in microbiology now offers mycology either as a major course, for example, at the University of Santo Tomas (MB 442—Mycology), UP Los Baños (VMCB 122—Veterinary Bacteriology and Mycology, PPTH 104—General Mycology, FBS 140—Forest Mycology), or as a taxonomy elective, for example, UP Diliman (BIO 112—Mycology), UP Baguio (BIO 103—Mycology), Bicol University (SE—Principles of Mycology). Through these course offerings, mycology can be further promoted as a major science discipline in the country.

The teaching of mycology and other science courses requires active learning strategies. This is in response to the directives of the Commission on Higher Education (CHED) which adopted the outcomes-based education (OBE). CHED through its Memorandum Order (CMO) No. 46 Series 2012 promulgated the policy standards for all private and public higher educational institutions in the Philippines which sets the minimum standard for the BS Biology program. It also focuses on the achievement of learning outcomes as demonstrated by the ability of graduates to effectively communicate, to work independently in multicultural and multidisciplinary teams, to demonstrate professional, social, and ethical responsibility, and to promote and preserve the "Filipino historical and cultural heritage" (dela Cruz, 2020). Some examples of active learning activities as applied to the teaching of undergraduate microbiology courses are outlined in the papers of dela Cruz (2020) and Mendoza (2020). An additional fun and engaging learning activity for students is the use of pick-up lines to facilitate learning of microbiology concepts (dela Cruz, 2014). Clapton (2015) stated that popular culture, such as pick-up lines which became widespread in the Philippines, can be used as a teaching tool to develop understanding of concepts and complex theories and as part of assessment strategies. For communicating scientific facts, particularly hot topics, to different audiences, illustrating editorial cartoons and writing editorial narratives (dela Cruz & Aril-dela Cruz, 2018) and creating infographics and info-posters (dela Cruz et al., 2021) could be ideal learning activities, even for a mycology course. The use of web-based identification guides (dela Cruz et al., 2012) and photoguides (Macabago & dela Cruz, 2012) can be used for species identification in a taxonomy course. Social media platforms such as Facebook can be a source of species records for ecological studies (dela Cruz & Olayta, 2022) while meta-analysis or correlation studies can be performed based on published literature (Sanvictores et al., 2022). Specimens available even within our own backyards can be easily collected and studied (dela Cruz & Eloreta, 2020; Estampador et al., 2022; Garcia et al., 2022). dela Cruz, Pangilinan et al. (2013) also suggested a class seminar activity, the "Meet a Microbiologist" program, to inspire students to pursue a career in microbiology. While higher education implements active and student-centered learning, this will only be successful if such strategy is also implemented at basic education levels. We believe that it is important to design learning activities for different learning levels as shown in the various activities with myxomycetes as outlined in the paper of Winsett et al. (2017, 2022).

In this chapter, we present different learning activities used in the teaching of fungi or mycology for different educational levels and educational institutions with varied facilities and resources. We initially presented three comic book materials that introduce fungi to young kids. Comics have been used to promote advocacies and share information and can raise enthusiasm among its readers (Rakower & Hallyburton, 2022; Saji et al., 2021; Scavone et al., 2019; Wang et al., 2018). We then outline learning activities that can be deployed for junior and senior high school students and for undergraduate students in both public and private universities across the country. While the list of learning activities presented here is not exhaustive and is applicable to other organisms or fields of study, we provide materials that can be easily deployed in the teaching of mycology in the country in the hope that this will inspire teachers to develop more active and student-centered activities for our learners.

2. Teaching fungi to kids in preschool and elementary levels

Children are naturally curious. Oftentimes, they ask the hardest questions. Our best reply is to provide them with age-appropriate answers that will fuel their interest and inspire creativity. But at times, we simply don't have the answers. Thus, letting them experience and interact with the natural world nurtures their natural curiosity. Here, we described three comic books that introduce fungi to kids. Developed and published by the Fungal Biodiversity, Ecogenomics and Systematics (FBeS) group at the Research Center for the Natural and Applied Sciences in the University of Santo Tomas, this comic book series introduces different groups of fungi, how to know them, and what are their uses (https://ustfungalbiodiversitylab.wordpress.com/learning-materials-for-kids/). Written in an engaging narrative and with colorful characters and images, these comic books can be read by kids. Various activities embedded within the comic book further facilitate interactions between the kids and the organisms they are reading about. These are just simple learning materials that can easily make kids learn and love fungi. We also described a simple learning activity with fungi that teachers can easily implement in their classrooms.

Title: **The Search for the Missing Lichens: The Adventures of Thallus Tanod**. Santiago, K. A. A., Gazo, S. M. T., dela Cruz, T. E. E. (2015). Fungal Biodiversity & Systematics Booklet I, Philippines, pp 20.

Synopsis: "The Search for the Missing Lichens" introduces the exciting adventures of Tanod Thallus. This comic book, the first of a series, provides relevant information on where to find and how to identify lichens. Readers will also get a glimpse of the important role lichens play in our environment. As kids learn about lichens, an activity page encourages them to explore their own backyard for nature's finest and most mysterious symbioses.

The story began when a friendly and adventurous tree, Tanod Thallus, discovered that his friends, namely Crustose (the paint-like lichen), Foliose (the leaf-like lichen), and Fruticose (the beard-like lichen), were missing (Fig. 14.1). These lichens used to inhabit the trunk of Tanod Thallus for years, and their disappearance is an enigma. To find his lichen friends, Tanod Thallus embarked on an exciting and fun-filled quest. But before he could set off for his adventure, Tanod Thallus needs to prepare things such as map and magnifying glass which are important for locating and observing lichens. He initially reached the plain where he found Crustose. His journey then led him to the forest where he met Foliose. At the mountain top, Tanod Thallus found Fruticose hanging on tree branches. The places where Tanod Thallus found his lichen friends are the usual habitats of these types of lichens. Through the adventure of Tanod Thallus, kids learn how to differentiate lichens from other organisms that also inhabit tree trunks like mosses. Kids also read trivia on the different uses of lichens.

Title: **The Adventures of Myxoman and Amoeboy**. Carascal, M. B., Pecundo, M. H., dela Cruz, T. E. E. (2016). Fungal Biodiversity & Systematics Booklet II, Philippines, pp 20.

Synopsis: The quest of Myxoman and his trusted sidekick, Amoeboy, introduces the amazing world of myxomycetes or slime molds to the young generation in a simple yet fun and interactive manner (Fig. 14.2). This comic book, the second in the series, provided information on how we can explore our environment for these tiny but extraordinary creatures. The comic book is packed with challenging activities, puzzles, and quizzes that encourage curiosity and understanding. Readers are also introduced to real images of myxomycetes.

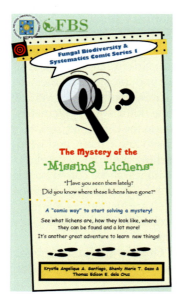  

FIGURE 14.1 The cover page and sample page of "The Mystery of the Missing Lichens". Kids will get up close with the three major types of lichens—crustose, foliose, and fruticose lichens and the habitats that support their growth.

2. Teaching fungi to kids in preschool and elementary levels 347

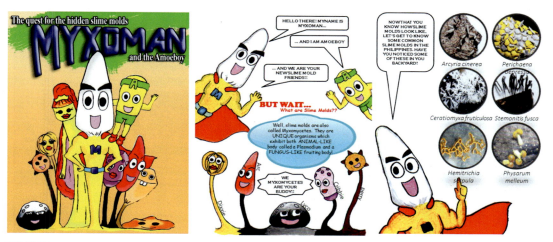

FIGURE 14.2 The cover page and sample page of "The Adventures of Myxoman and Amoeboy." Various species of myxomycetes serve as inspiration for the characters in the comic book.

The comic book feature as its main characters, Myxoman (based on the sporangium stage in the myxomycete life cycle) and his sidekick, Amoeboy (inspired from the amoeba stage of slime molds). Kids are also introduced to their myxomycete buddies, all are based on actual species of myxomycetes, namely, Dixxie [*Dictydium cancellatum* (Batsch) T. Macbr., now known as *Cribraria cancellata* (Batsch) Nann.-Bremek.], Arc [*Arcyria denudata* (L.) Wettst.], Cribbie [*Cribraria macrocarpa* (Schrad.) Pers.], Lyco [*Lycogala exiguum* Morgan], Perry [*Perichaena pedata* (Lister & G. Lister) G. Lister ex E. Jahn], and the Craetwins [*Craeterium leucocephalum* (Pers. ex J.F. Gmel.) Ditmar]. Together, they went on a quest to find the other hidden slime molds. These characters narrate the unique life cycle of myxomycetes and the habitats where they thrive. They also teach the readers how to find fruiting bodies of slime molds which they may encounter in their backyards. Myxoman and Amoeboy encourage kids to join them in several missions. These are engaging activities to learn more about myxomycetes. The kids also meet the story's main villain, Lord HD, the myxomycete habitat destroyer, whose main aim is to destroy all habitats of slime molds. Myxoman and Amoeboy, through their superpowers defeated Lord HD, and eventually met Princess Cinerea [*Arcyria cinerea* (Bull.) Pers.], the heiress to the kingdom of myxomycetes. The comic book is a creative gateway to introduce slime molds to children.

Title: **The TJ Mushketeers: A Journey to the Mushroom Land**. Carascal, M. B., Aril-dela Cruz, J. V., De Leon, A. M., dela Cruz, T. E. E. (2020). Fungal Biodiversity, Ecogenomics & Systematics Booklet III, Philippines, pp 20.

Synopsis: "The TJ Mushketeers — A Journey to the Mushroom Land" is written to introduce the ubiquitous but often neglected mushrooms (Fig. 14.3). Thanks to Super Mario and movies like The Smurfs and Trolls, mushroom-inspired animated characters are made known to the kids of this generation. As children read this comic book, they will be again introduced to colorful creatures based on mushrooms and will be more fascinated to know the real mushrooms that inspire these characters.

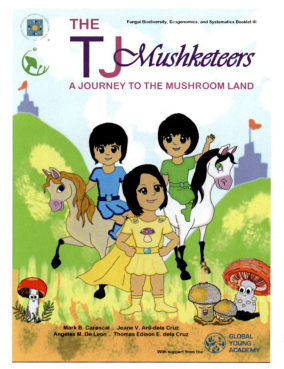 

FIGURE 14.3 The cover page and sample page of "The TJ Mushketeers — A Journey to the Mushroom Land." Kids will enjoy knowing about different species of mushrooms.

Tiffy, a TJ Mushketeer, brings with her two TJ Mushketeers-in-Training, the siblings Torrie and Tinnie, in an expedition to explore the Mushroom Land. With their curious, ready-for-adventure, and fun-loving nature, the three went on to encounter many interesting and surprising mushroom-inspired characters in the Agari Kingdom—the fairest mushroom of all, Rhoda Rhodotus [*Rhodotus palmatus* (Bull.) Maire], the captain of the guards, Fierce Amanita [*Amanita muscaria* (L.) Lam.], the members of the Council of Ministers [Doctor Puffy (*Lycoperdon perlatum* Pers.), Engineer Russell (*Russula emetica* (Schaeff.) Pers.), and Professor Billy (*Boletus edulis* Bull.)], and the ruler of the Agari Kingdom, Queen Angel. These characters explained what mushrooms are, their habitats, how they grow from tiny invisible spores to the fruiting bodies we are familiar with, and their many uses in the environment, in culinary, cosmetics, and medicines. They are also introduced to mushroom chef, Morchef Morchella [*Morchella esculenta* (L.) Pers.] and his sous-chef Marvin who assisted them in a hunt for edible mushrooms. On their way to explore the other roles of mushrooms, they met Phane [*Phanerochaete chrysosporium* Burds.] who explained the ability of fungi to degrade environmental pollutants. Meeting their Grandma Cion, the premier scientist of the kingdom, in her Mushroom Lab, opens more doors in understanding the different uses of mushrooms.

Older children will also learn while reading this comic book with engaging trivia and fun quizzes about mushrooms. Of course, what is adventure without mushroom hunting! The kids can use this comic book as pictorial guide to mushrooms. Younger children will also appreciate being read to by an adult. Therefore, we would like to encourage parents, older siblings, or caregivers to make reading an enriching experience by engaging the kids with questions that will foster imagination and appreciation for these fantastic organisms.

In addition to these comic books, we outline a simple class activity that kids both at the pre- and elementary schools can safely do at home or in their classroom.

Learning Activity: **Making Mushroom Spore Print**
Year Level: Preschool—Elementary School (6—12 years old)
Duration: 1 h, 1 class meeting
Learning Objective/s: At the end of the activity, the learner is expected to make and observe their own spore print.
Instructions: For this learning activity, the teachers will provide fresh fruiting bodies of mushrooms, specifically the Agarics. Alternatively, the learners can also bring their own mushroom specimens, preferably those found or collected in their own backyard. Under the supervision of the teacher, the learners will cut the stalk of the mushroom and the mushroom cap placed on either a white or black paper. It is important that the gills of the mushroom cap are directly above the paper. A wet cotton is placed on top of the mushroom cap. The setup is left overnight. The next day the learners can remove the mushroom cap and observe their spore prints (Fig. 14.4).
Assessment: At the end of the learning activity, the students can be tasked to illustrate the mushroom and/or record the color of the spores based on the spore print.

FIGURE 14.4 A sample spore print (white spores) from a brown mushroom.

3. Learning activities on fungi for high school students

Fungi can easily be introduced to high school students either as a unit or a chapter in the subject Biology under a regular secondary school curriculum or within a special elective subject such as Microbiology in a special science high school. In this section, we listed several learning activities designed for Junior (Grades 7–10 or 12–16 years old) to Senior (Grades 11–12 or 16–18 years old) High Schools. The activities presented here use locally available materials.

Learning Activity: **Observing Molds and Yeasts**
Year Level: Junior High School (Grades 7–10)
Duration: 1–2 h, 2 class meetings
Learning Objective/s: At the end of the activity, the student is expected to observe and document molds and yeasts from various natural sources.
Instructions: For observing yeasts, students will bring different sugar-rich natural substrates, for example, fruit peelings, nectar-filled flowers, and other similar substrata. These are cut into small pieces and transferred to sterile bottles with Glucose – Yeast Extract (GYE) broth. A balloon is then tied to seal the mouth of the culture bottle. The inoculated cultures are incubated at room temperature for 48–72 h. Gas formation as trapped in the balloon indicates the fermentation of the culture broth by naturally occurring yeasts. Students can then prepare a wet mount of the culture broth to observe for yeast cells. For observing molds, students will expose moistened food samples, e.g., bread, fruits, or fleshy vegetables, to air for at least 30 min. The exposed food sample is placed in a moist chamber, that is, a plastic container lined with moistened paper towel, and incubated at room temperature for 5–7 days. The students will regularly observe their moist chambers for the growth of filamentous molds. Similarly, a wet mount can be prepared with the growing molds and observe with an aid of a microscope. Students document their observations.
Assessment: At the end of the learning activity, the students will submit an illustration of molds and yeasts observed in their set-up. A rubric can be prepared to grade the student outputs.

Learning Activity: **Baiting Aquatic Fungi**
Year Level: Junior–Senior High School (Grades 7–12)
Duration: 1–2 h, 2 class meetings
Learning Objective/s: At the end of the activity, the student is expected to conduct baiting experiment for the observation of aquatic fungi.
Instructions: The sampling sites for this learning activity must be initially determined either by the teacher or the students. These would include rivers, streams, lakes, or ponds. In the chosen sampling site, three sampling stations will be established. For actively flowing aquatic habitats such as rivers and streams, we suggest collecting samples from upstream, midstream, and downstream. The teacher can also task the students to note any disturbance present in the sampling stations, e.g., any industrial or residential discharge of waste, agricultural plantations, etc. With a sterile bottle, water samples are collected and transported to the laboratory. To prepare the baits, 6 mm leaf discs will be cut from corn leaves with a single hole puncher. Together with another bait, sesame seeds, these are sterilized separately and later added on the collected water samples. We suggest adding 30 baits per petri plate of

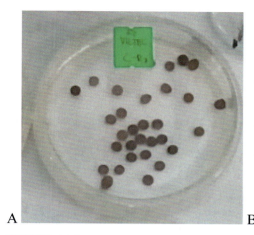
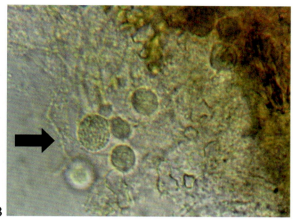

FIGURE 14.5 Sample baiting experiment for aquatic fungi. (A) corn leaf disks on collected water sample, (B) photomicrographs of chytrids (with arrow) colonizing corn leaf disk (brown area).

the water sample (Fig. 14.5). The baits are incubated at room temperature for 5 days and observed for the presence of aquatic fungi (Chytrids, Saprolegnias) with an aid of a microscope. The students are tasked to illustrate or photo document the aquatic fungi growing on their baits (Fig. 14.5). They are also expected to compute the % baits infected as the number of baits with aquatic fungi over the total number of baits multiplied by 100.

Assessment: At the end of the learning activity, the students will submit a written scientific report on their findings. Since the learning activity is designed for high school students, species identification of the observed aquatic fungi is not required, albeit encouraged.

Learning Activity: **Myxo Hunting**
Year Level: Junior — Senior High School (Grades 7—12)
Duration: 3 h, 2 class meetings
Learning Objective/s: At the end of the activity, the student is expected to learn basic skills in doing field collection, set up moist chamber cultures, and identify fruiting bodies of myxomycetes.

Instructions: Students are paired prior to the conduct of the activity. They are tasked to bring field materials, that is, a field notebook and pen for recording data, mobile phone with GPS app and camera for documentation, knife or twig cutter, glue, herbarium boxes, and labels. An orientation on safety during field collection is discussed by the faculty-in-charge with the students before any "myxo hunting" activity. If needed, a gratuitous permit will be secured from the local environment management agency or from appropriate local government units. It is also advised to conduct an ocular inspection of the sites by the faculty-in-charge to note the different vegetation types, climate conditions, and other relevant information as well as to determine the appropriate time for the field work. During the field collection, the faculty-in-charge will demonstrate how to find fruiting bodies of myxomycetes from the field area and to distinguish these from other organisms such as mosses, lichens, or tiny mushrooms. The preparation of herbarium specimens, that is, filling in the herbarium labels with relevant information and pasting of fruiting bodies on herbarium boxes, and field data, that is, documenting the GPS data and other relevant ecological parameters will also be demonstrated on site by the teacher. Relevant information for herbarium labels includes the specimen number, collection site, date of collection, name of the collector, substrate, and identity of the species. The students are given ample time to explore the sampling area for the collection of fruiting bodies of myxomycetes. Additionally, the students can also collect substrata where myxomycetes usually thrive, for example, ground leaf litter, twigs, dead barks, decayed inflorescences, etc. The collected substrata will be transported to the school laboratory and prepared as moist chamber cultures. This is done by cutting the materials into small pieces and placing in petri plates or plastic containers lined with filter paper or paper towel. The substrates are flooded with water overnight, and then, drained and placed in cabinet not directly exposed to sunlight. The moist chamber cultures are observed weekly for the fruiting bodies of myxomycetes until the eighth week. Substrates with identifiable fruiting bodies of myxomycetes will be transferred and glued to herbarium boxes as voucher specimens with informative labels. To aid with the initial identification of the myxomycetes, description of the fruiting bodies will be compared with taxonomic keys in published literature and with web-based electronic databases, e.g., http://slimemold.uark.edu. Students may also consult this website (https://philmyxos.wordpress.com/) for myxomycetes that are already recorded in the Philippines. Students are expected to document the myxomycetes they collected and observed in the moist chambers and their experiences during field and laboratory works.

Assessment: At the end of the learning activity, the students will submit a field report with pictures documenting their experiences and present a weekly observation of their moist chamber cultures.

Learning Activity: **Mushroom Treasure Hunt**
Year Level: Senior High School (Grades 11–12)
Duration: 3 h, 2 class meetings
Learning Objective/s: At the end of the activity, the student is expected to identify common mushrooms available in the local market and determine their habitats and economic values or health benefits.

Instructions: The teacher will provide the students with a mushroom treasure map (Fig. 14.6) and instruct them to visit any local market. The students will look for mushrooms sold in the market. They may interview the mushroom sellers to know the local names of the mushrooms and their knowledge about its local habitat. To document their mushrooms, the students will illustrate these on their treasure map. At home, the students will identify and do research to find more information about the economic values and health benefits of their mushroom treasures. The information they gathered will be recorded in a student worksheet (Fig. 14.6).

Assessment: At the end of the learning activity, the students will submit their accomplished treasure map and student worksheet for grading. For online deployment of courses, a discussion forum can be moderated by the teachers on the uses of mushrooms in food, medicine, and industry. Participation in this discussion forum can be graded.

The next learning activity can be built on the previous learning activities. For example, the mushrooms collected from the mushroom treasure hunt or the isolated fungi from baits can be sent to gene sequencing facilities such as the Philippine Genome Center to obtain fungal

3. Learning activities on fungi for high school students

A

Mushroom Treasures

B

Name: _____
Section: _____

	Mushroom 1	Mushroom 2	Mushroom 3	Mushroom 4
Scientific Name				
Local Name				
Habitat				
Economic Values				
Health Benefits				

FIGURE 14.6   (A) The mushroom treasure map and (B) the student worksheet for this activity.

DNA sequences. Alternatively, available fungal gene sequences from databases (NCBI Gen-Bank, UNITE) can be used as sample DNA sequences.

Learning Activity: **Molecular Identification of Fungi using DNA sequences**
Year Level: Senior High School (Grades 11–12)
Duration: 2–3 h, 2 class meetings
Learning Objective/s: At the end of the activity, the student is expected to BLAST search the assigned fungal DNA sequences, align the sequence with other related sequences which are downloaded from the NCBI GenBank or other databases, perform a phylogenetic analysis using the free software MEGA, and identify the fungal specimen based on their phylogenetic position and morphological characters.

Instructions: The students are initially grouped into teams of three members. They are provided with an instructional guide which contain information on the assigned unidentified fungus (i.e., source and relevant taxonomic data—colony and spore morphologies including pictures and photomicrographs of the specimens), and a detailed step-by-step instruction on analyzing gene sequences using the MEGA software. This activity requires the students to have access to a computer with internet connection. Each group will be assigned to an unidentified DNA sequence, which the students need to process (i.e., check for sequence quality and for assembly of the forward and reverse sequences to gain a consensus sequence using DNABaser). The consensus sequence is then subjected to BLAST search via the NCBI website (https://blast.ncbi.nlm.nih.gov/Blast.cgi) to obtain additional sequences from related taxa. Students follow the instructions on sequence alignment and phylogenetic analysis using MEGA. A phylogenetic tree will be generated after the process which the students interpret and use to identify the fungal specimen. To confirm the identity of the assigned fungi, the students need also to check if their molecular identification is supported by the morphological characteristics.

Assessment: At the end of the learning activity, the students will present a case study on the identification of their unknown fungal specimens using molecular data as supported by morphological characteristics. They will explain the different steps in the DNA sequence

III. Fungal conservation and management

354   14. Innovative learning activities for teaching mycology in the Philippines

analysis and give the significance of each step. A grading rubric will determine the points to be awarded to the students.

4. Innovative teaching strategies on fungi for undergraduate courses

Guerrero (2020) noted the different undergraduate research in his university which mirrors the low preference for mycological research topics in many universities. We believe that interest on fungi stems from a solid background information about these organisms and the deep appreciation of their uses and application. These can be covered within the syllabus of an undergraduate mycology course. A sample course syllabus in mycology offered in the College of Science at the University of Santo Tomas is presented in Fig. 14.7.

Interest of fungi can also be strengthened and sustained by providing learning activities that will surely engages undergraduate students as they learn about fungi and motivated them to pursue research on this topic. This led us to list down some of the learning activities we implemented in our respective classes. The list included some basic laboratory experiments that can easily be performed with easy-to-get, available resources and in limited laboratory facilities. We also included learning activities that can be implemented during online or face-to-face classes. We hope that this list serves as the first among the many learning activities that can be developed to promote active learning and engagement in the study of fungi in the Philippines.

Learning Activity: **Backyard Oomycetes**

Year Level: 1st year, preferably for General Mycology or General Biology/Microbiology

Duration: 1.5 h, 2 class meetings

Learning Objective/s: At the end of the activity, the student is expected to observe common oomycetes in our own backyards.

Instructions: For this activity, simple materials that are readily available in our own backyard will be used. To do this, the students will initially collect a handful of soil and few milliliters of water, for example, from stagnant water, lake, pond. For the soil samples, the student will place the soil in clean, wide-mouth, 50 mL glass jars, approximately 1 cm in depth. Then, the students slowly pour in distilled water over the soil sample until it is filled approximately 1 cm above the soil surface. Four to five pieces of baits, that is, insect wings, *Cleome* or *Phyllanthus* seeds, carabao grass blades, or sesame seeds, are then placed. Be sure that each bait will be placed separately in different jars. For the water samples, an ample amount of water will be placed in Petri plates. Make sure that the water will not reach the rim of the Petri plates. Similarly, four to five pieces of baits are added. All set-ups are incubated at room temperature. After 24—48 h or even longer, the students are expected to observe their baits under a microscope. This is done by carefully placing the baits on glass slides with forceps. For insect wings and grass blades, students will observe the periphery or margin of the baits, even its inner parts. For the seeds, they will observe its periphery. Students will search for aseptate or coenocytic hyphae, an indication of the presence of oomycetes. They will also observe for sporangia (or zoosporangia), a bag-like structure that contains motile zoospores. Observation of baits is done continuously for 7 days. The learning activity can be assigned to individuals or teams of students.

University of Santo Tomas
COLLEGE OF SCIENCE
España Blvd., Manila 1015 Philippines
Tel. Nos. (63-2) 7315728; (63-2) 4061611 local 8224
FAX: (63-2) 7409730
www.ust.edu.ph/science

BS MICROBIOLOGY
COURSE SPECIFICATIONS

Course Title:	MYCOLOGY LECTURE	*School Year:*	AY 2020-2021
Course Code:	MB 442	*Semester:*	2nd Term
Credit Units:	3	*Prerequisite Course/s:*	MB 441, MB441L

Course Facilitator:	Thomas Edison E. dela Cruz	*Class Schedule:*	MW (1MCR1) /TTH (1MCR2), 10:00 - 11:30 AM
Office:	Biological Sciences	*Room Assignment:*	Via UST Cloud Campus (Blackboard)
Contact No:	+632 3406 1611 local 8297	*Consultation Time:*	MT, 11:30 – 12:00 NN
E-mail:	thomas.delacruz@ust.edu.ph	*Consultation Room:*	virtual

Prepared by:	Prof. Dr. rer. nat. Thomas Edison E. dela Cruz	*Date Prepared:*	15 January 2021
Evaluated by:	A/Prof. Richard Thomas Pavia, PhD	*Dated Evaluated:*	22 January 2021
Approved by:	Prof. Rey Donne S. Papa, PhD	*Date Approved:*	26 January 2021

COURSE DESCRIPTION:	
Rationale	This 3-unit lecture course designed for B. Sc. Microbiology students deals with the study of fungi and other fungus-like organisms.
Focus	It covers the structure and function, growth, reproduction, physiology and genetics of the fungi and other fungus-like organisms such as myxomycetes (eumycetozoans) and straminopiles (oomycetes, labyrinthulomycetes). It also discusses their evolutionary relationship, diversity in form and structures, and ecology. The course also introduces the magnitude of applications and impacts of these organisms in the environment and in the human society.
Outcome	At the end of the course, the learners are expected to differentiate fungi and cther fungus-like organisms in terms of their morphology, physiology, reproductive biology, phylogenomics, and ecology.

MAPPING OF IILO WITH PILOs and CILOs:

	SEAL of Thomasian Education	Thomasian Graduate Attributes (ThoGA)	Program Intended Learning Outcomes (PILOs)	Course Intended Learning Outcomes (CILOs)
Institutional Outcomes: COMPETENCE, COMMITMENT AND COMPASSION	Servant Leader	Show leadership abilities to promote advocacies for life, freedom, justice, and solidarity in the service of the family, the local and global communities, the Church and the environment.	6. Engage in life-centered advocacies promoting social responsibility, preservation of the Filipino culture and heritage, data integrity and privacy, life-long learning, professional and ethical conduct, and Christian moral values.	1. Appraise the diverse roles and ecological and economic importance of fungi in nature and in the society (*cognitive domain*).
		Implement relevant projects and activities that speak of Christian compassion to the poor and the marginalized in order to raise their quality of life.	2. Apply basic knowledge in microbiology and utilize safe and responsible techniques and procedures to identify and solve problems related to the life sciences using current disciplinary and interdisciplinary principles.	2. Compare and contrast the major groups of fungi and fungus-like organisms in terms of morphology and life cycle (*cognitive domain*).
		Show respect for the human person, regardless of race, religion, age, and gender.	6. Engage in life-centered advocacies promoting social responsibility, preservation of the Filipino culture and heritage, data integrity and privacy, life-long learning, professional and ethical conduct, and Christian moral values.	1. Appraise the diverse roles and ecological and economic importance of fungi in nature and in the society (*cognitive domain*).
	Effective communicator and collaborator	Express myself clearly, correctly, and confidently in various environments, contexts, and technologies of human interaction.	7. Communicate effectively and accurately biological concepts and research outputs; and articulate the latest trends and developments in the biological sciences through written and oral methods using both English and Filipino language	4. Show the different nutritional requirements and integrate these to metabolic pathways of fungi (*psychomotor domain*).
		Work productively with individuals or groups from diverse cultures and demographics.	5. Work efficiently and effectively in individual and group-oriented activities in multi-disciplinary and multi-cultural teams.	2. Compare and contrast the major groups of fungi and fungus-like organisms in terms of morphology and life cycle (*cognitive domain*).

FIGURE 14.7 A sample syllabus for a general mycology lecture course offered to undergraduate students of biology and microbiology.

Assessment: At the end of the learning activity, the students will submit a detailed and labeled illustrations of the oomycetes observed in their own baits. Example of a typical oomycetes that can be observed with this learning activity is shown in Fig. 14.8.

COURSE ORGANIZATION:				
Intended Learning Outcome	Content Outline	Teaching-Learning Activities	Assessment Tasks	No. of Hours
	Class Orientation			1.5

1. To enumerate the different mycologists and appraise their discoveries and research that led to the development of fungal biology
2. To examine the different contributions of fungi to the environment and the society
	Chapter 1 Introduction to the Kingdom Fungi
	1. The Study of Fungi: Historical Perspectives
	2. Fungal studies in the Philippines
	3. Ecological Roles and Economic Importance
	  a. Fungi as food
	  b. Fungi in food processing and spoilage
	  c. Fungi in agriculture and environmental clean-up
	  d. Fungi and health including poisonous and hallucinogenic mushrooms
	  e. Biotechnological and industrial application of fungi
		Reading Assignment: Lange L. 2017. Microbiol Spectrum 5(1): FUNK-0007-2016.
		Lecture Discussion
		Video Presentation
		Tech Tool: PPT, video
			Journal Reading (Individual 5-min Essay)
			Online Test 1
				4.5

1. To identify and illustrate the unique features of the different major groups of true fungi 2. To compare and contrast the different fungal groups in terms of morphology and phylogeny	Chapter 2 The True Fungi (Kingdom Eumycota) 1. Classification, Nomenclature and Phylogeny 2. Phylum Chytridiomycota 3. Phylum Blastocladiomycota 4. Phylum Neocallimastigomycota 5. Phylum Zygomycota 6. Phylum Glomeromycota 7. Phylum Microsporidia 8. Phylum Ascomycota 9. Phylum Basidiomycota	Lecture Discussion Video Presentation Tech Tool: PPT, video, Padlet	Padlet Output (Peer Evaluation) Online Test 2	9.0
			Long Examination	1.5
1. To identify and illustrate the unique features of the different major groups of pseudofungi 2. To catalogue the key features that differentiate pseudofungi with the true fungi	Chapter 3 The "Pseudofungi" (Kingdoms Protozoa and Chromista) 1. Classification, Nomenclature and Phylogeny 2. Protozoan Pseudofungi: Myxostelida (Eumycetozoa), Dictyostelida, Labyrinthulida, Plasmodiophorida 3. Chromistan Pseudofungi: Hyphochytriomycota and Oomycota	Reading Assignment: Dagamac & dela Cruz. 2019. Philippine Journal of Systematic Biology 13 (2): 58-65. Lecture Discussion Video Presentation Tech Tool: PPT, video, Wizer worksheet	Group Worksheet Online Test 3	4.5

FIGURE 14.7 cont'd

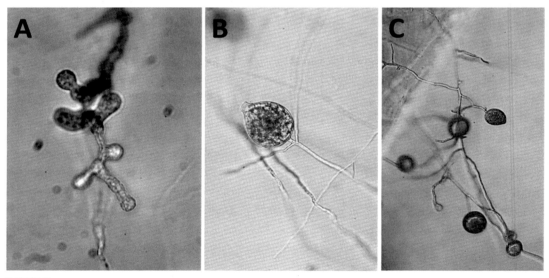

FIGURE 14.8 Typical Oomycetes. (A) Typical sporangium of *Pythium* spp. This group is often found in soil and appears on grass blades after 2–3 days. (B) Typical sporangium of *Phytophthora* spp. This group is also often found in soil and usually appears on grass blades and seeds after 2–3 days. (C) Branching pattern of typical oomycetes. Take note of the sporangiophores and the appearance of the hyphae.

Learning Activity: **Campus Mushrooms**
Year Level: 1st year, preferably for General Mycology or General Biology/Microbiology
Duration: 1–2 h, 2 class meetings
Learning Objective/s: At the end of the activity, the student is expected to identify mushrooms growing within the university campus and search for their potential nutraceutical benefits.

Instructions: The students will be tasked to explore the green spaces within the university campus for fruiting bodies of mushrooms. For this learning activity to be successful, it is important to determine the best time or period when mushroom fruiting bodies will be visibly growing. Ideal period would be during the rainy season, preferably at the beginning of the wet season. The students are tasked to collect and identify these mushrooms by comparing their fruiting body morphologies with descriptions from taxonomic keys and identification guides. Following identification, the students are then tasked to check through literature search if their collected mushrooms are edible. To check for the nutraceutical benefits of these edible mushrooms, the students are encouraged to test for these beneficial properties, for example, the presence of antioxidants, minerals, vitamins, etc. Alternatively, the students can simply do a journal search for the nutritional values of their collected edible mushrooms.

Assessment: At the end of the learning activity, the students will prepare an infographic or info-poster to raise awareness on the importance of mushrooms found in the campus. This will surely create a movement to conserve these mushrooms and promote its sustainable use.

Learning Activity: **Developing a Project Proposal for Drug Discovery with Fungi**
Year Level: 2nd year, preferably for Pharmaceutical Biology or Pharmacy
Duration: 3 h, 2–3 class meetings
Learning Objective/s: At the end of the activity, the student is expected to design a project for the isolation and testing of plant-associated fungi for drug discovery.

Instructions: The class will be initially divided into several groups with three to four members. Students are given ample time to do research of a local host plant with known medicinal value. They are then tasked to come up with their own techniques on how to isolate fungal endophytes from the host plant leaves. At this stage of the learning activity, the teacher is expected to demonstrate the proper way of conducting journal search to avoid using non–peer-reviewed articles from predatory publishers as references. Following their literature review, the students design a project proposal that includes details on the isolation of fungal endophytes from host medicinal plant, the production and extraction of bioactive secondary metabolites, and the testing of these metabolites for different bioactivities. A proposal oral defense can be conducted in class, with other members of the class serving as panel of evaluators. Grading will be based on the feasibility of their drug discovery program. While this may not be required for this learning activity, the students may opt to implement their isolation protocol and isolate fungal endophytes from their chosen host plant. Selected fungal isolates can then be grown on Potato Dextrose Broth (PDB) or Malt Extract Broth at room temperature for 4 weeks to produce secondary metabolites. The fungal cultures are extracted with organic solvent, e.g., ethyl acetate, and the crude culture extracts are tested for their antimicrobial activities using the standard Kirby–Bauer Antimicrobial Assay. Outputs of their experiments are documented and presented as evidence of the feasibility of the proposed study.

358   14. Innovative learning activities for teaching mycology in the Philippines

Assessment: At the end of the learning activity, the students will present a project proposal of their drug discovery strategy. A rubric will be used to grade the content, feasibility, and presentation of their proposed project.

Learning Activity: **Isolating Mangrove Fungi**
Year Level: 2nd—3rd year, preferably for General Mycology or General Microbiology
Duration: 3 h, 2—3 class meetings
Learning Objective/s: At the end of the activity, students are expected to isolate fungal endophytes from mangrove leaves and illustrate/identify the culturable mangrove fungal endophytes.
Instructions: Students will be grouped into four to five member teams. Each group will be tasked to collect mangrove leaves from a nearby coastal area. Students must select only healthy and symptomless leaves and place these in polythene bags. To isolate mangrove fungal endophytes, the students will initially cut leaf samples into 1 cm^2 and surface-sterilize by washing with running tap water for 10 min followed by successive soaking in 1% sodium hypochlorite for 3—4 min and 70% ethanol for 1 min, and finally rinsing three times with sterile distilled water. The leaf explants are placed on PDA plates supplemented with streptomycin sulfate. The students are tasked to monitor the petri plates for 7—14 days until fungal hyphae are visible on the edge of each leaf segment. The fungal hyphae are then reinoculated on fresh PDA medium for observation and illustration. Alternatively, the fungal isolates can be sent for outdoor sequencing and the gene sequences analyzed to confer species identity.
Assessment: At the end of the learning activity, the students submit an illustration of their isolated mangrove fungal endophytes.

Learning Activity: **Determining Secondary Metabolites of Mangrove Fungi**
Year Level: 2nd—3rd year, preferably for General Mycology or Microbial Physiology
Duration: 3 h, 3—4 class meetings
Learning Objective/s: At the end of the activity, students are expected to produce and extract secondary metabolites from mangrove fungal endophytes and determine its chromatographic fingerprint.
Instructions: Continuing from the previous learning activity, the students will select actively growing cultures of mangrove fungal endophytes. These will be inoculated on 250 mL PDB and incubated at room temperature for 14 days. The fungal culture broths are then filtered through muslin cloth and the filtrates extracted thrice with an equal amount of ethyl acetate. The organic component (upper phase) is then collected and evaporated using a rotary evaporator at $\leq 40°C$ until dried. To establish the chromatographic fingerprint of the fungal culture extracts, the students initially dissolve 10 mg of the crude culture extract in 1 mL anhydrous methanol. These are spotted and run-on thin layer chromatographic plate and visualized with UV. Students compute the different Rf values of the different spots corresponding to the different metabolites.
Assessment: At the end of the learning activity, the students submit a written laboratory report of their findings.

Learning Activity: **Determining the Bacteriostatic Activity of Mangrove Fungi**
Year Level: 2nd—3rd year, preferably for General Mycology or General Microbiology
Duration: 3 h, 2—3 class meetings

III. Fungal conservation and management

Learning Objective/s: At the end of the activity, students are expected to test fungal crude culture extracts for its bacteriostatic activities against test bacteria.

Instructions: The fungal culture extracts from the previous learning activity will be tested for its ability to inhibit the growth of test bacteria. Initially, a bacterial suspension of *Escherichia coli* and *Staphylococcus aureus* will be prepared with Nutrient Broth and standardized to 0.5 McFarland. Then, 100 µL of the bacterial suspension will be aseptically inoculated on a previously prepared Mueller-Hinton Agar (MHA) plate using a sterile L-rod spreader. The fungal culture extracts impregnated on filter paper disks are then placed over the inoculated MHA plates and incubated at 37°C for 24 h. Students are expected to measure and record the zones of inhibition, i.e., the diameter of the clearing zones around the paper disks.

Assessment: At the end of the learning activity, the students submit a written laboratory report. Alternatively, students can also be tasked to present their findings as a scientific poster.

Learning Activity: **Testing for Antagonistic Activity using Paired Culture Technique**
Year Level: 3rd year, preferably for General Mycology or Plant Pathology
Duration: 3 h, 2–3 class meetings
Learning Objective/s: At the end of the activity, the student is expected to test fungi for its ability to inhibit growth of other fungi.

Instructions: This learning activity will test the antagonistic activity of the fungus *Trichoderma harzianum* against the plant pathogenic fungus *Helminthosporium maydis*. Both fungi will be provided as live cultures by the faculty-in-charge. Initially, the test fungi are grown on Potato Dextrose Agar (PDA) plates for 3–5 days. Agar disks with a diameter of 6 mm will be cut with a flame-sterilized cork borer from the colony margin of the actively growing fungi. Then, these are inoculated on freshly prepared PDA plates at opposite sides and the plates are then sealed with parafilm and incubated at room temperature for 7 days. After incubation, the students determine the extent of inhibition with the scale listed in Table 14.1.

Assessment: At the end of the learning activity, the students will record their observation and photo document their paired cultures. They are expected to submit a written laboratory report which will be graded based on a grading rubric.

TABLE 14.1 Sample evaluation of the antagonistic activity of fungi.

Score	Description
1	Fungal antagonist overgrew the pathogen and completely covered the entire surface of the culture plate
2	Fungal antagonist grew on at least 2/3 of the surface of the culture plate while the remaining space or less is covered by the plant pathogen
3	Fungal antagonist and the plant pathogen colonized approximately half of the surface of the culture medium
4	Plant pathogen colonized at least two-third of the surface of the culture medium and appeared to tolerate the encroachment of the fungal antagonist
5	Plant pathogen completely overgrew the fungal antagonist and occupied the entire surface of the culture medium

Learning Activity: **Correlating Fruiting Body Morphology with Abiotic Factors**
Year Level: 3rd year, preferably for General Mycology or General Ecology
Duration: 3 h, 1–2 class meetings
Learning Objective/s: At the end of the activity, the student is expected to identify and measure the different fruiting body parts of the cosmopolitan mushroom *Schizophyllum commune* Fr. and correlate their morphometric data with various abiotic factors affecting their growth and development.

Instructions: Students are initially grouped in pairs or team of 3–4 members. They identify their sampling locality. It is advisable for the students to conduct an ocular inspection of their chosen sampling localities for the presence of *S. commune* prior to the actual conduct of the activity. Students are then tasked to collect fruiting bodies of *S. commune*. This is done by cutting the base of the fruiting bodies with a knife. The fruiting bodies are then wrapped in aluminum foil and placed in collecting basket to prevent any damage during transport. Students are expected to collect 30 individual fruiting bodies per substrate. It is also important for the students to record the GPS coordinates where the samples are collected and to describe the substrates where *S. commune* is growing. Different physico-chemical properties of the substrates and the sampling localities must also be determined and recorded, for example, bark pH and moisture content of the wood substrates and the temperature and relative humidity of the sampling area during the collection. For each of the fruiting body, the morphometric data is determined with the use of a ruler. The students will initially measure the lip angle distance, then followed by the claw width, the lamellae length, and the lamellae distance to apex (Fig. 14.9). All data are recorded in a table.

Assessment: At the end of the learning activity, the students will complete and submit the data sheet (see Table 14.2) and correlate the fruiting body morphometry with the different abiotic factors. This can be done with the correlation function in Microsoft Excel. Students may be tasked to give an oral presentation of their outputs in class.

Learning Activity: **Assessing Tolerance of Fungi to Environmental Pollutants**
Year Level: 3rd–4th year, preferably for General Mycology or General Ecology

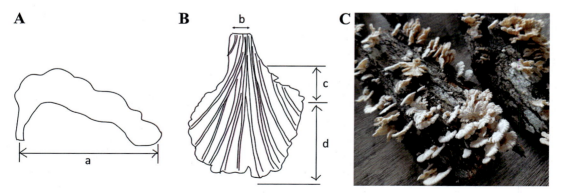

FIGURE 14.9 Morphometric measurement of the fruiting bodies of *Schizophyllum commune*. (A) lip, side view showing lip angle distance. (B) lip, front view, showing the claw width (b), lamellae length (c), and lamellae distance to apex (d). Owing to the uncanny similarity of the flower petal of *Corallorhiza* to the fruiting body morphology of *S. commune*, we adopted this morphometric measurement for our mushroom. Illustration is adopted from Freudenstein and Barrett (2014). (C) Fresh fruiting bodies of *S. commune* growing on fallen logs.

4. Innovative teaching strategies on fungi for undergraduate courses

TABLE 14.2 Student worksheet for the recording of the morphometric data of *S. commune*.

Substrate type	Sample no.	Morphometric measurement				Abiotic factors (Ave.)			
		Lip Angle	*Claw Width*	*Lamellae Length*	*Lamellae Distance*	*pH*	*Moisture*	*Temp*	*Humidity*
	1								
	2								
	3								
	4								
	5								
	etc								
	Mean								
	Correlation Coefficient[a]								

[a]*Interpretation: 1.00 (perfect correlation), 0.80–0.99 (very strong correlation), 0.60–0.79 (strong correlation), 0.40–0.59 (moderate correlation), 0.20–0.39 (weak correlation), 0.01–0.19 (very weak correlation), 0.00 (no correlation).*

Duration: 3 h, 2–3 class meetings

Learning Objective/s: At the end of the activity, the student is expected to test the ability of fungi to tolerate different heavy metals.

Instructions: In this learning activity, different species of filamentous fungi will be provided by the teacher as test organisms. These fungi will be initially grown on Potato Dextrose Agar plates for 7 days, from which agar disks are cut from the colony margin with a flame-sterilized cork borer. The fungal agar disks (6 mm in diameter) are inoculated at the center of PDA plates supplemented with 1 mM of different metal salts, for example, $CdCl_2$, $CuSO_4$, $PbSO_4$, $AsSO_4$, and Fe_2SO_4. The culture plates are incubated at 25°C for 12 days, during which the colony diameter is measured with a ruler every 2 days. The heavy metal tolerance potential of the fungal species, expressed as Tolerance Index, is calculated as the mean diameter of the fungal colony growing on PDA with heavy metal over the mean diameter of the fungal colony growing on PDA without heavy metal (Oladipo et al., 2018). Then, the heavy metal tolerance is rated as: 0.00–0.39 (very low tolerance), 0.40–0.59 (low tolerance), 0.60–0.79 (moderate tolerance), 0.80–0.99 (high tolerance), and 1.00–>1.00 (very high tolerance), with the higher values indicating higher tolerance by the fungi to the heavy metal (Oladipo et al., 2018). Following completion of the learning activity, it is important for the teacher to discuss the proper disposal of the culture plates and the dangers of heavy metal waste.

Assessment: At the end of the learning activity, the students will submit a student worksheet with their recorded values. The worksheet can be graded by the teacher.

Learning Activity: **Developing Value-Added Products with Mushrooms**

Year Level: 3rd–4th year, preferably for the course Bio-entrepreneurship

Duration: 3 h, 2–4 class meetings

Learning Objective/s: At the end of the activity, the student is expected to develop mushroom product and its marketing strategies and demonstrate integration of mycology into business.

III. Fungal conservation and management

Instructions: The students are initially grouped into teams of four to five members. They are tasked to improve or add value to a product of their choice by incorporating mushrooms. Students are then expected to do research on how the product will be developed. A prototype product can be made by the students which can be showcased during class-sponsored fairs or event. Additionally, the students may also conduct nutritional analysis and sensory evaluation of their developed mushroom-enriched product. They will also create packaging and other promotional materials for their product and design their own marketing strategies. Outputs of their project are presented as a case study paper.

Assessment: At the end of the learning activity, the students will submit a case study paper and present a product demonstration. A grading rubric will be used to assess the uniqueness of the proposed product and its packaging and the feasibility of its marketing strategies.

Learning Activity: **Mushroom Appreciation at Home**
Year Level: 4th year, preferably for General Mycology or General Ecology
Duration: 3 h, 2–4 class meetings
Learning Objective/s: At the end of the activity, the student is expected to cultivate mushrooms (basidiomycetes) in their home, demonstrate the usefulness of basidiomycetes in their daily home activities, and create a diary or scrapbook entry to show his/her appreciation of mushrooms.

Instructions: The learning activity is designed mainly for a small class and hence, the students will work individually. It starts with a lecture on the taxonomy of basidiomycetes and a video presentation on the cultivation of mushrooms. The students will also be provided with a step-by-step guide on mushroom cultivation which can be provided as a printed handout or can easily be deployed through online module of the university's learning management system. What makes this learning activity unique is its implementation as a home-based experiment, which lend itself to use in distance-education courses or during online learning. The students are expected to prepare or acquire mushroom spawns from local mushroom growers and take care of their own mushroom cultures at home. During the conduct of the learning activity, the students are also tasked to photo-document the growth of their mushrooms (Fig. 14.10) and to write an entry on their diary to record their learnings about basidiomycetes as they observe them growing and the challenges that they encountered caring for their "baby mushrooms." To cap the learning activity, students will demonstrate different uses of mushrooms, for example, through cooking, art, or in any engaging activity.

Assessment: At the end of the learning activity, the students will prepare and submit their diary or a scrapbook. A grading rubric will be used to evaluate their outputs. A sample diary entry is provided below.

A sample diary entry of a student.

"Since I was young, mushrooms have always been a gastronomic delight. My usual encounter with mushrooms is already on the plate - cooked and ready to be chomped down. But during the pandemic, I became more curious about how mushrooms grow as I watched the documentary film Fantastic Fungi: The Magic Beneath Us. The film illustrated how mycelium acts as the 'wood wide web' — connecting species and that mushrooms are only tip of the story.

This fascination then urged me to try to take care of oyster mushroom fruiting bags myself as 2022 started. As a beginner in the mycoworld, I learned that for them to thrive — a proper weaving of fresh air exchange, humidity and sunlight must be in place.

Each day, I am excited to wake up and commune with them: my day comprises of multiple misting visits … more so, having mushrooms as companion species during this pandemic

FIGURE 14.10 Sample photo of mushrooms being taken care by the students at home. *Photocredit: EM Wang, used with permission.*

taught me about the 'art of noticing'. Each time, it feels like encountering this life form in a new yet seemingly familiar way.

When nothing seems to be happening, they offer a random surprise: tiny pins suddenly emerge, then double in size in just a span of a day or two. In its magical unfolding, they grow into a thrilling harvest. Each harvest is a palpable testament that mushrooms are only the tips of a larger life process. A reminder of how each bountiful harvest is life in decay.

Overall, this growing experience is humbling. Each visit is an invitation to notice that I am not all in control. The unfolding reminds me of how our modern Anthropocentric (human-centric) lifestyle is too caught up in objectifying life forms and mechanically seeing them as products rather than as living beings. Encountering mushrooms in this context of relational care becomes an unexpected and poignant metabolic literacy. With this experience, I had a glimpse of why mycologist Merlin Sheldrake says that our close encounters with the fungi world increases our doses of awe and vulnerability. This newfound awe opens a journey for us humans into the world of taken-for-granted beauty, and that we are invited to witness the wonder of art, science, and life together." - EM Wang (BSc Biology Batch 2021, used with permission).

5. Safety when working with fungi

Fungi are generally safe to use in class. While susceptible individuals may show minor allergic reactions due to inhalation of fungal spores, handling mushrooms and other fungi inside classroom or in the field does not have any serious threat or health risks. However, it is still very important for teachers to remind their students of basic safety protocols when working with fungi. For example, prior to the conduct of any class excursion to collect specimens in the field, parental consent must be secured. When sampling natural parks or community forests, collecting and transport permits must be secured by the teachers from appropriate local government units or from the Department of Environment and Natural Resources. It is important to highlight compliance with local and national government laws and regulations among our students. During classroom activities with fungi, safety must still be taken into consideration. Proper protective covering such as laboratory gown or apron must be required for students. Though the learning activities presented in this chapter use nonpathogenic, nonharmful fungal specimens, it is still prudent to remind our students to treat these fungi or any microorganisms "as potentially harmful" to avoid carelessness during handling and experimentation. It is also equally important to teach the students proper waste disposal, for example, decontamination of any living fungal cultures prior to disposal, and to observe correct hygiene practices.

6. Concluding remarks

The teaching of mycology in developing countries like the Philippines needs not require highly sophisticated equipment and facilities. The use of locally available materials and fungal specimens and the study of local areas will provide our Filipino students with a "more realistic," "local feel," "direct connection," and "closer-to-home" (Philippine setting) learning experience. We also reiterate the need to develop engaging learning activities that promote the active learning of fungi and its many facets. With these innovative learning activities, appreciation of and studies on fungi increase which can significantly lead to the further development of the science of mycology in the Philippines.

References

Carascal, M. B., Aril-dela Cruz, J. V., De Leon, A. M., & dela Cruz, T. E. E. (2020). *The TJ Mushketeers: A journey to the mushroom Land. Fungal biodiversity* (p. 20). Ecogenomis & Systematics Booklet III.

Carascal, M. B., Pecundo, M. H., & dela Cruz, T. E. E. (2016). *The adventures of Myxoman and Amoeboy* (p. 20). Fungal Biodiversity & Systematics Booklet II.

Clapton, W. (2015). Pedagogy and pop culture: Pop culture as teaching tool and assessment practice. In F. Caso, & C. Hamilton (Eds.), *Popular culture and world politics: Theories, methods, pedagogies* (pp. 169–175). E-international Publication.

dela Cruz, T. E. E. (2014). "Pick-up lines": A fun way to facilitate learning microbiological concepts. *Journal of Microbiology & Biology Education, 15*(2), 299–300.

dela Cruz, T. E. E. (2020). Learning is fun! Developing an outcome-based, engaging learning activities in microbiology. In A. K. Raymundo, & G. T. Pawilen (Eds.), *Outcomes-based education (OBE) in microbiology* (pp. 31–41). National Academy of Science and Technology (NAST). Monograph Series 23.

dela Cruz, T. E. E., & Aril-dela Cruz, J. V. (2018). Communicating science through editorial cartoons in microbiology classrooms. *Journal of Microbiology & Biology Education, 19*(1). https://doi.org/10.1128/jmbe.v19i1.1496

dela Cruz, T. E. E., Dagamac, N. H. A., Torres, J. M. O., Santiago, K. A. A., & Yulo, P. R. J. (2013). Mycology. In V. L. Barraquio (Ed.), *Proceedings of the 22nd annual scientific meeting of the Philippine academy of microbiology* (pp. 61–68). ISSN: 2345-8453.

References

dela Cruz, T. E. E., & Eloreta, M. F. B. M. (2020). Myxomycetes in our backyard: Species exploration amidst the COVID-19 pandemic. *Antoninus Journal, 6* (COVID-19 special issue).

dela Cruz, T. E. E., & Olayta, C. O. M. (2022). Citizen taxonomy in social media: The use of Facebook for mapping species distribution of myxomycetes. *American Biology Teacher, 84*(4), 189–194.

dela Cruz, T. E. E., Paguirigan, J. A. G., & Aril-dela Cruz, J. V. (2021). Easy-to-do assessment tasks to create info-posters and infographics for communicating hot-button science issues. *The American Biology Teacher, 83*(7), 482–485.

dela Cruz, T. E. E., Pangilinan, M. V. P., Dagamac, N. H. A., Torres, J. M. O., Santiago, K. A. A., & Macabago, S. A. B. (2013b). The "meet a microbiologist" (MAM) program: A learning strategy for teaching microbiology and inspiring students. *The Asian Journal of Biology Education, 7*, 25–30.

dela Cruz, T. E. E., Pangilinan, M. V. B., & Litao, R. A. (2012). Printed identification key or web-based identification guide: An effective tool for species identification? *Journal of Microbiology & Biology Education, 13*(2), 180–182.

Dogma, I. J., Jr. (1975). Of Philippine mycology and lower fungi. *Kalikasan: the Philippine Journal of Biology, 4*, 69–105.

Estampador, F. R. F., Esteva, M. S. D. S., Garcia, N. F. G., Baylosis, D. C. G., Dagamac, N. H. A., & dela Cruz, T. E. E. (2022). A home-based experiment with myxomycetes for teaching basic ecological concepts. *SlimeMolds, 2*, V2A13.

Freudenstein, J. V., & Barrett, C. F. (2014). Fungal host utilization helps circumscribe leafless Coral root orchid species: An integrative analysis of *Corallorhiza odontorhiza* and *C. wisteriana*. *Taxon, 63*(4), 759–772.

Garcia, N. F. G., Baylosis, D. C. G., Esteva, M. S. D. S., Estampador, F. R. F., Dagamac, N. H. A, & dela Cruz, T. E. E. (2022). Assessment of natural and synthetic substrates as spore traps for myxomycetes. *SlimeMolds, 2*, V2A16.

Guerrero, J. J. G. (2020). Insights and prospects toward the undergraduate mycological researches of Bicol University. *Philippine Journal of Science, 149*(2), 405–413.

Macabago, S. A. B., & dela Cruz, T. E. E. (2012). Development of a myxomycete photoguide as a teaching tool for microbial taxonomy. *Journal of Microbiology & Biology Education, 13*(1), 67–69.

Mendoza, B. C. (2020). Learning is fun! Developing an outcome-based, engaging learning activities in microbiology. In A. K. Raymundo, & G. T. Pawilen (Eds.), *Outcomes-based education (OBE) in microbiology* (pp. 42–54). National Academy of Science and Technology (NAST). Monograph Series 23.

Oladipo, O. G., Awotoye, O. O., Olayinka, A., Bezuidenhout, C. C., & Maboeta, M. S. (2018). Heavy metal tolerance traits of filamentous fungi isolated from gold and gemstone mining sites. *Brazilian Journal of Microbiology, 49*(1), 29–37.

Rakower, J., & Hallyburton, A. (2022). Disease information through comics: A graphic option for health education. *Journal of Medical Humanities, 17*, 1–18.

Saji, S., Venkatesan, S., & Callender, B. (2021). Comics in the time of a pandemic: COVID-19, graphic medicine, and metaphors. *Perspectives in Biology and Medicine, 64*(1), 136–154.

Santiago, K. A. A., Gazo, S. M. T., & dela Cruz, T. E. E. (2015). *The search for the missing lichens: The adventures of Thallus Tanoa* (p. 20). Fungal Biodiversity & Systematics Booklet I.

Sanvictores, R.R., Gomez, M.C.C., Briones, S.L.R., Chang, W.T.M., Morales, A.M.A., Dagamac, N.H.A., dela Cruz, T.E.E (2022) Do fungal spore morphological traits correlate with allergenicity? *MycoAsia* 2022/05.

Scavone, P., Carrasco, V., Umpiérrez, A., Morel, M., Arredondo, D., & Amarelle, V. (2019). Microbiology can be comic. *FEMS Microbiology Letters, 366*(14), fnz171. https://doi.org/10.1093/femsle/fnz171

Wang, J. L., Acevedo, N., & Sadler, G. R. (2018). Using comics to promote colorectal cancer screening in the Asian American and Pacific Islander communities. *Journal of Cancer Education, 33*(6), 1263–1269.

Winsett, K. E., dela Cruz, T. E. E., & Wrigley de Basanta, D. (2017). Myxomycetes in education. The use of these organisms in promoting active and engaged learning. In S. L. Stephenson, & C. A. Rojas (Eds.), *Myxomycetes: Biology, Systematics, biogeography and Ecology* (pp. 389–412). Elsevier Academic Press.

Winsett, K. E., dela Cruz, T. E. E., & Wrigley de Basanta, D. (2022). Myxomycetes in education. The use of slime molds in promoting active and engaged learning. In S. L. Stephenson, & C. A. Rojas (Eds.), *Myxomycetes: Biology, Systematics, biogeography and Ecology* (pp. 533–563). Elsevier Academic Press.

Index

'*Note*: Page numbers followed by "f" indicate figures and "t" indicate tables.'

A

Acanthus ebracteatus, 21t–24t
Achlya ambisexualis, 69t–73t, 79t
Achlya Americana, 69t–73t
Achlya apiculata, 69t–73t
Achlya bisexualis, 69t–73t
Achlya flagellata, 69t–73t
Achlya proliferoides, 69t–73t, 79t
Achyla americana, 79t
Achyla bisexualis, 79t
Achyla flagellata, 79t
Acremonium kiliense, 36t–63t
Acremonium strictum, 36t–63t
Acrocordiopsis patilii, 36t–63t
Acrocordiopsis sphaerica, 36t–63t
Acrodictys liputii, 36t–63t
Acrodictys sacchari, 36t–63t
Acrodontium crateriforme, 36t–63t
Acrogenospora sphaerocephala, 36t–63t
Aegiceras corniculatum, 21t–24t
Aegiceras floridum, 21t–24t
Agriculture, biocontrol agents in, 245–253, 246t–250t
Agroforestry ecosystem, 92–93
Aigialus grandis, 36t–63t
Aigialus mangrovei, 36t–63t
Aigialus parvus, 36t–63t
Albugo candida, 79t
Albugo ipomoeae-panduratae, 79t
American Type Culture Collection (ATCC), 295
Aniptodera chesapeakensis, 36t–63t
Aniptodera inflatiascigera, 36t–63t
Aniptodera intermedia, 36t–63t
Aniptodera lignicola, 36t–63t
Aniptodera longispora, 36t–63t
Aniptodera mangrovei, 36t–63t
Annulatascus liputii, 36t–63t
Annulatascus velatisporus, 36t–63t
Antennospora nypae, 36t–63t
Antennospora nypensis, 36t–63t
Antennospora quadricornuta, 36t–63t
Antennospora salina, 36t–63t
Aphanomyces cladogamus, 79t

Aphanomyces helicoides, 69t–73t, 79t
Aphanomyces keratinophilus, 69t–73t, 79t
Aphanomyces laevis, 69t–73t, 79t
Apodachlya minima, 79t
Aquaphila albicans, 36t–63t
Arachidonic acid, 28
Arbuscular mycorrhizal fungi (AMF), 250–251
Arenariomyces trifurcatus, 36t–63t
Arthrinium phaeospermum, 36t–63t
Ascocratera manglicola, 36t–63t
Ascomycete, 93
Ascomycota, 26t, 35–63
Aspergilloides, 36t–63t
Aspergillosis, 215–218
 clinical management, 217–218
 clinical presentation, 217–218
 diagnostic strategies, 216–217
 epidemiology, 215–216
Aspergillus candidus, 36t–63t
Aspergillus flavus, 36t–63t
Aspergillus fumigatus, 36t–63t
Aspergillus fumisynnematus, 36t–63t
Aspergillus nidulans, 36t–63t
Aspergillus niger, 36t–63t
Aspergillus niveus, 36t–63t
Aspergillus novofumigatus, 36t–63t
Aspergillus ochraceus, 36t–63t
Aspergillus oryzae, 36t–63t
Aspergillus penicilloides, 36t–63t
Aspergillus restrictus, 36t–63t
Aspergillus sclerotiorum, 36t–63t
Aspergillus sydowii, 36t–63t
Aspergillus tamarii, 36t–63t
Aspergillus terreus, 36t–63t
Association of Systematic Biologists of the
 Philippines (ASBP), 7
Astrosphaeriella mangrovei, 36t–63t
Astrosphaeriella papillata, 36t–63t
Astrosphaeriella stellata, 36t–63t
Astrosphaeriella striatispora, 36t–63t
Astrosphaeriella tornata, 36t–63t
Avicennia alba, 21t–24t

368 Index

Avicennia marina, 21t–24t
Avicennia officinalis L., 21t–24t
Avicennia rumphiana, 21t–24t

B

Bacteriostatic activity of mangrove fungi, 358
Bactrodesmium longisporum, 36t–63t
Bactrodesmium stilboideum, 36t–63t
Baiting Aquatic Fungi, 350
Basidiomycete, 93
Basidiomycota, 35–63
Batanes Protected Landscape and Seascape, 123
Bathyascus grandisporus, 36t–63t
Biatriospora marina, 36t–63t
Bicol University (BU), 4
Bioactivities, 285, 286t–287t
Biodiversity, 1
Biodiversity Conservation Society of the Philippines (BCSP), 7
Biological control agents (BCAs), 245
Biological resource centers (BRCs)
  microbial culture collections (MCC), 294–302, 295f
    American Type Culture Collection (ATCC), 295
    Castellani's procedure, 307–308
    cryopreservation methods, 309
    freeze-drying, 308
    functions and types of, 294–295
    fungal resources of, 305–309, 306t
    history of, 296
    lyophilization, 308
    oil overlay, 306–307
    opportunities and challenges in, 310–312
    Philippine Network of Microbial Culture Collections (PNMCC, Inc.), 301–302, 303t–304t
    short-term preservation, 306
    sporulating filamentous fungi, 308
    University of the Philippines Culture Collection (UPCC), 299–300
  mycological herbarium (MH), 313–317
    challenges in, 317–319
    collections, 315–316
    fungal herbarium specimens, preservation methods for, 316
    opportunities, 317–319
  valuable biological materials (VBMs), 293–294
Biology Teachers Association of the Philippines (BIOTA), 7
Bioluminescence fungi, 137–139, 138f
  in Philippines, 143
  in Southeast Asia, 139–143, 140t–141t
    Indonesia, 142
    Malaysia, 139–141
    Singapore, 141, 142f

uses and applications of, 143–144
Blastocladia angusta, 36t–63t
Blastocladia globosa, 36t–63t
Blastocladia gracilis, 36t–63t
Blastocladia incrassata, 36t–63t
Blastocladia pringsheimii, 36t–63t
Blastocladia ramosa, 36t–63t
Blastocladia sparrowii, 36t–63t
Boerlagiomyces grandisporus, 36t–63t
Bremia lactucae, 79t
Bruguiera cylindrica, 21t–24t
Bruguiera gymnorhiza, 21t–24t
Bruguiera parviflora, 21t–24t
Bruguiera sexangula, 21t–24t
Buenavista Protected Landscape (BPL), 125
  Initao-Buenavista protected landscape and seascape, 125
  macrofungi composition in, 125

C

Camposporium fusisporum, 36t–63t
Camposporium quercicola, 36t–63t
Campus Mushrooms, 357
Canarium ovatum, 4
Candelabrum brocchiatum, 36t–63t
Candida albicans, 4
Candida famata, 36t–63t
Candida guilliermondii, 36t–63t
Candidiasis
  candidal paronychia, 220–221
  diagnostic strategies, 220
  epidemiology, 218–220
  invasive candidiasis, 222–223
  mucocutaneous candidiasis, 221
  vulvovaginal candidiasis, 221–222
Carinispora nypae, 36t–63t
Caryospora callicarpa, 36t–63t
Caryospora minima, 36t–63t
Caryosporella rhizophorae, 36t–63t
Castellani's procedure, 307–308
Catanduanes Watershed Forest Reserve, 122
Catenaria anguillulae, 36t–63t
Cavinti Underground River and Cave Complex (CURCC), 126–127
Central Luzon State University (CLSU), 3–4
Ceriops decandra, 21t–24t
Ceriops tagal, 21t–24t
Ceriops zippeliana, 21t–24t
Chloridium cylindrosporum, 36t–63t
Chytridium lagenaria, 36t–63t
Chytridium oedogoniarum, 36t–63t
Chytridium palmelloidea, 36t–63t
Chytridium schenkii, 36t–63t

Index

369

Chytriomyces annulatus, 36t–63t
Chytriomyces appendiculatus, 36t–63t
Chytriomyces tabellariae, 36t–63t
Cirrenalia tropicalis, 36t–63t
Cladochytrium cladosporioides, 36t–63t
Cladochytrium replicatum, 36t–63t
Cladochytrium sphaerospermum, 36t–63t
Clavatospora bulbosa, 36t–63t
Coelomomyces indiana, 36t–63t
Coelomomyces quadrangulatus, 36t–63t
Coelomomyces stegomyiae, 36t–63t
Colletotrichum falcatum, 175
Colletotrichum fructicola, 36t–63t
Colletotrichum queenslandicum, 36t–63t
Colletotrichum siamense, 36t–63t
Colletotrichum tropicale, 36t–63t
Convention on Biological Diversity, 271–272
Cordana abramovii, 36t–63t
Coriolaceae, 119
Coronopapilla mangrovei, 36t–63t
Crop loss studies, 194–196
   maize banded leaf, 195
   maize stenocarpella disease complex, 196
   multiple rice fungal diseases, 195
   perennial crops, 196
   rice blast, 194, 194t
   rice sheath blight, 195
   sheath blight, 195
Cryopreservation methods, 309
Cryptococcus, 36t–63t
Cucullosporella mangrovei, 36t–63t
Curvularia intermedia, 36t–63t
Cylindrochytridium johnstonii, 36t–63t
Cytospora rhizophorae, 36t–63t

D

Dactylaria africana, 36t–63t
Dactylaria halioptrepha, 36t–63t
Dactylaria haliotrepha, 36t–63t
Dactylaria longidentata, 36t–63t
Dendryphiella salina, 36t–63t
De novo fatty acid biosynthesis, 75–76, 76f
Diaporthe siamensis, 36t–63t
Dictyochaeta curvispora, 36t–63t
Dictyochaeta plovercovensis, 36t–63t
Dictyomorpha dioica, 36t–63t
Dictyosporium heptasporum, 36t–63t
Dictyuchus anomalus, 69t–73t
Didymella aptrootii, 36t–63t
Digitodesmium bambusicola, 36t–63t
Dinagat Island Watershed, 122–123
Diplophlyctis asteroidea, 36t–63t
Diplophlyctis complicata, 36t–63t

Dipterocarps, 91
DNA sequencing, 255–256
Docosahexaenoic acid (DHA), 28

E

Edible mushrooms
   bioactivities, 285, 286t–287t
   choice of mushroom species, 283
   different indigenous substrata, mycelial growth on, 283
   different physicochemical parameters, mycelial growth at, 283–284
   ethnomycology, 272–273
   mass production, 285
   mushroom species, choice of, 283
   mycochemicals, 285
   pure mycelial culture, isolation to, 283
   spawn inoculum preparation, 284
   species listing of, 273–275, 274f
   in vitro culture of, 275–285, 276t–281t
Eicosapentaenoic acid (EPA), 28
Ellisembia adscendens, 36t–63t
Ellisembia vaginata, 36t–63t
Endolichenic fungi (ELF), 156–157
Endomycorrhizal fungi, 91–92
Engyodontium album, 36t–63t
Entophlyctis confervae-glomeratae, 36t–63t
Enumeration of Philippine Fungi, 2
Ethnomycology, 3–4, 272–273
Eupenicillium javanicum, 36t–63t
Excoecaria agallocha, 21t–24t
Export commodities, pests preventing movement with, 332–333

F

Flora de Filipinas, 2
Fluminicola bipolaris, 36t–63t
Forest formation, 90–92
   beach forest, 92
   forest over limestone/karst forest, 91–92
   fungal diversity and mycological collection, 94–128, 95f, 96t–118t
      forest and watershed reservation, 119–123
      mycological collections, 127–128
      national parks and wildlife sanctuary, 125–127
      protected landscape and seascape, 123–125
      wetlands and peatlands, 127
   peat swamp forest, 92
   tropical lower montane, 91
   tropical lowland evergreen forest, 90–91
   tropical semievergreen rainforest, 91
   tropical upper montane, 91
Forest fungi, 92–94

370 Index

Forest/watershed reservation, 119–123
  Catanduanes Watershed Forest Reserve, 122
  Dinagat Island Watershed, 122–123
  La Mesa Watershed Reservation (LMWR), 121
  Malagos Watershed Reservation, 122
  Mount Makiling Forest Reserve (MMFR), 121–122
  Nueva Vizcaya Watershed, 120
  Palali-Mamparang Mountain Range (PMMR), 120
  San Fernando Forest, 119
  Sierra Madre Mountain Range, 119
  Subic Watershed Forest Reserve, 120
Foxfire, 138
Freeze-drying, 308
Fungal Biodiversity, Ecogenomics and Systematics
    (FBeS), 3
Fungal diversity, 94–128
Fungal endophytes, 29, 35–63
Fungal growth requirements, 94
Fungal plant diseases, epidemiology of
  Aspergillus ear rots, 189–190
  climate change, 190, 205–206
    maize, 205
    perennial crops, 206
    rice, 205
  crop loss studies, 194–196
    maize banded leaf, 195
    maize stenocarpella disease complex, 196
    multiple rice fungal diseases, 195
    perennial crops, 196
    rice blast, 194, 194t
    rice sheath blight, 195
    sheath blight, 195
  disease-yield loss modeling, 204
  empirical and mechanistic models, 190
  fusarium, 189–190
  Maize stenocarpella disease complex, 193
  perennial crops, 193–194
  plant disease modeling, 194t, 197–204
    maize, 203
    perennial crops, 203–204
    rice blast, 197–202, 200t–201t
    rice sheath blight, 202–203
  rice blast, 189, 191
  rice brown spot, 189, 192
  rice sheath blight, 189, 192
  survey studies, 196–197
Fungal plant pathogens of quarantine
  controlling entry practices, exit and epidemics,
      331–334
    export commodities, pests preventing movement
        with, 332–333
    pest epidemics, preventing, 333
    pest incursions, preventing, 331–332

current policy, 325–327
disease epidemics, 333–334
fungal pathogens with high impact in, 328–331
  downy mildew on corn, 329–330
  panama disease of banana, 328–329
  phytophthora on coconut, 328
history, 325–327
plant pest, 333–334
plant quarantine protocols, 325–327
protocols, 327
  domestic transport, 327
  importation, 327
South American leaf blight (SALB) on rubber,
    330–331
Fusarium chlamydosporum, 36t–63t
Fusarium oxysporum, 36t–63t
Fusarium proliferatum, 36t–63t
Fusarium solani, 36t–63t

G
Gaddang indigenous communities, 272–273
Ganodermataceae, 119
Geographic information system (GIS), 190
Glomeromycota, 93
Gonapodya polymorpha, 36t–63t
Gonapodya prolifera, 36t–63t
Graphidacea, 150–151
Guignardia mangiferae, 36t–63t

H
Haematonectria haematococca, 36t–63t
Haliphthoros milfordensis, 69t–73t, 79t
Haliphthoros philippinensis, 69t–73t, 79t
Halocyphina villosa, 36t–63t
Halomassarina thalassiae, 36t–63t
Halophytophthora batemanensis, 69t–73t
Halophytophthora exoprolifera, 69t–73t
Halophytophthora porrigovesica, 69t–73t
Halophytophthora vesicula, 69t–73t
Halorosellinia oceanica, 36t–63t
Halosarpheia heteroguttulata, 36t–63t
Halosarpheia lotica, 36t–63t
Halosarphiea marina, 36t–63t
Heavy metals, 238–242, 239t–241t
Helicascus kanaloanus, 36t–63t
Helicorhoidion nypicola, 36t–63t
Helicosporium gigasporum, 36t–63t
Hinulugang Taktak Protected Landscape, 124
Horae Physicae Berolinensis, 2
Hortaea werneckii, 36t–63t
Hydea pygmea, 36t–63t
Hydrocarbons, 256–258, 257t
Hypoxylon oceanicum, 36t–63t

I

International Rice Research Institute (IRRI), 191, 195
Ityorhoptrun verruculosum, 36t—63t

J

Jahnula seychellensis, 36t—63t

K

Kallichroma tethys, 36t—63t
Kirschsteiniothelia elaterascus, 36t—63t

L

Labyrinthulomycota, 29
Labyrinthulomycota, 26t
Lagenidium giganteum, 69t—73t, 79t
Lagenidium humanum, 79t
Lagenidium oophilum, 79t
Lagenidium pygmaeum, 69t—73t, 79t
La Mesa Watershed Reservation (LMWR), 121
Lasiodiplodia theobromae, 36t—63t, 175
Leptolegniella keratinophilum, 79t
Leptomitus lacteus, 69t—73t, 79t
Lichens
  bioactive secondary metabolites, 148—149
  definition of, 147—148
  elements, 148
  multiple microbial partners, 148f
  natural habitats, 149f
  Philippines, 150
    antimicrobial activities, 153—155
    antioxidant activities, 155
    biological activities of, 154t
    cytotoxic activities, 155
    endolichenic fungi (ELF), 156—157
    herbicidal activities, 156
    natural product research, 151—156, 153f
    taxonomic diversity of, 150—151, 151f, 152t
Lichtheimia ramosa, 36t—63t
Lignincola laevis, 36t—63t
Lignincola longirostris, 36t—63t
Lignincola nypae, 36t—63t
Lignincola tropica, 36t—63t
Lineolata rhizophorae, 36t—63t
Linocarpon angustatum, 36t—63t
Linocarpon appendiculatum, 36t—63t
Linocarpon bambusicola, 36t—63t
Linoleic acid, 28
Linolenic acid, 28
Local journals, 7, 8t—10t
Lophiostoma bipolare, 36t—63t
Lophiostoma mangrovei, 36t—63t
Lulworthia grandispora, 36t—63t
Lumnitzera littorea, 21t—24t
Lumnitzera racemosa, 21t—24t

M

Macroalgae, 63—64
Macrofungi, 91—92
Malagos Watershed Reservation, 122
Manglicolous fungi, 18
Mangrove ecosystems, oomycetes in, 65—74, 68f, 79t
Mangrove forests, 35—63, 36t—63t
Mangrove fungi, 18—19, 358
Mangrove leaves
  applications, 25—26
  decaying, 26—28
    applications, 27, 28t
    diversity, 26
    fungi genera isolation, 26, 26t, 27f
  diversity, 20
  fungal endophytes, 20, 21t—24t
    colonies and conidia of, 20, 25f
Mangrovispora pemphii, 36t—63t
Marasmiellus palmivorus, 36t—63t
Marine/marine-derived fungi, 64—65
Marine oomycetes
  biotechnological applications of, 74—75
  de novo fatty acid biosynthesis, 75—76, 76f
  fatty acid production and profiling, 76—80, 77f
  in mangrove ecosystems, 65—74, 68f, 79t
Marine yeasts, 34—65
Marinosphaera mangrovei, 36t—63t
Massarina thalassioidea, 36t—63t
Mass production, 285
Mean colony extension rates (MCER), 67
Microbial consortia, 242
Microbial culture collections (MCC), 294—302, 295f
  American Type Culture Collection (ATCC), 295
  Castellani's procedure, 307—308
  cryopreservation methods, 309
  freeze-drying, 308
  functions and types of, 294—295
  fungal resources of, 305—309, 306t
  history of, 296
  lyophilization, 308
  oil overlay, 306—307
  opportunities and challenges in, 310—312
  Philippine Network of Microbial Culture Collections
    (PNMCC, Inc.), 301—302, 303t—304t
  short-term preservation, 306
  sporulating filamentous fungi, 308
  University of the Philippines Culture Collection
    (UPCC), 299—300
Mine tailings contamination, 192
Molecular Identification of Fungi using DNA
  sequences, 353
Monodictys levis, 36t—63t
Monodictys monilicellularis, 36t—63t
Monotosporella microaquatica, 36t—63t

372   Index

Morosphaeria ramunculicola, 36t—63t
Morosphaeria velatispora, 36t—63t
Mount Arayat National Park, 125—126
Mount Banahaw and San Cristobal Protected
     Landscape (MBSCPL), 124
Mount Makiling Forest Reserve (MMFR), 121—122
Mt. Hamiguitan Range Wildlife Sanctuary
     (MHRWS), 127
Mt. Palaypalay-Mataas na Gulod Protected
     Landscape (MPMGPL), 123
Mueller-Hinton Agar (MHA), 359
Museum of Natural History (MNH), 2—3
Mushroom Treasure Hunt, 352
Mycelia, 138—139
Mycena illuminans, 139—141
Mycena manipularis, 143
Mycena pruinoso-viscida, 139—141
Mycological herbarium (MH), 313—317
     challenges in, 317—319
     collections, 315—316
     fungal herbarium specimens, preservation methods
         for, 316
     opportunities, 317—319
Mycological Society of the Philippines (MSP), 5,
     6t—7t
Mycology
     safety when working with fungi, 363—364
     teaching
         high school students, learning activities on,
             350—354, 351f, 353f
         Mushketeers, 347
         Myxoman and Amoeboy, adventures of, 346
         preschool and elementary levels, kids in, 345—349
         undergraduate courses, 354—363, 355f
Mycoremediation, 236—237, 237f
     heavy metals, 238—242, 239t—241t
     hydrocarbons, 256—258, 257t
Mycorrhizas, 93
Mycosis
     aspergillosis, 215—218
         clinical management, 217—218
         clinical presentation, 217—218
         diagnostic strategies, 216—217
         epidemiology, 215—216
     candidiasis
         candidal paronychia, 220—221
         diagnostic strategies, 220
         epidemiology, 218—220
         invasive candidiasis, 222—223
         mucocutaneous candidiasis, 221
         vulvovaginal candidiasis, 221—222
     chemoprophylaxis, 228
     other fungal infections, 223—228

     clinical management, 225—228
     clinical presentation, 225—228
     diagnostic strategies, 224—225
     epidemiology, 223—224
     overview of, 213—215
     subcutaneous mycosis, 215
MYKOVAM, 238—242
Myristic acid, 28
Myxo Hunting, 351
Myzocytium megastomum, 69t—73t, 79t
Myzocytium proliferum, 69t—73t, 79t

N

Neotropical rainforest, 90—91
Neptunella longirostris, 36t—63t
Nitrogen, 94
Nowakowskiella elegans, 36t—63t
Nowakowskiella haemisphaerospora, 36t—63t
Nueva Vizcaya Watershed, 120
Nypa fruticans, 21t—24t

O

Oedogoniomyces lymnaeae, 36t—63t
Oil overlay, 306—307
Oleaginous organisms, 28
Oleic acid, 28
Olpidiopsis karlingiae, 79t
Olpidiopsis luxurians, 69t—73t, 79t
Olpidiopsis pythii, 79t
Olpidium allomycetos, 36t—63t
Olpidium decipiens, 36t—63t
Olpidium gregarium, 36t—63t
Olpidium pendulum, 36t—63t
Olpidium sparrowii, 36t—63t
Oncobasidium theobromae, 179
Ophioceras dolichostomum, 36t—63t
Organochloride pesticides, 243—245
Osbornia octodonta, 21t—24t
Oxybiodegarable (OBD), 255
Oxydothis nypicola, 36t—63t

P

Paecilomyces formosus, 36t—63t
Paecilomyces javanicus, 36t—63t
Paecilomyces lilacinus, 36t—63t
Paecilomyces persicinus, 36t—63t
Paecilomyces victoriae, 36t—63t
Paired Culture Technique, 359
Palali-Mamparang Mountain Range (PMMR), 120
Palmitic acid, 28
Panama disease of banana, 328—329
Parmeliaceae, 150—151
Parmotrema gardneri, 155

Passeriniella savoryellopsis, 36t—63t
Pathogenic fungi, 93
Peat swamp forest, 92
Pemphis acidula, 21t—24t
Penicillium canescens, 36t—63t
Penicillium chrysogenum, 36t—63t
Penicillium citreonigrum, 36t—63t
Penicillium citrinum, 36t—63t
Penicillium janthinellum, 36t—63t
Penicillium purpurascens, 36t—63t
Penicillium purpurogenum, 36t—63t
Penicillium rubrum, 36t—63t
Penicillium rugulosum, 36t—63t
Penicillium spinulosum, 36t—63t
Periconia prolifica, 36t—63t
Pestalotiopsis adusta, 36t—63t
Pestalotiopsis cocculi, 36t—63t
Pestalotiopsis microspora, 36t—63t
Pest epidemics, preventing, 333
Pesticide-polluted soil, 243
Pesticides, bioremediation agents for, 242—245, 243f, 244t
Pest incursions, preventing, 331—332
Pest Management Council of the Philippines (PMCP) Inc., 5
Phaeoisaria clematidis, 36t—63t
Phaeosphaeriopsis musae, 36t—63t
Phanerochaete chrysosporium, 255
Phialophora verrucosa, 36t—63t
Phialophorophoma litoralis, 36t—63t
Philippine Academy of Microbiology (PAM), 5
Philippine National Collection of Microorganisms (PNCM), 2—3, 5—7
Philippine Network on Microbial Culture Collection (PNMCC) Inc., 5—7
Philippine Phytopathological Society (PPS), 5
Philippines
  bioluminescence fungi in, 143
  forest formation in, 90—92
  fungal research in, 2—4
  higher education and research institutions, 2—4
  lichens, 150
    antimicrobial activities, 153—155
    antioxidant activities, 155
    biological activities of, 154t
    cytotoxic activities, 155
    endolichenic fungi (ELF), 156—157
    herbicidal activities, 156
    natural product research, 151—156, 153f
    taxonomic diversity of, 150—151, 151f, 152t
  mangrove fungi, 18—19
Philippines, environmental mycology in
  agriculture, biocontrol agents in, 245—253, 246t—250t

mycoremediation, 236—237, 237f
  heavy metals, 238—242, 239t—241t
  hydrocarbons, 256—258, 257t
pesticides, bioremediation agents for, 242—245, 243f, 244t
plastic degrading fungi, 253—256, 254t
Southeast Asia, growing environmental concern in, 235—236
Philippine Society for Microbiology (PSM), 5
Phlyctidium anatropum, 36t—63t
Phlyctochytrium planicorne, 36t—63t
Phlyctorhiza endogena, 36t—63t
Phoma lingam, 36t—63t
Phoma nebulosa, 36t—63t
Phomatospora berkeleyi, 36t—63t
Phomopsis mangrovei, 36t—63t
Phomopsis pittospori, 36t—63t
Photobiont, 147—148
Phyllosticta musarum, 194
Phylogenetic analysis, 29
Phylum Ascomycota, 165f, 166—177, 167t—173t
Phylum Basidiomycota, 177—179, 178t—179t
Phylum Blastocladiomycota, 179—180, 181t
Phylum Chytridiomycota, 179—180, 181t
Phylum Mucoromycota, 179—180, 181t
Physoderma maydis, 179
Phytophthora elongata, 69t—73t
Phytophthora insolita, 69t—73t
Phytophthora meadii, 79t
Phytophthora nicotianae, 79t
Phytophthora on coconut, 328
Phytophthora palmivora, 79t
Phytophthora parasitica, 79t
Phytophthora phaseoli, 79t
Phytopythium dogmae, 69t—73t
Phytopythium kandeliae, 69t—73t
Phytopythium leanoi, 69t—73t
Pichia angusta, 36t—63t
Plant diseases, 194t, 197—204
  Colletotrichum falcatum, 175
  Lasiodiplodia theobromae, 175
  maize, 203
  perennial crops, 203—204
  Phylum Ascomycota, 165f, 166—177, 167t—173t
  Phylum Basidiomycota, 177—179, 178t—179t
  Phylum Blastocladiomycota, 179—180, 181t
  Phylum Chytridiomycota, 179—180, 181t
  Phylum Mucoromycota, 179—180, 181t
  plant pathogenic fungi, genera of, 164—165, 165f
  rice blast, 197—202, 200t—201t
  rice sheath blight, 202—203
Plant pathogenic fungi, genera of, 164—165, 165f
Plant pathogens, 20

374   Index

Plant pest, 333–334
Plant quarantine protocols, 325–327
Plasmopara viticola, 79t
Plastic degrading fungi, 253–256, 254t
Plectospira gemmifera, 79t
Pleurophragmium bitunicatum, 36t–63t
Polyethylene (PE), 253
Polyethylene terephthalate (PET), 253
Polyporaceae, 93, 119
Polypropylene (PP), 253
Polystyrene (PS), 253
Polyurethane (PUR), 253
Potato Dextrose Agar (PDA), 359
Potato Dextrose Broth (PDB), 357
Project Proposal for Drug Discovery with Fungi, 357
Protocols, 327
    domestic transport, 327
    importation, 327
Pseudocersosporella, 36t–63t
Pseudohalonectria longirostrum, 36t–63t
Pseudoperonospora cubensis, 79t
Pseudospiropes cubensis, 36t–63t
Pythium aphanidermatum, 79t
Pythium arrhenomanes, 79t
Pythium debaryanum, 79t
Pythium echinulatum, 79t
Pythium gracile, 69t–73t
Pythium monospermum, 69t–73t, 79t
Pythium proliferum, 69t–73t, 79t
Pythium torulosum, 69t–73t, 79t

Q
Quezon Protected Landscape (QPL), 124–125
Quintaria lignatilis, 36t–63t

R
Relative humidity (RH), 191
Rhizidiomyces apophysatus, 69t–73t
Rhizidiomyces hirsutus, 69t–73t
Rhizoclosmatium globosum, 36t–63t
Rhizophila marina, 36t–63t
Rhizophora apiculata, 21t–24t
Rhizophora mucronata, 21t–24t
Rhizophora stylosa, 21t–24t
Rhizophydium carpophilum, 36t–63t
Rhizophydium pollinis-pini, 36t–63t
Rhizopus microsporus, 36t–63t
Rhizosiphon multiporum, 36t–63t
Rice blast, 189, 191
Rice brown spot, 189, 192
Rice sheath blight, 189, 192
Roussoëlla minutella, 36t–63t
Rozella allomycis, 36t–63t

Rozella cladochytrii, 36t–63t
Rozella myzoctii, 36t–63t

S
Saagaromyces ratnagiriensis, 36t–63t
Saccardoella minuta, 36t–63t
Saccharomyces cerevisiae, 256
Salisapilia bahamensis, 69t–73t
Salisapilia elongata, 69t–73t
Salisapilia epistomium, 69t–73t
Salispilia bahamensis, 69t–73t
Salispilia elongata, 69t–73t
Salispilia epistomium, 69t–73t
Salispina hoi, 69t–73t
Salispina lobata, 69t–73t
Salispina spinosa, 69t–73t
Salsuginea ramicola, 36t–63t
San Fernando Forest, 119
Saprolegnia diclina, 79t
Saprolgenia declina, 69t–73t
Saprophytic basidiomycetes, 92–93
Sarvoryella paucispora, 36t–63t
Savoryella aquatica, 36t–63t
Savoryella lignicola, 36t–63t
Savoryella longispora, 36t–63t
Scanning Electron Microscope (SEM), 254–255
Scedosporium aurantiacum, 36t–63t
Schizochytrium aggregatum, 69t–73t
Scopulariopsis brumptii, 36t–63t
Scyphiphora hydrophylacea, 21t–24t
Seagrasses, 63–64
Seaweeds, 34
Shining wood, 138
Sierra Madre Mountain Range, 119
Siphonaria variabilis, 36t–63t
Sonneratia alba, 21t–24t
Sonneratia caseolaris, 21t–24t
Sonneratia ovalis, 21t–24t
South American leaf blight (SALB) on rubber, 330–331
Southeast Asia, growing environmental concern in, 235–236
Spiropes caaguazuense, 36t–63t
Sporidesmiella hyalosperma, 36t–63t
Sporidesmium paludosum, 36t–63t
Sporoschisma juvenile, 36t–63t
Stearic acid, 28
Stenocarpella macrospora, 193
Subcutaneous mycosis, 215
Subic Watershed Forest Reserve, 120
Swampomyces triseptatus, 36t–63t

T

Taal Volcano Protected Landscape (TVPL), 124
Talaromyces macrosporus, 36t–63t
Tetraploa aristata, 36t–63t
Thalassogena sphaerica, 36t–63t
Thraustochytrids, 26–28, 27f
  isolation of, 29
Thraustochytrium globosum, 69t–73t
Thraustochytrium proliferum, 69t–73t
Thraustochytrium roseum, 69t–73t
Thraustotheca clavata, 79t
Tiarosporella paludosa, 36t–63t
Tirisporella beccariana, 36t–63t
Torpedospora radiata, 36t–63t
Toxigenic fungi, 179
Traditional knowledge, 271–272
Trametes cubensis, 36t–63t
Trematosphaeria mangrovis, 36t–63t
Trichocladium achrasporum, 36t–63t
Trichocladium alopallonellum, 36t–63t
Trichoderma aureoviride, 36t–63t
Trichoderma polysporum, 36t–63t
Tricladium lunderi, 36t–63t
Tritirachium oryzae, 36t–63t
Tropical lowland evergreen forest, 90–91
Tropical semievergreen rainforest, 91
Tylopocladium terricola, 36t–63t

U

University of Santo Tomas (UST), 3
University of the Philippines (UP), 2–3
Usnea philippina, 151–152

V

Valsa brevispora, 36t–63t
Valuable biological materials (VBMs),
  293–294
Vanakripa gigaspora, 36t–63t
Verruculina enalia, 36t–63t
Verticillium nigrescens, 36t–63t
Vibrissea nypicola, 36t–63t

W

Water management, 191
Wildlife Conservation Society of the Philippines
  (WCSP), 7
Woroninella dolichi, 179

X

Xylariaceae, 119
Xylaria cubensis, 36t–63t
Xylocarpus granatum, 21t–24t
Xylomyces, 36t–63t

Y

Yeasts, 34–35
  aquatic habitats, 34

Z

Zalerion maritimum, 36t–63t
Zalerion varia, 36t–63t
Zoophagus insidians, 69t–73t
Zygomycota, 35–63

Printed in the United States
by Baker & Taylor Publisher Services